American Mathematical Society

Colloquium Publications
Volume 34

The Location of Critical Points of Analytic and Harmonic Functions

J. L. Walsh

American Mathematical Society
Providence, Rhode Island

2000 *Mathematics Subject Classification.* Primary 30-XX.

Library of Congress Cataloging-in-Publication Data

Walsh, J. L. (Joseph Leonard), 1895–
The location of critical points of analytic and harmonic functions.
p. cm. — (American Mathematical Society Colloquium publications, ISSN 0065-9258 ; v. 34)
New York, N.Y., American Mathematical Society, 1950.
Includes bibliography.
ISBN 978-0-8218-4643-8 (alk. paper)
1. Functions. 2. Harmonic functions. I. Title. II. Colloquium publications (American Mathematical Society) ; v. 34.

QA1 .A5225 vol. 34
517.5 50012177

∞ The paper used in this book is acid-free and falls within the guidelines
established to ensure permanence and durability.
Visit the AMS home page at http://www.ams.org/

10 9 8 7 6 5 4 13 12 11 10 09

PREFACE

Commencing with Gauss, numerous mathematicians have contributed to the study in the plane of the complex variable of the geometric relations between the zeros of a polynomial and those of its derivative. More generally there have been studied the relations between the zeros and poles of an arbitrary rational function and the zeros of its derivative, and an elaborate body of material has evolved which applies also to the critical points of an arbitrary analytic function, of Green's function, and of other harmonic functions. It is the purpose of the present volume to assemble and unify a large portion of this material, to make it available for study by the beginner or by the specialist, and for reference.

Results concerning even polynomials and rational functions are far too numerous to be completely treated here. The omission is not serious, for a general survey of the entire field of the geometry of zeros of polynomials has recently been written by Marden [1949]*, to which the reader may refer for broader perspective in that field. The present material has been chosen to emphasize (i) the determination of regions which are free from critical points (or alternately, which contain all critical points), rather than to study mere enumeration of critical points in a given region, and (ii) results for polynomials and rational functions which can be extended to and furnish a pattern for the cases of more general analytic functions and of harmonic functions. Our main problem, then, is the approximate determination of critical points—approximate not in the sense of computation which may be indefinitely refined, but in the sense of geometric limitation of critical points to easily constructed regions, preferably bounded by lines and circles. The point sets shown to contain the critical points are naturally defined in terms of the zeros of a given polynomial, in terms of the zeros and poles of a given rational or more general analytic function, and in terms of suitable level curves of a given harmonic function. The theorem of Lucas is typical in this field both as to general content and method of proof, and occupies a central position in the entire theory. Although there are close connections with topology, our methods are mainly study of a field of force, use of algebraic inequalities, analytic geometry, geometry of circles and plane curves, circle transformations, potential theory, and conformal mapping. So far as concerns rational functions, the methods are largely elementary, as seems to be in keeping with the nature of the problems.

We use the term *critical point* to include both zeros of the derivative of an analytic function and points where the two first partial derivatives of a harmonic function vanish. It is hardly necessary to emphasize the importance of critical points as such: 1) they are notable points in the behavior of the function, and in particular for a harmonic function are the multiple points of level curves and

* Dates in square brackets refer to the Bibliography

their orthogonal trajectories (curves of steepest descent); for an analytic function $f(z)$ they are the multiple points of the loci $|f(z)| = \text{const}$ and $\arg [f(z)] = \text{const}$; 2) for an analytic function $f(z)$ they are the points where the conformality of the transformation $w = f(z)$ fails; 3) they are conformal invariants, of importance in numerous extremal problems of analytic functions and in the study of approximation by rational and other functions; 4) they are precisely the positions of equilibrium in a potential field of force due to a given distribution of matter or electricity; 5) they are stagnation points in a field of velocity potential. We obtain incidentally in the course of our work numerous results concerning the direction of the force or velocity corresponding to a given potential, but such results are secondary and are frequently implicit, being left to the reader for formulation.

The term *location* of critical points suggests the term *locus*, and ordinarily we determine an actual locus of critical points under suitable restrictions on the given function. In the former part of the book we take pains to indicate this property, but in the latter part leave to the reader the detailed discussion. Of course our entire problem, of determining easily found regions free from (or containing) critical points, is clearly a relative one, and it is a matter of judgment how far the theory should be developed. Convenience, simplicity, and elegance are our criteria, but the theory admits of considerable further development, especially in the use of algebraic curves of higher degree.

A large part of the material here set forth has not been previously published, except perhaps in summary form. Thus Chapter V considers in some detail the critical points of rational functions which possess various kinds of symmetry. Chapters VII–IX present a unified investigation of the critical points of harmonic functions, based on the study of a field of force due to a spread of matter on the boundary of a given region, a boundary which may consist of a finite number of arbitrary Jordan curves. The introduction and use of special loci called *W-curves* adds unity and elegance to much of our entire discussion.

For help with manuscript and proof, the author is indebted to: Dorothy W. Allen, Esther Blauer, Edward Block, Edward Carterette, Barbara P. Hunicke, Robert Kendall, Marion B. Morris, and Elizabeth S. Walsh. The author is also grateful for support from the Milton Fund of Harvard University, and to the American Mathematical Society for accepting the book in its Colloquium Series.

J. L. WALSH

Feb. 1949.

TABLE OF CONTENTS

Preface iii

Chapter I. Fundamental Results 1

§1.1. Terminology. Preliminaries 1

§1.1.1. Point set terminology 1

§1.1.2. Function-theoretic preliminaries 2

§1.2. Gauss's Theorem 5

§1.3. Lucas's Theorem 6

§1.3.1. Statement and proof 6

§1.3.2. Complements 8

§1.4. Jensen's Theorem 9

§1.4.1. Proof 10

§1.4.2. Complements 11

§1.5. Walsh's Theorem 13

§1.5.1. Preliminaries 13

§1.5.2. Proof 15

§1.6. Lemniscates 17

§1.6.1. Level curves 17

§1.6.2. Lemniscates and their orthogonal trajectories 18

§1.6.3. Behavior at infinity 21

Chapter II. Real Polynomials 24

§2.1. Polynomials with real zeros 24

§2.2. Jensen's Theorem, continued 27

§2.2.1. Special cases 28

§2.2.2. A region for non-real critical points 29

§2.3. Number of critical points 31

§2.4. Reality and non-reality of critical points 32

§2.4.1. Sufficient conditions for non-real critical points 32

§2.4.2. Sufficient conditions for real critical points 33

§2.4.3. Real critical points, continued 35

§2.5. Jensen configuration improved 38

§2.6. W-curves 40

§2.6.1. General results 41

§2.6.2. Real polynomials 44

§2.6.3. Zeros on horizontal segments 49

§2.7. Non-real polynomials 52

§2.7.1. Jensen circles 52

§2.7.2. Infinite sectors 55

Chapter III. Polynomials, Continued 58

§3.1. Infinite circular regions as loci 58

§3.1.1. Exterior of a circle 58

§3.1.2. Half-planes 60

§3.2. A characteristic property of circular regions 62

§3.3. Critical points of real polynomials as centers of circles 63

§3.4. Lucas polygon improved 68

§3.5. The Lucas polygons for a polynomial and its derivative 71

§3.5.1. Relations of convex sets 71

§3.5.2. Dependence of zeros of polynomial on those of its derivative 73

§3.6. Symmetry in the origin 75

§3.6.1. Ordinary symmetry 75

§3.6.2. Multiple symmetry in O 78
§3.7. Circular regions and symmetry 80
§3.8. Higher derivatives 83
§3.9. Further results 87
CHAPTER IV. RATIONAL FUNCTIONS 89
§4.1. Field of force 89
§4.1.1. Fundamental theorem 89
§4.1.2. Transformations of the plane 92
§4.1.3. Stereographic projection 94
§4.2. Bôcher's Theorem 97
§4.2.1. Proof 98
§4.2.2. Locus of critical points 99
§4.2.3. Specializations 101
§4.2.4. Circular regions as loci 103
§4.2.5. Converse 105
§4.3. Concentric circular regions as loci 106
§4.4. The cross-ratio theorem 109
§4.4.1. Geometric locus 109
§4.4.2. Discussion of locus 113
§4.4.3. Locus of critical points 115
§4.4.4. Generalizations 120
§4.5. Marden's Theorem 121
CHAPTER V. RATIONAL FUNCTIONS WITH SYMMETRY 123
§5.1. Real rational functions 123
§5.1.1. Real zeros and poles 123
§5.1.2. Non-real zeros 125
§5.1.3. Regions as loci 128
§5.1.4. Regions as loci, continued 131
§5.1.5. W-curves 135
§5.2. Zeros and poles concyclic 138
§5.2.1. Zeros and poles interlaced 139
§5.2.2. Extensions 141
§5.2.3. Critical points near a given zero 145
§5.2.4. Continuation 148
§5.2.5. Zeros and poles on prescribed arcs 152
§5.3. Hyperbolic plane 156
§5.3.1. Analog of Lucas's Theorem 157
§5.3.2. Extensions 159
§5.3.3. Analog of Jensen's Theorem 163
§5.3.4. Circular regions as loci 165
§5.3.5. Analog of Bôcher's Theorem 169
§5.3.6. Continuation 170
§5.3.7. NE half-planes as loci 172
§5.4. Elliptic plane 176
§5.4.1. Analog of Bôcher's Theorem 176
§5.4.2. Circular regions as loci 178
§5.5. Symmetry in the origin 181
§5.5.1. Regions bounded by concentric circles 181
§5.5.2. Regions bounded by equilateral hyperbolas 183
§5.5.3. Circular regions as loci of zeros and poles 186
§5.5.4. Multiple symmetry in O 189
§5.5.5. Multiple symmetry in O, continued 191
§5.5.6. Polynomials and multiple symmetry in O 193
§5.6. Skew symmetry in O 194

§5.6.1. Sectors containing zeros and poles 194
§5.6.2. W-curves 197
§5.7. Symmetry in z and $1/z$ 202
§5.7.1. Polynomials in z and $1/z$ 202
§5.7.2. General rational functions 206
§5.8. Miscellaneous results 208
§5.8.1. Combinations of symmetries 208
§5.8.2. Circular regions and symmetry in a line 209
§5.8.3. Circular regions and real polynomials 212
CHAPTER VI. ANALYTIC FUNCTIONS 217
§6.1. Entire and meromorphic functions 217
§6.2. The hyperbolic plane 221
§6.2.1. Analytic and meromorphic functions 221
§6.2.2. Blaschke products 223
§6.3. Functions with multiplicative periods 224
§6.3.1. Uniform functions 224
§6.3.2. Non-uniform functions 227
§6.4. Simply periodic functions 231
§6.4.1. General theorems 232
§6.4.2. Functions with symmetry 234
§6.5. Doubly periodic functions 235
§6.6. General analytic functions 236
§6.7. Analytic functions and harmonic functions 240
CHAPTER VII. GREEN'S FUNCTIONS 241
§7.1. Topology 241
§7.1.1. Level curves 241
§7.1.2. Variable regions 243
§7.1.3. A formula for Green's function 245
§7.1.4. Level curves and lemniscates 247
§7.2. Geometry of level loci 249
§7.2.1. Analog of Lucas's Theorem 249
§7.2.2. Center of curvature 252
§7.3. Symmetry in axis of reals 255
§7.3.1. Analog of Jensen's Theorem 255
§7.3.2. Number of critical points 257
§7.4. Analog of Walsh's Theorem 259
§7.4.1. Numbers of critical points 259
§7.4.2. Numbers of critical points, alternate treatment 262
§7.4.3. Symmetry 263
§7.4.4. Inequalities on masses 264
§7.4.5. Circles with collinear centers 266
§7.5. Doubly-connected regions 266
CHAPTER VIII. HARMONIC FUNCTIONS 269
§8.1. Topology 269
§8.1.1. Level loci 269
§8.1.2. A formula for harmonic measure 271
§8.1.3. Number of critical points 274
§8.2. Analog of Bôcher's Theorem 275
§8.3. A cross-ratio theorem 278
§8.4. Hyperbolic plane 279
§8.5. Other symmetries 283
§8.6. Periodic functions 286
§8.7. Harmonic measure, Jordan region 288
§8.7.1. Finite number of arcs 288

§8.7.2. Infinite number of arcs 290
§8.8. Hyperbolic geometry and conformal mapping 292
§8.8.1. Doubly-connected region 292
§8.8.2. Case $p > 2$ 293
§8.9. Linear combinations of Green's functions 294
§8.9.1. Simply-connected region 295
§8.9.2. Regions of arbitrary connectivity 296
§8.9.3. Multiply-connected regions 299
§8.10. Harmonic measure of arcs 300
§8.10.1. Doubly-connected regions 300
§8.10.2. Arbitrary connectivity 301
§8.10.3. Approximating circular regions 302
Chapter IX. Further Harmonic Functions 304
§9.1. A general field of force 304
§9.2. Harmonic measure of arcs 307
§9.2.1. A single arc 307
§9.2.2. Several arcs 308
§9.2.3. Linear combinations 310
§9.3. Arcs of a circle 312
§9.3.1. Disjoint arcs 313
§9.3.2. Abutting arcs 315
§9.3.3. Poisson's integral 317
§9.3.4. Poisson's integral as potential of a double distribution 319
§9.3.5. Harmonic functions and rational functions 322
§9.4. A circle as partial boundary 324
§9.4.1. Harmonic measure 324
§9.4.2. Linear combinations 327
§9.4.3. Harmonic measure, resumed 329
§9.4.4. Symmetry 330
§9.4.5. Symmetry, several arcs 333
§9.4.6. Case $p > 1$ 337
§9.5. Green's functions and harmonic measures 339
§9.5.1. A numerical example 339
§9.5.2. General linear combinations 341
§9.6. Limits of boundary components 344
§9.6.1. Hyperbolic geometry 344
§9.6.2. Comparison of hyperbolic geometries 346
§9.7. Methods of symmetry 348
§9.7.1. Reflection in axis 349
§9.7.2. Reflection in point 353
§9.7.3. Reflection of an annulus, Green's function 356
§9.7.4. Reflection of an annulus, harmonic measure 358
§9.8. Topological methods 360
§9.8.1. Simple-connectivity 360
§9.8.2. Higher connectivity 362
§9.8.3. Contours with boundary values zero 364
§9.9. Assigned distributions of matter 365
§9.9.1. Simple layers 365
§9.9.2. Superharmonic functions 366
§9.9.3. Double layers 370

Bibliography 377

Index 381

CHAPTER I

FUNDAMENTAL RESULTS

§1.1. Terminology. Preliminaries. We shall be concerned primarily with the plane of the complex variable $z = x + iy$; the plane may be either the *finite plane* (i.e. plane of finite points) or the *extended plane* (the finite plane with the adjunction of a single point at infinity). The extended plane is often studied by stereographic projection onto the sphere (*Neumann sphere*), on which the point at infinity is no longer exceptional. When we study primarily polynomials, we ordinarily deal with the finite plane except when we use an auxiliary inversion in a circle; when we deal with more general rational functions, or with harmonic functions, and with the non-integral linear transformation of the complex variable, we usually operate in the extended plane. We frequently identify a value of z and the point representing it; thus the term *non-real point* refers to a point representing a non-real value of z.

In the finite plane the term *circle* means simply circumference; in the extended plane the use of the term is ordinarily broadened to include straight line.

§1.1.1. Point set terminology. A *neighborhood* of a finite point is the interior of a circle with that point as center. A neighborhood of the point at infinity is the exterior of a circle of the finite plane, point at infinity included.

If S is a given point set, the point P is an *interior point* of S if some neighborhood of P contains only points of S, is a *boundary point* of S if every neighborhood of P contains both points of S and points not in S, and is an *exterior point* if some neighborhood of P contains no point of S. The point P is a *limit point* of S if every neighborhood of P contains an infinity of points of S. A set S is *open* if every point of S is an interior point of S; a set S is *closed* if it contains its limit points.

The *closure* of a set S consists of S plus its limit points; the closures of sets R, S, T are respectively denoted by $\bar{R}$, $\bar{S}$, $\bar{T}$.

A *Jordan arc* is the image of the closed line segment $0 \leqq x \leqq 1$ under a one-to-one continuous transformation, where continuity may be interpreted either in the finite plane or on the sphere, the latter being equivalent to the extended plane. A *Jordan curve* is similarly the image of a circumference under a one-to-one continuous transformation.

A *region* is an open set of which any two points can be joined by a Jordan arc consisting wholly of points of the set. A *closed region* is the closure of a region, and need not properly be a region, but the term *region* may be somewhat loosely applied to include closed region, and even arcs and points as degenerate closed regions. A *Jordan region* is a region bounded by a Jordan curve.

A *component* of a closed set S is a closed subset which cannot itself be separated, but which can be separated from all other points of S, by a Jordan curve disjoint from S.

A *Jordan configuration* is a point set consisting of a finite number of Jordan arcs.

§1.1.2. Function-theoretic preliminaries. A *polynomial in z of degree n* is a function which can be written as $a_0z^n + a_1z^{n-1} + \cdots + a_n$, $a_0 \neq 0$. A function $f(z)$ is *analytic* at a finite point z_0 if it possesses a continuous derivative at every point of a neighborhood of z_0, and is analytic at infinity if $f(1/z)$ is analytic at the origin.* A function is analytic in a region if it is analytic at every point of that region, and is assumed single-valued unless otherwise specified. A function $f(z)$ is *meromorphic* in a region if it is analytic there except perhaps for poles. A finite *critical point* of an analytic function $f(z)$ is a point z_0 at which the derivative $f'(z)$ vanishes (this notation for derivative will be used consistently); the *multiplicity* of z_0 as a critical point of $f(z)$ is the multiplicity of z_0 as a zero of $f'(z)$. The point at infinity is a critical point of an analytic function $f(z)$, and of order k, if the origin is a critical point of the function $f(1/z)$, and of order k. If an analytic function $f(z)$ has a pole at infinity, that point is not a critical point; if $f(z)$ is analytic there, the point at infinity is a critical point of order k if and only if $f'(z)$ has a zero at infinity of order $k + 2$.

A few well known theorems are of central importance in the sequel.

Principle of Argument. *Let R be a region whose boundary B consists of a finite number of mutually disjoint Jordan curves, and let the function f(z) be meromorphic in R, continuous on R + B in the closed neighborhood of B, different from zero on B. As z traces B in the positive sense (i.e. so that the region R is situated to the left of the forward moving observer), the net increase in* arg $[f(z)]$ *is* 2π *times the number of zeros minus the number of poles of f(z) interior to R.*

An application of this principle is

Rouché's Theorem. *Let the functions* $\varphi(z)$ *and* $\psi(z)$ *be analytic interior to a region R whose boundary B consists of a finite number of mutually disjoint Jordan curves, and let the functions* $\varphi(z)$ *and* $\psi(z)$ *be continuous in the closed region* $\bar{R}$, *with the relation* $|\varphi(z)| > |\psi(z)|$ *on B. Then the functions* $\varphi(z)$ *and* $\varphi(z) + \psi(z)$ *have the same number of zeros in R.*

Let z_0 be a point of R at which at least one of the functions $\varphi(z)$ and $\varphi(z) + \psi(z)$ vanishes; if z_0 is a zero of $\varphi(z) + \psi(z)$ of order m but not a zero of $\varphi(z)$, then z_0 is a zero of $f(z) \equiv [\varphi(z) + \psi(z)]/\varphi(z)$ of order m; if z_0 is a zero of $\varphi(z)$ of order n but not a zero of $\varphi(z) + \psi(z)$, then z_0 is a pole of $f(z)$ of order n; if z_0 is a zero of $\varphi(z) + \psi(z)$ of order m and a zero of $\varphi(z)$ of order n, then z_0 is respectively a zero of $f(z)$ of order $m - n$, or a non-zero point of analyticity of $f(z)$, or a pole of $f(z)$ of order $n - m$ according as we have $m > n$, or $m = n$, or $m < n$.

* We assume $f(1/z)$ to be defined for $z = 0$ so as to be continuous there if possible; a similar remark applies below at other isolated singularities.

Thus the number of zeros of $\varphi(z) + \psi(z)$ in R minus the number of zeros of $\varphi(z)$ in R equals the number of zeros of $f(z)$ minus the number of poles in R. On the other hand, as z traces each component of B the point $w = f(z)$ remains interior to the circle $|w - 1| = 1$, so the net total change in arg w is zero; the number of zeros of $f(z)$ minus the number of poles of $f(z)$ in R is zero.

HURWITZ'S THEOREM. *Let R be a region in which the functions $f_n(z)$ and $f(z)$ are analytic, continuous in $\bar{R}$, with $f(z)$ different from zero on the boundary B of R, and let the sequence $f_n(z)$ converge uniformly to $f(z)$ on $R + B$. Then for n greater than a suitably chosen N, the number of zeros of $f_n(z)$ in R is the same as the number of zeros of $f(z)$ in R.*

In the proof we assume, as we may do, that B consists of a finite number of mutually disjoint Jordan curves. If $\delta\,(> 0)$ is chosen so that $|f(z)| > \delta$ on B, we need merely choose N_δ so that $n > N_\delta$ implies $|f_n(z) - f(z)| < \delta$ on B, and apply Rouché's Theorem.

Hurwitz's Theorem applies not merely to the given region R, but also to a neighborhood $N(z_0)$ whose closure lies in R of an arbitrary zero z_0 of $f(z)$ in R, provided $f(z)$ does not vanish on the boundary of $N(z_0)$. If z_0 is a zero of $f(z)$ of order m, and if $N(z_0)$ contains no zero of $f(z)$ other than z_0, then n greater than a suitably chosen N_δ implies that $N(z_0)$ contains precisely m zeros of $f_n(z)$. It follows that if no function $f_n(z)$ vanishes identically, the limit points in R of the zeros of the functions $f_n(z)$ in R are precisely the zeros of $f(z)$ in R.

In particular, under the hypothesis of Hurwitz's Theorem with $f(z)$ not identically constant, the sequence $f_n'(z)$ converges uniformly to $f'(z)$ on any closed subset of R, so if z_0 is a finite or infinite critical point of $f(z)$ in R of order m, and if $N(z_0)$ is a neighborhood of z_0 to which B and all other critical points of $f(z)$ are exterior, then n greater than a suitably chosen N_δ implies that $N(z_0)$ contains precisely m critical points of $f_n(z)$; if no $f_n(z)$ is identically constant, the limit points in R of the critical points of the $f_n(z)$ are precisely the critical points of $f(z)$ in R; if $f'(z)$ and $f_n'(z)$ are continuous in $R + B$ and different from zero on B, and if $f_n'(z)$ converges uniformly in $R + B$, then for suitably large index the function $f_n(z)$ has precisely the same total number of critical points in R as does $f(z)$.

The zeros of an algebraic equation are continuous functions of the coefficients: if the variable polynomial $P_k(z)$ of bounded degree approaches the fixed polynomial $P(z)$ (not identically constant) in the sense that coefficients of $P_k(z)$ approach corresponding coefficients of $P(z)$, then each zero of $P(z)$ is approached by a number of zeros of $P_k(z)$ equal to its multiplicity; all other zeros of $P_k(z)$ become infinite. More explicitly, let

$$P(z) \equiv a_0 z^n + a_1 z^{n-1} + \cdots + a_n$$

be a polynomial not identically constant, and let

$$P_k(z) \equiv a_{0k} z^n + a_{1k} z^{n-1} + \cdots + a_{nk}$$

be a variable polynomial such that $a_{mk} \to a_m$ $(m = 0, 1, 2, \cdots, n)$ *as* k *becomes infinite. Let* $\alpha_1, \alpha_2, \cdots, \alpha_p$ *be the distinct zeros of* $P(z)$, *and let* $\epsilon > 0$ *be arbitrary,* $|\alpha_i - \alpha_j| > 2\epsilon$, $|\alpha_j| + \epsilon < 1/\epsilon$. *Then there exists* M_ϵ *such that* $k > M_\epsilon$ *implies that each neighborhood* $|z - \alpha_j| < \epsilon$ *contains precisely a number of zeros of* $P_k(z)$ *equal to the multiplicity of* α_j *as a zero of* $P(z)$, *and all other zeros of* $P_k(z)$ *lie exterior to the circle* $|z| = 1/\epsilon$.

The sequence $P_k(z)$ converges uniformly to the function $P(z)$ in any closed bounded region:

$$|P(z) - P_k(z)| \leqq |a_0 - a_{0k}| \cdot |z|^n + |a_1 - a_{1k}| \cdot |z|^{n-1} + \cdots + |a_n - a_{nk}|,$$

hence in the closed region $|z| \leqq 1/\epsilon$; the conclusion follows from Hurwitz's Theorem applied simultaneously to $|z| \leqq 1/\epsilon$ and to all the regions $|z - \alpha_j| \leqq \epsilon$. This method of proof is used by Bieberbach-Bauer [1928].

If we modify the hypothesis here so as to admit a polynomial $P(z)$ which is identically constant but not identically zero, the proof requires no modification, and shows that $k > M_\epsilon$ implies that all zeros of $P_k(z)$ lie exterior to $|z| = 1/\epsilon$.

A real function $u(x, y)$ or $u(z)$ is *harmonic* in a finite region if there it is continuous together with its first and second partial derivatives, and satisfies Laplace's equation. The function $u(z)$ is harmonic at a finite point if it is harmonic throughout a neighborhood of that point, and is harmonic at infinity if $u(1/z)$ is harmonic at the origin. If $u(x, y)$ is harmonic in a region R, it possesses a conjugate $v(x, y)$ there, which is single-valued if R is simply connected; the function $f(z) \equiv u(x, y) + iv(x, y)$ is then analytic in R. A finite *critical point* of $u(x, y)$ is a point at which $\partial u/\partial x$ and $\partial u/\partial y$ vanish, that is, a critical point of $f(z)$; it is sufficient if the directional derivatives of $u(x, y)$ in two essentially distinct (i.e. not the same nor opposite) directions vanish; the point at infinity is a critical point of $u(x, y)$ if and only if it is a critical point of $f(z)$. The *order* of a critical point of $u(x, y)$ is its order as a critical point of $f(z)$. If $z_0 = x_0 + iy_0$ is a finite critical point of $u(x, y)$ of order m, then all partial derivatives of $u(x, y)$ of orders $1, 2, \cdots, m$ vanish there, but not all partial derivatives of order $m + 1$.

If a function $f(z)$ is analytic at infinity, its derivative $f'(z)$ has there a zero of order at least two, and consequently if $u(x, y)$ is harmonic at infinity its first and second order partial derivatives vanish there; if $f'(z)$ has a zero there of order k (> 2), then both $f(z)$ and $u(x, y)$ have critical points there, of order $k - 2$, and conversely. A finite or infinite critical point z_0 of an analytic or harmonic function of order m retains the property of being a critical point of order m under one-to-one conformal transformation of the neighborhood of z_0, even if a finite z_0 is transformed to infinity or an infinite z_0 is transformed to a finite point.

If a sequence of functions $u_k(x, y)$ harmonic in a region R converges uniformly in R to the function $u(x, y)$ harmonic but not identically constant in R, if (x_0, y_0) is a finite or infinite critical point of $u(x, y)$ in R of order m, and if $N(x_0, y_0)$ is a neighborhood of (x_0, y_0) whose closure lies in R and contains no other critical point of $u(x, y)$, then for k sufficiently large precisely m critical points of $u_k(x, y)$ lie in $N(x_0, y_0)$, each critical point being counted according to its multiplicity. The proof may be conveniently given by means of Hurwitz's Theorem, for if the functions $v(x, y)$ and $v_k(x, y)$ conjugate to $u(x, y)$ and $u_k(x, y)$ are suitably chosen, the functions $u_k(x, y) + iv_k(x, y)$ are analytic in $N(x_0, y_0)$ and converge uniformly there to the function $u(x, y) + iv(x, y)$.

We state for reference without proof (which may be given by inequalities derived from Poisson's integral)

HARNACK'S THEOREM. *Let $u_n(x, y)$ be a monotonically increasing sequence of functions harmonic in a region R. Then either $u_n(x, y)$ becomes infinite at every point of R or the sequence converges throughout R, uniformly on any closed set interior to R.*

§1.2. Gauss's Theorem. We commence our study of the location of critical points by considering the simplest non-trivial functions, namely polynomials. Rolle's Theorem of course applies to real polynomials as to any real function possessing a derivative, and informs us that between two zeros of the function lies at least one zero of the derivative. Beyond Rolle's Theorem, the first general result concerning the zeros of the derivative of an arbitrary polynomial seems to be due to Gauss [1816]:

GAUSS'S THEOREM. *Let $p(z)$ be the polynomial $(z - \alpha_1)(z - \alpha_2) \cdots (z - \alpha_n)$, and let a field of force be defined by fixed particles situated at the points $\alpha_1, \alpha_2, \cdots, \alpha_n$, where each particle repels with a force equal to the inverse distance. At a multiple zero of $p(z)$ are to be placed a number of particles equal to the multiplicity. Then the zeros of the derivative $p'(z)$ are, in addition to the multiple zeros of $p(z)$, precisely the positions of equilibrium in this field of force.*

The logarithmic derivative of $p(z)$ is

$$(1) \qquad \frac{p'(z)}{p(z)} = \frac{1}{z - \alpha_1} + \frac{1}{z - \alpha_2} + \cdots + \frac{1}{z - \alpha_n}.$$

The conjugate of $1/(z - \alpha_k)$ is a vector whose direction (including sense) is the direction from α_k to z and whose magnitude is the reciprocal of the distance from α_k to z, so this vector represents the force at the variable point z due to a single fixed particle at α_k. Every multiple zero (but no simple zero) of $p(z)$ is a zero of $p'(z)$; every other zero of $p'(z)$ is by (1) a position of equilibrium in the field of force; every position of equilibrium is by (1) a zero of $p'(z)$. Gauss's Theorem is established.

As a simple example here, we choose $p(z) = (z - \alpha_1)^{m_1}(z - \alpha_2)^{m_2}$, $m_1 m_2 \neq 0$. Except for α_1 and α_2, the only zero of $p'(z)$ is given by

$$\frac{m_1}{z - \alpha_1} + \frac{m_2}{z - \alpha_2} = 0, \qquad \frac{z - \alpha_1}{\alpha_2 - z} = \frac{m_1}{m_2},$$

so z is the point $(m_2\alpha_1 + m_1\alpha_2)/(m_1 + m_2)$ which divides the segment $\alpha_1\alpha_2$ in the ratio $m_1 : m_2$.

The particles of Gauss's Theorem may be termed of *unit mass*; more generally we may consider particles of arbitrary (not necessarily integral) positive mass, repelling with a force equal to the quotient of the mass by the distance. Thus we have the

COROLLARY. *If particles of positive masses $\mu_1, \mu_2, \cdots, \mu_n$ are placed at the respective points $\alpha_1, \alpha_2, \cdots, \alpha_n$, then the positions of equilibrium in the resulting field of force are precisely the zeros of the derivative of $p(z) = \prod_{k=1}^{n} (z - \alpha_k)^{\mu_k}$, except that α_k is also a zero of $p'(z)$ if we have $\mu_k > 1$.*

Gauss's Theorem is of such central importance in the sequel that if a polynomial $p(z)$ is given we often automatically set up the corresponding field of force, and identify a zero of $p(z)$ with a fixed particle.

§1.3. Lucas's Theorem. If all the zeros of a polynomial $p(z)$ are real, it follows from Rolle's Theorem that each interval J of the axis of reals bounded by successive zeros of $p(z)$ contains at least one zero of the derivative $p'(z)$. If the respective multiplicities of the zeros of $p(z)$ are $m_1, m_2, \cdots, m_k$, whose sum is n, the corresponding multiplicities of the same points as zeros of $p'(z)$ are the positive numbers among the set $m_1 - 1, m_2 - 1, \cdots, m_k - 1$, whose sum is $n - k$. There are $k - 1$ intervals J each containing in its interior at least one zero of $p'(z)$, which has a totality of $n - 1$ zeros, so each interval J contains in its interior precisely one zero of $p'(z)$. All zeros of $p'(z)$ lie in the smallest interval of the axis of reals which contains the zeros of $p(z)$. This consequence of Rolle's Theorem can be generalized to apply to an arbitrary polynomial.

§1.3.1. Statement and proof.

LUCAS'S THEOREM. *Let $p(z)$ be a polynomial* whose zeros are $\alpha_1, \alpha_2, \cdots, \alpha_n$, and let Π be the smallest convex set on which those zeros lie. Then all zeros of the derivative $p'(z)$ also lie on Π. No zero of $p'(z)$ lies on the boundary of Π unless it is a multiple zero of $p(z)$, or unless all the zeros of $p(z)$ are collinear.*

Let pegs be placed in the z-plane at the points $\alpha_1, \alpha_2, \cdots, \alpha_n$; let a large rubber band be stretched in the plane so as to to include all the pegs. If the rubber band is allowed to contract so as to rest only on the pegs, and if it remains taut, it will fit over the pegs in the form of the boundary of Π. Thus if $p(z)$ has but one distinct zero, Π coincides with that zero; if the zeros of $p(z)$ are collinear, Π is a line segment; in any other case Π is the closed interior of a rectilinear polygon, called the *Lucas polygon* for $p(z)$ or for the points α_j. In every case the set Π

* Here and in corresponding places throughout the present work, the qualifying adjective *non-constant* is tacitly understood.

is convex in the sense that if two points belong to Π, so also does the line segment joining them.

The set Π can also be defined as the point set common to all closed half-planes each containing all the α_k.

If z_0 is a point exterior to Π, there exists a line L separating z_0 and Π; in fact, L may be chosen as the perpendicular bisector of the shortest segment joining z_0 to Π. In the field of force set up by Gauss's Theorem, all particles lie on one side of L; the force at z_0 due to each particle has a non-vanishing component perpendicular to L directed toward the side of L on which z_0 lies. Consequently z_0 cannot be a position of equilibrium. Moreover z_0 cannot be a multiple zero of $p(z)$, hence cannot be a zero of $p'(z)$.

If z_0 is a boundary point of Π not a zero of $p(z)$, and if not all the zeros of $p(z)$ are collinear, denote by L the line containing the side of the boundary of Π on which z_0 lies. Then one of the two half-planes bounded by L contains points

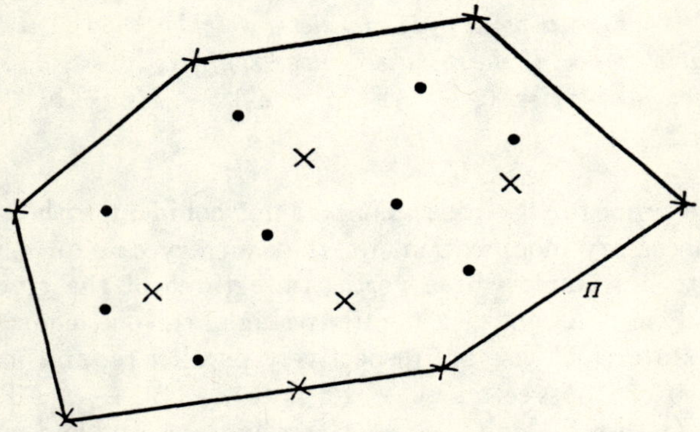

Fig. 1 illustrates §1.3.1 Lucas's Theorem

α_k in its interior while the other half-plane does not. Again the force at z_0 due to all the particles has a non-vanishing component perpendicular to L, so z_0 cannot be a position of equilibrium nor a zero of $p'(z)$. The proof is complete.

Although Lucas's Theorem follows at once from that of Gauss, there seems to be no evidence that it was stated by Gauss. It is credited to F. Lucas [1874]. Since his proof, as D. R. Curtiss [1922] has remarked, it has been "discovered and rediscovered, proved and reproved in most of the languages of Europe—and all the proofs are substantially the same".

In the present work we shall find numerous analogs and generalizations of Lucas's Theorem, many of them proved by this same method of considering at a point outside of a certain locus the total force in the field set up by Gauss's Theorem, and showing that this total force cannot be zero. In particular we shall make frequent use of the fact that a point z_0 cannot be a position of equilibrium under the action of several forces if those forces are represented by vectors having their initial points in z_0 and either their terminal points in an angle less than π

with vertex z_0 or their terminal points in an angle π with vertex z_0 and at least one terminal point interior to that angle.

§1.3.2. Complements. If the positions of the zeros of a variable polynomial $p(z)$ are fixed, and if the multiplicities of those zeros are permitted to take all possible values, the totality of critical points of $p(z)$ forms a countable point set S. Lucas's Theorem utilizes the positions of the zeros of a given polynomial but not their multiplicities, and essentially asserts that S lies in Π. However, we now prove that Π *is the closure of* S, in other words, if we wish to assign only a *closed* set as one in which the critical points lie, Lucas's Theorem is the best possible result [Fejér, Toeplitz, 1925; the corresponding result for real polynomials with non-real zeros (slightly weaker than §1.4.2 Theorem 1) had been previously proved (1920a) by this same method by the present writer]:

THEOREM 1. *Let Π be the closed interior of the Lucas polygon for the points $\alpha_1, \alpha_2, \cdots, \alpha_n$. Let z_0 be a point of Π, and let a neighborhood $N(z_0)$ of z_0 be given. Then for a suitable choice of the positive integers $\mu_1, \mu_2, \cdots, \mu_n$, the derivative $p'(z)$ of the polynomial $p(z) = (z - \alpha_1)^{\mu_1} (z - \alpha_2)^{\mu_2} \cdots (z - \alpha_n)^{\mu_n}$ has a zero in $N(z_0)$.*

We phrase the proof for the case that z_0 is not collinear with any two of the points α_k; the necessary modifications for the contrary case are obvious.* Then z_0 lies interior to a triangle whose vertices are three of the given points α_k, say α_1, α_2, and α_3. A non-degenerate triangle can be constructed whose sides directed counterclockwise are respectively parallel to, and indeed suitable positive multiples of, the vectors $z_0 - \alpha_1$, $z_0 - \alpha_2$, $z_0 - \alpha_3$. Thus we have $m_1'(z_0 - \alpha_1) + m_2'(z_0 - \alpha_2) + m_3'(z_0 - \alpha_3) = 0$, where m_1', m_2', and m_3' are positive, whence with $m_k' = m_k/(z_0 - \alpha_k)(\bar{z}_0 - \bar{\alpha}_k)$

$$\frac{m_1}{\bar{z}_0 - \bar{\alpha}_1} + \frac{m_2}{\bar{z}_0 - \bar{\alpha}_2} + \frac{m_3}{\bar{z}_0 - \bar{\alpha}_3} = 0,$$

where m_1, m_2, and m_3 are positive, If the positive rational numbers r_1, r_2, r_3 approach respectively m_1, m_2, m_3, and the positive rational numbers $r_4, r_5, \cdots, r_n$ approach zero, then a zero z of the function

$$\frac{r_1}{z - \alpha_1} + \frac{r_2}{z - \alpha_2} + \cdots + \frac{r_n}{z - \alpha_n} \tag{1}$$

approaches z_0, and hence for suitable choice of $r_1, r_2, \cdots, r_n$ this zero of (1) lies in $N(z_0)$. If m is the least common multiple of the denominators of the latter numbers r_k, we need merely set $\mu_k = mr_k$ to complete the proof.

---

* It is not sufficient to assume here merely that z_0 is an interior point of Π; for instance if $p(z)$ is the polynomial $z^4 - 1$, the point $z = 0$ is interior to Π but not interior to a triangle whose vertices are zeros of $p(z)$.

The Corollary to Gauss's Theorem (§1.2) yields an extension of Lucas's Theorem:

COROLLARY. *If* Π *is the closed interior of the Lucas polygon for the points* $\alpha_1, \alpha_2, \cdots, \alpha_n$, *then* Π *contains the zeros of the derivative of the function*

$$(z - \alpha_1)^{\mu_1}(z - \alpha_2)^{\mu_2} \cdots (z - \alpha_n)^{\mu_n},$$

where all the μ_k *are positive. No zero of the derivative lies on the boundary of* Π *unless it is a point* α_k *with* $\mu_k > 1$, *or unless all the points* α_k *are collinear.*

Another general property of the zeros of the derivative of a polynomial is expressed in

THEOREM 2. *The zeros of a polynomial* $p(z)$ *of degree greater than unity and the zeros of its derivative* $p'(z)$ *have the same center of gravity.*

The center of gravity of the zeros of

$$p(z) = z^n + a_1 z^{n-1} + \cdots + a_n$$

is $-a_1/n$, as is the center of gravity of the zeros of

$$p'(z) = nz^{n-1} + (n - 1)a_1 z^{n-2} + \cdots + a_{n-1}.$$

The significance of Theorem 2 lies especially in the fact that any line L through the center of gravity of a finite set of points bounds two closed half-planes each containing one or more of those points; either all the points lie on L, or at least one point lies interior to each of the half-planes.

Under the conditions of Lucas's Theorem, the center of gravity of the α_k lies in Π, and is an interior point of Π unless all the α_k are collinear. Any point z_0 of Π has the property that any line through z_0 bounds two closed half-planes each containing one or more of the α_k; this property is shared by no point exterior to Π. If z_0 is an interior point of Π, then at least one α_k lies on each side of every line through z_0.

§1.4. Jensen's Theorem. For a real polynomial $p(z)$, non-real zeros occur in pairs of conjugate imaginaries; this symmetry may enable us to improve Lucas's Theorem as applied to $p(z)$. Using as diameter the segment joining each pair of conjugate imaginary zeros of $p(z)$, we construct a circle, which we shall call a *Jensen circle*. We shall prove [Jensen, 1913; proof due to Walsh, 1920a]:

JENSEN'S THEOREM. *Each non-real zero of the derivative* $p'(z)$ *of a real polynomial* $p(z)$ *lies on or within a Jensen circle for* $p(z)$. *A non-real point* z_0 *not a multiple zero of* $p(z)$ *nor interior to a Jensen circle for* $p(z)$ *is a zero of* $p'(z)$ *only if* $p(z)$ *has no real zeros and* z_0 *lies on all Jensen circles.*

§1.4.1. Proof. For convenience we first establish the

Lemma. *If two unit particles are situated at the points $+i$ and $-i$, the corresponding force is horizontal at every point on the unit circle C, has a vertical component directed away from the axis of reals at every non-real point exterior to C, and has a vertical component directed toward the axis of reals at every non-real point interior to C.*

For a given point z we replace the two particles at $+i$ and $-i$ by a double particle α equivalent to them in the sense that the force at z due to the two

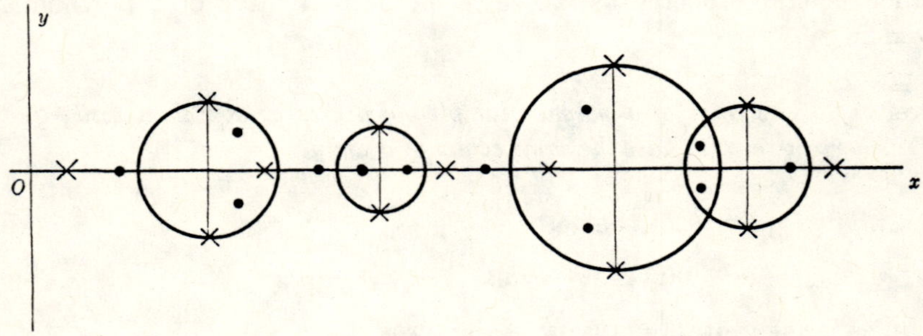

Fig. 2 illustrates §1.4 Jensen's Theorem

original particles is equal to the force at z due to the double particle. Then α is defined by the equation

$$\frac{1}{\bar{z}+i}+\frac{1}{\bar{z}-i}=\frac{2}{\bar{z}-\bar{\alpha}},$$

whence $\alpha = -1/z$; as a matter of fact, α is the harmonic conjugate of z with respect to $+i$ and $-i$. We may write

$$\arg \alpha = \pi - \arg z, \ |\alpha| = 1/|z|;$$

it follows that the half lines Oz and $O\alpha$ are equally inclined to and (if z is not real) on the same side of the axis of reals, and that $|\alpha|$ is equal to, less than, or greater than unity according as $|z|$ is equal to, greater than, or less than unity. The conclusion follows, and applies in essence to the field of force due to any pair of unit particles in conjugate imaginary points, if C now denotes the circle for which the particles lie in diametrically opposite points.

We proceed to the proof of Jensen's Theorem, by means of the field of force set up by Gauss's Theorem. At a point z_0 exterior to all Jensen circles but not on the axis of reals, it follows from the Lemma that the force due to each pair of particles situated at conjugate imaginary zeros of $p(z)$ has a non-vanishing vertical component directed away from the axis of reals. The force at z_0 due to each particle at a real zero of $p(z)$ also has such a component, so z_0 cannot be a position of equilibrium. Moreover z_0 cannot be a multiple zero of $p(z)$, hence cannot be a zero of $p'(z)$.

If the non-real point z_0 is not a multiple zero of $p(z)$ and is not interior to any Jensen circle but is a zero of $p'(z)$, then by the Lemma the force at z_0 due to either a pair of particles at conjugate imaginary zeros of $p(z)$ or a particle on the axis of reals can have no vertical component directed toward the axis of reals; since z_0 is a position of equilibrium such a force can have no vertical component directed away from the axis of reals, so z_0 lies on every Jensen circle and $p(z)$ has no real zeros.

When $p(z)$ is real with some non-real zeros, Jensen's Theorem may enable us to cut down the set Π assigned to the zeros of $p'(z)$ by Lucas's Theorem; yet neither of these theorems includes the other.

A fairly obvious consequence of the Lemma is the

COROLLARY. *If $p(z)$ is a real polynomial with no real zeros, then no non-real point interior to all the Jensen circles for $p(z)$ is a zero of $p'(z)$.*

Indeed, at a non-real point z_0 interior to all the Jensen circles the force due to each pair of zeros of $p(z)$ has a vertical component directed toward the axis of reals, so z_0 is not a position of equilibrium.

A special situation deserves mention here: *if $p(z)$ is a polynomial of degree four whose zeros lie at the vertices of a rectangle, the zeros of $p'(z)$ lie at the center of the rectangle, and at the intersection of the two circles whose diameters are the longer sides of the rectangle.* If the rectangle is a square, all zeros of $p'(z)$ lie at the center, for a suitable rotation and translation transforms $p(z)$ into a constant multiple of $z^4 - A$, where A is real. In the general case, it is a consequence of symmetry that the force at the center of the rectangle is zero; moreover, at an intersection z_0 of the two circles mentioned, it follows from the Lemma that the force due to each pair of particles situated at the extremities of a longer side of the rectangle is parallel to the shorter sides; by the symmetry these two forces are equal in magnitude and opposite in sense, so z_0 is a zero of $p'(z)$.

§1.4.2. Complements. If the fixed points $\alpha_1, \bar{\alpha}_1, \alpha_2, \bar{\alpha}_2, \cdots, \alpha_n, \bar{\alpha}_n$ are assigned as non-real zeros of unrestricted multiplicities of a real polynomial $p(z)$ which may have real zeros arbitrary in position and multiplicity, Jensen's Theorem describes a certain point set L known to contain all critical points. Namely L consists of the axis of reals, the interiors of all Jensen circles, the points α_k and $\bar{\alpha}_k$, and possible points z_0 which lie on all Jensen circles. Except for these points z_0, the set L is the actual *locus* of critical points of $p(z)$, as we proceed to demonstrate. In this sense Jensen's Theorem cannot be improved. We thus establish [compare Walsh, 1920a, where a weaker theorem is given] a result stronger than the exact analog of §1.3.2 Theorem 1.

THEOREM 1. *Let non-real numbers $\alpha_1, \bar{\alpha}_1, \alpha_2, \bar{\alpha}_2, \cdots, \alpha_n, \bar{\alpha}_n$ be given, and let*

z_0 be a point interior to one of the corresponding Jensen circles. Then there exists a real polynomial $p(z)$ with zeros in all points α_k and $\bar{\alpha}_k$ and with no other non-real zeros, for which z_0 is a critical point.

It is clear that an arbitrary real point is a multiple zero and hence a critical point of a suitably chosen $p(z)$; this conclusion applies also to each of the points α_k and $\bar{\alpha}_k$. Let the non-real point z_0 lie for definiteness interior to the Jensen circle for the pair α_1 and $\bar{\alpha}_1$, and let z_0 be distinct from the α_k and $\bar{\alpha}_k$. It follows from the Lemma that the vector

$$\frac{1}{\bar{z}_0 - \bar{\alpha}_1} + \frac{1}{\bar{z}_0 - \alpha_1}$$

has a non-vanishing component directed toward the axis of reals. Consequently if the positive integers $m_1, m_2, \cdots, m_n$ are suitably chosen, with m_1 large in comparison with $m_2, \cdots, m_n$, the vector

$$A = \frac{m_1}{\bar{z}_0 - \bar{\alpha}_1} + \frac{m_1}{\bar{z}_0 - \alpha_1} + \frac{m_2}{\bar{z}_0 - \bar{\alpha}_2} + \frac{m_2}{\bar{z}_0 - \alpha_2} + \cdots + \frac{m_n}{\bar{z}_0 - \bar{\alpha}_n} + \frac{m_n}{\bar{z}_0 - \alpha_n}$$

also has a non-vanishing component directed toward the axis of reals. Let the line of action of the vector A (with initial point z_0) cut the axis of reals in the point β, and let the integer m be chosen so that the force at z_0 due to a $(2m)$-fold particle at β is in magnitude greater than A.

The particles introduced in Gauss's Theorem are especially useful and simple because the force at a point z due to a unit particle at a point ζ is in magnitude, direction, and sense the vector $\zeta' z$, where ζ' is the inverse of ζ in the unit circle whose center is z. This use of inversion throughout the present work frequently effects a simplification in our geometric configuration, as in the case at hand. Invert in the unit circle whose center is z_0, denote by C' the image of the axis of reals, and by β' the image of β. If $z_0\alpha$ is the vector $A/(2m)$, the point α lies interior to the circle C' on the line segment (z_0, β'). A suitably chosen chord $\beta_1' \beta_2'$ of C' is bisected by α, and the sum of the vectors $z_0\beta_1'$ and $z_0\beta_2'$ is twice the vector $z_0\alpha$, hence is equal to A/m. If β_1 and β_2 denote the inverses of β_1' and β_2' in the unit circle whose center is z_0, it now follows that the total force at z_0 due to m-fold particles at each of the real points β_1 and β_2 is $-A$, so z_0 is a critical point of the polynomial $(z - \beta_1)^m(z - \beta_2)^m(z - \alpha_1)^{m_1}(z - \bar{\alpha}_1)^{m_1} \cdots (z - \alpha_n)^{m_n} \cdot (z - \bar{\alpha}_n)^{m_n}$; if β_1' or β_2' coincides with z_0, we simply drop the corresponding factor $(z - \beta_1)^m$ or $(z - \beta_2)^m$ here; Theorem 1 is established.

We shall discuss in more detail in Chapter II the possibility of improving Jensen's Theorem by restricting the multiplicities of various zeros of $p(z)$ and the positions of the real zeros.

Throughout our discussion of §§1.3 and 1.4 we have used the language of fields of force, as did Gauss, Lucas, and later writers, and as we shall continue to do. Any analytic function or its conjugate can be interpreted as defining such a field. It is entirely in keeping to study fields of force in a subject closely related to harmonic functions and potential theory. However, as Bôcher [1904] says in this connection, "the proofs of the theorems which we have here deduced from mechanical intuition can readily be thrown, without essentially modifying their character, into purely algebraic form. The mechanical problem must nevertheless be regarded as valuable, for it suggests not only the theorems but also the method of proof."

§1.5. Walsh's Theorem. We have already (§1.2) indicated that the zeros of the derivative of the polynomial $p(z) = (z - \alpha_1)^{m_1} (z - \alpha_2)^{m_2}$ are α_1 (if $m_1 > 1$), α_2 (if $m_2 > 1$), and the point $(m_2\alpha_1 + m_1\alpha_2)/(m_1 + m_2)$; it follows that if the zeros of $p(z)$ are allowed to vary slightly from these positions, those of $p'(z)$ also vary but slightly. A broader and more precise result can be obtained [1920, 1921]:

WALSH'S THEOREM. *Let the closed interiors of the circles $C_1 : | z - \alpha_1 | = r_1$ and $C_2 : | z - \alpha_2 | = r_2$ be the respective loci of m_1 and m_2 independent zeros of a variable polynomial $p(z)$ of degree $m_1 + m_2$. Then the locus of the zeros of the derivative $p'(z)$ consists of the closed interior of C_1 (if $m_1 > 1$), the closed interior of C_2 (if $m_2 > 1$), and the closed interior of the circle*

$$C: \left| z - \frac{m_2\alpha_1 + m_1\alpha_2}{m_1 + m_2} \right| = \frac{m_2 r_1 + m_1 r_2}{m_1 + m_2}. \tag{1}$$

Any of these circles C_1, C_2, or C which is exterior to the other two circles contains respectively the following number of zeros of $p'(z)$: $m_1 - 1$, $m_2 - 1$, 1.

If $m_k = 1$, we omit the closed interior of C_k, $k = 1$ or 2, in *enumerating* the regions composing the locus of the zeros of $p'(z)$, but parts or all of these regions may actually belong to that locus.

§1.5.1. Preliminaries. In the proof we shall make use of several lemmas. The first of these is equivalent to a lemma due to Laguerre [1878], although his statement was expressed in algebraic form and thereby loses some of the simplicity and elegance due to the mechanical interpretation. By a *circular region* we understand the closed interior of a circle, the closed exterior of a circle, or a closed half-plane; as degenerate cases we include here a single point and the entire plane, for which the following property is trivial:

LEMMA 1. *If the point P lies exterior to a circular region R containing m unit particles, the corresponding resultant force at P is equivalent to the force at P due to m coincident unit particles located at a suitably chosen point of R. This conclusion persists if P lies on the boundary of R, provided no particle coincides with P.*

The force at P due to a unit particle at any point Q is in magnitude, direction, and sense the vector $Q'P$, where Q' denotes the inverse of Q in the unit circle whose center is P. To replace m particles Q by m coincident particles, we replace the m vectors $Q'P$ by m coincident equivalent vectors, and hence replace their initial points Q' by the center of gravity of the m points Q'. Denote by R' the circular region which is the inverse of R in the unit circle whose center is P; since P is either exterior to or on the boundary of R, it follows that R' is the closed interior of a circle or a closed half-plane, which is convex. Since all the points Q lie in R, all the points Q' lie in R', as does their center of gravity Q_0'. The inverse Q_0 of Q_0' lies in R, as we were to prove. The point Q_0 depends on P and on the given particles, but not on R; thus if each of a number of circular regions R contains the given particles but does not contain P (assumed not at a given particle) in its interior, then each of the regions R contains Q_0.

In Lemma 1 with P exterior to R, the point Q_0 cannot lie on the boundary of R unless Q_0' lies on the boundary of R', and unless all the given m particles coincide on the boundary of R. With P on the boundary of R, the point Q_0 lies on the boundary of R when and only when all the given m particles lie on the boundary of R.

In this lemma we do not exclude the possibility that if R is infinite a particle may lie at infinity, and is then defined to exert a zero force at P. If a given particle does lie at infinity, we may consider in effect in the lemma m finite particles equivalent to n $(> m)$ finite or infinite coincident particles, so far as concerns the force exerted at P.

The following lemma is closely related to prior results of H. Bohr and Minkowski on convex sets:

LEMMA 2. *If the points z_1 and z_2 vary independently and have as respective loci the closed interiors of the circles $C_1 : | z - \alpha_1 | = r_1$ and $C_2 : | z - \alpha_2 | = r_2$, then the locus of the point $z_0 = (m_2 z_1 + m_1 z_2)/(m_1 + m_2)$ which divides the segment (z_1, z_2) in the constant ratio $m_1 : m_2$ (where $m_1 > 0$, $m_2 > 0$) is the closed interior of the circle C given by* (1).

Since we have $| z_1 - \alpha_1 | \leqq r_1$, $| z_2 - \alpha_2 | \leqq r_2$, we also have

$$\left| \frac{m_2 z_1 + m_1 z_2}{m_1 + m_2} - \frac{m_2 \alpha_1 + m_1 \alpha_2}{m_1 + m_2} \right| = \left| \frac{m_2(z_1 - \alpha_1)}{m_1 + m_2} + \frac{m_1(z_2 - \alpha_2)}{m_1 + m_2} \right|$$
$$\leqq \frac{m_2 r_1 + m_1 r_2}{m_1 + m_2},$$

so z_0 lies on or within C. Reciprocally, if z_0 is given on or within C:

$$\left| z_0 - \frac{m_2 \alpha_1 + m_1 \alpha_2}{m_1 + m_2} \right| \leqq \frac{m_2 r_1 + m_1 r_2}{m_1 + m_2},$$

we define z_1 and z_2 by the equations

$$(2) \qquad \frac{z_1 - \alpha_1}{r_1} = \frac{z_2 - \alpha_2}{r_2} = \left(z_0 - \frac{m_2 \alpha_1 + m_1 \alpha_2}{m_1 + m_2} \right) \Big/ \frac{m_2 r_1 + m_1 r_2}{m_1 + m_2},$$

from which we deduce that $| z_1 - \alpha_1 | \leqq r_1$, $| z_2 - \alpha_2 | \leqq r_2$, and $z_0 = (m_2z_1 + m_1z_2)/(m_1 + m_2)$, so z_0 belongs to the locus and Lemma 2 is established.

The center of C divides the segment (α_1, α_2) in the ratio $m_1 : m_2$. If r_1 and r_2 are equal, all three circles have the same radius, and any line parallel to the line of centers cutting one circle cuts all three at the same angle. If r_1 and r_2 are unequal, the point $\alpha_0 = (r_2\alpha_1 - r_1\alpha_2)/(r_2 - r_1)$ is an external center of similitude for any pair of the three circles, in the sense that a suitable stretching or shrinking of the plane with α_0 fixed and without change of direction transforms any of those circles into any other. Any line through α_0 cuts those three circles at the same angle, and this is true even if the circles are all oriented counterclockwise. If C_1 and C_2 have common external tangents, these lines are tangent to C. If C_2 lies interior to C_1, C separates C_1 and C_2. Whenever z_0 lies on the boundary of its locus, the points z_1 and z_2 lie on the boundary of their loci, and the line joining these three points passes through α_0 (is parallel to the line $\alpha_1\alpha_2$ if $r_1 = r_2$) and cuts the oriented circles C_1, C_2, and C at z_1, z_2, and z_0 at the same angle; equations (2) are satisfied.

If z_1 and z_2 have as loci not the closed interiors of C_1 and C_2 but the closed segments in those closed regions of a line L through α_0 (or parallel to $\alpha_1\alpha_2$ if we have $r_1 = r_2$), the corresponding locus of z_0 is the portion of L in the closed interior of C.

Any point z_0 common to the closed interiors of C_1 and C_2 must lie in the closed interior of C, for we may set $z_1 = z_2 = z_0$; any such point which lies on C must also lie on C_1 and C_2 at a point where the three circles are tangent (if the three circles do not coincide).

§1.5.2. Proof. We are now in a position to prove the theorem. Suppose a point z exterior to C_1 and C_2 to be a position of equilibrium in the field of force set up in Gauss's Theorem. The force at z due to the m_1 particles on or within C_1 is equivalent to the force at z due to m_1 particles coinciding at some point z_1 on or within C_1, and the force at z due to the m_2 particles on or within C_2 is equivalent to the force at z due to m_2 particles coinciding at some point z_2 on or within C_2. Consequently (§1.2) z divides the segment (z_1, z_2) in the ratio $m_1 : m_2$ and hence lies on or within C. Therefore the three circles C_1, C_2, C contain on or within them all positions of equilibrium; they contain also all multiple zeros of $p(z)$, for even if $m_1 = 1$ or $m_2 = 1$ or both, each possible multiple zero of $p(z)$ lies in C_1, C_2, or C.

If we have $m_1 = 1$ and if z lies on or within C_1 but exterior to C_2 and C, then by the reasoning as given z can be neither a position of equilibrium nor a multiple zero of $p(z)$, hence is not a point of the locus. If we have $m_2 = 1$ and if z lies on or within C_2 but exterior to C_1 and C, then z is not a point of the locus. If z lies on or within both C_1 and C_2, it lies on or within C, and may be either a position of equilibrium or a multiple zero of $p(z)$. Thus every possible zero of $p'(z)$ belongs to the locus of the zeros of $p'(z)$ as described in the theorem.

Any point z on or within C divides the segment (z_1, z_2) in the ratio $m_1 : m_2$ for suitable choice of z_1 and z_2 in their assigned loci, so any such point belongs to the locus of zeros of $p'(z)$. If m_k is greater than unity, $k = 1$ or 2, each point on or within C_k may be a multiple zero of $p(z)$ and belongs to this locus. Consequently the locus of the zeros of $p'(z)$ is as stated in the theorem.

To prove the remainder of the theorem we use the *method of continuity* used by Bôcher [1904] in a similar situation. Allow the m_k zeros of $p(z)$ whose locus is the closed interior of C_k to move continuously remaining in their locus, and to coalesce at α_k, $k = 1, 2$. In the final position the numbers of zeros of $p'(z)$ are respectively $m_1 - 1$, $m_2 - 1$, and 1 at the centers of C_1, C_2, C. Throughout the motion, the zeros of $p'(z)$ move continuously, none can enter or leave the closed interior of a circle C_1, C_2, C which lies exterior to the other two circles, so there has been no change in the number of zeros of $p'(z)$ on or within such a circle due to the motion. The proof is now complete.

In connection with Walsh's Theorem, we study for later use the conditions that a point z_0 with $p'(z_0) = 0$ can be on the boundary of its locus, using the

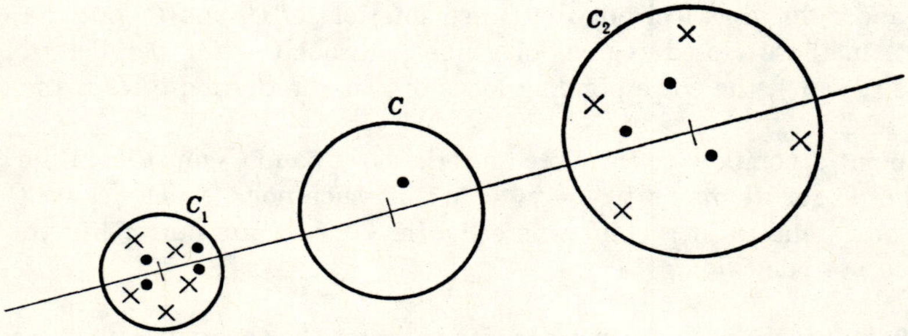

Fig. 3 illustrates §1.5 Walsh's Theorem

methods already developed in proving the theorem and the remarks already made concerning boundary points under the conditions of Lemmas 1 and 2. If z_0 lies on C exterior to C_1 and C_2, the m_1 zeros of $p(z)$ on or within C_1 coincide at some point z_1 on C_1, the m_2 zeros of $p(z)$ on or within C_2 coincide at some point z_2 on C_2, and (2) is satisfied. If z_0 lies on C interior to C_1 with $m_1 = 1$ but is exterior to C_2, the m_2 zeros of $p(z)$ on or within C_2 coincide at some point z_2 on C_2, and if z_1 denotes the zero of $p(z)$ on or within C_1, we have (2) satisfied. If z_0 lies on C_1 but is exterior to C and C_2, then we must have $m_1 > 1$, and z_0 is a multiple zero of $p(z)$. If z_0 lies on C_2 but is exterior to C and C_1, then we have $m_2 > 1$ and z_0 is a multiple zero of $p(z)$. If z_0 not a multiple zero of $p(z)$ lies on both C_1 and C exterior to C_2, the m_2 zeros of $p(z)$ on or within C_2 coincide at some point z_2 on C_2, and (2) is satisfied so far as concerns z_2; if we have $m_1 = 1$, the zero z_1 of $p(z)$ on or within C_1 must lie on C_1 so that (2) is satisfied; if we have $m_1 > 1$, the zeros of $p(z)$ on or within C_1 must lie on C_1, but surprisingly an infinity of positions of these zeros are possible. If z_0 lies on C_1 and C_2, then z_0 lies interior to C except in the case where all three circles either are tangent at a single point, or coincide; in this excepted case z_0 is on the boundary of the locus of Lemma 2, and z_0 must be a multiple zero of $p(z)$.

We add one further remark concerning the proof. A necessary and sufficient condition that a point P be a position of equilibrium in the field of force is that P be the center of gravity of the inverses of *all* the particles with respect to the unit circle whose center is P. The theorem is readily proved by use of this fact.

The theorem applies when C_1 or C_2 degenerates to a point, and thus may be useful in paring down the corners of the Lucas polygon, as well as being useful if C_1 and C_2 are relatively far apart.

In the theorem the circles C_1, C_2, C if not mutually exterior may contain more than $m_1 - 1$, $m_2 - 1$, 1 critical points respectively; indeed if those circles are all relatively large, each of them may contain both α_1 and α_2 and hence may contain all the zeros of both $p(z)$ and $p'(z)$.

By the use of Lucas's Theorem, or directly by the method already used, it is readily shown that if the locus of m_k zeros of $p(z)$ of degree $m_1 + m_2$ is no longer the closed interior of C_k but is the closed segment in that closed region of a line L through α_0 (or parallel to $\alpha_1\alpha_2$ if we have $r_1 = r_2$), $k = 1, 2$, then the locus of critical points consists of those segments (insofar as we have $m_k > 1$) plus the segment of L in the closed interior of C; compare §2.1.

In the theorem, it is admissible to choose C_1 and C_2 identical, in which case C coincides with them. We thus obtain a new formulation, which is equivalent to Lucas's Theorem in the sense that either proposition is readily proved from the other:

If the open or closed interior of a circle contains all zeros of a polynomial, it contains all zeros of the derived polynomial.

§1.6. Lemniscates.

In order to study the level curves for a polynomial or its real and pure imaginary components, we first consider such curves for analytic functions in general.

§1.6.1. Level curves.

Let the function $w = f(z) = u(x, y) + iv(x, y)$ be analytic but not identically constant throughout the neighborhood of (x_0, y_0). The locus $L_1 : u(x, y) = u(x_0, y_0)$ can be considered as the image in the z-plane of a line $u =$ const in the plane of $w = u + iv$. If $f'(z)$ is different from zero at the point $z_0 = x_0 + iy_0$, then the transformation $w = f(z)$ maps a neighborhood of z_0 in the z-plane smoothly onto a region containing $w_0 = f(z_0)$ in the w-plane, so in the neighborhood of z_0 the locus L_1 consists of a single analytic arc passing through z_0. If $f'(z)$ has an m-fold zero at z_0, we have the development $f(z) - w_0 = w - w_0 = (z - z_0)^{m+1}[a_0 + a_1(z - z_0) + \cdots]$, $a_0 \neq 0$, from which we have by two applications of the implicit function theorem, for a suitable choice of the $(m + 1)$st roots and of the coefficients b_k and c_k,

$$(w - w_0)^{1/(m+1)} = (z - z_0)[b_0 + b_1(z - z_0) + \cdots], \qquad b_0 \neq 0,$$

$$z - z_0 = c_1(w - w_0)^{1/(m+1)} + c_2(w - w_0)^{2/(m+1)} + \cdots, \qquad c_1 \neq 0;$$

this representation is valid for sufficiently small $w - w_0$, and for a suitably chosen $(m + 1)$st root. The transformation $w = f(z)$ maps a neighborhood of z_0 in the z-plane uniformly onto a region containing w_0 in the Riemann surface for $(w - w_0)^{1/(m+1)}$ over the w-plane, and in the map angles at z_0 are multiplied $(m + 1)$-fold. Thus in the neighborhood of z_0 the locus L_1 consists of $m + 1$ analytic arcs through z_0 intersecting each other at z_0 at successive angles of $\pi/(m + 1)$. The locus L_2: $v(x, y) = v(x_0, y_0)$ is likewise the image in the z-plane of a line $v =$ const in the w-plane. If we have $f'(z_0) \neq 0$, then in the neighborhood of z_0 the locus L_2 consists of a single analytic arc, orthogonal to L_1 at z_0; if $f'(z)$ has an m-fold zero at z_0, then in the neighborhood of z_0 the locus L_2 consists of $m + 1$ analytic arcs through z_0 cutting each other at z_0 in successive angles of $\pi/(m + 1)$ and bisecting the angles between successive arcs of L_1 at z_0. Of course the functions $u(x, y)$ and $v(x, y)$ cannot have maxima or minima at (x_0, y_0), and the neighborhood of (x_0, y_0) is divided by L_1 into $2(m + 1)$ regions in which we have alternately $u(x, y) > u(x_0, y_0)$ and $u(x, y) < u(x_0, y_0)$, where m is the multiplicity of z_0 as a zero of $f'(z)$ if $f'(z_0) = 0$, and otherwise $m = 0$. Geometrically in three dimensions the surfaces representing $u(x, y)$ and $v(x, y)$ have inclined tangent planes at the points for which $(x, y) = (x_0, y_0)$ if $f'(z_0) \neq 0$, and have saddle points (perhaps multiple) if $f'(z_0) = 0$.

Suppose the function $w = f(z)$ analytic at z_0 but not identically constant in the neighborhood of z_0; the locus $\lambda_1 : | f(z) | = | f(z_0) |$ in that neighborhood is the image in the z-plane of the circle (perhaps degenerate) $| w | = | f(z_0) |$, and the locus $\lambda_2 : \arg [f(z)] = \arg [f(z_0)]$ is the image in the z-plane of the half-line $\arg w = \arg [f(z_0)]$. If we have $f'(z_0) \neq 0$, $f(z_0) \neq 0$, these loci λ_1 and λ_2 consist in the neighborhood of z_0 each of an analytic Jordan arc through z_0, and the two loci are orthogonal there. If $f'(z)$ has an m-fold zero at z_0 but $f(z_0) \neq 0$, each of the loci consists in the neighborhood of z_0 of $m + 1$ analytic Jordan arcs through z_0 making successive angles $\pi/(m + 1)$ at z_0; the arcs of λ_1 bisect the angles between successive arcs of λ_2 at z_0. If we have $f(z_0) = 0$, the locus λ_1 consists in the neighborhood of z_0 of the single point z_0, and λ_2 consists of $m + 1$ analytic Jordan arcs terminating in z_0 and intersecting there at successive angles of $2\pi/(m + 1)$, where m is the multiplicity of z_0 as a zero of $f'(z)$ if $f'(z_0) = 0$ and otherwise $m = 0$. The functions $| f(z) |$ and $\arg [f(z)]$ have no maxima nor minima at z_0, except that $| f(z) |$ has a minimum if $f(z_0) = 0$, so geometrically in three dimensions the surfaces representing these functions (with $f(z_0) \neq 0$) have inclined tangent planes except at a point corresponding to a critical point z_0 of $f(z)$, and at such a point the surfaces have saddle points.

Of course if we set $F(z) = e^{f(z)} = e^{u+iv}$, then for $F(z)$ the loci $\lambda_1 : | F(z) | = e^u =$ const $\neq 0$ and $\lambda_2 : \arg [F(z)] = v =$ const are precisely the loci L_1 and L_2 for $f(z)$. Reciprocally, if we set $\varphi(z) = \log [f(z)] = \log | f(z) | + i \arg | f(z) |$, then for $\varphi(z)$ the loci $L_1 : \Re[\varphi(z)] =$ const and $L_2 : \Im[\varphi(z)] =$ const are precisely the loci λ_1 (except perhaps for the locus $| f(z) | = 0$) and λ_2 for $f(z)$.

§1.6.2. Lemniscates and their orthogonal trajectories. [Lucas, 1879, 1888].

If the points $\alpha_1, \alpha_2, \cdots, \alpha_n$ and the number $\mu(> 0)$ are fixed, the locus

$$(1) \qquad | p(z) | \equiv | (z - \alpha_1)(z - \alpha_2) \cdots (z - \alpha_n) | = \mu$$

is called a *lemniscate*, and the points α_j are called its *poles*. The set of loci (1), where μ varies, is a family of lemniscates. A unique locus $| p(z) | = | p(z_0) |$ of the family passes through an arbitrary point z_0 distinct from the α_j. Equation (1) can also be written

$$\prod_{k=1}^{n} [(z - \alpha_k)(\bar{z} - \bar{\alpha}_k)] = \mu^2,$$

so (1) is an algebraic curve of degree $2n$, not necessarily irreducible. The terms in x and y of highest degree can be written $(x^2 + y^2)^n$, so the curve is by definition an *n-circular* $(2n)$-*ic*. The special case $| z^2 - 1 | = 1$ is the Bernoullian lemniscate, which is a bicircular quartic.

Some properties both of the individual locus and of the family can be deduced at once from the general properties of level curves (§1.6.1) and from the principle of maximum modulus for an analytic function. Thus we have $p(\alpha_k) = 0$, and $| p(z) |$ becomes infinite as z becomes infinite, so the locus (1) separates each point α_k from the point at infinity. No point of the locus (1) can lie interior to a Jordan curve consisting wholly of points of (1); each such Jordan curve must contain at least one point α_k in its interior. Consequently the locus (1) consists of a finite number of bounded Jordan curves, each containing in its interior one or more of the points α_k; these Jordan curves are mutually exterior except that a multiple point of (1) may belong to several such curves; at such a point z_0 we have $p'(z_0) = 0$, and the tangents to the branches of (1) are equally spaced. Since each multiple point of (1) is a critical point of $p(z)$, no more than $n - 1$ lemniscates of the family possess multiple points. If the locus (1) consists of k Jordan curves, mutually disjoint except for multiple points of (1), then (1) separates the plane into precisely $k + 1$ regions, of which k are finite and characterized by the inequality $| p(z) | < \mu$ and one of which is infinite and defined by the inequality $| p(z) | > \mu$.

Obviously the two loci $| p(z) | = \mu_1 \ (> 0)$ and $| p(z) | = \mu_2 \ (> \mu_1)$ cannot intersect, and the latter separates the former from the point at infinity; each point of the former must lie interior to one of the Jordan curves which compose the latter. Thus when μ is small and positive, the lemniscate (1) consists of small Jordan curves, one about each of the distinct points α_k; the general theory of schlicht functions shows that these curves are convex for μ sufficiently small. As μ increases, these Jordan curves increase monotonically in size, but (as topological considerations show) the number of the Jordan curves in a lemniscate does not change unless a limiting locus has a multiple point and thus passes through a zero of $p'(z)$. If the locus $| p(z) | = \mu_1$ consists of k Jordan curves, and passes through zeros of $p'(z)$ of total multiplicity m, then for μ smaller than μ_1 but sufficiently near to μ_1 the locus (1) also consists of k Jordan curves, and these are mutually disjoint, but for μ larger than μ_1 and sufficiently near to μ_1 the locus (1) separates the plane into m fewer regions than does $| p(z) | = \mu_1$ and hence consists of $k - m$ mutually disjoint Jordan curves. If q zeros of $p'(z)$ lie in the region $|p(z)| > \mu$ and none on (1), then (1) consists of precisely $q + 1$ Jordan curves. For sufficiently large μ, the locus (1) consists of a single analytic Jordan curve, which we shall show (§1.6.3) to be approximately circular in shape.

The loci

$$\arg [p(z)] = \text{const} \tag{2}$$

likewise form a family covering the plane; through each point other than a point α_j passes a (geometrically) unique locus (2). In the neighborhood of α_j, the locus (2) consists of m analytic Jordan arcs terminating in α_j, where m is the multiplicity of α_j as a zero of $p(z)$; in the neighborhood of the point at infinity the locus (2) consists of n analytic Jordan arcs terminating at infinity. Only a finite number of the loci (2) have multiple points other than points α_j or the point at infinity, namely those loci passing through zeros of $p'(z)$. Except for those loci, which are not more than $n - 1$ in number, each locus (2) consists of n analytic Jordan arcs, mutually disjoint except for the point at infinity and except for multiple points α_j, each arc extending from a point α_j to the point at infinity. Each exceptional locus also consists of n Jordan arcs each extending from a point α_j to the point at infinity; each of these n arcs is analytic except perhaps in the zeros of $p'(z)$ different from the α_j, and the interiors of the arcs are mutually disjoint except for such points; at an m-fold zero of $p'(z)$ not an α_j, precisely $m + 1$ arcs of (2) cross, and successive tangents to them make angles of $\pi/(m + 1)$.

Whenever a locus (2) cuts a locus (1), except perhaps at a zero of $p'(z)$, the two loci cut orthogonally. At a zero of $p'(z)$ other than a point α_j, the set of tangents to one locus bisects the angles between successive tangents to the other locus.

In the w-plane, the normal to the curve $|w| = \text{const}$ in the direction of increasing $|w|$ has the direction $\arg w$. If we consider the w-plane and z-plane to be related by the transformation $w = p(z)$, then except perhaps at a critical point of the transformation the normal to the locus (1) has the direction

$$\arg w + \arg\left(\frac{dz}{dw}\right) = \arg \frac{p(z)}{p'(z)} = -\arg \frac{p'(z)}{p(z)};$$

indeed this conclusion is general, and depends on the analyticity but not the polynomial character of $p(z)$. Thus except at a critical point of $p(z)$, *the normal to the locus* (1) *in the direction of increasing* $|p(z)|$ *is the direction of the force in the field set up by Gauss's Theorem. The lines of force are the loci* (2), *sensed from the points* α_j *to infinity*, and at each point not a critical point represent the direction and sense of the force. Of course it follows here that the normal at z to the locus (1) (which is the tangent at z to the locus (2)) extended in the sense of decreasing $|p(z)|$ must intersect the Lucas polygon Π (§1.3), if z is exterior to Π; each force exerted at z by a particle at α_j is represented by a vector $\alpha_j'z$, where α_i' lies in the angle subtended at z by Π, so the sum of the vectors is a vector $\alpha' z$, where α' also lies in that angle. All critical points lie on or within Π, and the normals to the locus (1) and tangents to the locus (2) at a critical point also intersect Π.

The locus (2) is not always a complete algebraic curve, as is illustrated by the example $p(z) = z$, for which the locus (2) is a half-line. However, the two following equations are equivalent to each other:

$$\arg [p(z)] = \varphi \text{ or } \varphi + \pi, \qquad \Re[e^{-i\varphi} p(z)] = 0;$$

the latter equation represents an algebraic curve of degree n, which need not be irreducible.

§1.6.3. Behavior at infinity. Lucas [1879, 1888] studied the behavior of the loci (1) and (2) not merely at finite points but also in the neighborhood of the point at infinity. In order to set forth his results according to modern standards of rigor, we consider [Walsh, 1936] the behavior of the polynomial $p(z)$ itself in that neighborhood. Broadly considered, $p(z)$ behaves there approximately like the polynomial $(z - \alpha_0)^n$, where α_0 is the center of gravity of the points α_k. More explicitly, we prove

THEOREM 1. *Let there be given the polynomial* $p(z) = (z - \alpha_1)(z - \alpha_2) \cdots (z - \alpha_n)$. *For* $|A|$ *sufficiently large, the zeros of the polynomial*

$$p(z) - A \tag{3}$$

lie in the neighborhoods of the zeros of the polynomial $(z - \alpha_0)^n - A$, *where* $\alpha_0 = (\alpha_1 + \alpha_2 + \cdots + \alpha_n)/n$ *is the center of gravity of the* α_k. *That is to say, let* $\epsilon(>0)$ *be given; there exists* M_ϵ *such that* $|A| > M_\epsilon$ *implies that every zero* z *of* (3) *satisfies an inequality* $|z - (\alpha_0 + A^{1/n})| < \epsilon$, *where* $A^{1/n}$ *is a suitably chosen n-th root of* A.

By means of Theorem 1 we shall establish the

COROLLARY. *When* z *becomes infinite, a zero* α *of the equation in* α

$$p(z) = (z - \alpha)^n \tag{4}$$

approaches the point α_0. *That is to say, let* $\epsilon > 0$ *be given; there exists* R_ϵ *such that* $|z| > R_\epsilon$ *implies that equation* (4) *has a zero* α *in the region* $|\alpha - \alpha_0| < \epsilon$.

Of course $|p(z)|$ becomes infinite when and only when $|z|$ becomes infinite, so the Corollary is a consequence of Theorem 1. Indeed, suppose $\epsilon > 0$ is given; if R_ϵ is suitably chosen, the inequality $|z| > R_\epsilon$ implies $|p(z)| > M_\epsilon$. If z_0 is arbitrary, with $|z_0| > R_\epsilon$, we have $|p(z_0)| > M_\epsilon$, and Theorem 1 implies that any zero of the polynomial $p(z) - p(z_0)$ satisfies the inequality

$$|z_0 - (\alpha_0 + [p(z_0)]^{1/n})| < \epsilon,$$

for a suitably chosen n-th root, whence for a suitably chosen zero α of (4) with $z = z_0$, we have $|\alpha - \alpha_0| < \epsilon$, as we were to prove.

In the proof of Theorem 1, two lemmas are convenient:

LEMMA 1. *For sufficiently small values of* ζ *and a suitable branch of the function* $\log(1 - \zeta)$ *we have*

$$|\log(1 - \zeta) + \zeta| \leqq m|\zeta^2|,$$

where the constant m *is independent of* ζ.

Here we need merely set

$$(5)\qquad \begin{aligned} \log(1 - \zeta) &= -\zeta - \zeta^2/2 - \zeta^3/3 - \cdots, & |\zeta| < 1, \\ \log(1 - \zeta) + \zeta &= -\zeta^2[\tfrac{1}{2} + \zeta/3 + \zeta^2/4 + \cdots], & |\zeta| < 1. \end{aligned}$$

When we have $|\zeta| < r(<1)$, the square bracket has a meaning and its modulus has a bound m independent of ζ.

LEMMA 2. *For sufficiently small values of* β *we have*

$$|e^\beta - 1| \leqq m_1|\beta|,$$

where the constant m_1 *is independent of* β.

We write

$$e^\beta - 1 = \beta[1 + \beta/2! + \beta^2/3! + \cdots];$$

the series in brackets converges for all values of β, uniformly for all β in modulus less than a given bound, so the sum of the series is bounded uniformly for all β less than a given bound.

In the proof of Theorem 1 we choose $\alpha_0 = 0$, a choice which involves no loss of generality. The zeros of the polynomial (3) are formally zeros of the equation

$$(6)\qquad \frac{1}{n}\sum_1^n \log\left(1 - \frac{\alpha_k}{z}\right) = \log\frac{A^{1/n}}{z}.$$

If z is a zero of (3), and if $|z|$ is sufficiently large, the first member of (6) can be defined uniquely by means of (5), and we can consequently define $A^{1/n}$ from (6) by means of the exponential function. Thanks to the choice $\alpha_0 = 0$ we may write (6) as

$$\frac{1}{n}\sum_1^n\left[\log\left(1 - \frac{\alpha_k}{z}\right) + \frac{\alpha_k}{z}\right] = \log\frac{A^{1/n}}{z}.$$

It follows from Lemma 1 that for $|z|$ sufficiently large, and in any case greater than every $|\alpha_k|$, we have

$$\left|\log\frac{A^{1/n}}{z}\right| \leqq \frac{m'}{|z|^2},$$

where m' is a suitably chosen constant independent of z, and hence it follows from Lemma 2 that we have, again for $|z|$ sufficiently large,

$$\left|\frac{A^{1/n}}{z} - 1\right| \leqq m_1 \left|\log \frac{A^{1/n}}{z}\right| \leqq \frac{m_1 m'}{|z|^2},$$

$$|A^{1/n} - z| \leqq m_1 m'/|z|. \tag{7}$$

If now ϵ is given, there exists m_2 such that $|z| > m_2$ implies both (7) and the inequality $m_1 m'/|z| < \epsilon$. There exists M_ϵ such that $|p(z)| > M_\epsilon$ implies $|z| > m_2$, so $|A| > M_\epsilon$ implies that any zero of (3) satisfies ($\alpha_0 = 0$)

$$|z - (\alpha_0 + A^{1/n})| < \epsilon,$$

and Theorem 1 is established.

We return now to the lemniscates (1) and the Corollary. When μ is sufficiently large, the equation $|p(z)| = \mu$ is the equation $|(z - \alpha)^n| = \mu$, where α is near α_0; otherwise expressed, the locus (1) lies within a distance ϵ of the circle $|z - \alpha_0| = \mu^{1/n}$, and hence has approximately the shape of a large circle whose center is α_0.

Likewise the locus

$$\arg [p(z)] = \varphi, \tag{8}$$

where φ is constant, is for large $|z|$ approximately the locus $\arg [(z - \alpha)^n] = \varphi$, where α is near α_0; the points of (8) are within a distance ϵ of the locus $\arg (z - \alpha_0) = (\varphi + 2k\pi)/n$, $k = 0, 1, 2, \cdots, n - 1$; consequently the locus (8) has as asymptotes the n half-lines $\arg (z - \alpha_0) = (\varphi + 2k\pi)/n$. The locus $\arg [p(z)]^2 = 2\varphi$ has as asymptotes the $2n$ half-lines

$$\arg (z - \alpha_0) = (\varphi + k\pi)/n, \qquad k = 0, 1, 2, \cdots, 2n - 1.$$

In the special case $n = 2$ with $\alpha_1 \neq \alpha_2$, each line of force (8) is part of an algebraic curve of degree two through α_1 and α_2 whose asymptotes are orthogonal and meet in the point $(\alpha_1 + \alpha_2)/2$; these algebraic curves are therefore equilateral hyperbolas with that point as center, including the degenerate curve consisting of the line $\alpha_1\alpha_2$ and the perpendicular bisector of the segment $\alpha_1\alpha_2$. Of course this same fact is a consequence of the results of §1.6.2; the lines of force in the field due to particles at the zeros of the polynomial $p(z)$ are the loci $\arg [p(z)] = \text{const}$; in the case $p(z) \equiv z^2 + 1$ we have $\tan \arg [p(z)] = 2xy/(x^2 - y^2 + 1)$, so the lines of force are arcs of the equilateral hyperbolas $x^2 - y^2 + 2Cxy + 1 = 0$ with centers in $z = 0$ passing through the points $z = i$ and $-i$; the degenerate equilateral hyperbola $xy = 0$ is not excluded.

CHAPTER II

REAL POLYNOMIALS

§2.1. Polynomials with real zeros. In the present chapter we shall consider the critical points of real polynomials, and consequently Jensen's Theorem (§1.4) dominates our entire discussion. However, we first treat the special case of polynomials all of whose zeros are real. In this case it follows from Rolle's Theorem, as we have pointed out in §1.3, that the critical points are also all real. A general result [Walsh, 1922] on the location of those critical points is

THEOREM 1. *Denote by I_k the interval $a_k \leqq z \leqq b_k$, $k = 1, 2, \cdots, n$. Let the critical points of $\prod_1^n (z - a_k)$ be c_j, with $c_j \leqq c_{j+1}$, and the critical points of $\prod_1^n (z - b_k)$ be d_j, with $d_j \leqq d_{j+1}$. Then if the intervals I_j are the respective loci of the points α_j, the locus of the k-th critical point (in algebraic magnitude) of the variable polynomial $p(z) = \prod_1^n (z - \alpha_j)$ consists of the interval $I_k': c_k \leqq z \leqq d_k$, $k = 1, 2, \cdots, n - 1$. Any interval I_k' which is disjoint from every other distinct interval I_j' contains a number of zeros of $p'(z)$ equal to the multiplicity of c_k as a critical point of $\prod_1^n (z - a_m)$.*

We have the usual formula for the zeros of $p'(z)$ not also zeros of $p(z)$:

$$\frac{p'(z)}{p(z)} = \sum_{k=1}^{n} \frac{1}{z - \alpha_k} = 0,$$

whence by differentiation

$$\frac{\partial z}{\partial \alpha_j} \sum_{k=1}^{n} \frac{1}{(z - \alpha_k)^2} - \frac{1}{(z - \alpha_j)^2} = 0, \qquad \frac{\partial z}{\partial \alpha_j} > 0,$$

and this inequality holds unless z coincides with some α_k. If z coincides with some α_k, but does not coincide with α_j, then z must be a multiple zero of $p(z)$, and remains unchanged if α_j varies and the other zeros of $p(z)$ remain unchanged. If z coincides with α_j, then z must also be a multiple zero of $p(z)$. As α_j now increases, say from the fixed point α_j^0, the remaining zeros of $p(z)$ being unchanged, one zero of $p'(z)$ also increases from α_j^0, by Rolle's Theorem; any other zeros of $p'(z)$ at α_j^0 remain unchanged. Thus in every case, *each zero z of $p'(z)$ increases or remains unchanged as α_j increases, the other zeros α_k of $p(z)$ being fixed.*

Since the zeros of $p'(z)$ are ordered according to algebraic magnitude, the k-th zero of $p'(z)$ varies continuously with each α_j, and in the same sense (if at all) as α_j. Thus the locus of the k-th zero of $p'(z)$ can be found by considering first the k-th zero of $p'(z)$ with $\alpha_j = a_j$ for all j, and then by continuously moving the points α_j to the right one at a time, while they trace their respective loci. The k-th zero of $p'(z)$ increases continuously and monotonically during this process, starting from the point c_k, and reaching but never going beyond the point d_k. The theorem follows, the last part by the method of continuity (§1.5.2).

Several special cases of this theorem are of interest.

COROLLARY 1. [for $k = 1$, Nagy, 1918; in part due to Laguerre]. *Let α_1 be algebraically the least zero of the polynomial $p(z)$ with only real zeros; denote the multiplicity of α_1 by k and the degree of $p(z)$ by n. Let all other zeros of $p(z)$ lie in the interval $(\alpha_1 <)\alpha_2 \leqq z \leqq \alpha_m$. Then at least one zero of $p'(z)$ lies in the interval*

$$\frac{(n-k)\alpha_1 + k\alpha_2}{n} \leqq z \leqq \frac{(n-k)\alpha_1 + k\alpha_m}{n},$$

and no zero of $p'(z)$ lies in the interval $\alpha_1 < z < [(n-k)\alpha_1 + k\alpha_2]/n$.

Corollary 1 is established by choosing I_1 as the null interval $z = \alpha_1$, locus of k zeros of $p(z)$, and I_2 as the interval $\alpha_2 \leqq z \leqq \alpha_m$, locus of $n - k$ zeros of $p(z)$.

COROLLARY 2. *Let $P(z)$ be a polynomial of degree n, whose zeros (assumed all real) are denoted by $a_1, a_2, \cdots, a_n$, with $a_j \leqq a_{j+1}$, and let the zeros of the derivative $P'(z)$ be denoted by $b_1, b_2, \cdots, b_{n-1}$, with $b_j \leqq b_{j+1}$. If the respective loci of the points $\alpha_1, \alpha_2, \cdots, \alpha_n$ are the intervals $a_k - h \leqq z \leqq a_k + h$, then the respective loci of the critical points of the polynomial $\prod_1^n (z - \alpha_k)$ are the intervals $b_k - h \leqq z \leqq b_k + h$.*

COROLLARY 3. *Let $P(z)$ be a polynomial of degree n, whose zeros $a_1, a_2, \cdots, a_n$ are all real, with $a_j \leqq a_{j+1}$, and let the zeros of $P'(z)$ be $b_1, b_2, \cdots, b_{n-1}$, with $b_j \leqq b_{j+1}$. If the respective loci of the points $\alpha_1, \alpha_2, \cdots, \alpha_n$ are the intervals $a_k - \lambda|\gamma - a_k| \leqq z \leqq a_k + \lambda|\gamma - a_k|$, where λ is positive and γ is real with $\gamma \leqq a_1$ or $\gamma \geqq a_n$, then the locus of the critical points of the polynomial $p(z) = \prod_1^n (z - \alpha_k)$ consists of the intervals I_k': $b_k - \lambda|\gamma - b_k| \leqq z \leqq b_k + \lambda|\gamma - b_k|$. If we have $\lambda \leqq 1$, the locus of the k-th critical point (in algebraic magnitude) of $p(z)$ is the interval I_k'.*

In Corollary 3 the two sets of points

$$a_k - \lambda|\gamma - a_k|, \qquad a_k + \lambda|\gamma - a_k|, \tag{1}$$

are respectively the initial and terminal points of the intervals which are the loci of the α_k, and the substitution $w - \gamma = (1 + \lambda)(z - \gamma)/(1 - \lambda)$ or $w - \gamma = (1 - \lambda)(z - \gamma)/(1 + \lambda)$ according as we have $\gamma \leqq a_1$ or $\gamma \geqq a_n$ transforms the former set into the latter. That is to say, the polynomial $\prod_1^n (z - \alpha_k)$ whose zeros are the initial points of the given intervals is found by a linear integral transformation with γ as fixed point from the polynomial whose zeros are the terminal points of the given intervals, so the critical points of the one polynomial are found by that same transformation from the critical points of the other. If we have $\lambda < 1$, all the points a_j, b_j, and the intervals I_k' lie on one side of γ, in the order indicated by their subscripts, and the intervals I_k' of Corollary 3 are precisely the intervals I_k' of Theorem 1; Corollary 3 follows from Theorem 1. If we have $\lambda > 1$, the point γ separates the two sets of points (1); the notation I_k' of Corollary 3 does not agree with that of Theorem 1; nevertheless the interval I_1' or I_{n-1}' (notation of Corollary 3) according as we have $\gamma \geqq a_n$ or $\gamma \leqq a_1$ contains all critical points of $p(z)$; Corollary 3 is a consequence of Theorem 1. The case $\lambda = 1$ is exceptional, for here either the original substitution or its inverse degenerates, but the geometric situation is simple, for γ is either the common initial or the common terminal point of all the given intervals and also of the intervals I_k'; the intervals I_k' of Corollary 3 are the respective intervals I_k' of Theorem 1, and Corollary 3 remains valid.

In Corollary 3 the assumption that γ does not separate any pair of zeros of $P(z)$ is essential, for under the appropriate substitution as defined above, each significant initial point of an interval (locus of an α_k) is transformed into the terminal point of that same interval; but if γ separates a pair of the α_k, that substitution transforms some initial points into terminal points and some terminal points into initial points. For application of Theorem 1 we must determine the final locus of critical points of $p(z)$ as a set of intervals whose initial points are the critical points of $p(z)$ when all the α_k are as far to the left as possible and whose terminal points are the critical points of $p(z)$ when all the α_k are as far to the right as possible.

Corollary 2 is a limiting case as $\gamma \to \infty$ of Corollary 3, and both corollaries can be extended (§3.3) to the case of circular regions rather than line intervals. For the case of precisely two distinct intervals in the hypothesis, these two corollaries are analogous to and can be proved from Walsh's Theorem (§1.5).

COROLLARY 4. *Let α_1 be a zero of order k_1 of the polynomial $p(z)$, let k_2 other zeros of $p(z)$ lie in the interval $I_2 : \alpha_2 \leqq z \leqq \beta_2$, and let all the remaining k_3 zeros lie in the interval $I_3 : \alpha_3 \leqq z \leqq \beta_3$. Then all zeros of $p'(z)$ not at α_1 nor in the intervals I_2 and I_3 lie in the intervals $I_1' : c_2 \leqq z \leqq d_2$ and $I_2' : c_3 \leqq z \leqq d_3$, where c_2 and c_3 ($c_2 \leqq c_3$) are the zeros of the polynomial defined as*

$$(z - \alpha_1)^{-k_1+1}(z - \alpha_2)^{-k_2+1}(z - \alpha_3)^{-k_3+1}$$

times the derivative of $(z - \alpha_1)^{k_1}(z - \alpha_2)^{k_2}(z - \alpha_3)^{k_3}$, and where d_2 and d_3 ($d_2 \leqq d_3$) are the zeros of the polynomial defined as $(z - \alpha_1)^{-k_1+1}(z - \beta_2)^{-k_2+1}(z - \beta_3)^{-k_3+1}$ times the derivative of $(z - \alpha_1)^{k_1}(z - \beta_2)^{k_2}(z - \beta_3)^{k_3}$. An interval I_j' disjoint from the other interval I_k', from α_1, and from the intervals I_2 and I_3 contains precisely one critical point of $p(z)$.

The order of the points α_1, c_k, and d_k on the axis of reals depends on the order of the points α_k and β_j. Nevertheless a direct proof of Corollary 4 including all cases is readily given by the method of proof of Theorem 1 and by the remark that if k unit particles lie in an interval I of the axis of reals, the corresponding force at a real point z exterior to I is equal to the force at z due to k *coincident* unit particles in I. Corollary 4 does not describe the complete locus of critical points, which obviously consists of I_1', I_2', I_2 (if $k_2 > 1$), I_3 (if $k_3 > 1$), and the point α_1 (if $k_1 > 1$). In effect, Corollary 4 reduces the approximate solution of the equation $p'(z) = 0$ to the solution of a quadratic, just as Corollary 1 reduces that approximate solution to the solution of a linear equation. The case $k_3 = 0$ of Corollary 4 is Corollary 1 in slightly more general form.

§2.2. Jensen's Theorem, continued. We now resume the study of Jensen's Theorem. In the field of force due to two unit particles at the points i and $-i$, it is proved in §1.6.3 *that the lines of force are equilateral hyperbolas through those points and with centers in the origin.* A direct proof of this fact is not difficult. The force at an arbitrary point $z = x + iy$ is

$$\frac{1}{\bar{z} + i} + \frac{1}{\bar{z} - i} = \frac{2\bar{z}}{\bar{z}^2 + 1} = \frac{2\bar{z}(z^2 + 1)}{(z^2 + 1)(\bar{z}^2 + 1)},$$

whose slope when computed yields the differential equation for the lines of force:

$$(1) \qquad \frac{dy}{dx} = \frac{y(x^2 + y^2 - 1)}{x(x^2 + y^2 + 1)} \quad \text{or} \quad x^2y^2d\left(\frac{x^2 - y^2 + 1}{xy}\right) = 0.$$

The solution is then the family of equilateral hyperbolas

$$(2) \qquad x^2 - y^2 + 1 = Cxy,$$

including the limiting curve $xy = 0$; these curves pass through the points $(0, \pm 1)$, and have the origin as center. A unique curve of the family passes through an arbitrary point of the plane different from the points $(0, \pm 1)$.

In the use of equilateral hyperbolas, we recall the following elementary property: *Let H be an equilateral hyperbola with vertices V and V', let the points P and P' be symmetric in the conjugate axis of H, and let C be the circle whose diameter is the segment PP'. Then V and V' lie on, interior to, or exterior to C according as P and P' lie on, interior to, or exterior to H.* The *interior* of the hyperbola $b^2x^2 - a^2y^2 + a^2b^2 = 0$ is defined as the set $b^2x^2 - a^2y^2 + a^2b^2 < 0$, and the *exterior* as the set $b^2x^2 - a^2y^2 + a^2b^2 > 0$. If H is the hyperbola $x^2 - y^2 + a^2 = 0$, and P the point (x_0, y_0), then C is the circle $(x - x_0)^2 + y^2 - y_0^2 = 0$, and the condition that V: $(0, a)$ and V': $(0, -a)$ lie on, interior to, or exterior to C is $x_0^2 + a^2 - y_0^2 = 0$, <0, or >0, which is the condition that P and P' lie on, interior to, or exterior to H. Here C is the Jensen circle for P and P'.

Let us restate Jensen's Theorem in the form: *Let $p(z)$ be a real polynomial, and let z be a non-real critical point not a multiple zero of $p(z)$. Then either* (i) *z lies on all Jensen circles and $p(z)$ has no real zeros, or* (ii) *z lies interior to at least one Jensen circle and $p(z)$ has at least one real zero, or* (iii) *z lies interior to at least one Jensen circle and exterior to at least one Jensen circle.* We have thus an equivalent statement [essentially due to Curtiss, 1920] of Jensen's Theorem: *Let $p(z)$ be a real polynomial, let z be a non-real critical point not a multiple zero* *of $p(z)$, and let H be the equilateral hyperbola whose vertices are z and $\bar{z}$. Then either all zeros of $p(z)$ lie on H or some zeros of $p(z)$ lie in the interior and some in the exterior of H.*

§2.2.1. Special cases. In order to prepare a deeper investigation into the situation of Jensen's Theorem, we study two special cases under that theorem.

THEOREM 1. *In the field of force due to unit particles at the points $+i$ and $-i$, the locus of non-real points at which the lines of action of the force pass through the point $(\alpha, 0)$, $\alpha \neq 0$, is the circle (excepting points of the axes)*

$$(x + 1/\alpha)^2 + y^2 = 1 + 1/\alpha^2. \tag{3}$$

This is the unique circle passing through the points $+i$ and $-i$ whose tangents at those points pass through $(\alpha, 0)$.

The condition of the theorem is by (1) precisely the condition

$$\frac{y(x^2 + y^2 - 1)}{x(x^2 + y^2 + 1)} = \frac{y}{x - \alpha},$$

which for $y \neq 0$ reduces to (3).

Under the conditions of Theorem 1 the force in the half-plane $x > 0$ due to the particles at $+i$ and $-i$ has a component toward the right, and in the half-plane $x < 0$ has a component toward the left. If we have $\alpha > 0$, then on the arc of the circle (3) in the half-plane $x > 0$ the force is directed toward $z = \alpha$ and on the arc in the half-plane $x < 0$ is directed away from α. There follows the

COROLLARY. *All non-real critical points of the polynomial $(z^2 + 1)^{m_1}(z - \alpha)^{m_2}$, $m_1 > 0$, $m_2 > 0$, lie on the arc of (3) contained in the closed interior of the unit circle.*

This result, like Jensen's Theorem, for fixed α provides a geometric locus for non-real critical points which is independent of the (variable) multiplicities of the zeros of $p(z)$. A related result which does depend on those (fixed) multiplicities but where now α is allowed to vary is

THEOREM 2. *The non-real critical points other than $+i$ and $-i$ of the polynomial $p(z) = (z^2 + 1)^k(z - \alpha)^{n-2k}$, where α is real, lie on the circle whose center is the origin and radius $[(n - 2k)/n]^{1/2}$.*

We eliminate α from the equations found by separating real and pure imaginary parts of the equation

$$(z^2 + 1)^{-k+1}(z - \alpha)^{-n+2k+1}p'(z)$$
$$= (n - 2k)[(x + iy)^2 + 1] + 2k(x + iy)(x + iy - \alpha) = 0,$$

which yields the equation $y = 0$ with the equation

$$n(x^2 + y^2) = n - 2k. \tag{4}$$

When α is large and positive, one of the critical points of $p(z)$ other than $\pm i$ or α lies near the origin (as α becomes positively infinite, such a critical point approaches the origin), and the other lies near the point $2k\alpha/n$ (the center of gravity of these two critical points is $k\alpha/n$, by the equation for $p'(z)$). Both critical points must then be real, by the symmetry. As α decreases continuously, the former critical point moves to the right and the latter to the left, for the correspondence between α and z is analytic, and z cannot change sense or leave the axis of reals except when two critical points coincide; the two critical points coalesce at $[(n - 2k)/n]^{1/2}$, when $\alpha = [(n^2 - 2kn)/k^2]^{1/2}$; this latter value need not be greater than unity. As α continues to decrease, the critical points move on the circle already determined, remaining conjugate imaginary, and when $\alpha = 0$ the critical points are $\pm i[(n - 2k)/n]^{1/2}$. The behavior of the critical points as α continues to decrease is found from symmetry.

Another interpretation of the behavior of the critical points is found by considering the force at a real point z due to the particles involved. The force due to the k-fold particles at $+i$ and $-i$ is $2kz/(z^2 + 1)$, a function which is zero when z is zero, is positive for positive z, has the derivative $2k$ at $z = 0$, has a maximum for $z = 1$, and approaches zero as z becomes infinite. The force at z due to the $(n - 2k)$-fold particle at $z = \alpha$ has the magnitude $(n - 2k)/|z - \alpha|$, and is directed away from α. When α is large and positive, the graphs of those two functions intersect near O and also in a large value of z. When α moves to the left, the former intersection moves to the right and the latter to the left. The intersections coincide for $\alpha = [(n^2 - 2kn)/k^2]^{1/2}$, and when α continues farther to the left but remains positive, the graphs have no significant intersection.

§2.2.2. A region for non-real critical points. We have shown (§1.4.2 Theorem 1) that if the non-real zeros of a real polynomial are prescribed, Jensen's Theorem cannot be improved except (a) by restricting the degree of the polynomial or (b) by restricting the positions of the real zeros, or both. We proceed to consider such improvements.

Theorem 2 is especially suggestive in the further study of critical points. We prove [Walsh, 1920a for $k = 1$; Nagy, 1922 for arbitrary k (the original method is applicable)]:

THEOREM 3. *If $p(z)$ is a real polynomial of degree n whose zeros are all real except for k-fold zeros at $+i$ and $-i$, then all non-real critical points of $p(z)$ except $+i$ and $-i$ lie on or within the circle whose center is the origin and radius $[(n - 2k)/n]^{1/2}$.*

It will be convenient to prove two lemmas.

LEMMA 1. *If the force at a point P due to m variable particles on a line L not through P has the direction of a line λ at P, then the force is not greater than it would be if all m particles were concentrated at the intersection of L and λ.*

Denote by L' the circle which is the inverse of L in the unit circle Γ whose center is P. The force at P due to a particle at Q is in magnitude, direction, and sense the vector $Q'P$, where Q' is the inverse of Q in Γ. The force at P due to the m given particles is equivalent to the force at P due to an m-fold particle on λ situated at the center of gravity Q_0' of the inverses Q' of the m given particles. Since the points Q lie on L, their inverses Q' lie on L', and Q_0' lies on or within L' and on λ. The greatest force $Q_0'P$ is exerted if Q_0' lies at the intersection (other than P) of L' and λ, which implies that all m particles are concentrated at the intersection of L and λ.

LEMMA 2. *Let A be an open circular arc interior to the unit circle with end-points $+i$ and $-i$, cutting the axis of reals in the point β. Then the force at a point z_0 of A due to unit particles at $+i$ and $-i$ passes through the point $\alpha = 2\beta/(1 - \beta^2)$, and this force increases monotonically in magnitude as z moves monotonically on A from β to $+i$.*

We identify A with an arc of circle (3). The fact that the force at z_0 passes through the point α follows from Theorem 1 by setting $x = \beta$, $y = 0$ in (3) and solving for α. We have shown (§1.4.1) that the force at z_0 due to the particles at $+i$ and $-i$ is equivalent to that at z_0 due to a double particle at the point $-1/z_0$, and when we choose $z_0 = \beta$, *the point $-1/z_0$ turns out to be the second intersection of the circle* (3) *with the axis of reals*; for with $y = 0$, the product of the abscissas of the two intersections is -1. Moreover, when z_0 moves along a circle so also does the point $-1/z_0$; when $z_0 = \pm i$ we have $-1/z_0 = z_0$; thus if z_0 moves along A, the point $-1/z_0$ moves along the circle (3). The distance $|z_0 + 1/z_0|$ is a chord of the circle (3) on a line through α, and is greatest for z_0 on A when $z_0 = \beta$; the force is then least for $z_0 = \beta$ and increases monotonically as z_0 moves monotonically along A from $z = \beta$ to $z = +i$.

We are now in a position to prove Theorem 3, and our method is to show that if z is a non-real point of an arc A which does not intersect the circle (4), then the force F_1 at z due to the k-fold particles at $+i$ and $-i$ is greater than any force F_0 at z acting in the same direction as F_1 but opposite in sense and due to $n - 2k$ real particles. For the moment concentrate these $n - 2k$ particles at the point α corresponding to the arc A and denote by F_2 their force at z; as z moves along A from β to the point $+i$, this force F_2 decreases monotonically in magnitude (because the distance $|z - \alpha|$ increases monotonically) and always has the direction from α to z; the force F_1 increases monotonically in magnitude (Lemma 2) and always has the direction from z to α; if F_2 were greater than F_1 for the position $z = \beta$, there would be a position of equilibrium at some intermediate point z of A between β and $+i$, which by Theorem 2 is not the case. Since at $z = \beta$ the force F_2 is not greater than F_1, at every non-real point z of A the force F_1 is greater than F_2, and since by Lemma 1 the force F_2 is greater than or equal to the force F_0 with the same line of action as but opposite sense to F_1, where F_0 is due to $n - 2k$ arbitrarily placed particles on the axis of reals, it follows that no non-real point z of A can be a critical point of $p(z)$ if A does not intersect the circle (4). Precisely the same method applies if A intersects the circle (4), so far as concerns points z of A exterior to (4), and this completes the proof.

§2.3. Number of critical points. As a complement to Jensen's Theorem we have

THEOREM 1. *Let $p(z)$ be a real polynomial, let the real points α and β ($\alpha < \beta$) lie exterior to all Jensen circles for $p(z)$, and let α and β be neither zeros nor critical points of $p(z)$. Denote by K the configuration consisting of the segment $\alpha\beta$ together with the closed interiors of all Jensen circles intersecting that segment. If K contains k zeros of $p(z)$, then K contains $k - 1$, k, or $k + 1$ zeros of $p'(z)$.*

If the force in the field set up by Gauss's Theorem is directed to the left at α and to the right at β, or is in identical directions (to the right or left) at α and β, or is directed to the right at α and to the left at β, then K contains respectively $k - 1$, k, or $k + 1$ zeros of $p'(z)$.

We establish Theorem 1 by considering the variation in the direction of force as we trace a Jordan curve C near to K, lying in the complement of K, and containing K in its interior. For definiteness we may by way of illustration choose C as the image of a circle $|w| = R(>1)$ under the conformal map of the complement of K in the z-plane onto the region $|w| > 1$ of the w-plane so that the two points at infinity correspond to each other. Let C intersect the axis of reals in the point α' to the left of α and the point β' to the right of β; we choose C so close to K that no zero or critical point of $p(z)$ lies on C or between C and K, and that no point of a Jensen circle for $p(z)$ lies in either of the intervals $\alpha'\alpha$ or $\beta\beta'$. For definiteness let us treat the case that the force at α is directed toward the left and that at β toward the right; the forces at α' and β' are then respectively toward left and right. We commence at the point β' to trace C and to consider on C the variation in the direction of the force; we constantly use the fact (§1.4.1) that at a point not on the axis of reals and exterior to all Jensen circles the force has a non-vanishing component directed away from the axis of reals. The force at β' is directed horizontally toward the right, and as we proceed along C in the counterclockwise sense, the force varies first so as to make a positive acute angle with the positive direction of the axis of reals. Between β' and α', the direction-angle of the force varies between zero and π; when we reach α', the force is directed horizontally toward the left and has increased its angle by π. As we continue to trace C, the direction-angle of the force becomes slightly greater than π, remains between π and 2π at points between α' and β', and is again directed horizontally toward the right when we reach β'. The total change in direction of the force is thus 2π, and is the increase in the argument of the conjugate of $p'(z)/p(z)$. Consequently the increase in arg $[p'(z)/p(z)]$ as z traces C is -2π; the number of zeros of $p(z)$ interior to C is k, so by the Principle of Argument (§1.1.2) the number of zeros of $p'(z)$ interior to C is $k - 1$. This same method can be used in every case, whatever the directions of the forces at α and β, as the reader will note, and this completes the proof of the theorem.

One special case here deserves explicit mention:

COROLLARY. *Let (α, β) with $\alpha < \beta$ be a closed interval of the axis of reals exterior to all Jensen circles and containing no zero of $p(z)$. If the force at α acts toward the right and that at β toward the left, the interval contains precisely one zero of $p'(z)$. If the forces at α and β have the same sense, the interval contains no zero of $p'(z)$. The force at α cannot act toward the left and that at β toward the right.*

The Corollary and a somewhat less general theorem were proved by the present writer [1920a] by a different method; an intermediate theorem was proved by Nagy [1922], using that same method.

§2.4. Reality and non-reality of critical points. Certain conditions [Walsh, 1920a] on the positions of zeros of a real polynomial of given degree can be obtained which are respectively sufficient for the non-reality and reality of particular critical points.

§2.4.1. Sufficient conditions for non-real critical points. The following theorem is suggested by §2.2.1 Theorem 2 and §2.2.2 Theorem 3:

THEOREM 1. *Let $p(z)$ be a real polynomial of degree $n(>2)$ with simple zeros at $+i$ and $-i$ and all its other zeros in the interval $0 \leqq z \leqq (n^2 - 2n)^{1/2}$, but not all these real zeros concentrated at the point $(n^2 - 2n)^{1/2}$; then $p(z)$ has precisely two non-real critical points.*

No real critical point z_0 of $p(z)$ lies to the left of the algebraically least real zero of $p(z)$, for if we assume the reverse, then the force at z_0 due to the particles at $+i$ and $-i$ equals in magnitude the force at z_0 due to the remaining $n - 2$ particles z_j, with $z_0 < z_j \leqq (n^2 - 2n)^{1/2}$:

$$\frac{2z_0}{z_0^2 + 1} = \sum_{j=1}^{n-2} \frac{1}{z_j - z_0} > \frac{n - 2}{(n^2 - 2n)^{1/2} - z_0}, \qquad 0 > [n^{1/2} z_0 - (n - 2)^{1/2}]^2,$$

which is impossible.

Each interval of the axis of reals bounded by two zeros of $p(z)$ contains at least one critical point, by Rolle's Theorem; to prove the theorem it is sufficient to show that no such interval contains more than one critical point. In the (assumed) contrary case, we move continuously to the left the zero of $p(z)$

which is the terminal point of that interval and cause it to coincide with the initial point of that interval while keeping all other zeros of $p(z)$ constant; at least two critical points become non-real; for when a j-fold zero and an m-fold zero of $p(z)$ coalesce, they do so at a $(j + m - 1)$-fold critical point. We reach a contradiction by showing that no non-real point (x, y) interior to the unit circle and of ordinate numerically less than $\frac{1}{2}$ can be a critical point, assuming that at least one zero of $p(z)$ whose abscissa is less than x lies in the interval of the theorem. Let (x, y) lie in the first quadrant; the vertical component of the force at (x, y) due to a particle at the zero just mentioned is not less than the corresponding component for a particle at the origin: $y/(x^2 + y^2)$. If (x, y) is a critical point, the vertical component of the force at (x, y) due to the particles at $+i$ and $-i$ must then be numerically greater than $y/(x^2 + y^2)$:

$$\frac{2y(1 - x^2 - y^2)}{(x^2 + y^2)^2 + 2(x^2 - y^2) + 1} > \frac{y}{x^2 + y^2}, \qquad 4y^2 - 1 > 3(x^2 + y^2)^2,$$

which is impossible with $y < \frac{1}{2}$. Theorem 1 is established. Under the conditions of Theorem 1, each open finite interval of the axis of reals bounded by zeros of $p(z)$ and containing no zero of $p(z)$ thus contains precisely one zero of $p'(z)$.

The method used is not sufficiently powerful to prove or disprove the corresponding result where the given polynomial of degree n has zeros of multiplicity k at $+i$ and $-i$, and the remaining zeros real and in the interval $0 \leqq z \leqq [(n^2 - 2kn)/k^2]^{1/2}$, but the method does apply in extended form and proves the precise analog of Theorem 1 in certain cases, for instance if we have $k \leqq 4$. However, Theorem 1 cannot be extended here without restriction on k and n, as is shown by the following counterexample, constructed by Mr. R. E. Chamberlin. We set $f(z) = z(z^2 + 1)^{10}(z - 97)^{979}$, and the critical points of $f(z)$ not multiple zeros of $f(z)$ are the zeros of $f_1(z) = 1000z^3 - 2037z^2 + 980z - 97$; we have $f_1(0) < 0$, $f_1(\frac{1}{2}) > 0$, $f_1(1) < 0$, $f_1(2) > 0$, so $f_1(z)$ has precisely three real zeros; we also have the inequality

$$[n(n - 2k)]^{1/2}/k = 98.99 > 97.$$

§2.4.2. Sufficient conditions for real critical points.

THEOREM 2. *Let $p(z)$ be a real polynomial with k-fold zeros at $+i$ and $-i$, m zeros whose abscissas are greater than or equal to $[m(m + 2k)/k^2]^{1/2}$, n zeros whose abscissas are less than or equal to $-[n(n + 2k)/k^2]^{1/2}$, and with no other zeros. Let the unit circle C be exterior to all other Jensen circles for $p(z)$. Then $p'(z)$ has precisely one zero in the interval I: $-[n/(n + 2k)]^{1/2} \leqq z \leqq [m/(m + 2k)]^{1/2}$, and no non-real zeros other than $+i$ and $-i$ on or within the unit circle C.*

The degenerate cases here are $m \neq 0$, $n = 0$, and all the real zeros of $p(z)$ concentrated at $[m(m + 2k)/k^2]^{1/2}$, in which case $p'(z)$ has a double zero at $[m/(m + 2k)]^{1/2}$; and $m = 0$, $n \neq 0$, and all the real zeros of $p(z)$ concentrated at $-[n(n + 2k)/k^2]^{1/2}$, in which case $p'(z)$ has a double zero at $-[n/(n + 2k)]^{1/2}$.

In either of these two cases we make the convention that only one zero of $p'(z)$ shall be assigned to the interval I. Henceforth we exclude these degenerate cases.

The force at the point $[m/(m+2k)]^{1/2}$ is directed toward the right, for otherwise the force at that point due to the particles at $+i$ and $-i$ is less than the force at that point due to the m particles, all coincident at $[m(m+2k)/k^2]^{1/2}$:

$$\frac{2k}{\left(\frac{m}{m+2k}\right)^{1/2}+\left(\frac{m+2k}{m}\right)^{1/2}} < \frac{m}{\left(\frac{m(m+2k)}{k^2}\right)^{1/2}-\left(\frac{m}{m+2k}\right)^{1/2}},$$

$$m < m,$$

which is absurd. We use here the fact that a pair of conjugate imaginary particles of abscissa not less than $[m(m+2k)/k^2]^{1/2}$ can be replaced by an equivalent double particle whose abscissa has that same property, without altering the force at $[m/(m+2k)]^{1/2}$. Similarly the force at the point $-[n/(n+2k)]^{1/2}$ is directed toward the left. Consequently the interval I contains at least one zero of $p'(z)$. Suppose for the moment that no real zero of $p(z)$ lies on or within C; if the force at the point $+1$ is directed toward the right, and the force at -1 toward the left, it follows from §2.3 Theorem 1 that C contains precisely one zero of $p'(z)$ distinct from $+i$ and $-i$, which is thus real and in I; no matter what the directions of the forces at ± 1, the only zeros of $p'(z)$ interior to C are real, and precisely one zero lies in I. As real zeros of $p(z)$ are now moved continuously toward O but satisfying the given hypothesis, the other zeros of $p(z)$ being fixed, no real critical point can enter or leave the interval I, and it is a consequence of §2.3 Theorem 1 that no real critical point interior to C can become non-real. Theorem 2 is established.

If in Theorem 2 we omit the requirement that C be exterior to all other Jensen circles, it is no longer true that C can contain on or within it no non-real zeros. As an illustration we choose $k = 4$, $m = 2$, $n = 0$, with the zeros of $p(z)$ other than $+i$ and $-i$ in $(5/4)^{1/2} \pm 21^{1/2}i/6$. At the point $z_0 = 5^{1/2}/3 + 2i/3$ on both Jensen circles for $p(z)$ the total force is zero, so z_0 is a critical point.

Of course if m and n are given, the intervals assigned to the zeros of $p(z)$ are the largest that suffice for the reality of the critical points interior to C. For instance, if we allow an abscissa smaller than $[m(m+2k)/k^2]^{1/2}$ for the m zeros, we can concentrate them at a point $z(>0)$ with such an abscissa, and remove the n zeros (if any), by assigning numerically large abscissas, to such a distance that their influence in the field of force is negligibly small. Then $p(z)$ has two non-real critical points interior to C.

A number of the previous results of §§2.2–2.4 when applied to the particular case of simple zeros yields the following:

Let $p(z)$ be a polynomial of degree n with simple zeros at the points $+i$ and $-i$, and with all the remaining $n-2$ (>0) zeros real and—

1). *coincident at* $[n(n-2)]^{1/2}$, *then $p(z)$ has a double critical point at* $[(n-2)/n]^{1/2}$.

2). *coincident, then all non-real critical points lie on the circle whose center is the origin and radius* $[(n-2)/n]^{1/2}$.

3). *with abscissas unrestricted, then all non-real critical points lie in the closed interior of that circle.*

4). *with abscissas not less than* $[n(n-2)]^{1/2}$, *all critical points are real and except in case* 1) *precisely one critical point lies in the closed interval*

$$(0, [(n-2)/n]^{1/2}).$$

5). *with abscissas non-negative but not greater than* $[n(n-2)]^{1/2}$, *then except in case* 1) *there are precisely two non-real critical points.*

§2.4.3. Real critical points, continued. Even though Theorem 2 does not extend without restriction to the case of m non-real zeros, an extension is possible in the case $k = 1$:

THEOREM 3. *Let* $p(z)$ *be a real polynomial with simple zeros at* $+i$ *and* $-i$, m *zeros whose abscissas are not less than* $[m(m+2)]^{1/2}$, n *zeros whose abscissas are not greater than* $-[n(n+2)]^{1/2}$, *and with no other zeros; then* $p'(z)$ *has precisely one zero in the interval* $-[n/(n+2)]^{1/2} \leqq z \leqq [m/(m+2)]^{1/2}$ *and no non-real zero on or within the unit circle* C.

As in Theorem 2, the degenerate cases may arise, are included in the theorem by a suitable convention, and are henceforth in the proof excluded. The entire discussion concerning Theorem 2 is valid here with $k = 1$, and is merely to be supplemented.

We shall now prove that under no circumstances consistent with our hypothesis can any point of C except $+1$ and -1 be a critical point. First suppose a point of C in the first quadrant to lie on or within one of the Jensen circles pertaining to the m given zeros. At such a point (x, y) the horizontal component of the force due to the two particles at $+i$ and $-i$ is greater than the horizontal component of the force due to the m particles; for the contrary implies (we replace a pair of non-real particles by an equivalent double particle)

$$\frac{2}{2x} \leqq \frac{2}{2\{[m(m+2)]^{1/2}-1\}} + \frac{m-2}{[m(m+2)]^{1/2}-1},$$

$$x \geqq \frac{[m(m+2)]^{1/2}-1}{m-1} > 1, \tag{1}$$

which is impossible. The proof just given is also valid for the point $z = 1$; if that point is on or within one of the Jensen circles belonging to the m zeros, it is not a zero of $p'(z)$.

Second, it is conceivable that a point $P:(x, y)$ in the first quadrant and on C should lie exterior to all the Jensen circles pertaining to the m zeros, and yet should lie interior to one or more of the Jensen circles pertaining to the n zeros, and should be a position of equilibrium. We shall prove the impossibility of this, roughly, as follows. Such a zero of $p'(z)$ must be near the point $z = 1$, for as we move upward from that point along C the horizontal force at P due to the particles at $+i$ and $-i$ increases and eventually exceeds the force at P due to the m particles. The latter force is always of such nature that the total force $\mathfrak{F}$ at P due to the m particles and the two particles at $+i$ and $-i$ is inclined steeply to the horizontal, so steeply in fact that for the force at P due to the n particles to be inclined at that same angle, P must be quite near the center of the corresponding Jensen circle, which turns out to be impossible. We proceed to the details of the proof.

Denote by Q_0 the point $[m(m+2)]^{1/2}$, by P_0 the reflection of P in the vertical line through Q_0, and by P_1 the reflection of P in the axis of reals. Any pair of the m particles which are conjugate imaginary can without altering the corresponding force at P be replaced by a double equivalent particle which lies in the closed acute-angled infinite sector S with vertex Q_0 bounded by a half-line from Q_0 through P_0 and an infinite segment of the axis of reals. Invert the entire

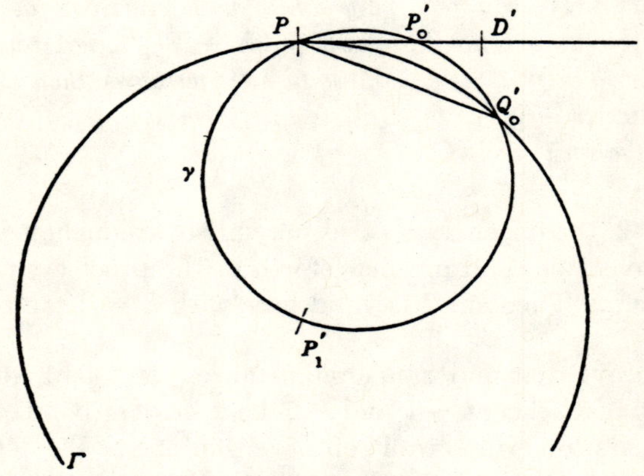

Fig. 4 illustrates §2.4.3 Theorem 3

configuration in the unit circle whose center is P, and denote the images of P_0, P_1, and Q_0 by P_0', P_1', and Q_0'; the image of the axis of reals is a circle Γ whose center P_1' lies vertically below P, and the image of the line $P_0Q_0P_1$ is the circle γ through P, Q_0', and P_1'. The point P_0' lies on γ on the horizontal line through P. The image of S is a lens-shaped region S' bounded by an arc of each of the two circles γ and Γ terminated by P and Q_0', and S' is contained in the closed segment T of γ bounded by the line segment PQ_0' and an arc of γ. The force at P due to the m particles is represented by an m-fold vector from a point M of T to P. The force at P due to the particles at $+i$ and $-i$ is represented by m times a certain horizontal vector PD' directed to the right, and the total force $\mathfrak{F}$ is represented by m times the vector MD'. The point D' lies to the right of P_0', for we have shown by the use of (1) that if one pair of the m particles exerts a horizontal force at P (which implies that P is on the corresponding Jensen circle), and if the others of the m particles are at Q_0, then the total force $\mathfrak{F}$ has a non-zero component toward the right; here $\mathfrak{F}$ is m times the vector MD'

to D' from a point M of the segment $Q_0'P_0'$. The point of contact of a tangent from D' to γ cannot lie between Q_0' and P_0' below the line PP_0'; for the force at P due to the particles at $+i$ and $-i$ is $1/x$, the length of PD' is $1/mx$, and the inverse D of D' in the unit circle whose center is P is at a distance mx horizontally to the right of P; we study the original figure (i.e. before the original inversion), and note the distances:

$$\begin{aligned} \overline{P_0Q_0^2} &= \{[m(m+2)]^{1/2} - x\}^2 + y^2, \\ P_0P &= 2\{[m(m+2)]^{1/2} - x\}, \\ P_0D &= 2[m(m+2)]^{1/2} - (m+2)x, \end{aligned}$$

from which we write the condition that the point of tangency below PP_0 of the circle through P and D tangent to the line P_0Q_0 should not lie interior to the segment P_0Q_0:

$$\overline{P_0Q_0^2} \leqq P_0P \cdot P_0D,$$

$$\{[m(m+2)]^{1/2} - x\}^2 + 1 - x^2 \leqq 2\{[m(m+2)]^{1/2} - x\} \cdot \{2[m(m+2)]^{1/2} - (m+2)x\},$$

$$0 \leqq 2(m+2)\{x^2 - (m+3)[m/(m+2)]^{1/2}x + m(m+3)^2/4(m+2)\} + 3m(m+2) - 1 - m(m+3)^2/2;$$

the absolute minimum of this second member occurs for

$$x = (m+3)[m/(m+2)]^{1/2}/2,$$

which is greater than unity, and the minimum for $0 \leqq x \leqq 1$ occurs for $x = 1$ and has a value which is positive. It follows that *the slope of* $\mathfrak{F}$ *(assumed to have a horizontal component to the left) is numerically least when all the* m *particles are concentrated at* Q_0.

If all the m particles are located at $[m(m+2)]^{1/2}$, and if $\mathfrak{F}$ has a horizontal component toward the left, the slope of $\mathfrak{F}$ at the point P on C is numerically equal to

$$\frac{mxy}{[m(m+2)^3]^{1/2}\,x - mx^2 - m(m+2) - 1},$$

and a simple algebraic manipulation shows that this expression for $0 < x < 1$ is not less than

$$\frac{my}{[m(m+2)^3]^{1/2} - m - m(m+2) - 1}.$$

Consider now the (ξ, η)-plane and the force due to two unit particles at $+i$ and $-i$ respectively. *At any point* P *of the first quadrant, the slope of the line of force is numerically less than that of* OP. Indeed, the force at P is (§1.4) equivalent to the force at P due to a double particle situated at some point Q on the half-line defined as the reflection in the η-axis of the half-line from O through P. The slope of PQ is numerically less than that of OP.

For the point $P:(x, y)$ previously considered, if $-\mu(<0)$ is the slope of the line of action of the force due to the n particles whether real or non-real, we have

$$\mu \leqq y/[n(n+2)]^{1/2} < y.$$

Thus if P is a position of equilibrium we must have

$$y > \frac{my}{[m(m+2)^3]^{1/2} - m - m(m+2) - 1}, \qquad [m(m+2)]^{1/2} > m+1,$$

which is impossible.

We have therefore proved that *no point of the circle C except $+1$ or -1 can be a critical point of $p(z)$*, and that $+1$ is not a critical point if it lies on or within one of the Jensen circles pertaining to the m zeros of $p(z)$, nor is -1 a critical point if it lies on or within one of the Jensen circles pertaining to the n zeros of $p(z)$.

If the Jensen circle of any pair of the m zeros of $p(z)$ contains or intersects C, continuously move those zeros toward the right along the circle joining them with the point -1, and continue this motion until their Jensen circle cuts the axis of reals slightly to the right of the point $+1$, and so that no critical point lies on the axis of reals between their Jensen circle and $+1$. During this motion the force at -1 is constant, and by (1) the force at $+1$ has not changed in sense, so there is no change in the number of critical points on and within C. Similarly move any pair of the n zeros whose Jensen circle cuts or contains C, keeping the force constant in the point $+1$. In the final position, the forces at the points $+1$ and -1 are in the same direction as were the forces in the initial position, never having changed sense; a final force is zero when and only when the initial force is zero. In the final position, which is of course the initial position so far as concerns those of the m and n zeros of $p(z)$ whose Jensen circles contain no point of C, the circle C contains on or within it one, two, or three critical points according as the forces at $+1$ and -1 are both, one, or neither directed away from the origin. The forces at the points $[m/(m+2)]^{1/2}$ and $-[n/(n+2)]^{1/2}$ are initially and finally directed away from the origin, by the proof of Theorem 2, so it is clear that the critical points on and within C are initially and finally all real, with precisely one critical point in the interval of the theorem, and not more than one each (the same initially and finally) in the intervals $-1 < z < -[n/(n+2)]^{1/2}$, $[m/(m+2)]^{1/2} < z < +1$. The phraseology as given can be made more precise by the reader to include the case that $+1$ or -1 is a critical point, and this completes the proof of the theorem.

§2.5. Jensen configuration improved. We now resume the situation of §2.2.1 Theorem 1, and proceed to determine regions which are free from critical points and depend on the positions but not the multiplicities of the zeros or the degree of the given polynomial. A simple result [Walsh, 1946a] is

THEOREM 1. *Let $p(z)$ be a real polynomial with precisely one pair $(\alpha, \bar{\alpha})$ of non-real zeros. Let z_1 be the algebraically least and z_2 the algebraically greatest of the real zeros of $p(z)$. Let A_k $(k = 1, 2)$ be the circular arc bounded by α and $\bar{\alpha}$ of angular measure less than π which is tangent to the lines αz_k and $\bar{\alpha} z_k$ (A_k is the line segment joining α and $\bar{\alpha}$ if $z_k = (\alpha + \bar{\alpha})/2$). Then all non-real zeros of $p'(z)$ lie in the closed lens-shaped region R bounded by A_1 and A_2. No non-real point of A_k other than α or $\bar{\alpha}$ can be a zero of $p'(z)$ unless all real zeros of $p(z)$ coincide at z_k.*

Let z_0 be a non-real point not in R but interior to the Jensen circle for α and $\bar{\alpha}$. The line of action L of the force at z_0 due to the particles at α and $\bar{\alpha}$ cuts the axis of reals at a point exterior to the closed interval $z_1 \leqq z \leqq z_2$, by §2.2.1 Theorem 1, for the intersection of a (variable) arc such as A_1 with the axis of

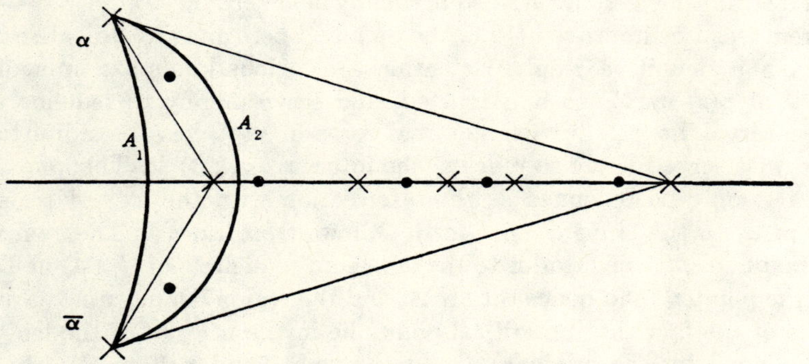

Fig. 5 illustrates §2.5 Theorem 1

reals varies monotonically with and in the same sense as z_1. All real zeros of $p(z)$ lie on one side of L, the force exerted at z_0 by the particles at those real zeros cannot have the direction of L, so the total force at z_0 cannot be zero. Moreover z_0 cannot be a multiple zero of $p(z)$, so z_0 cannot be a critical point of $p(z)$. Similarly a non-real boundary point z_0 of R cannot be a position of equilibrium unless all real zeros of $p(z)$ lie on L; if these real zeros lie on L they all coincide at z_k if z_0 lies on A_k.

In Theorem 1 there are at most two non-real zeros of $p'(z)$ distinct from α and $\bar{\alpha}$; if there are two such zeros, all real zeros of $p'(z)$ lie in the interval

$$z_1 \leqq z \leqq z_2 .$$

Theorem 1 admits of extension to the case of several pairs of non-real zeros; we prove for instance

THEOREM 2. *Let $p(z)$ be a real polynomial with precisely two pairs of non-real zeros: $\alpha_1, \bar{\alpha}_1, \alpha_2, \bar{\alpha}_2$, and with no real zeros. Let the Jensen circle C_j $(j = 1, 2)$*

corresponding to α_1 *and* $\bar{\alpha}_1$ *intersect the axis of reals in points* z_1' *and* z_1'', *with* $z_1' < z_1''$, *and suppose we have* $z_1'' < z_2'$. *Let the tangent to the circle* $z_1'\alpha_2\bar{\alpha}_2$ *at* α_2 *cut the axis of reals in* z_1 *and the tangent to the circle* $z_1''\alpha_2\bar{\alpha}_2$ *at* α_2 *cut the axis of reals in* z_2, *and let* A_k $(k = 1, 2)$ *be the circular arc bounded by* α_1 *and* $\bar{\alpha}_1$ *of angular measure less than* π *which is tangent to the lines* $\alpha_1 z_k$ *and* $\bar{\alpha}_1 z_k$. *Then all non-real zeros of* $p'(z)$ *on or within* C_1 *lie in the closed lens-shaped region* R *bounded by* A_1 *and* A_2.

If z_0 is any non-real point interior to C_1, and if β denotes the intersection with the axis of reals of the tangent at α_2 to the circle $z_0\alpha_2\bar{\alpha}_2$, then the force at z_0 due to the two particles at α_2 and $\bar{\alpha}_2$ has the line of action $z_0\beta$. No matter where z_0 may lie on or interior to C_1, the corresponding point β lies in the closed interval $z_1 \leqq \beta \leqq z_2$. If z_0 lies interior to C_1 but exterior to R, the line of action L of the force at z_0 due to the particles at α_1 and $\bar{\alpha}_1$ cuts the axis of reals exterior to the interval $z_1 \leqq z \leqq z_2$, so z_0 cannot be a position of equilibrium. Of course z_0 cannot be a multiple zero of $p(z)$, so z_0 cannot be a zero of $p'(z)$.

Theorem 2 can be improved (§2.6.2 below), and here aims to indicate a method rather than a definitive result. Extensions of various kinds are immediate: 1) real zeros of $p(z)$ may also be admitted, and serve merely to redefine the significant interval z_1z_2; 2) further non-real zeros of $p(z)$ may be admitted, and these likewise serve merely to redefine the interval z_1z_2; 3) in Theorem 2 itself, the point z_1 may be redefined as the intersection with the axis of reals of the tangent at α_2 to the circle $\alpha_1\alpha_2\bar{\alpha}_2$, for it follows from Lucas's Theorem that no critical points of $p(z)$ lie interior to the left-hand semicircle of C_1; 4) in Theorem 2 itself, the point z_2 (and hence the arc A_2 and the region R) may now be redefined by virtue of the fact that no critical points lie in the interior of the lens-shaped region bounded by the original A_2 and the right-hand half of C_1; the new z_2 may be chosen as the intersection with the axis of reals of the tangent at α_2 to the circle through α_2, $\bar{\alpha}_2$, and the intersection of the original A_2 with the axis of reals; indeed, this process of revising the arc A_2 may be continued indefinitely, or the position of the limiting arc may be computed.

We emphasize the fact that in Theorem 2 we assert only that the *non-real* zeros of $p'(z)$ on or within C_1 lie in R; if the multiplicities of α_1 and $\bar{\alpha}_1$ as zeros of $p(z)$ are large, and those of α_2 and $\bar{\alpha}_2$ are relatively small, a real zero of $p'(z)$ lies near the center of C_1, and need not lie in R.

The method here indicated can be applied in essence even if Jensen circles intersect or lie one interior to another. In every case, it may be noticed, this method yields a point set which cannot be further improved by the use of Lucas's Theorem.

§2.6. W-curves. The circular arc A_k used in §2.5 Theorem 1 (and a similar remark applies to the arc A of §2.2.1 Theorem 1) may be interpreted as composing, together with a suitable segment of the axis of reals, the locus of positions of equilibrium in the field of force due to three particles of variable mass, namely equal particles at α and $\bar{\alpha}$ and a single particle at z_k, where all possible positive masses integral or not, are contemplated. Such a locus is a natural object of research when for a given geometric configuration we study regions free from critical points, regions depending on the positions of the zeros but not depending on the multiplicities of the zeros or on the degree of a given polynomial; §2.5 Theorem 1 is perhaps the simplest non-trivial illustration of a more general situation that we proceed to investigate in detail. We prove

THEOREM 1. *If* α $(\geqq 0)$ *is fixed, but* k *and* m *are variable positive integers, the closure of the locus of the critical points of the polynomial*

$$p(z) = (z^2 + 1)^k (z - \alpha)^m$$

consists of the segment $0 \leqq z \leqq \alpha$ *plus the arc of the circle* (*line if* $\alpha = 0$)

$$(1) \qquad \alpha x^2 + 2x + \alpha y^2 = \alpha$$

in the closed interior of the unit circle.

In the logarithmic derivative of $p(z)$ we set $k = 1$ and study z as an analytic function of the variable m, $0 \leqq m \leqq \infty$, although only values of z corresponding to rational values of m belong to the original locus. We omit the limiting case $\alpha = 0$. Then we have

$$(2) \qquad (m + 2)z^2 - 2\alpha z + m = 0, \qquad \frac{\partial z}{\partial m} = \frac{1 + z^2}{2(\alpha - mz - 2z)}.$$

When m is positive and near zero, the two values of z are real and near zero and α respectively. As m increases, the former value increases and the latter decreases algebraically until they coincide at the point $z = [(1 + \alpha^2)^{1/2} - 1]/\alpha$ when we have $m = (1 + \alpha^2)^{1/2} - 1$. As m continues to increase, the two values of z move on the curve (1) found by eliminating m from the first of equations (2) split into real and pure imaginary components. When m becomes infinite, these two values of z approach $+i$ and $-i$ respectively. The locus of critical points is the bow-and-arrow shaped configuration mentioned. The circle (1) (compare §2.2.1 Theorem 1) passes through the points $\pm i$, and is tangent at those points to the lines $(\pm i, \alpha)$ respectively.

§2.6.1. General results. We now formulate a result [Walsh, 1946a] which is extremely broad in its implications; the special situation just discussed, which is inherently that of §2.5 Theorem 1, is typical. The totality of zeros of a polynomial is said to possess *symmetry* if it is invariant under a given set of transformations of the plane. We consider various kinds of symmetry, such as symmetry in the axis of reals, in the origin O, in a circle C, or p-fold symmetry in O, in the sense that the totality of zeros is invariant under rotation about O through an angle of $2\pi/p$; combinations of these symmetries are not excluded. If a symmetry is given and also the zeros of $p(z)$ (perhaps with suitable multiplicities) which exhibit that symmetry, we shall use the term *group* to indicate any irreducible subset of these zeros which (multiplicities considered) possess the prescribed symmetry. For instance if the symmetry is defined by reflection in the axis of reals, a group is a real point or a pair of conjugate imaginary points; if the symmetry is p-fold symmetry about O, a group is either O or the set of vertices of a regular p-gon whose center is O. Two groups, being irreducible, are either disjoint or identical.

Corresponding to any two groups of points or of zeros of a given polynomial $p(z)$, we use the term *W-curve* to indicate *the closure of the locus of the critical points of a variable polynomial whose zeros retaining the prescribed symmetry are those respective groups of arbitrary integral multiplicities*. Thus Theorem 1 may be summarized in the statement that for symmetry in the axis of reals the W-curve of a real zero α and a pair of conjugate imaginary zeros $+i$ and $-i$ consists of that portion of the axis of reals and of the circle (1) in the closed Lucas polygon for the three points α, $+i$, and $-i$.

It follows from this definition that *the W-curve for two groups G_1 and G_2 is precisely G_1 plus G_2 plus the set of points at which the forces due to the two groups are the same in direction but opposite in sense*. A finite number of points may be exceptional here, namely points at which the force due to G_1 or G_2 is zero and thus has properly no direction. If the forces at a point z_0 due to G_1 and G_2 have the same direction but are opposite in sense, then for suitable multiplicities (not necessarily integral) of the groups of particles of G_1 and G_2, the point z_0 is a position of equilibrium; if a neighborhood of z_0 is given, then for suitably chosen integral multiplicities of those groups a position of equilibrium lies in that neighborhood (by the continuity of the zeros of a polynomial as functions of the coefficients), so z_0 lies on the W-curve. Conversely, if z_0 lies on the W-curve but does not belong to G_1 or G_2, then in an arbitrary neighborhood of z_0 lie positions of equilibrium for the field of force due to G_1 plus G_2 with suitable multiplicities; consequently if the forces at z_0 due to G_1 and G_2 have directions, those directions must be the same with the forces in opposite senses.

If the polynomials $p_1(z)$ and $p_2(z)$ have simple zeros at the respective points of G_1 and G_2 and no other zeros, the W-curve is precisely the locus defined by

$$\arg \frac{p_1'(z)}{p_1(z)} = \arg \frac{p_2'(z)}{p_2(z)} - \pi, \qquad \arg \frac{p_1(z)p_2'(z)}{p_1'(z)p_2(z)} = \pi.$$

Each arc of a W-curve is bounded, and commences in a zero of $p_1(z)p_2'(z)$ and terminates in a zero of $p_1'(z)p_2(z)$. All the zeros of those two polynomials belong to the W-curve. The multiple points of the W-curves are the multiple zeros of those two polynomials, and the critical points of their quotient; compare §1.6.1. The W-curve pertaining to G_1 and G_2 lies in the Lucas polygon pertaining to the zeros G_1 and G_2, hence lies in the Lucas polygon pertaining to the zeros of $p(z)$ if $p(z)$ has $p_1(z)$ and $p_2(z)$ as factors. The W-curve is not necessarily a complete algebraic curve, but is part of the algebraic curve

$$\Re\left[\frac{p_1(z)p_2'(z)}{p_1'(z)p_2(z)}\right] = 0.$$

At every point of this algebraic curve, the lines of action if defined of the two forces due to the two groups of particles are identical, the W-curve consists of the closure of that part of the algebraic curve on which those forces are opposite in sense.

In the situation of Lucas's Theorem, non-trivial symmetry is not required and we may consider the groups as single points; the W-curve of two groups consists of the line segment joining the two points and thus is precisely *an arc of a line of force common to the two groups* (points); the W-curves as such need not be introduced here because of this identity. In various other cases of symmetry, notably p-fold symmetry about O (see §3.6), each W-curve is composed of arcs of a line of force common to two groups. Whenever the two groups G_1 and G_2 have a common arc of a line of force on which the senses of the two forces are opposite, this arc belongs to the corresponding W-curve. The importance of W-curves in general is illustrated by a fundamental result:

THEOREM 2. *Let a symmetry be given, and a polynomial $p(z)$ whose zeros possess that symmetry. Let S be the set of all W-curves for all pairs of groups of zeros of $p(z)$. Let R be the set of all points which can be joined to the point at infinity by Jordan arcs not intersecting S. Then R contains no zero of $p'(z)$.*

The totality of zeros of $p(z)$ is the sum of a number of groups, each with a suitable multiplicity. If a point z_0 is remote from S, the forces at z_0 due to the usual particles at the individual zeros of $p(z)$ are represented by vectors terminating in z_0 and having their initial points in the acute angle subtended at z_0 by the Lucas polygon for $p(z)$; we replace these vectors by other vectors, each representing the force at z_0 due to a group of particles; these new vectors, found respectively by adding certain of the original vectors, terminate in z_0 and have their initial points in the same acute angle whose vertex is z_0. If z_1 is a point of R, and if the bounded Jordan arc J in R joins z_0 to z_1, then as z moves from z_0 along J to z_1, the point z never lies in S, and the vectors representing the forces at z due to the various groups of particles always have their terminal points in z and their initial points in a suitably chosen *acute angle* whose vertex is z, for the set S is precisely the locus on which two of the vectors have the same direction but opposite senses, including possible zero vectors. Thus when z is at z_1 the vectors lie in a suitably chosen acute angle whose vertex is z_1, so z_1 is not a position of equilibrium. The point z_1 is not a multiple zero of $p(z)$, hence is not a zero of $p'(z)$.

Under the conditions of Theorem 2, let a point z_1 of the boundary of R be a critical point but not a multiple zero of $p(z)$; then z_1 lies on at least one W-curve, and the forces at z_1 due to the various groups of particles are represented by vectors terminating in z_1 and with initial points in a closed sector with vertex z_1 and angle π; no such vector can have its initial point interior to this sector, so all vectors are collinear; thus the groups of particles fall into two classes according to the sense of the forces they exert, and z_1 lies simultaneously on the W-curves for all pairs of groups, each pair consisting of one group from each class.

Non-trivial symmetry is not essential in Theorem 2, and in defining W-curves for a given polynomial the groups of zeros may then be chosen arbitrarily with due reference to the multiplicities; but the use of non-trivial symmetry when it exists has the effect of decreasing the number of loci used, as compared with choosing individual zeros, has the effect of simplifying the nature of the loci obtained without symmetry, and also yields sharper results; compare §2.7. Theorem 2 is in essence a generalization of Lucas's Theorem, for in the latter no symmetry is required, and the W-curves can be considered as line segments joining pairs of zeros of $p(z)$.

As an entirely general remark, it is clear that if a symmetry is given, and point sets $S_1, S_2, \cdots, S_m$ on which groups of zeros of a polynomial may lie, and if the totality of W-curves corresponding to pairs of arbitrary groups (including pairs of groups both on one set S_j) lies in a point set S, then no critical points of $p(z)$ lie in the region composed of all points which can be joined to the point at infinity by Jordan arcs not intersecting S. We apply this remark in §2.6.3. This conclusion persists for the functions of §1.2, Corollary.

§2.6.2. Real polynomials. For a real polynomial symmetry in the axis of reals is involved, and a group is either a real point or a pair of non-real points symmetric in the axis of reals. The W-curve corresponding to two real points is the line segment joining them; the W-curve of one real point and a pair of conjugate imaginary points is determined in §2.6 Theorem 1; we proceed to study the W-curve corresponding to two pairs of non-real points.

The W-curve corresponding to the two pairs of points $\pm ia$ and $\pm ib$, $0 < a < b$, consists of the two segments of the axis of imaginaries $a^2 \leqq y^2 \leqq b^2$ plus the origin, for it will be noted that any point of either segment $a^2 < y^2 < b^2$ or the origin is a position of equilibrium in the field of force due to pairs of particles at the prescribed points, of positive but not necessarily integral mass. No point other than those mentioned can be a position of equilibrium, for a point not on the interval $y^2 \leqq b^2$ of the axis of imaginaries is exterior to the Lucas polygon for the two particles, and a point of the intervals $0 < y^2 < a^2$ is non-real and interior to both Jensen circles, hence not a position of equilibrium. Henceforth we treat in detail only the case of pairs of particles not having the same abscissa.

Let pairs of particles be placed respectively at the points $z_0 = x_0 + iy_0$, $\bar{z}_0 = x_0 - iy_0$, and $z_1 = x_1 + iy_1$, $\bar{z}_1 = x_1 - iy_1$, $x_0 < x_1$, $y_0 > 0$, $y_1 > 0$. The W-curve is part of the algebraic curve found by equating the slope of the force at (x, y) due to one pair of particles and the slope of the force due to the other pair:

$$(3) \qquad \frac{y[(x - x_0)^2 + y^2 - y_0^2]}{(x - x_0)[(x - x_0)^2 + y^2 + y_0^2]} = \frac{y[(x - x_1)^2 + y^2 - y_1^2]}{(x - x_1)[(x - x_1)^2 + y^2 + y_1^2]}.$$

At an arbitrary point of this curve the two forces have the same direction (if defined) and either the same sense or opposite senses; the point does not belong to the W-curve if the senses are the same, and does belong to the W-curve if the senses are opposite. The algebraic curve (3) degenerates into the axis of reals and the curve

$$(4)\quad \begin{aligned} F(x, y) \equiv (x - x_1)\,&[(x - x_0)^2 + y^2 - y_0^2][(x - x_1)^2 + y^2 + y_1^2] \\ &- (x - x_0)\,[(x - x_0)^2 + y^2 + y_0^2][(x - x_1)^2 + y^2 - y_1^2] = 0, \end{aligned}$$

which is seen by inspection to be a bicircular quartic symmetric in Ox and passing through the points z_0, $\bar{z}_0$, z_1, and $\bar{z}_1$. Moreover we have $F(x_0, 0) > 0$, $F(x_1, 0) > 0$, $F(+\infty, 0) < 0$, $F(-\infty, 0) < 0$, so the curve cuts the axis of reals in an odd number of points to the left of x_0, in an even number of points (at most two) between x_0 and x_1, and in an odd number of points to the right of x_1.

The W-curve lies in the closed Lucas polygon for the four given points. At a point between the lines $x = x_0$ and $x = x_1$, the force due to the particles z_0 and $\bar{z}_0$ has a component toward the right, and the force due to the particles z_1 and $\bar{z}_1$ has a component toward the left; such a point if a point of (4) must lie in the Lucas polygon and be a point of the W-curve. Thus the W-curve consists precisely of that part of (4) plus the axis of reals which lies in the closed Lucas polygon; the curve (4) cuts the Lucas polygon precisely in the four points z_0, $\bar{z}_0$, z_1, $\bar{z}_1$. The W-curve lies also in the point set consisting of the closed Jensen circles plus the segment $x_0 \leqq x \leqq x_1$.

The W-curve can also be considered as the closure of the locus of positions of equilibrium of pairs of particles at z_0, $\bar{z}_0$, and z_1, $\bar{z}_1$ of respective multiplicities unity and m, where m takes all positive rational values:

$$(5)\qquad \frac{1}{z - z_0} + \frac{1}{z - \bar{z}_0} + \frac{m}{z - z_1} + \frac{m}{z - \bar{z}_1} = 0, \qquad m > 0.$$

In other words, the W-curve is the image in the z-plane of the infinite segment $0 \leqq m \leqq \infty$ under the transformation (5). An arbitrary value of m corresponds to three values of z, not necessarily all distinct. The value $m = 0$ corresponds to $z = z_1$ and $\bar{z}_1$ as well as $z = x_0$, and the value $m = \infty$ corresponds to $z = z_0$ and $\bar{z}_0$ as well as $z = x_1$. Thus the W-curve joins the set z_0, $\bar{z}_0$, x_1 to the set z_1, $\bar{z}_1$, x_0, but not necessarily the points of one set to the respective points of the other. To each value of z corresponds a unique m; for given m, the equation determining z is a real cubic which then can have no non-real multiple values of z; if the locus (5) has a multiple point z it must correspond to a single value of m and to a multiple zero of a single equation (5), and must be real.

The bicircular quartic (4) *cannot degenerate.* If a bicircular quartic degenerates, it must consist of two circles, and (4) is symmetric in the axis of reals. The curve (4) cannot consist of a circle counted twice, for the W-curve lies in the Lucas polygon and passes through the points z_0, $\bar{z}_0$, z_1, $\bar{z}_1$. The curve (4) cannot consist of two circles not intersecting the axis of reals, or mutually symmetric in and tangent to each other on the axis of reals, or intersecting each other on that axis, for the curve cuts the axis in an odd number of points to the left of x_0 and in an odd number of points to the right of x_1. The curve (4) cannot consist of two circles with their centers on Ox and cutting each other in two non-real points, for the curve (4) together with the axis of reals is the locus of z in (5) where now m takes all real values; to each z corresponds a unique m, and the determination of z from m depends on the solution of a real cubic, which cannot have non-real multiple zeros. The curve (4) cannot consist of two circles tangent to each other on Ox and having their centers on Ox for (§§2.6.1, 1.6.1) at a multiple point of (3) the tangents to the various branches of the curve in order make equal angles with each other. The curve (4) cannot consist of two circles which do not intersect and which have their centers on Ox, as we now indicate. If the curve (4) consists of two such circles, an arc A of one of them joining z_1 and $\bar{z}_1$ in the closed Lucas polygon and lying in the closed interior of the Jensen circle for z_1 and $\bar{z}_1$ belongs to the W-curve. Let the line tangent to A at z_1 intersect Ox in the point β, necessarily finite. Then the line of action of the force at every point of A due to the particles at z_1 and $\bar{z}_1$ passes through β, so by the definition of the W-curve the line of action of the force at every point of A due to the two particles at z_0 and $\bar{z}_0$ also passes through β, so A is an arc of a circle through z_0 and $\bar{z}_0$, which is impossible.

The non-degenerate bicircular quartic can have at most one finite double point; if there were two such double points, a circle through them and through a third finite point of the curve would intersect the curve in the projective plane in a total of at least five finite and four infinite points (the circular points at infinity counted twice), which is impossible.

The entire curve (4) can be plotted by points. Choose an arbitrary point β on Ox, and let C_k be the circle through z_k and $\bar{z}_k$ tangent to the line βz_k at z_k. At a point z (if any) of intersection of C_0 and C_1, the forces due to the pairs of particles z_0, $\bar{z}_0$ and z_1, $\bar{z}_1$ both have the line of action $z\beta$, so z is a point of (4). At a possible point z of intersection of the two Jensen circles for those pairs of particles the two forces are horizontal, and a finite point β of the kind just mentioned does not exist, but the point z is a point of (4); even if this point z lies on Ox it is a point of (4), as follows from (4) itself. In fact if any point z' of (4) lies on Ox, the two tangents at the two points z_k to the two circles through z', z_k, and $\bar{z}_k$ meet at a finite or infinite point β of Ox, for the corresponding fact is true of points z on (4) and near to z' but not on Ox; of course the two circles through z', z_k, and $\bar{z}_k$ are tangent at z'. Every point of the curve (4) other than a point of intersection of the two Jensen circles can thus be obtained by the construction mentioned, and such points of intersection are themselves readily constructed.

Let C denote the circle through the points z_0, $\bar{z}_0$, z_1, $\bar{z}_1$, and let the tangent to C at z_0 meet Ox in the point β_0. Then *the line* $\beta_0 z_1$ *is tangent to the curve* (4) *at* z_1. A geometric proof can be given by studying the directions of forces at a point of (4) near z_1, but requires some delicacy as to the infinitesimal distances involved. However, differentiation of (4) yields the equation

$$D_x y \,|_{(x_1,y_1)} = \frac{y_1[(x_1 - x_0)^2 + y_1^2 - y_0^2]}{(x_1 - x_0)[(x_1 - x_0)^2 + y_1^2 + y_0^2]};$$

the second member is the slope of the line of force at (x_1, y_1) due to the particles at z_0 and $\bar{z}_0$, hence is the slope of the line $\beta_0 z_1$.

The methods and results of §2.5 are also of interest in considering the shape and position of the W-curve. In addition to the notation just used, let the tangent to C at z_1 meet Ox in the point β_1, let A_0 be the circular arc $z_0\bar{z}_0$ to the right of the line $x = x_0$ tangent to the line $z_0\beta_1$ at z_0, and let A_1 be the circular arc $z_1\bar{z}_1$ to the left of the line $x = x_1$, tangent to the line $z_1\beta_0$ at z_1. Then *the non-real part of the W-curve lies exterior to the circles of which A_0 and A_1 are arcs*. We treat in detail merely the case that z_0 lies to the left and z_1 to the right of the vertical diameter of C, so that A_0 and A_1 are arcs less than π, and show that a non-real point $z = x + iy$ in the trapezoid $\bar{z}_0\bar{z}_1 z_1 z_0$ to the left of A_0 is not a point of the W-curve. Denote by B_k the circular arc $z_k z \bar{z}_k$, and by α_k the intersection with Ox of the tangent to B_k at z_k. The line of action of the force at z due to the particles at z_k and $\bar{z}_k$ passes through α_k. By noting the relation between A_k and B_k we have $x_0 < x < \alpha_0 < \beta_1$, and either $\alpha_1 < x$ or $\alpha_1 > \beta_1$; consequently we have $\alpha_0 \neq \alpha_1$, so z is not a position of equilibrium nor a point of the W-curve. The omitted cases are similarly treated. The arc A_k is tangent to the curve (4) at z_k and $\bar{z}_k$.

We proceed to study the variation of z as a function of m in (5). If the Jensen circles for the two pairs of points are mutually exterior, the W-curve consists of the segment $x_0 \leqq x \leqq x_1$ plus a Jordan arc joining the two points z_0 and $\bar{z}_0$ lying in the closed interior of their Jensen circle, plus a Jordan arc joining the two points z_1 and $\bar{z}_1$, lying in their Jensen circle. Of course the two arcs intersect Ox in double (not triple!) points of the W-curve, hence intersect Ox orthogonally. Whenever the W-curve consists of the segment $x_0 \leqq x \leqq x_1$ plus two mutually disjoint Jordan arcs J_0 and J_1 joining z_0 and $\bar{z}_0$, z_1 and $\bar{z}_1$ respectively, that is to say, whenever the curve (4) cuts Ox at two distinct points in the interval

$$x_0 \leqq x \leqq x_1,$$

the variation of z appears from (5). When m is small but positive, the three values of z are respectively: real but near x_0, and on J_1 near z_1 and $\bar{z}_1$. As m increases, each point z moves on the W-curve without changing sense, the two points on J_1 until they coincide at the intersection a_1 of J_1 with Ox, and the real z to the right on Ox. As m continues to increase, one of the points z moves to the right from a_1 and the other to the left, both on Ox, and the third point continues real and moving to the right. As m increases still further, the two of these points moving in opposite directions coincide at the intersection a_0 of J_0 with Ox, then move on J_0 in opposite senses; the remaining point z continues to the right on Ox. As m becomes infinite, the respective points z approach z_0, $\bar{z}_0$, and x_1.

If the interval $x_0 \leqq x \leqq x_1$ lies interior to both Jensen circles, and since (§1.4.1) no non-real position of equilibrium can lie interior to both Jensen circles, the W-curve consists of the segment $x_0 \leqq x \leqq x_1$ plus two disjoint Jordan arcs joining z_0 and z_1 and $\bar{z}_0$ and $\bar{z}_1$ respectively and not intersecting Ox. Whenever the W-curve consists of that segment plus two such disjoint Jordan arcs, that is to say, when no point of the curve (4) lies on the segment $x_0 \leqq x \leqq x_1$, then as m increases from zero to infinity, the three points z remain distinct and move monotonically on separate Jordan arcs of the W-curve, from z_1 to z_0, from $\bar{z}_1$ to $\bar{z}_0$, and from x_0 to x_1.

If the W-curve has a triple point β, that is to say if the curve (4) has a double point on the segment $x_0 \leqq x \leqq x_1$, as is the case for instance when the four points $z_0, \bar{z}_0, z_1, \bar{z}_1$ lie at the vertices of a square, then the W-curve consists of the segment $x_0 \leqq x \leqq x_1$ plus Jordan arcs from the points $z_0, \bar{z}_0, z_1, \bar{z}_1$ to β, mutually disjoint except for β. The tangents to the two arcs to the curve (4) at β make angles of $\pi/3$ and $-\pi/3$ with Ox. As m increases from zero, the three points z start at the points $z_1, \bar{z}_1, x_0$, move along the arcs $z_1\beta$, $\bar{z}_1\beta$ of (4) and along the segment $x_0 \leqq x \leqq \beta$ monotonically until they all coincide at β, and then move monotonically along the arcs βz_0, $\beta\bar{z}_0$, and the segment $\beta \leqq x \leqq x_1$ until as m becomes infinite these points approach $z_0, \bar{z}_0$, and x_1.

Let us investigate more closely the condition for a triple point of the W-curve. We choose temporarily $x_0 = 0$, $y_0 = 1$, and set $x_1 = \alpha$, $y_1 = b(>0)$. Equation (5) is then equivalent to the condition

$$z^3 - \frac{2+m}{1+m}\alpha z^2 + \frac{\alpha^2 + b^2 + m}{1+m} z - \frac{m\alpha}{1+m} = 0,$$

and this equation has a triple zero when and only when we have

$$\frac{m\alpha}{1+m} = \left[\frac{(2+m)\alpha}{3(1+m)}\right]^3, \qquad \frac{\alpha^2 + b^2 + m}{3(1+m)} = \left[\frac{(2+m)\alpha}{3(1+m)}\right]^2,$$

conditions which are equivalent to

$$(6) \qquad \alpha^2 = \frac{27m(m+1)^2}{(m+2)^3}, \qquad b^2 = \frac{m(2m+1)^3}{(m+2)^3}.$$

Computation of the derivatives in (6) shows that those two functions of m increase monotonically with $m(>0)$. Thus if b is given, m is uniquely determined from (6) and hence α is uniquely determined; if α is given, m is uniquely determined and so also is b. In particular if we have $b = 1$, we have $m = 1$, $\alpha = 2$; if we have $b < 1$, we have $m < 1$, $\alpha < 2$; if we have $b > 1$, we have $m > 1$, $\alpha > 2$. From (6) we may write

$$\alpha^2 + b^2 = \frac{8m^2 + 7m}{m+2},$$

an equation which can be solved for m and which by elimination of m from (6) gives an algebraic equation connecting α and b directly.

Let b be fixed, with α variable and not necessarily defined by (6); we consider the shapes of the corresponding W-curves. It is to be noticed that these curves vary continuously with α. When α is large, the bicircular quartic (4) cuts the axis of reals at precisely two distinct points in the interval $0 \leqq x \leqq \alpha$. As α

decreases, there is a particular value α_0 of α defined by (6) for which the curve (4) has a double point in that interval. As α continues to decrease beyond α_0, the curve (4) does not cut the axis at all in that interval, for the curve (4) can have no double point for values of α less than α_0, and for small values of α we have shown that the curve does not cut the axis in the interval $0 \leqq x \leqq \alpha$.

We return now to the previous arbitrary values of x_0, y_0, x_1, y_1, with $x_0 < x_1$, $y_0 > 0$. Let us set $y_1/y_0 = b$, and let α be the value defined by (6). *If we have* $(x_1 - x_0)/y_0 > \alpha$, *the curve* (4) *cuts the segment* $x_0 \leqq x \leqq x_1$ *in two distinct points; if we have* $(x_1 - x_0)/y_0 = \alpha$, *the curve* (4) *has a double point on that segment; if we have* $(x_1 - x_0)/y_0 < \alpha$, *the curve* (4) *does not cut that segment.*

We add one further remark concerning the shape of the curve (4) in a particular case. *If the Jensen circles for the two pairs of points are tangent but not equal, the curve* (4) *cuts the axis of reals at their point of tangency and at a point on the segment* $x_0 \leqq x \leqq x_1$ *interior to the larger Jensen circle.* Let us choose the origin as the point of tangency: $y_0 = -x_0$, $y_1 = x_1$, and suppose $x_1 > -x_0$; the intersections of (4) with the axis of reals are given by $x = 0$ and

$$f(x) \equiv x^3 - 4(x_0 + x_1)x^2 + 2(x_0^2 + 5x_0x_1 + x_1^2)x - 4x_0x_1(x_0 + x_1) = 0.$$

Then we have $f(0) = -4x_0x_1(x_0 + x_1)$, which is positive, and

$$f(x_1) = -x_1[(x_0 - x_1)^2 + x_0^2],$$

which is negative.

§2.6.3. Zeros on horizontal segments. The discussion of §2.6.1 applies to all real polynomials, but can be made more precise in special situations, as we now indicate. For later application we state

LEMMA 1. *Let the horizontal line segments* S_1 *and* S_2 *be mutually symmetric in the axis of reals, and let the distance of each from the axis of reals be not less than* $2^{-1/2}$ *times the length of* S_1. *Then the two closed regions bounded by* S_1 *and* S_2 *and by the non-real arcs of the W-curve for their end-points contain the non-real arcs of the W-curve for any two pairs of points on* S_1 *and* S_2.

The W-curve for the end-points of S_1 and S_2 consists of a segment of the axis of reals plus two arcs which do not intersect the axis of reals, by virtue of our restriction on the distance of S_1 from the axis of reals, so that W-curve together with S_1 and S_2 bounds two disjoint closed regions. Of course the W-curve lies in the rectangle whose opposite sides are S_1 and S_2.

We return to equation (4) in form (3) with the factors y omitted and with $y_1 = y_0$. If x_0, y_0, and x are considered fixed, $x_0 < x < x_1$, $y^2 < y_0^2$, we compute from (4) the derivative $\partial y^2/\partial x_1$; it follows that if we have either

$$(x - x_1)^2[(x - x_1)^2 + 2y^2 - 4y_0^2] + y^4 - y_0^4 < 0 \text{ or } 2y_0^2 > (x - x_1)^2$$

we have also $\partial y^2/\partial x_1 < 0$. If we choose first x_0 and x_1 as the abscissas of the end-points of S_1, and then allow x_1 to move to the left, the part of the W-curve not on the axis moves farther from the axis of reals, monotonically on each vertical line; if we allow x_0 to move to the right, it follows by symmetry that the part of the W-curve not on the axis moves farther from the axis of reals. Since the W-curve for any two pairs of points on S_1 and S_2 lies in the rectangle whose opposite sides are S_1 and S_2, the conclusion follows.

By Theorem 2 and the Lemma we have at once

THEOREM 3. *Let the horizontal line segments S_1 and S_2 be mutually symmetric in the axis of reals, and let the distance of each from that axis be not less than $2^{-1/2}$ times the length of S_1. Let $p(z)$ be a real polynomial whose zeros lie on S_1 and S_2. Then all non-real critical points of $p(z)$ lie in the two regions bounded by S_1 and S_2 and by the non-real arcs of the W-curve for the end-points of S_1 and S_2.*

Theorem 3 is not valid if we omit the restriction between the length of S_1 and its distance from the axis of reals, for if we have $2y_0^2 < (x_0 - x_1)^2$ it follows that for points near (x_0, y_0) on the W-curve the relation $\partial y^2/\partial x_1 > 0$ holds, so the conclusion of Lemma 1 is false.

The precise extension of Theorem 3 to the case where we assume no relation between the length of S_1 and the distance from the axis of reals is algebraically more difficult, and will not be treated here. We shall establish, however, a related result, in the proof of which we use

LEMMA 2. *Let C be the circle through the points $+i, -i, \xi + i, \xi - i$, with $\xi > 0$, and let β be the intersection with the axis of reals of the tangent to C at $\xi + i$. Then we have $\beta \geqq 2^{3/2}$, and the minimum value $\beta = 2^{3/2}$ is assumed only for $\xi = 2^{1/2}$.*

The equation of C can be written $(x - \xi/2)^2 + y^2 = 1 + (\xi/2)^2$, whence $\beta = \xi + 2/\xi$, and the conclusion follows. A similar proof yields

LEMMA 3. *Under the conditions of Lemma 2, let δ be the distance from β to C. Then δ decreases as ξ increases.*

We have $\delta = \xi/2 + 2/\xi - (1 + \xi^2/4)^{1/2}$, and we find $d\delta/d\xi < 0$, in each of the two cases $\xi \geqq 2$, $0 < \xi < 2$. We now establish

THEOREM 4. *Let S_1 be the segment $x_0 \leqq x \leqq x_1$, $y = 1$, and let S_2 be the reflection of S_1 in the axis of reals. Let A_0 be the circular arc $(x_0 + i, x_0 - i)$ less than π tangent to the line $(x_0 + i, x_0 + 2^{3/2})$ at $x_0 + i$, and let A_1 be the circular arc $(x_1 + i, x_1 - i)$ less than π tangent to the line $(x_1 + i, x_1 - 2^{3/2})$ at $x_1 + i$. If $p(z)$ is a real polynomial whose zeros lie on S_1 and S_2, then all non-real critical points of $p(z)$ lie in the closed curvilinear quadrilateral $S_1A_0S_2A_1$.*

It is sufficient to consider, as we do, a non-real point $z = x + iy$ interior to the circle of which A_0 is an arc, and interior to the rectangle $\bar{z}_0\bar{z}_1z_1z_0$, $z_0 = x_0 + i$,

$z_1 = x_1 + i$; let x' be arbitrary, $x_0 < x' < x$, let A denote the arc $(x' + i, z, x' - i)$, and let α denote the intersection with the axis of reals of the tangent to A at $x' + i$. The arc congruent to and with the same orientation as A tangent to A_0 on the axis of reals has a tangent at a point of the line $y = 1$ which cuts the axis of reals to the left of the point $x_0 + 2^{3/2}$, by Lemma 3; hence the tangent to A at the point $x' + i$ cuts the axis of reals to the left of the point $x_0 + 2^{3/2}$ but to the right of the point x. If we now have $x < x'' < x_1$, let B denote an arc $(x'' + i, x'' - i)$ not greater than π of the circle $(z, x'' + i, x'' - i)$; then B may or may not be the arc $(x'' + i, z, x'' - i)$. It follows from Lemma 2 that the tangent to B at the point $x'' + i$ if not parallel to the axis of reals cuts the axis of reals either to the left of the abscissa x or to the right of the point $x_0 + 2^{3/2}$, according as z does or does not lie on B. The force at z due to each pair of particles at zeros of $p(z)$ has a component orthogonal to the line through z and $x_0 + 2^{3/2}$, and the sense of this component is independent of the pair of particles considered. Thus z is not a position of equilibrium and Theorem 4 is established. If A_0 and A_1 intersect, the interior of the quadrilateral $S_1A_0S_2A_1$ is interpreted as the sum of the interiors of two curvilinear triangles.

Theorems 3 and 4 can be extended by admitting also real zeros of $p(z)$; here we need certain properties of the W-curve for a real point and a pair of conjugate imaginary points. Let α and β be real, and let $A(\alpha, \beta)$ denote the circular arc less than π bounded by the points $\beta + i$ and $\beta - i$, tangent at those points to the lines from those points to α; if we have $\alpha = \beta$, the arc $A(\alpha, \beta)$ is the line segment $(\beta + i, \beta - i)$. The W-curve for a particle at α and a pair of particles at $\beta \pm i$ consists (§2.6 Theorem 1) of the arc $A(\alpha, \beta)$ plus a segment of the axis of reals. It is clear that $A(\alpha, \beta)$ varies monotonically with α in the sense that if α moves to the right the intersection of $A(\alpha, \beta)$ with every line $y = c$, $-1 < c < +1$, also moves to the right. It is also true, as we now show, that $A(\alpha, \beta)$ varies monotonically with β in the sense that if β moves to the right the intersection of $A(\alpha, \beta)$ with every line $y = c$, $-1 \leqq c \leqq +1$, also moves to the right. We choose $\alpha = 0$, $\beta > 0$, so the distance from α to $A(\alpha, \beta)$ is δ of Lemma 3, where now $\xi = 2/\beta$, so δ increases with β. The monotonic variation of $A(\alpha, \beta)$ follows. We are now in a position to prove the extension of Theorem 3:

THEOREM 5. *Let S_1 be the segment $x_0 \leqq x \leqq x_1$, $y = 1$, with $x_1 - x_0 \leqq 2^{1/2}$, and let S_2 be symmetric to S_1 in the axis of reals. Let S_0 be the segment $\alpha_0 \leqq x \leqq \alpha_1$ of the axis of reals. Let Π_1 denote the closed interior of the two regions bounded by S_1, S_2, and the non-real arcs of the W-curve for the pairs $x_0 \pm i$ and $x_1 \pm i$; let Π_2 denote the closed interior of the Jordan curve formed by S_1, S_2, $A(\alpha_0, x_0)$, $A(\alpha_1, x_1)$. If the zeros of the real polynomial $p(z)$ lie on S_0, S_1, and S_2, then all non-real critical points of $p(z)$ lie in $\Pi_1 + \Pi_2$.*

Any arc $A(\alpha, \beta)$ with $\alpha_0 \leqq \alpha \leqq \alpha_1$ and $x_0 \leqq \beta \leqq x_1$ lies in Π_2, for moving α to the left until it coincides with α_0 and β to the left until it coincides with x_0 moves all points of $A(\alpha, \beta)$ on each line $y = c$, $-1 < c < +1$, to the left, and moving α to the right until it coincides with α_1 and β to the right until it coincides with x_1 moves all points of $A(\alpha, \beta)$ on each line $y = c$ to the right. Thus the point set $\Pi_1 + \Pi_2$ contains all non-real arcs of W-curves for all pairs of groups of points of $p(z)$, so Theorem 5 follows from Theorem 2.

If C denotes the circle through the end-points of S_1 and S_2, and if α_k' denotes the intersection with the axis of reals of the tangent to C at $x_k + i$, then the non-real arcs of the W-curve of Theorem 5 lie (§2.6.2) in the curvilinear quadrilateral formed by S_1, S_2, $A(\alpha_1', x_0)$, $A(\alpha_0', x_1)$. Thus in Theorem 5 all non-real critical points of $p(z)$ lie in the curvilinear quadrilateral whose sides are S_1, S_2, $A(\alpha_0'', x_0)$, $A(\alpha_1'', x_1)$, where $\alpha_0'' = \min(\alpha_0, \alpha_1')$, $\alpha_1'' = \max(\alpha_1, \alpha_0')$. In Theorem 3 with S_1 the line segment $x_0 \leqq x \leqq x_1$, $y = 1$, all non-real critical points of $p(z)$ lie in the curvilinear quadrilateral whose sides are S_1, S_2, $A(\alpha_1', x_0)$, $A(\alpha_0', x_1)$; in the limiting case $(x_1 - x_0)^2 = 2$, we have by Lemma 2 the relations $\alpha_1' = x_0 + 2^{3/2}$, $\alpha_0' = x_1 - 2^{3/2}$, so this limiting case is included also in Theorem 4.

An extension of Theorem 4 is

Theorem 6. *Let S_1 be the segment $x_0 \leqq x \leqq x_1$, $y = 1$, let S_2 be the reflection of S_1 in the axis of reals, and let S_0 be the segment $\alpha_0 \leqq x \leqq \alpha_1$ of the axis of reals. Let Π denote the closed interior of the curvilinear quadrilateral whose sides are S_1, S_2, and the arcs $A(\alpha_0', x_0)$ and $A(\alpha_1', x_1)$, where $\alpha_0' = \min(\alpha_0, x_0 + 2^{3/2})$, $\alpha_1' = \max(\alpha_1, x_1 - 2^{3/2})$. If $p(z)$ is a real polynomial whose zeros lie on S_0, S_1, and S_2, then all non-real critical points of $p(z)$ lie in Π.*

All W-curves for $p(z)$ lie in the set Π_1 composed of Π plus a suitable segment of the axis of reals, whether W-curves for two pairs of non-real zeros, for a pair of non-real zeros and one real zero, or for two real zeros; thus Π_1 contains all critical points of $p(z)$, and Π contains all non-real critical points.

Of course, under the conditions of Theorem 5 all non-real points of the W-curves for two pairs of non-real zeros lie in the closed rectangle with S_1 and S_2 as opposite sides, so all non-real critical points of $p(z)$ not in Π_2 lie in that rectangle. A similar remark applies to Theorem 6.

§2.7. Non-real polynomials. Some of the methods we have developed for the study of real polynomials apply as well in the study of non-real polynomials [compare Walsh, 1920a].

§2.7.1. Jensen circles.

Theorem 1. *Let $p_1(z)$ be a real polynomial, and $p_2(z)$ a polynomial of positive degree whose zeros are not all real but lie in the closed lower half-plane $y \leqq 0$. Then the critical points of the polynomial $p(z) \equiv p_1(z)p_2(z)$ in the open upper half-plane $y > 0$ not multiple zeros of $p_1(z)$ lie in the open Jensen circles for $p_1(z)$.*

At a point z_0 of the upper half-plane not a multiple zero of $p_1(z)$ and not interior to a Jensen circle, the force due to the zeros of $p_1(z)$ has no downward vertical component, and the force due to the zeros of $p_2(z)$ has a non-vanishing upward component, so z_0 is not a position of equilibrium. Of course z_0 cannot be a multiple zero of $p_1(z)p_2(z)$, so z_0 cannot be a critical point. If we modify the hypothesis by requiring the zeros of $p_2(z)$ to lie in the half-plane $y \leqq b$ (>0), then the critical points of $p_1(z)p_2(z)$ in the complementary half-plane $y > b$ not multiple zeros of $p_1(z)$ lie in the open Jensen circles. Theorem 1 can be sharpened by the methods of §2.5; we give merely a simple illustration:

COROLLARY 1. *Let the zeros of the polynomial $p_1(z)$ of positive degree lie both in the closed lower half-plane $y \leqq 0$, and in the closed infinite sector bounded by half-lines from $+i$ through real points z_1 and z_2, with $z_1 < z_2$. Let A_k denote the circular arc less than π bounded by $+i$ and $-i$ tangent at $+i$ to the line $z_k i$, and let R denote the closed lens-shaped region bounded by A_1 and A_2. Then no critical point of the polynomial $(z^2 + 1)^k p_1(z)$ lies exterior to R in the open upper half-plane.*

If z_0 is a point of the open upper half-plane in the open unit circle but not in R, the line of action of the force at z_0 due to the particles at $+i$ and $-i$ cuts the axis of reals in a point outside of the interval $z_1 \leqq z \leqq z_2$. All zeros of $p_1(z)$ lie on one side of this line of action, so z_0 cannot be a critical point.

COROLLARY 2. *Under the conditions of Theorem 1, no point of the axis of reals not a multiple zero of $p(z)$ can be a critical point of $p(z)$. The open upper half of the interior of any Jensen circle exterior to all other Jensen circles contains precisely one critical point of $p(z)$. Any maximal region R of the upper half-plane consisting wholly of the interior points of the upper half of each of m distinct Jensen circles contains precisely m critical points of $p(z)$ not multiple zeros of $p(z)$.*

Here the number m is to be chosen as large as possible.

The boundary C of R is a Jordan curve which consists of a segment of the axis of reals together with various connecting arcs of Jensen circles. We modify this boundary so as to replace the segment of the axis of reals near each real zero z_0 of $p(z)$ by a small semicircle with center z_0, so chosen that the force at each point of the semicircle has a non-vanishing component vertically upward, and that no critical point of $p(z)$ lies between that semicircle and the axis of reals; if z_0 lies at a corner of C, we replace a small segment of the axis of reals plus a small circular arc of C by a small arc α of a circle whose center is z_0, again so that the force at every point of α has a component vertically upward and so that α cuts off no critical point other than z_0. We modify C also in the neighborhood of each non-real zero z_0 of $p(z)$ on C by replacing an arc of C by a small neighboring circular arc exterior to R but so that the modification C' of C is a Jordan region containing in its interior precisely those zeros of $p(z)$ in the open upper half-plane the interiors of whose Jensen circles belong to R.

At every point of the axis of reals not a zero of $p(z)$ the force has a non-vanishing component vertically upward, and this conclusion holds for each point of C'. Thus when z traces C', the net change in the direction of the force at z is zero, the net change in arg $[p'(z)/p(z)]$ is zero, so the number of zeros of $p'(z)$ interior to C' is equal to the number of zeros of $p(z)$ interior to C'; if the multiplicities of the latter are $n_1, n_2, \cdots, n_m$, those same points are critical points of multiplicities $n_1 - 1, n_2 - 1, \cdots, n_m - 1$, and there remain m other critical points interior to C' and hence in R.

THEOREM 2. *Let the zeros of the polynomial $p_1(z)$ lie in the closed half-plane $y \leqq -1$. Then no critical point of the polynomial $(z^2 + 1)^k p_1(z)$ other than $-i$ lies in the closed lower half of the unit circle.*

At a point z_0 other than $-i$ in the closed lower half of the unit circle, the force due to the pair of particles at $+i$ and $-i$ has no component vertically downward; the force at z_0 due to the remaining particles has a non-vanishing component vertically upward, so z_0 is not a position of equilibrium nor a critical point.

The corollaries to Theorem 1 apply to Theorem 2. We add a further

COROLLARY. *Under the hypothesis of Theorem 2, let all zeros of $p_1(z)$ lie in the infinite sector with vertex $-i$ opposite to the sector $z_1(-i)z_2$, z_1 and z_2 real, $z_1 < 0 < z_2$, and let A_j be the circle through $+i$ and $-i$ tangent to the line $z_j(-i)$ at $-i$. Then no critical point lies in the open lower half-plane interior to A_1 or A_2.*

At a point z_0 of the open lower half-plane interior to A_1 or A_2 but exterior to the unit circle, the line of action of the force due to equal particles at $+i$ and $-i$ cuts the axis of reals outside of the interval $z_1 \leqq z \leqq z_2$, and all zeros of $p_1(z)$ lie below this line, so z_0 cannot be a critical point. Theorem 2 and the Corollary extend to the case that the zeros of $p_1(z)$ lie in the closed half-plane $y \leqq b$, where b is negative.

The following theorem with its corollaries represents a sharpening of results due to Ancochea [1945] and Montel [1945], as well as to the present writer [1920a].

THEOREM 3. *If every zero of a polynomial $p(z)$ not on or within a circle C is one of a pair symmetric with respect to C, then all critical points of $p(z)$ not on or within C lie in the closed Jensen circles constructed on the segments joining pairs of symmetric zeros as diameters. No point of a Jensen circle exterior to all other Jensen circles and exterior to C is a critical point unless it is a multiple zero of $p(z)$.*

Choose C as the unit circle. If the point z_0 is exterior to C and to the Jensen circle Γ for the pair of symmetric zeros z_1 and z_2, the force $\mathfrak{F}$ at z_0 due to unit particles at z_1 and z_2 has a component along Oz_0 directed away from O. If z_0 lies on the same side as O of the perpendicular bisector β of the segment z_1z_2, the force $\mathfrak{F}$ is equal to the force at z_0 due to two coincident particles at a point ζ interior to Γ, where if α denotes the center of Γ, the angles $z_0\alpha O$ and $O\alpha\zeta$ are equal. The circle constructed on $O\alpha$ as diameter passes through the intersections of C and Γ, so z_0 is exterior to this circle and the angle $Oz_0\alpha$ is numerically less than $\pi/2$; the angle $Oz_0\zeta$ is likewise numerically less than $\pi/2$, so $\mathfrak{F}$ has a component in the direction and sense Oz_0. If now z_0 is on β or separated by β from O, let Oz_0 cut β in the point β_0; the angle $O\beta_0\alpha$ is numerically less than $\pi/2$, and hence the angle $Oz_0\zeta$ is numerically less than $\pi/2$, so $\mathfrak{F}$ has a component in the direction and sense Oz_0.

If the point z_0 is exterior to C and to all the Jensen circles, or exterior to C and on a Jensen circle but exterior to all other Jensen circles and not a zero of $p(z)$, then the force at z_0 due either to each pair of particles, or to a single particle on or within C, or the total force due to all particles, has a non-vanishing component in the direction and sense Oz_0; thus z_0 is not a critical point.

Theorem 3 is analogous to Theorem 1, and indeed Theorem 1 is the limiting case of Theorem 3 as the radius of C becomes infinite; and the Corollaries to Theorem 1 apply here in suitably modified form. A further remark is the

COROLLARY. *Let the points z_1 and z_2 be mutually inverse in the circle C, let β denote the perpendicular bisector of the segment joining them, and let Γ denote the circle constructed on the segment z_1z_2 as diameter. If the zeros of the polynomial $p_1(z)$ of positive degree lie in the closed interior of C, then no critical points of the polynomial $(z - z_1)^k(z - z_2)^k p_1(z)$ lie in the closed interior of Γ on β or between β and C. The open half H of the interior of Γ bounded by β and containing no point of C contains precisely one critical point.*

Denote by O the center of C and by α the center of Γ. If z_0 is a point interior to Γ between β and C, the force at z_0 due to unit particles at z_1 and z_2 is equal to the force at z_0 due to a double particle at a point ζ exterior to Γ so chosen that the angles $z_0\alpha O$ and $O\alpha\zeta$ are equal. If the line ζz_0 cuts $O\alpha$ in ξ, the angle $O\xi z_0$ is obtuse, so the force at z_0 due to the particles at z_1 and z_2 has a non-vanishing component in the direction and sense Oz_0. The force at z_0 due to the particles at the zeros of $p_1(z)$ likewise has such a component, so z_0 is not a critical point. It may be verified directly that no point of Γ or of β exterior to C can be a critical point; indeed no point of C not a multiple zero of $p_1(z)$ can be a critical point, either in Theorem 3 or the Corollary.

The remainder of the Corollary follows by the method of continuity, if we continuously vary the zeros of $p_1(z)$ in the closed interior of C so as to remain in the closed region, and cause them to coincide at z_1 or z_2 interior to C. During the motion no critical point can enter or leave H, and in the final configuration H contains precisely one critical point.

§2.7.2. Infinite sectors. Under certain conditions the methods already developed determine infinite sectors free from critical points. For convenience we first establish the

LEMMA. *Let the point P lie in the open sector S with vertex O and angular opening α less than π. Then the force at P due to a pair of unit particles symmetric in O and situated in the complement of S is equivalent to the force at P due to two coincident unit particles situated in the sector Σ with vertex O and angular opening $2\pi - 2\alpha$, not containing P, whose sides are the reflections of the half-line OP in the sides of S. If we have $\alpha \geqq \pi/2$, the force at P due to a set of m unit particles symmetric in O and situated in the complement of S is equivalent to the force at P due to an m-fold particle Q which lies in every closed half-plane containing Σ and not containing P.*

The force at P due to two particles symmetric in O situated on a line L is equivalent to the force at P due to two coincident particles situated on the reflection in L of the half-line OP. The given particles lie in the closed double sector complementary to the point set consisting of S plus the reflection of S in O. Thus for each pair of the given particles the equivalent double particle lies in Σ; a possible given particle may also lie at O. The sector Σ is of angular opening not greater than π, if we have $\alpha \geqq \pi/2$; any half-plane containing Σ but not containing P contains each double particle equivalent to the given pairs of particles and also contains O, and consequently (§1.5.1 Lemma 1) contains the m-fold particle Q equivalent to the m given particles; the particle Q is finite.

The Lemma applies essentially also to a point P on the boundary of S not coinciding with one of the given particles, but in this situation the total force at P is zero if P coincides with O, and may be zero otherwise if all the given particles lie on the boundary of S.

A number of results are now immediately available.

THEOREM 4. *Let R be a closed double sector with vertex O and angular opening β not greater than $\pi/2$, and let the zeros of a polynomial $p_1(z)$ be symmetric in O and lie in R.*

1). *All critical points of $p_1(z)$ lie in R.*

2). *If all zeros of the polynomial $p_2(z)$ lie in the closure of the sector S_1, one of the two sectors S_1 and S_2 with vertex O complementary to R, then the other (open) sector S_2 contains no critical point of $p_1(z)p_2(z)$.*

3). *If we have $\beta = \pi/2$, and if all zeros of $p_1(z)$ lie on the boundary of R, so do all critical points of $p_1(z)$.*

To prove 1), we merely notice that the force at a point P exterior to R due to particles at the given zeros of $p_1(z)$ is not zero, for there exists a half-plane not containing P but containing the sector Σ of the Lemma, and we notice that P cannot be a multiple zero of $p_1(z)$. To prove 3), it is sufficient to apply 1) in succession to each of the four sectors into which the boundary of R separates the plane. We shall later (§3.6) resume the study of polynomials whose zeros are symmetric in a point.

In the Lemma, as the half-line OP rotates about O through a given angle, the sector Σ rotates about O in the reverse sense through that same angle; when P approaches a boundary point P_0 of S, a side of Σ approaches the half-line OP_0. Thus Σ remains of constant size and in the complement of S; as OP rotates through the closure of S, the sector Σ rotates through the complement of S, first with one side of Σ coinciding with one side of S, then with the other side of Σ coinciding with the other side of S.

Under the hypothesis of 2), when P lies in S_2 the sector Σ lies in the complement of S_2, and the two sectors S_1 and Σ lie in a suitably chosen closed half-plane depending on P but not containing P. Thus the total force at P due to particles at the zeros of both $p_1(z)$ and $p_2(z)$ is equivalent to the force at P due to a finite multiple particle in that half-plane, and P cannot be a position of equilibrium or a critical point.

Part 2) can be somewhat generalized:

THEOREM 5. *Let the zeros of the polynomial $p_1(z)$ be symmetric in the point O and lie in the closed sectors $\varphi_0 \leqq \arg z \leqq \varphi_1$ and $\varphi_0 + \pi \leqq \arg z \leqq \varphi_1 + \pi$, where we have $\varphi_0 \leqq 0 \leqq \varphi_1, \varphi_1 - \varphi_0 < \pi/2$. Let the zeros of the polynomial $p_2(z)$ of positive degree lie in the closed sector $\theta_0 + \pi \leqq \arg z \leqq 2\pi$, where we have $0 \leqq \theta_0 \leqq \varphi_1$. Then the closed sector $S_0 : 2\varphi_1 - \theta_0 \leqq \arg z \leqq 2\varphi_0 + \pi$ contains no critical points of $p_1(z)p_2(z)$, except perhaps on its boundary in the cases $\varphi_0 = 0, \varphi_1 = \theta_0$.*

The origin is not a critical point unless it is a multiple zero or unless $\theta_0 = 0$, for the force at O due to the particles at the zeros of $p_1(z)$ vanishes, and the particles at the zeros of $p_2(z)$ lie in a sector of angular opening $\pi - \theta_0$ with vertex O. Except in the cases $\varphi_0 = 0$ and $\varphi_1 = \theta_0$, the sector S_0 is disjoint (except for the vertex O) to the sectors assigned to the zeros of $p_1(z)$. Let now $z_0 \neq 0$ be a point of the sector S_0 but $\arg z_0 \neq \varphi_1$, $\arg z_0 \neq \pi$. Let Σ be the sector which is the locus of the double particles equivalent so far as concerns the force at z_0 to pairs of particles symmetric in O situated in the sectors assigned to the zeros of $p_1(z)$. When $\arg z_0$ has the value $2\varphi_1 - \theta_0$, the sector Σ (considered positive) has its terminal side on the half-line $\arg z = \theta_0$, and when $\arg z_0$ increases, Σ rotates clockwise. When $\arg z_0$ has the value $2\varphi_0 + \pi$, the initial side of Σ is $\arg z = \pi$. In every admissible position of $\arg z_0$, both Σ and the sector $\theta_0 + \pi \leqq \arg z \leqq 2\pi$ lie in a suitably chosen half-plane depending on z_0 but not containing z_0. Thus z_0 cannot be a critical point.

The case $\varphi_0 = 0, \varphi_1 = \theta_0$ is part 2) of Theorem 4.

CHAPTER III

POLYNOMIALS, CONTINUED

The present chapter makes a deeper study of the critical points of polynomials in general, and of some polynomials with special properties.

§3.1. Infinite circular regions as loci. We proceed to prove several analogs [due to the present writer, 1921] of Walsh's Theorem (§1.5) where now either the exterior of a circle or a half-plane may be the locus of a number of zeros of a given polynomial.

§3.1.1. Exterior of a circle. We shall use the analog of §1.5.1 Lemma 2:

LEMMA 1. *Let the fixed circular regions* $| z - \alpha_1 | \leqq r_1$ *and* $| z - \alpha_2 | \geqq r_2$ *be the respective loci of the variable points* z_1 *and* z_2. *Let* m_1 *and* m_2 *be positive constants, with* $r = (m_1 r_2 - m_2 r_1)/(m_1 + m_2)$ *positive. Then the locus of the point*

$$z = \frac{m_2 z_1 + m_1 z_2}{m_1 + m_2} \tag{1}$$

is the circular region $| z - \alpha | \geqq r$, *with*

$$\alpha = \frac{m_2 \alpha_1 + m_1 \alpha_2}{m_1 + m_2}.$$

By the given conditions we have

$$z - \alpha = \frac{m_1(z_2 - \alpha_2) + m_2(z_1 - \alpha_1)}{m_1 + m_2},$$

$$| z - \alpha | \geqq \frac{m_1 r_2 - m_2 r_1}{m_1 + m_2} = r, \tag{2}$$

so the point z lies in the prospective locus.

If z is now given with $| z - \alpha | \geqq r$, we choose z_1 to satisfy the equation $z_1 - \alpha_1 = -(z - \alpha)r_1/| z - \alpha |$, whence we have $| z_1 - \alpha_1 | = r_1$; we choose z_2 to satisfy (1), whence

$$z_2 - \alpha_2 = \frac{m_1 + m_2}{m_1}(z - \alpha) + \frac{m_2}{m_1}\frac{z - \alpha}{| z - \alpha |} r_1,$$

$$| z_2 - \alpha_2 | \geqq \frac{m_1 + m_2}{m_1} r + \frac{m_2}{m_1} r_1 = r_2, \tag{3}$$

so z is a point of the locus.

The point $z_0 = (r_2 \alpha_1 + r_1 \alpha_2)/(r_1 + r_2)$ is an internal center of similitude for the pairs (C_1, C_2) and (C_1, C) in the sense that a suitable stretching or shrinking of the plane with z_0 fixed followed by a rotation about z_0 through the angle π carries one circle of a pair into the other, and is an external center of similitude for the pair (C_2, C), in the sense that a suitable stretching or shrinking of the plane with z_0 fixed and without rotation carries one circle into the other; here C_1, C_2, and C indicate circles according to the subscripts already used. Any line through z_0 cutting the circles cuts C and C_2 at the same angle, and cuts C_1 at the supplementary angle, if all three circles are considered oriented in the same sense. It follows from (2) that whenever z lies on the boundary of its locus, so also do z_1 and z_2; the line passing through these three points passes through z_0 and cuts C and C_2 at the same angle and C_1 at the supplementary angle; we have

$$\frac{z_1 - \alpha_1}{r_1} = -\frac{z_2 - \alpha_2}{r_2} = -\frac{z - \alpha}{r}. \tag{4}$$

If we consider a new locus problem, where the assigned loci of z_1 and z_2 are now the closed segments in the circular regions C_1 and C_2 of a line L through z_0 cutting C_1, then the locus of z as defined by (1) is the closed segment of L in the circular region C.

In the limiting case $r = 0$ of Lemma 1, the circle C is the null circle z_0, and whenever z lies at z_0 the points z_1 and z_2 in their proper loci satisfying (1) are not uniquely determined; it is necessary and sufficient for (1) that z_1 and z_2 lie on the boundaries of their respective loci on a line through z_0, with the first equation of (4) satisfied.

Even if r is not positive, if z is given the derivation of (3) remains valid, and shows that in this case the locus of z is the entire plane.

If we modify the hypothesis of Lemma 1 so as to assign the closed exteriors of circles as the respective loci of both z_1 and z_2, the locus of z is the entire plane, for we can leave the locus of z_2 unchanged as the closed exterior of a circle of radius r_2, and choose a new locus for z_1, namely a finite circular region of radius r_1 contained in the original locus with the property $m_1 r_2 - m_2 r_1 = 0$; here the locus of z is the entire plane, so the original locus is the entire plane.

THEOREM 1. *Let the circular regions* $C_1 : |z_1 - \alpha_1| \leqq r_1$ *and* $C_2 : |z_2 - \alpha_2| \geqq r_2$ *be the respective loci of* m_1 *and* m_2 *zeros of a variable polynomial* $p(z)$ *of degree* $m_1 + m_2$. *Then the locus of the zeros of* $p'(z)$ *consists of* C_1 (*if* $m_1 > 1$) *and* C_2 (*if* $m_2 > 1$) *together with the region*

$$C: \left| z - \frac{m_1 \alpha_2 + m_2 \alpha_1}{m_1 + m_2} \right| \geqq \frac{m_1 r_2 - m_2 r_1}{m_1 + m_2}.$$

If C_1 *has no point in common with the region* $C_2 + C$, *those two regions contain respectively* $m_1 - 1$ *and* m_2 *zeros of* $p'(z)$.

Each point z of the region C is a possible critical point of $p(z)$, for we need merely choose the m_1 and m_2 zeros assigned to the regions C_1 and C_2 as coinciding at respective points z_1 and z_2 in those regions, with (1) satisfied. Each point z in a region C_1 or C_2 with $m_1 > 1$ or $m_2 > 1$ is a possible multiple zero of $p(z)$ and hence a possible critical point. Each point common to the regions C_1 and C_2 is also a possible multiple zero, but belongs to the region C, for in (1) we may choose $z = z_1 = z_2$. Thus every point of the alleged locus is a possible critical point.

Conversely, if z is a critical point of $p(z)$ it is either in a region C_k with $m_k > 1$ or in both C_1 and C_2 and thus in C, or a position of equilibrium not in a region C_k with $m_k > 1$; in the latter case the force at z due to m_j particles in each of the regions C_j is equal to the force at z due to m_j coincident particles in that region, so z lies in the region C. The remainder of Theorem 1 follows by the method of continuity.

If z is a critical point of $p(z)$ not a multiple zero of $p(z)$ nor in a closed region C_k with $m_k > 1$ but is on the boundary of C, then z is a position of equilibrium in the field of force, and all the zeros of $p(z)$ in the region C_j are concentrated at a point z_j on the circle C_j, with (1) and (4) satisfied. The line through z, z_1, and z_2 passes through z_0 and cuts C and C_2 at the same angle and C_1 at the supplementary angle.

In Theorem 1 we do not exclude the case that C_1 degenerates to a point: $r_1 = 0$. In this case the circle C is found by shrinking C_2 toward α_1 in the ratio $(m_1 + m_2):m_1$. If we choose also $\alpha_1 = \alpha_2$ we have [Alexander, 1915, for the case $m_1 = 1$]:

COROLLARY. *If $z = \alpha_1$ is a zero of multiplicity m_1 of the polynomial $p(z)$ whose other zeros are m_2 in number and lie in the closed region $|z - \alpha_1| \geqq r_2$, then all critical points of $p(z)$ other than α_1 lie in the closed region $|z - \alpha_1| \geqq m_1r_2/(m_1 + m_2)$.*

§3.1.2. Half-planes. Here again we use a geometric

LEMMA 2. *With $z = x + iy$, let the circular regions $C_1: |z - \alpha| \leqq r$ with $\alpha = \alpha_1 + i\alpha_2$ and $C_2: x \geqq 0$ be the respective loci of the variable points z_1 and z_2. Let m_1 and m_2 be positive constants. Then the locus of the point z defined by* (1) *is the circular region*

$$C: x \geqq \frac{m_2(\alpha_1 - r)}{m_1 + m_2}. \tag{5}$$

If $z_1 = x_1 + iy_1$ and $z_2 = x_2 + iy_2$ lie in the assigned regions, we have from (1)

$$x = \frac{m_2 x_1 + m_1 x_2}{m_1 + m_2},$$

with $x_1 \geqq \alpha_1 - r$, and $x_2 \geqq 0$, so (5) is satisfied. Conversely, if z is given with (5) satisfied, we write $z_1 = \alpha - r$, $z_2 = x_2 + iy_2 = (m_1 + m_2)z/m_1 - m_2 z_1/m_1$, whence $|z_1 - \alpha| \leqq r$, $x_2 \geqq 0$. The proof is complete.

We remark that if z lies on the boundary of its locus C, the points z_1 and z_2 which satisfy (1) are uniquely defined on the boundaries of their respective loci with $z_1 = \alpha - r$. The line through z, z_1, z_2 passes through the point $\alpha - r$, and cuts the three circles C_1, C_2, C at the same angle.

THEOREM 2. *With $z = x + iy$, let the circular regions $C_1:|z - \alpha| \leqq r$ with $\alpha = \alpha_1 + i\alpha_2$ and $C_2:x \geqq 0$ be the respective loci of m_1 and m_2 zeros of a variable polynomial $p(z)$ of degree $m_1 + m_2$. Then the locus of the zeros of $p'(z)$ consists of C_1 (if $m_1 > 1$) and C_2 (if $m_2 > 1$), together with the region C defined by* (5). *If the region C_1 has no point in common with the region $C_2 + C$, those two regions contain respectively $m_1 - 1$ and m_2 zeros of $p'(z)$.*

We omit the proof of Theorem 2, but point out that if z is a zero of $p'(z)$ not a multiple zero of $p(z)$ nor in a closed region C_k with $m_k > 1$ but lies on the boundary of C, then z is a position of equilibrium and all the zeros of $p(z)$ in the region C_k are concentrated at a point z_k on the circle C_k, with (1) satisfied and $z_1 = \alpha - r$; the line through z, z_1, z_2 cuts the three circles C_1, C_2, C at the same angle.

In Theorem 2 we do not exclude the case $r = 0$, for which C is a half-plane found by shrinking C_2 toward α in the ratio $(m_1 + m_2):m_1$.

Still a third case deserves mention here:

LEMMA 3. *Let the closed half-planes $C_1:x \geqq a_1$ and $C_2:x \geqq a_2$ be the respective loci of the points $z_1 = x_1 + iy_1$ and $z_2 = x_2 + iy_2$. Then the locus of the point z defined by* (1), *where m_1 and m_2 are positive constants, is the half-plane*

$$C:x \geqq \frac{m_2 a_1 + m_1 a_2}{m_1 + m_2}. \tag{6}$$

The inequalities $x_1 \geqq a_1$, $x_2 \geqq a_2$ obviously imply (6). Conversely, if z is given with (6) satisfied, we need merely set

$$z_1 - a_1 = z_2 - a_2 = z - \frac{m_2 a_1 + m_1 a_2}{m_1 + m_2};$$

then z_1 and z_2 are in their proper loci, and (1) is satisfied. Here the points z_1 and z_2 are not uniquely defined by the mere fact that z is on the boundary of the region C; indeed if z is given on the boundary of that region we may choose $x_1 - a_1 = x_2 - a_2 = 0$ and then choose y_1 arbitrarily with y_2 determined from (1); the points z_1 and z_2 are in their proper loci with (1) satisfied. Even if z_1 and z_2 are not uniquely determined from the fact that z lies on the boundary of its locus, nevertheless for any possible choice of z_1 and z_2 these points also lie on the boundaries of their loci, and the line zz_1z_2 naturally cuts the three parallel lines C_1, C_2, C at the same angle.

If we modify the hypothesis of Lemma 3 so that the regions C_1 and C_2 are half-planes whose boundaries are not parallel or whose boundaries are parallel but bound their regions on opposite sides, the locus of z is the entire plane. In either of these cases, let z be an arbitrary point of the plane. There exists a line L through z such that an infinite segment S_1 of L lies in C_1 and an infinite segment S_2 of L disjoint from S_1 lies in C_2. There exist points z_1 and z_2 on S_1 and S_2 respectively such that (1) is satisfied.

THEOREM 3. *Let the closed half-planes $C_1\colon x \geqq a_1$ and $C_2\colon x \geqq a_2$ be the respective loci of m_1 and m_2 zeros of a variable polynomial $p(z)$ of degree $m_1 + m_2$. Then the locus of the zeros of $p'(z)$ consists of C_1 (if $m_1 > 1$) and C_2 (if $m_2 > 1$) together with the region C defined by* (6).

We omit the proof of Theorem 3. If z is a zero of $p'(z)$ not a multiple zero of $p(z)$ nor in a closed region C_k with $m_k > 1$ but lies on the boundary of C, then z is a position of equilibrium and all the zeros of $p(z)$ in the region C_k are concentrated at a point z_k on the boundary of that region, with (1) satisfied. The points z_k are not uniquely determined, but they are collinear with z and the line zz_1z_2 cuts the three parallel lines C_1, C_2, C at the same angle.

We have mentioned in §3.1, as in §1.5, details concerning the positions of the points z, z_1, and z_2 partly for the sake of their intrinsic interest and partly for use in §3.3 below.

§3.2. A characteristic property of circular regions. Circular regions have appeared frequently in the foregoing development, and will appear frequently in the sequel. To some extent this is due to their simplicity and convenience, but also to their intrinsic properties. The property of invariance under inversion has shown itself useful, as will the property of invariance under linear transformation. Another property [Walsh, 1922] is also striking:

THEOREM 1. *Let C be a closed point set of the plane with the general property that the force at an arbitrary point P not in C due to two unit particles in C is equivalent to the force at P due to two unit coincident particles at some point of C. Then C is a circular region.*

If P is the point z and the two given particles are at z_1 and z_2, the equivalent coincident particles are at z_0, where we have by taking conjugates

$$\frac{1}{z - z_1} + \frac{1}{z - z_2} = \frac{2}{z - z_0}, \qquad (z_1, z_2, z, z_0) = 2;$$

here and in the sequel we define cross-ratio by the equation

$$(w_1, w_2, w_3, w_4) \equiv \frac{(w_1 - w_2)(w_3 - w_4)}{(w_2 - w_3)(w_4 - w_1)}. \tag{1}$$

Thus z_0 is the harmonic conjugate of z with respect to z_1 and z_2, is concyclic with those three points, and is separated from z by z_1 and z_2. If z is not in C, but z_1 and z_2 are in C, the harmonic conjugate z_0 of z with respect to z_1 and z_2 lies in C. Then the harmonic conjugate z_0' of z with respect to z_0 and z_1 lies in C, as does the harmonic conjugate z_0'' of z with respect to z_0 and z_2. We continue this process indefinitely—a convenient interpretation is found by taking z at infinity, so that the harmonic conjugate of z with respect to any two finite points is the mid-point of their segment. From the closure of the set C it follows that the entire arc z_1z_2 not containing z of the circle z_1z_2z belongs to C. Consequently, for an arbitrary circle Γ of the plane the intersection of Γ with C must consist of no point, a single point, a closed arc, or the entire circle Γ.

If C degenerates to a single point or the entire plane, the conclusion of Theorem 1 is true. In any other case there must be at least one point P of the plane not in C, and some neighborhood of P consists wholly of points not in C. The set C contains at least two points and hence contains no isolated point, but has an infinite number of boundary points.

Let Γ be the circle (or straight line) through three distinct boundary points z_1, z_2, z_3 of C. Then every point of Γ is a point of C; otherwise an open arc α of Γ, say bounded by z_1 and z_2, contains no points of C, but the remaining points of Γ including z_3 belong to C. Let the sequence of points ζ_k each exterior to C approach z_3; then the harmonic conjugate of ζ_k with respect to z_1 and z_2 belongs to C and approaches a point ζ_0 of α, the harmonic conjugate of z_3 with respect to z_1 and z_2, so ζ_0 belongs to C contrary to the definition of α.

Let P be a point not belonging to C, and denote by Γ_0 the circular region not containing P bounded by Γ. The set C contains the intersections with Γ of any circle through P intersecting Γ, so C contains Γ_0. If C contains a point Q exterior to Γ_0, the harmonic conjugate of ζ_k with respect to z_1 and Q belongs to C, as does the harmonic conjugate of z_3 with respect to z_1 and Q, and indeed C contains every point of the arc z_1Q (not containing z_3) of the circle z_3z_1Q. The interior points of this arc z_1Q lie exterior to Γ_0. Similarly if z lies on Γ near z_1, C contains every point of the arc zQ (not containing z_3) of the circle z_3zQ. Thus a suitably chosen neighborhood of z_1 consists wholly of points of C, contrary to our assumption that z_1 is a boundary point of C. This contradiction shows that no point Q not in Γ_0 can be a point of C, and completes the proof of Theorem 1.

Theorem 1 is stated and proved for the particle equivalent to *two* given particles. It is clear that a corresponding result holds for the particle equivalent to k given particles, where k is arbitrary but fixed.

Theorem 1 can be regarded as a converse to §1.5.1 Lemma 1.

§3.3. Critical points of real polynomials as centers of circles. Walsh's Theorem (§1.5) may be interpreted as using certain geometric loci concerned with the proportional division of segments to define the locus of critical points of a polynomial; it can also be interpreted in the case $r_1 = r_2$, α_1 and α_2 real, as the comparison of a given polynomial with a suitably chosen real polynomial: equal circles are constructed with centers in the points α_1 and α_2; the closed interiors of these circles are assigned as the respective loci of m_1 and m_2 zeros of a polynomial $p(z)$; the zeros of $p'(z)$ then have as their loci the closed interiors of circles with the same radius, whose centers are the critical points of the polynomial $(z - \alpha_1)^{m_1}(z - \alpha_2)^{m_2}$. When the theorem is expressed in this form, a natural extension exists [Walsh, 1924]:

THEOREM 1. *Let the circles $C_1, C_2, \cdots, C_n$ have collinear centers $\alpha_1, \alpha_2, \cdots, \alpha_n$ respectively, and the common radius r. Let the closed interiors of these circles be the respective loci of $m_1, m_2, \cdots, m_n$ zeros of the variable polynomial $p(z)$ which has no other zeros. Denote by $\alpha_1', \alpha_2', \cdots, \alpha_k'$ the distinct zeros of the derivative of $(z - \alpha_1)^{m_1}(z - \alpha_2)^{m_2} \cdots (z - \alpha_n)^{m_n}$, of the respective multiplicities $q_1, q_2, \cdots, q_k$. Then the locus of the zeros of $p'(z)$ consists of the closed interiors of the circles $C_1', C_2', \cdots, C_k'$ whose centers are the points $\alpha_1', \alpha_2', \cdots, \alpha_k'$, and whose common radius is r. Any circle C_j' which has no point in common with the other circles C_i' contains on or within it precisely q_j zeros of $p'(z)$; any set of circles C_j' disjoint from the other circles C_i' contains a number of zeros of $p'(z)$ equal to the sum of the corresponding numbers q_j.*

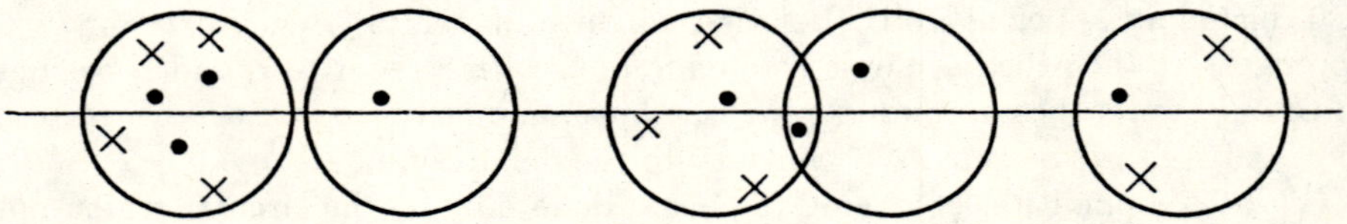

Fig. 6 illustrates §3.3 Theorem 1

We assume, as we may do with no loss of generality, that the points α_j lie on the axis of reals. We assume also $n > 2$.

It follows from Lucas's Theorem that all zeros of $p'(z)$ lie in the oval region bounded by segments of the two common external tangents to all the circles C_j and by the extreme semicircles of the C_j. It follows by a simultaneous translation of all the points α_j that every point on or within a circle C_j' belongs to the locus of critical points. To determine the locus it is sufficient to determine all boundary points; we proceed to show that all boundary points lie on the circles C_j'. Any point in the closed interiors of two of the given circles C_j lies interior to a circle C_j' and is an interior point of the locus. Any point in the closed interior of a circle C_j whose closed interior is the locus of several zeros of $p(z)$ may be a multiple zero of $p(z)$ and hence is a point of the locus; any point on such a circle C_j lies on a circle C_i'. In order to determine the locus completely, it remains to study an arbitrary point z on the boundary of the locus and either exterior to all the C_j, or in the closed interior of but one such circle, whose closed interior is the locus of only one zero of $p(z)$.

In a general way, our proof consists of fixing a point z on the boundary of the locus of the zeros of $p'(z)$, and replacing the various particles determining the field of force by equivalent multiple particles, by use of §1.5 Lemmas 1 and 2 and §3.1 Lemmas 1 and 2. The special geometric properties of the circles involved will then enable us to show that whenever z lies on the boundary of its locus, it lies on a circle C_j'. We proceed with the details of the proof.

If a point z on a common tangent to the circles C_j is a zero of $p'(z)$ but not a multiple zero of $p(z)$, then as follows from Lucas's Theorem, all zeros of $p(z)$ must also lie on this tangent; consequently z lies on a circle C_j'.

We study the usual field of force (§1.2), and suppose a point z of the locus of zeros of $p'(z)$ to lie on the boundary of that locus, but not on a common tangent to the circles C_j, and either exterior to all the C_j or in the closed interior of but one C_j, namely a circle whose closed interior is the assigned locus of but one zero of $p(z)$. For the present assume z exterior to all the C_j; the horizontal line L through z cuts each C_j in precisely two distinct points, and cuts all C_j at the same angle. We replace the m_j particles on or within C_j by an equivalent m_j-fold particle in that closed region so far as concerns the force exerted at z (§1.5.1 Lemma 1). We then replace these two multiple particles ζ_1 and ζ_2 in the two given circular regions say C_1 and C_2 which intersect L nearest and to the right of z by a single equivalent particle ζ_{12}; here the defining equation is of the form

$$\frac{m_1}{z-\zeta_1}+\frac{m_2}{z-\zeta_2}=\frac{m_1+m_2}{z-\zeta_{12}},$$

and when we transform the point z to infinity by a linear transformation of the z-plane to the W-plane, say $W = 1/(z - Z)$, $W_1 = 1/(z - \zeta_1)$, $W_2 = 1/(z - \zeta_2)$, $W_{12} = 1/(z - \zeta_{12})$, we have precisely the situation of §1.5 Lemma 2. The locus of ζ_{12} is thus the closed interior of a circle C_{12} which is cut by L at the same angle as are the circles C_j; the two intersections of C_{12} with L lie on L to the right of z between the extreme intersections of C_1 and C_2 with L.

We now replace the $(m_1 + m_2)$-fold particle ζ_{12} and the m_3-fold particle ζ_3 in the next nearest circular region say C_3, if any, intersecting L to the right of z, by an equivalent $(m_1 + m_2 + m_3)$-fold particle ζ_{23} so far as concerns the force exerted at z; the locus of ζ_{23} is the closed interior of a circle C_{23} which is cut by L in two distinct points at the same angle as are the circles C_j; the two intersections of C_{23} with L lie on L to the right of z between the extreme intersections of C_{12} and C_3 with L. We proceed in this manner as long as possible; in the next-to-the-last stage we have two multiple particles ξ_1 and ξ_2 whose respective loci are the closed interiors of circles Γ_1 and Γ_2, which are replaced by a single equivalent multiple particle ξ_0 whose locus is the closed interior of a circle Γ_0; all the circles Γ_1, Γ_2, Γ_0, C_j are proper circles cut by L in two distinct finite points to the right of z and at the same angle; the intersections of Γ_0 with L lie between the extreme intersections of the circles C_j with L to the right of z. We deal likewise with the particles assigned to the closed interiors of the circles C_j which intersect L to the left of z, in order of nearness to z replacing pairs of particles by single equivalent multiple particles; the last step replaces two particles η_1 and η_2 whose respective loci are the closed interiors of circles Δ_1 and Δ_2, by an equivalent particle η_0 whose locus is the closed interior of a circle Δ_0; all the circles Δ_1, Δ_2, Δ_0, Γ_1, Γ_2, Γ_0, C_j are proper circles cut by L in two distinct points and at the same angle; the intersections of Δ_0 with L lie between the extreme intersections of the circles C_j with L to the left of z.

Since z is on the boundary of its locus, and since the loci Γ_0 and Δ_0 depend continuously on z, it follows that z is on the boundary of the locus Σ of positions of equilibrium when ξ_0 and η_0 are now considered independent of z and to have as loci the closed interiors of Γ_0 and Δ_0 ; otherwise a sufficiently small change in z but arbitrary in direction would cause only a small change in the loci Γ_0, Δ_0, and Σ, and z would still belong to the locus Σ and to the locus of the zeros of $p'(z)$, contrary to our original hypothesis on z. By the proof of Walsh's Theorem (§1.5.2) it follows that the particles ξ_0 and η_0 must lie on the boundaries of their proper regions (z is exterior to Γ_0 and Δ_0) and that the line $\xi_0 z \eta_0$ cuts Γ_0 and Δ_0 at the same angle. The circles Γ_0 and Δ_0 are neither interior to the other and are not internally tangent, so there is but one line through an arbitrary point of the circle Σ which cuts those directed circles at the same angle; the line L through z is known to have this property, so ξ_0 and η_0 lie at the intersections of L with Γ_0 and Δ_0, intersections so chosen that the angles L makes with those circles directed counterclockwise are equal. It follows (§1.5.1) from the fact that ξ_0 lies on L and Γ_0 that ξ_1 and ξ_2 lie at the intersections of L with Γ_1 and Γ_2, again intersections so chosen that the angles L makes with those directed circles are equal to the angles L makes with Γ_0 and Δ_0. Continuation of this reasoning shows that each particle of multiplicity m_j representing the m_j particles assigned to the closed interior of C_j must lie at that intersection of C_j with L, so chosen that the angles L makes with the directed circles C_j at those intersections are all equal; then z is a position of equilibrium for these positions of the particles and consequently z lies on a circle C_j'.

It remains to consider a point z on the boundary of the locus of the zeros of $p'(z)$ but on or interior to a circle, say C_1, whose closed interior is the prescribed locus of but a single zero ζ_1 of $p(z)$; we assume that z is not on or interior to any of the circles $C_2, \cdots, C_n$, for every point in the closed interiors of two of the given circles is interior to a circle C_j', hence not a boundary point of the locus of zeros of $p'(z)$. The point z does not lie on a common tangent to the circles C_j. Let L be the horizontal line through z. Again we replace the m_j particles assigned to the closed interior of C_j by an equivalent m_j-fold particle in that same circular region without altering the corresponding force at z. We treat the particles in the regions C_j $(j > 1)$ intersecting L to the right of z precisely as before, and also those in the regions C_j $(j > 1)$ intersecting L to the left of z, and then we replace the two resulting particles by a single equivalent particle ξ_0. The locus of ξ_0 does not contain z, and may be the closed interior or exterior of a circle Γ_0 or a closed half-plane bounded by a line Γ_0 ; in any case the line L cuts Γ_0 in two distinct points not necessarily finite but distinct from z, at precisely the same angle as that in which L cuts all the C_j, where Γ_0 and the C_j are directed in the positive sense with respect to the significant loci. The point z is a position of equilibrium in the field of force due to ξ_0 and a unit particle on or within C_1. We omit further details; we apply as necessary the results of §1.5 or those of

§3.1; the case of §3.1.2 Lemma 3 and Theorem 3 cannot arise, nor can the case of indeterminate position of particles mentioned in §1.5.2. In any case it follows as before that z must lie on a circle C_j'.

One new phenomenon can occur here, however; in the study of the locus of all positions of equilibrium in the field due to particles ξ_0 and ζ_1, considered as independent variables in the loci bounded by fixed circles Γ_0 and C_1, the point z may coincide with z_0, in the notation of §3.1.1 Lemma 1 and Theorem 1, where we have $r = 0$; it may still be true that a small displacement of z in a suitable direction causes z to move exterior to the locus of positions of equilibrium in the field due to ξ_0 and ζ_1 restricted to their loci for the variable z, so z may still be on the boundary of the locus of critical points of $p(z)$. The points ξ_0 and ζ_1 in whose field z is a position of equilibrium are then not uniquely defined by the fact that z lies at ζ_0, but may lie on the boundary of their loci on any line L_1 through z_0 cutting Γ_0 and C_1. However, the point z is a position of equilibrium when ξ_0 and ζ_1 are chosen *arbitrarily* on such a line L_1 and on the circles Γ_0 and C_1 in such positions that the directed angles at ξ_0 and ζ_1 between line and circles are equal, the circles being oriented according to the loci which they bound. In particular we may choose L_1 as L; as before it then follows that z lies on a circle C_j'. Incidentally, this phenomenon effectively arises if z lies on two distinct circles C_j' interior to C_1; then (by Theorem 1, whose proof is not yet complete) the point z lies on the boundary of its locus but the points ξ_0 and ζ_1 are not uniquely determined, nor are the zeros of the corresponding polynomial $p(z)$ if z is a critical point of $p(z)$. We proceed to complete the proof of Theorem 1.

The locus Δ' of the zeros of $p'(z)$ is closed, because the locus of the zeros of $p(z)$ is closed and bounded. The locus Δ' is bounded, it contains the closed interiors of the circles C_j', and each boundary point of Δ' is a point of a circle C_j'. Each point exterior to all the C_j' can be joined to the point at infinity by a curve cutting no circle C_j' and thus containing no boundary point of Δ'. If S is generally a closed bounded set and a point P can be joined to the point at infinity by a Jordan arc disjoint from the boundary of S, then P is not a point of S. Hence Δ' consists precisely of the closed interiors of the C_j'. The remainder of the theorem follows by the method of continuity (§1.5.2).

A slight modification of the proof given yields the

COROLLARY. *Let the circles $C_1, C_2, \cdots, C_n$ with centers $\alpha_1, \alpha_2, \cdots, \alpha_n$ have the common external center of similitude α. Let the closed interiors of these circles be the respective loci of $m_1, m_2, \cdots, m_n$ zeros of the variable polynomial $p(z)$ which has no other zeros. Denote by $\alpha_1', \alpha_2', \cdots, \alpha_k'$ the distinct zeros of the derivative of $(z - \alpha_1)^{m_1}(z - \alpha_2)^{m_2} \cdots (z - \alpha_n)^{m_n}$, of respective multiplicities $q_1, q_2, \cdots, q_k$, and denote by C_j' the circle whose center is α_j' which has α as common external center of similitude with each C_i. Then the locus of the zeros of $p'(z)$ consists of the closed interiors of the circles C_j'. Any circle C_j' which is exterior to the other circles C_i' contains on or within it precisely q_j zeros of $p'(z)$; any set of closed interiors of* *circles C'_j disjoint from the closed interiors of the remaining C'_j contains a number of zeros of $p'(z)$ equal to the sum of the corresponding numbers q_j.*

In this Corollary it is not intended to require that α shall lie exterior to the circles C_j, although that is the most interesting situation, we require merely that a suitable stretching of the plane with α as center and without rotation carries any C_j into any other C_j.

§3.4. Lucas polygon improved. The Lucas polygon, as a point set known to contain all critical points of a given polynomial $p(z)$, is improved by §3.1.1 Corollary to Theorem 1; a deleted neighborhood of an arbitrary zero of $p(z)$ can be assigned which is known to contain no critical point of $p(z)$. Since no critical point other than a multiple zero of $p(z)$ can lie on the boundary of the Lucas polygon (assumed non-degenerate), it follows that no critical point lies in a certain strip inside the polygon and bounded by a side of the polygon and by a line parallel to that side. We proceed to make this conclusion more precise under certain conditions:

THEOREM 1. *Let $p(z)$ be the polynomial $(z^2 + 1)^k p_1(z)$ of degree $2k + m$, where the m zeros of $p_1(z)$ lie in the circular region $C: | z - a | \leqq r$, where a is real. Then all critical points of $p(z)$ lie in the points $+i$ and $-i$ (if we have $k > 1$), in the region C (if we have $m > 1$), and on the bounded point set separated from infinity by points of the curve*

$$(1) \quad [(m + 2k)(x^2 + y^2) - 2akx + m]^2 - 4k^2r^2x^2 + 4[a^2k^2 - k^2r^2 - m(m + 2k)]y^2 = 0.$$

Let us denote by E the point set which is the locus of positions z of equilibrium in the field of force due to k-fold particles at $+i$ and $-i$, and to an m-fold particle α in C. It follows from Lucas's Theorem that E is bounded, and it follows from the functional dependence of z on α that z is not on the boundary of its locus unless α is on the boundary of its locus. We determine the boundary of E by choosing $\alpha = a + re^{i\theta}$. Then z satisfies the equation

$$(2) \quad \frac{k}{z - i} + \frac{k}{z + i} + \frac{m}{z - \alpha} = 0, \qquad \alpha = a + re^{i\theta},$$

$$e^{i\theta} = \frac{2kz^2 - 2akz + m(z^2 + 1)}{2krz};$$

this second member multiplied by its conjugate and set equal to unity gives precisely the curve (1), a bicircular quartic symmetric in the axis of reals.

If now z is a critical point of $p(z)$ distinct from i and $-i$ and not in C if we have $m > 1$, then z is a position of equilibrium in the field due to k-fold particles at i and $-i$ and to an m-fold particle in C, so z lies in E but E lies in the bounded point set separated from the point at infinity by (1), so Theorem 1 is established.

Theorem 1 is of primary interest in the case $a > r$, to which we now restrict ourselves. The region C and the curve (1) lie to the right of the axis of imaginaries. At a point z of the axis of reals, the force due to the particles at the points $+i$ and $-i$ is horizontal, so if z satisfies (2) the point α must also be real. Thus the only intersections z of (1) with the axis of reals are positions of equilibrium in the field with real α. The smallest such z (if any) corresponds to the largest value of α, namely $\alpha = a + r$, by §2.2.1. *We denote this value by* z_0, *the smaller real zero (if any) of* (2) *with* $e^{i\theta} = 1$. To an arbitrary z corresponds by (2) a unique value of α; for only one α do the two values of z coincide, namely $\alpha = [m(m + 2k)/k^2]^{1/2}$, so the curve has at most one double point, necessarily real, $z = [m/(m + 2k)]^{1/2}$.

It also follows from §2.2.1 that the real part of z defined by the general equation (2) need not vary monotonically with the real part of α. To find the nearest point of the locus (1) to the axis of imaginaries, we find the points of (1) at which we have $dx/dy = 0$; at a possible multiple point of (1) the tangents are equally spaced, for (1) is the image of C under the analytic transformation from α to z, so the point of (1) nearest to the axis of imaginaries is not a multiple point of (1). Direct differentiation of (1) with $dx = 0$, $dy \neq 0$, and suppression of the factor y yields

$$(m + 2k)[(m + 2k)(x^2 + y^2) - 2akx + m] + 2[a^2k^2 - k^2r^2 - m(m + 2k)] = 0;$$

substitution into (1) of the corresponding value of y^2 yields

$$\begin{aligned} &(m + 2k)^2[-a^2k^2 + m(m + 2k)]x^2 \\ &\qquad + 2ak(m + 2k)[a^2k^2 - k^2r^2 - m(m + 2k)]x \\ &\qquad\qquad - [a^2k^2 - k^2r^2 - m(m + 2k)][a^2k^2 - k^2r^2] = 0, \end{aligned} \tag{3}$$

an equation which may or may not have positive zeros.

Equation (3) is of significance in connection with the vertical tangents of the curve (1) at points not on the axis of reals when and only when the lines represented by (3) intersect the circle represented by the equation preceding (3) in points not on the axis of reals; when this condition is satisfied we denote by x_1 the smaller zero of (3) corresponding to a vertical line cutting the circle but not tangent to it; when this condition is not satisfied we set $x_1 = +\infty$.

Some points of the set E (but not of C) lie to the left of the line $x = a - r$, for with $\alpha = a - r$ two positions of equilibrium given by (2) lie in the strip $0 < x < a - r$. Thus we have the

COROLLARY. *Under the conditions of Theorem* 1 *with* $0 < r < a$, *no critical points of* $p(z)$ *lie in the region* $0 < x < \min [z_0, x_1]$.

This corollary enables us to pare down the set assigned by Lucas's Theorem, along any side of the Lucas polygon, unless all zeros of the given polynomial

$p(z)$ lie on that side, unless that side passes through more than two distinct zeros of $p(z)$, or unless the side passes through precisely two zeros of $p(z)$ and those are of unequal multiplicities.

We can be more explicit regarding the shape of the curve (1), still assuming $r < a$. If we have $a - r > \alpha_0 = [m(m + 2k)/k^2]^{1/2}$ and if r is small, the curve (1) consists of two mutually exterior Jordan curves each symmetric in the axis of reals, and the set E consists of their closed interiors. The extreme intersections of (1) with the axis of reals correspond to the values of z in (2) with $\alpha = a + r$, and the other intersections correspond to $\alpha = a - r$. As the circular region C now increases continuously and monotonically, the Jordan regions composing E increase continuously and monotonically, and remain disjoint (for (1) has no double point in the real plane except on the axis of reals) until the circle C passes through the point α_0. Here the two intermediate intersections of (1) with the axis of reals coincide, and as C continues to enlarge further those intersections disappear; the set E is the closed interior of a single Jordan curve.

On the other hand, if we commence with a small value of r and with $a + r < \alpha_0$, the curve (1) consists of two mutually exterior Jordan curves not cutting the axis of reals, and E is the sum of their closed interiors. When the region C increases continuously and monotonically, so does the set E, and when the circle C passes through α_0 these two Jordan curves have a single point in common. When C continues to increase further, E becomes and remains the closed interior of a single Jordan curve.

Under the hypothesis $r < a$, the curve (1) cannot consist of two nested Jordan curves, for in the case $a + r < \alpha_0$, the curve does not cut the axis of reals; in the case $a - r > \alpha_0$ the curve consists of two mutually exterior Jordan curves whose intersections with the axis of reals are the four values of z in (2) for $\alpha = a - r$ and $\alpha = a + r$; in the case $a - r < \alpha_0 < a + r$ the curve cuts the axis of reals in precisely two points, corresponding to $\alpha = a + r$ in (2).

In the case $a - r > \alpha_0$, the curve (1) consists of two mutually exterior Jordan curves and it follows that (1) has no vertical tangent to the left of the line $x = z_0$; the existence of such a tangent would imply the existence of a circle cutting one of these Jordan curves in at least four finite points and the other Jordan curve in at least two such points, and also cutting (1) in the circular points at infinity each counted twice, which is impossible. Thus in the case $a - r > \alpha_0$, the conclusion of the Corollary may be replaced by the statement that *no critical points of $p(z)$ lie in the region* $0 < x < z_0$. It follows by a limiting process that this statement applies also in the case $a - r = \alpha_0$.

We add a further remark in the case $a^2k^2 - k^2r^2 - m(m + 2k) \geqq 0$, or $a^2 - r^2 \geqq \alpha_0^2$; whenever (x, y) lies on the curve (1) we have

$$[(m + 2k)(x^2 + y^2) - 2akx + m]^2 - 4k^2r^2x^2 \leqq 0.$$

It is a consequence of (2) that the circle whose diameter is the segment joining the positions of equilibrium corresponding to $\alpha = a + r$ (we must have $a + r > \alpha_0$) is

$$(4) \qquad (m + 2k)(x^2 + y^2) - 2akx + m - 2krx = 0.$$

If we have $a - r > \alpha_0$, the circle whose diameter is the segment joining the positions of equilibrium corresponding to $\alpha = a - r$ is

$$(5) \qquad (m + 2k)(x^2 + y^2) - 2akx + m + 2krx = 0;$$

if we have $a - r = \alpha_0$, equation (5) represents a null circle and if we have $a - r < \alpha_0$, has no locus. Every point on circle (5) lies interior to the circle (4) except in the degenerate case $r = 0$. It follows that in the case $a^2k^2 - k^2r^2 - m(m + 2k) \geqq 0$, the curve (1) lies in the closed interior of the circle (4) and in the closed exterior of the circle (5) if any. The curve (1) is tangent to these circles (insofar as the circles are non-degenerate) at points of intersection with them on the axis of reals.

Of course in the case $a^2k^2 - k^2r^2 - m(m + 2k) = 0$, the curve (1) is identical with the circle (4), and in the case $a^2k^2 - r^2k^2 - m(m + 2k) < 0$ equation (5) has no locus and the curve (1) lies in the closed exterior of the circle (4).

§3.5. The Lucas polygons for a polynomial and its derivative. According to Lucas's Theorem, all the zeros of the derivative $p'(z)$ of a given polynomial $p(z)$ lie in the smallest closed convex set Π containing the zeros of $p(z)$. If Π' denotes the smallest closed convex set containing the zeros of $p'(z)$, it is thus obvious that Π contains Π'. Moreover no vertex of Π' other than a multiple zero of $p(z)$ can lie on the boundary of Π unless Π is a line segment. What other relations exist between Π and Π'? It is clear for instance that Π' is identical with Π when and only when each vertex of Π is a multiple zero of $p(z)$.

If $p'(z)$ is given, the integral $p(z)$ is determined merely to within an additive constant C, and for each value of C there exists a corresponding $p(z)$ and a corresponding convex set $\Pi(C)$ which necessarily contains Π'. The convex point set Π^0 which is the intersection of the $\Pi(C)$ thus contains Π'; under what conditions are Π^0 and Π' identical? We proceed to discuss these questions, using the notation just introduced.

§3.5.1. Relations of convex sets.

THEOREM 1. *If $p'(z)$ is a given polynomial of degree one or two, then Π^0 and Π' are identical.*

If we have $p'(z) = (z - \alpha)^{n-1}$, we may choose $p(z) \equiv (z - \alpha)^n/n$; here Π' reduces to the point $z = \alpha$, as does $\Pi(C)$ for this particular $p(z)$, so indeed Π^0 and Π' are identical whenever $p'(z)$ has but one distinct zero.

If $p'(z)$ is of degree two and has distinct zeros, it is essentially no loss of generality to choose $p'(z) = z^2 - 1$, whence $p(z) = z^3/3 - z + C$. For $C = 2/3$, the set $\Pi(C)$ becomes $-2 \leqq z \leqq 1$ and for $C = -2/3$ becomes $-1 \leqq z \leqq 2$, so Π' and Π^0 are both the segment $-1 \leqq z \leqq 1$.

THEOREM 2. *If $p'(z)$ is a polynomial of degree three with all zeros collinear, then Π^0 and Π' are identical.*

We choose $p'(z) = z(z - 1)(z - \alpha)$ with $\alpha > 1$, but only obvious modifications need be made in our discussion if we have $\alpha = 1$. There exists a number C so that if $p(z) = z^4/4 - (1 + \alpha)z^3/3 + \alpha z^2/2 + C$, we have $p(0) < 0$, $p(1) > 0$, $p(\alpha) < 0$; indeed the condition $p(1) > 0$ is $C > (1 - 2\alpha)/12$, and this last expression is negative; the condition $p(\alpha) < 0$ is $C < \alpha^3(\alpha - 2)/12$; the relation $1 - 2\alpha < \alpha^3(\alpha - 2)$ is the obvious inequality $(\alpha^2 - 1)(\alpha - 1)^2 > 0$; for a negative value of C which satisfies the relations $(1 - 2\alpha)/12 < C < \alpha^3(\alpha - 2)/12$, the polynomial $p(z)$ has only real zeros, so Π^0 contains no point not on the axis of reals. If we set $C = 0$, we have $p(z) \equiv z^2[3z^2 - 4(1 + \alpha)z + 6\alpha]/12$, whose zeros other than $z = 0$ are either real and positive or non-real with real part $2(1 + \alpha)/3$, so Π^0 contains no point $z < 0$. By symmetry it follows that Π^0 contains no point $z > \alpha$, whence Π^0 and Π' coincide.

THEOREM 3. *There exists a polynomial $p'(z)$ of degree three such that Π^0 and Π' are not identical.*

We choose $p'(z) = z^3 + 1$, whose zeros -1, ω, and $\bar{\omega}$, with $\omega = \frac{1}{2} + 3^{1/2}i/2$, possess three-fold symmetry about the origin. We note that the choice $p(-1) = 0$ gives $p(z) = (z + 1)^2(z^2 - 2z + 3)/4$, which has the zeros $z = -1, -1, 1 \pm 2^{1/2}i$; the points ω and $\bar{\omega}$ are of course interior to the Lucas polygon for $p(z)$. Assume that Π^0 and Π' are identical; we shall reach a contradiction. The points $\beta_n = \frac{1}{2} + 1/n$, $n = 1, 2, 3, \cdots$, are exterior to Π', yet there exists C_n such that β_n is exterior to $\Pi(C_n)$. A subsequence of these values C_n, say C_{n_k}, approaches a finite or infinite limit C_0. This limit cannot be infinite, for (§1.6.3) if $| C_n |$ is large, the region $\Pi(C_n)$ is approximately a large square with center O. The zeros of $p_{n_k}(z) \equiv z^4/4 + z + C_{n_k}$ approach the respective zeros of $p_0(z) \equiv z^4/4 + z + C_0$, which are finite; a side of the polygon bounding $\Pi(C_{n_k})$ approaches a segment of the line $\Re(z) = \frac{1}{2}$; at least two zeros of $p_{n_k}(z)$ approach finite limits on the line $\Re(z) = \frac{1}{2}$. Two zeros of $p_{n_k}(z)$ cannot approach respectively the points ω and $\bar{\omega}$, for then $p_0(z)$ would have ω and $\bar{\omega}$ as double zeros and -1 would not be a zero of $p'(z)$. One zero of $p_{n_k}(z)$ cannot approach ω or $\bar{\omega}$, as is seen by symmetry from the specific example already considered. Two zeros of $p_{n_k}(z)$ cannot approach points of the line $\Re(z) = \frac{1}{2}$ other than ω and $\bar{\omega}$, for in such a case $p_0(z)$ could not have ω and $\bar{\omega}$ on its Lucas polygon bounding $\Pi(C_0)$, by Lucas's Theorem. Thus we have a contradiction, and the theorem is proved.

THEOREM 4. *There exists a polynomial $p'(z)$ of degree four with real zeros such that Π^0 and Π' are not identical.*

We choose here $p'(z) = (z^2 - 1)^2$, and the proof is similar to that of Theorem 3. The set Π' is the segment $-1 \leqq z \leqq 1$. Again we assume Π^0 and Π' identical, and shall reach a contradiction. Set $\beta_n = i/n$, $n = 1, 2, 3, \cdots$, so that β_n must be exterior to some $\Pi(C_n)$; the numbers C_n have a finite limit point C_0, for (§1.6.3) when $| C_n |$ is large the region $\Pi(C_n)$ is approximately a large regular pentagon whose center is 0. If the subsequence C_{n_k} approaches C_0, a side of the polygon bounding $\Pi(C_{n_k})$ approaches a segment of the axis of reals. The polynomial $p_0(z)$ corresponding to the constant C_0 must have all its zeros in the closed lower half-plane. The points $z = +1$ and -1 cannot both be zeros of $p_0(z)$, for then $p_0(z)$ would have at least six zeros. It follows from Rolle's Theorem by study of the possible intervals involved that $p_0(z)$ cannot have only real zeros. At least one of the points $+1$ and -1 thus lies on a side of the Lucas polygon and is a zero of $p'(z)$ but not a multiple zero of $p_0(z)$. Thus $p_0(z)$ cannot exist, and this contradiction completes the proof.

§3.5.2. Dependence of zeros of polynomial on those of its derivative. In connection with the present topic, the study of the zeros of $p(z)$ when those of $p'(z)$ are given, several other results deserve mention.

THEOREM 5. *Let $p'(z)$ be a polynomial of degree n all of whose zeros are real and non-positive. If the integral $p(z)$ is so chosen that $p(z_0) = 0$, where z_0 is positive, then $p(z)$ has no zeros other than z_0 in the sector $| \arg z | < 2\pi/(n + 1)$.*

For the present we choose $n > 2$, and set $p'(z) = (z - \alpha_1') \cdots (z - \alpha_n')$,

$$p(z) = \int_{z_0}^{0} p'(z)\, dz + w(z), \quad w(z) = \int_0^z p'(z)\, dz;$$

the first of these integrals is to be taken over a segment of the axis of reals, and is negative; the second of these integrals is to be taken over the line segment joining O to z, where we choose $0 < \arg z = \theta < 2\pi/(n + 1) \leqq \pi/2$, and $w(z)$ is to be investigated further. We have $dw/dz = p'(z)$, $\arg (dw) = \theta + \sum \arg (z - \alpha_k')$, from which it follows that $\arg (dw)$ never decreases as $| z |$ increases. If Γ denotes the image in the w-plane of the line segment Oz, then as w traces Γ commencing at $w = 0$, *the tangent to Γ always rotates in the counterclockwise sense. We shall prove that as w traces Γ commencing at $w = 0$, the curvature of Γ never increases.* We write

$$w(re^{i\theta}) = \int_0^{re^{i\theta}} p'(re^{i\theta})e^{i\theta}\, dr,$$

$$\frac{dw}{dr} = p'(re^{i\theta})e^{i\theta}, \qquad \frac{| dw |}{dr} = | p'(re^{i\theta}) |.$$

Each factor $| z - \alpha_k' |$ increases as r increases, so $| dw |/dr$ increases. If we set $z = x + i\lambda x$, we have $\arg (z - \alpha_k') = \tan^{-1}[\lambda x/(x - \alpha_k')]$, $d[\arg(z - \alpha_k')]/dx = -\alpha_k'\lambda/[(x - \alpha_k')^2 + \lambda^2 x^2]$, and this last member decreases or remains unchanged as x increases. Consequently $d[\arg p'(z)]/dr$ decreases or remains unchanged as

r increases, as does the curvature $d[\arg(dw)]/|dw|$; the curvature actually decreases as r increases unless we have $p'(z) \equiv z^n$.

So long as the relation $0 < \arg(dw) < 2\pi$ persists, the arc Γ cannot cut the positive half of the axis of reals $\Re(w) > 0$. If Γ does cut the positive half of the axis of reals under these conditions, we obviously have $p'(z) \not\equiv z^n$, and the arc Γ necessarily cuts itself and forms a loop Γ_0, a convex Jordan curve, analytic except for a vertex A. The largest circle C_0 whose interior lies interior to Γ_0 is tangent to Γ_0 in at least two distinct points, and has no less curvature than Γ_0 at those points of tangency. As a circle C moves from the position of C_0 remaining doubly tangent to Γ_0, the interior of C remaining interior to Γ_0 and both points of tangency moving along Γ_0 away from A, C approaches a position in which the two points of tangency are coincident, and at this point of tangency the curvature of Γ_0 is not less than that of C_0, or of Γ_0 at the points of tangency of C_0 and Γ_0; this contradicts the strongly monotonic character of the curvature of Γ.

The inequality $0 < \theta < 2\pi/(n+1)$ implies $0 < \arg(z - \alpha_k') < 2\pi/(n+1)$, $\theta < \arg(dw) \leqq (n+1)\theta < 2\pi$, so $w(z)$ cannot be positive and $p(z)$ cannot vanish. Of course $p(z)$ cannot vanish other than in the point z_0 on the positive half of the axis of reals, by Rolle's Theorem.

The exceptional case $n = 2$ can be treated directly; it is sufficient to consider $p'(z) = z^2 + z$, whence $p(z) = (z^3 - z_0^3)/3 + (z^2 - z_0^2)/2$. The zeros of $p(z)$ other than z_0 are given by

$$4z = -2z_0 - 3 \pm(-12z_0^2 - 12z_0 + 9)^{1/2},$$

from which it can be verified at once that we cannot have $|\arg z| < 2\pi/3$. The theorem is trivial in the cases $n = 0$ and $n = 1$. This completes the proof.

Except in the case that $p'(z)$ is a constant multiple of z^n, this proof shows that no zero of $p(z)$ lies on a half-line $\arg z = \pm 2\pi/(n+1)$, $n \not\equiv 1$.

In Theorem 5 it is not sufficient merely to assume that the zeros of $p'(z)$ lie in the closed left half-plane, as is shown by setting $p'(z) = z^2 + b^2$, $b > 0$; in addition to the zero $z = z_0$, the integral $p(z)$ has the zeros

$$z = \tfrac{1}{2}[-z_0 \pm 3^{1/2}i(z_0^2 + 4b^2)^{1/2}],$$

for which we have $|\arg z| < 2\pi/3$.

The number $2\pi/(n+1)$ that occurs in Theorem 5 can be replaced by no larger number, for if we set $p'(z) \equiv z^n$, we have $p(z) = (z^{n+1} - z_0^{n+1})/(n+1)$, whose zeros are the points $e^{2\pi ki/(n+1)}z_0$, $k = 0, 1, 2, \cdots, n$. Indeed, if $p'(z)$ satisfying the prescribed conditions is arbitrary, for large z_0 the zeros of $p(z)$ are (§1.6.3) approximately the points $\alpha + e^{2\pi ki/(n+1)}(z_0 - \alpha)$, where α is the center of gravity of the points α_k'; again it appears that the number $2\pi/(n+1)$ is the best possible.

COROLLARY. *Under the hypothesis of Theorem* 5, *no zero of* $p(z)$ *lies in the region*

$|z| < z_0$; *unless* $p'(z)$ *is a constant multiple of* z^n, *no zero of* $p(z)$ *other than* z_0 *lies on the circle* $|z| = z_0$.

For $0 < r < z_0$ we obviously have $|re^{i\theta} - \alpha_k'| \leqq r - \alpha_k'$, $|p'(re^{i\theta})| \leqq p'(r)$, whence for $0 < |z| \leqq z_0$

$$|p(z) - p(0)| = \left| \int_0^z p'(re^{i\theta})\, d(re^{i\theta}) \right| \leqq \int_0^{|z|} p'(r)\, dr \leqq \int_0^{z_0} p'(z)\, dz = -p(0).$$

For these integrals the first inequality is a strong inequality except when θ is zero or $p'(z)$ is a multiple of z^n; the second inequality is strong unless we have $|z| = z_0$. Of course we have $p(0) < 0$, so the conclusion follows.

Both Theorem 5 and the Corollary extend under suitable conditions to polynomials $p'(z)$ whose zeros are not necessarily collinear.

§3.6. Symmetry in the origin. We proceed now to study polynomials whose zeros have special properties of symmetry, first symmetry in the origin O and then extensions to p-fold symmetry in O.

§3.6.1. Ordinary symmetry. We have already (§2.7.2) commenced the study of polynomials whose zeros are symmetric in O. The fundamental result here is [Walsh, 1933]

THEOREM 1. *Let* $p(z)$ *be a polynomial whose zeros are symmetric in the origin* O.

1). *If all the zeros of* $p(z)$ *lie in a closed double sector with vertex* O *and angle not greater than* $\pi/2$, *then all zeros of* $p'(z)$ *lie in that sector.*

2). *If all the zeros of* $p(z)$ *lie in the closed interior of a non-degenerate equilateral hyperbola whose center is* O, *then all zeros of* $p'(z)$ *also lie in that closed interior, except for a simple zero at* O *itself.*

3). *If all the zeros of* $p(z)$ *lie in the closed exterior of a non-degenerate equilateral hyperbola whose center is* O, *all the zeros of* $p'(z)$ *lie in that same closed region.*

4). *If all the zeros of* $p(z)$ *lie on an equilateral hyperbola whose center is* O, *then all the zeros of* $p'(z)$ *also lie on that hyperbola except for a simple zero at* O. *Any open finite arc of the hyperbola bounded by two zeros of* $p(z)$ *and containing neither such a zero nor* O *contains in its interior precisely one (a simple) zero of* $p'(z)$.

In the field of force due to two particles symmetric in O, the lines of force are (§§1.6.3 and 2.2) equilateral hyperbolas with center O passing through those particles, including the degenerate curve consisting of two perpendicular lines through O, one of them through the given particles. Theorem 1 is readily proved by the study of this field of force for all pairs of particles; compare §2.7.2 for the proof of 1).

A second method of proof is perhaps more interesting, and will be discussed in detail. We can write $p(z)$ in the form

$$p(z) = z^k \prod_1^m (z^2 - \alpha_j^2), \quad \alpha_j \neq 0,$$

so the function $[p(z)]^2$ is a polynomial in z^2 of the same degree as $p(z)$:

$$P(w) \equiv [p(z)]^2, w = z^2. \tag{1}$$

Under the transformation $w = z^2$ the totality of zeros of $p(z)$ corresponds to the totality of zeros of $P(w)$, and reciprocally. The totality of zeros of $P'(w)$ corresponds to the totality of zeros of $p(z)$ and $p'(z)$, and reciprocally except that $z = 0$ is always a zero of $p'(z)$ (obviously a position of equilibrium if not a zero of $p(z)$) unless $z = 0$ is a simple zero of $p(z)$:

$$d[P(w)] /dw = p'(z) \cdot p(z)/z.$$

Under the transformation $w = u + iv = z^2 = (x + iy)^2$, the line $Au + Bv = C$ in the w-plane is the image of $A(x^2 - y^2) + 2Bxy = C$, so the properties of the images can be read off directly. An arbitrary straight line in the w-plane corresponds in the z-plane to an equilateral hyperbola whose center is the origin and reciprocally; here a straight line through $w = 0$ corresponds to a degenerate equilateral hyperbola, namely two perpendicular lines through O. Angles are unchanged by the transformation except that angles at the origin $w = 0$ in the w-plane are twice the corresponding angles at the origin $z = 0$ in the z-plane

Thanks to the properties of the transformation $w = z^2$, the proof of Theorem 1 is immediate. In case 1), the image of the given double sector S is a simple sector S' in the w-plane with vertex $w = 0$ and angle not greater than π; since all zeros of $p(z)$ lie in S, all zeros of $P(w)$ lie in S', as do all zeros of $P'(w)$, by Lucas's Theorem; consequently all zeros of $p'(z)$, including a possible zero at $z = 0$, lie in S. The remaining parts of Theorem 1 are proved in precisely this same manner, by use of Lucas's Theorem. In case 2), the given closed interior of a hyperbola (§2.2) corresponds to a closed half-plane in the w-plane not containing O. In case 3), the given closed exterior of a hyperbola (§2.2) corresponds to a closed half-plane in the w-plane for which $w = 0$ is an interior point. In case 4), the given hyperbola corresponds to a straight line of the w-plane; as a degenerate case here, the given curve may consist of two perpendicular lines, which correspond to a straight line in the w-plane through $w = 0$; here any simple open segment of one of the given lines bounded by zeros of $p(z)$, containing no such zero and not containing O, contains a single zero of $p'(z)$; but $w = 0$ may be a zero of $P'(w)$ and then $z = 0$ is a multiple zero of $p'(z)$; moreover $z = 0$ may be a zero of $p(z)$ and unless a multiple zero of $p(z)$ is not a zero of $p'(z)$.

As a remark complementary to Theorem 1, we notice that in case 1) no boundary point P of the given sector other than O can be a zero of $p'(z)$ unless P is a multiple zero of $p(z)$ or unless all zeros of $p(z)$ lie on the line OP, except when the given sector has the angle $\pi/2$ and all zeros of $p(z)$ lie on the boundary. Likewise in cases 2) and 3) no point of the hyperbola is a zero of $p'(z)$ unless it is a multiple zero of $p(z)$ or unless all zeros of $p(z)$ lie on the curve. In case 1) the conclusion is false without the restriction that the angle α of the sector be not greater than $\pi/2$, as we illustrate by the example

$$p(z) = (z - a - ib)(z - a + ib)(z + a - ib)(z + a + ib), \qquad 0 < a < b,$$

where the zeros of $p(z)$ lie at the vertices of a rectangle and hence (§1.4.1) the zeros of $p'(z)$ other than at the center lie at the intersections of the circles constructed on the longer sides as diameters; these zeros lie exterior to the closed double sector $-b/a \leqq \tan[\arg z] \leqq b/a$ containing the zeros of $p(z)$.

Theorem 1, if applied repeatedly to a given polynomial $p(z)$, shows that *the smallest point set S convex with respect to the family F of equilateral hyperbolas with center O and which contains all zeros of* $p(z)$ *also contains all zeros of* $p'(z)$ *except perhaps O*, which is always a zero of $p'(z)$ unless it is a simple zero of $p(z)$. This is an exact analog of Lucas's Theorem.

The requirement that S be convex with respect to the family F of hyperbolas with center O is to be interpreted as follows: identify each point P with its reflection in O; through each pair of distinct points P and Q then passes a unique curve of the family F; whenever P and Q belong to S, so also shall the finite arc PQ of this unique curve of the family. If P: (x_1, y_1) and Q: (x_2, y_2) are given, this curve has the equation

$$\begin{vmatrix} x^2 - y^2 & xy & 1 \\ x_1^2 - y_1^2 & x_1 y_1 & 1 \\ x_2^2 - y_2^2 & x_2 y_2 & 1 \end{vmatrix} = 0,$$

which is non-trivial by virtue of the distinctness of P and Q. The hyperbola degenerates when and only when the lines OP and OQ are either identical or orthogonal; in the latter case the arc PQ consists of two broken lines each with a right angle at O and which have been identified; this arc PQ corresponds to a single line segment in the w-plane. Of course all the facts concerning convexity with respect to the family F in the z-plane may be interpreted as a transformation of, and proved from, the facts concerning ordinary convexity in the w-plane.

In more usual language, the set S may be the closed interiors of two disjoint curvilinear polygons; it is bounded by arcs of equilateral hyperbolas of the family F, arcs included among the W-curves for the groups of zeros of $p(z)$. The set S is the image in the z-plane of the Lucas polygon for $P(w)$ in the w-plane.

As an application of these remarks on convexity we have the

COROLLARY. *Let* $p(z)$ *be a polynomial whose zeros are symmetric in O. If all the zeros of* $p(z)$ *lie in the closed interior* [*exterior*] *of the hyperbola H of eccentricity not greater than* $2^{1/2}$ [*not less than* $2^{1/2}$] *whose center is O, so also do all zeros of* $p'(z)$ *other than O.*

If the zeros of $p(z)$ lie in the closed interior of $H: b^2x^2 - a^2y^2 = a^2b^2$ with $a \geqq b > 0$, the zeros of $P(w)$ defined by (1) lie in the w-plane in a region the equation of whose boundary is found from the equation of H by substituting $2x^2 = (u^2 + v^2)^{1/2} + u$, $2y^2 = (u^2 + v^2)^{1/2} - u$:

$$\left(u - \frac{a^2 + b^2}{2}\right)^2 \Big/ \frac{(a^2 - b^2)^2}{4} - \frac{v^2}{a^2 b^2} = 1; \tag{2}$$

the curve (2) is the image in the w-plane of both H and the hyperbola $a^2x^2 - b^2y^2 = a^2b^2$; of course (2) is a line $u = a^2$ if we have $a = b$, for which case the corollary has been proved; otherwise the intercepts on the axis of reals of the hyperbola (2) are $u = a^2$ and b^2. The entire closed interior of H is a point set connected with the vertices of H: $z = \pm a$, and its image is connected with the image of those vertices: $w = a^2$, so the image of the closed interior of H is the closed interior of the right-hand branch of (2), which is convex. The conclusion follows from Lucas's Theorem applied to $P(w)$.

If the zeros of $p(z)$ lie in the closed exterior of H this same method and formulas are valid, now with $b \geqq a$; we take $b > a$, for the case $b = a$ is included in Theorem 1. The closed exterior of H is a point set connected with $z = 0$, and its image is connected with $w = 0$, so the image of the closed exterior of H is the closed interior of the left-hand branch of (2), which is convex. The Corollary is established. It is also true that under the conditions of the Corollary, no zero of $p'(z)$ other than a multiple zero of $p(z)$ can lie on H unless H degenerates or unless H is equilateral and all zeros of $p(z)$ lie on H.

Any theorem concerning the zeros of the derivative of a polynomial can be applied to $P(w)$ in the w-plane and by transformation yields a theorem concerning $p(z)$ in the z-plane. For instance Jensen's Theorem in the w-plane, interpreted as a theorem on a polynomial whose zeros are symmetric in an *arbitrary* line, yields a theorem in the z-plane involving the symmetry of points with respect to a line or to an equilateral hyperbola and involving certain bicircular quartics. As a single example of this general method we prove

THEOREM 2. *Let all zeros of a polynomial $p(z)$ of degree n be symmetric in O, and let O be a k-fold zero, while all other zeros of $p(z)$ lie in the closed interior of an equilateral hyperbola H. Then all zeros of $p'(z)$ other than O lie in the closed interior of the equilateral hyperbola H' found by shrinking H toward O in the ratio $(n:k)^{1/2}$.*

By suitable orientation of the z-plane we take H as $x^2 - y^2 = a^2$, and under the transformation $w = z^2$ the image of H is $u = a^2$. In the w-plane the polynomial $P(w)$ defined by (1) has a zero of multiplicity k at O, and has its remaining $n - k$ zeros in the half-plane $u \geqq a^2$. All zeros of $P'(w)$ other than $w = 0$ lie in the half-plane $u \geqq ka^2/n$, by §3.1.2 Theorem 2, so all zeros of $p'(z)$ other than $z = 0$ lie in the closed interior of the hyperbola H':$x^2 - y^2 = ka^2/n$.

§3.6.2. Multiple symmetry in O. In the discussion of §3.6.1, equilateral hyperbolas with center O play a leading role; they are both 1) the lines of force in the z-plane due to a pair of particles having the prescribed symmetry (hence W-curves), and 2) the images in the z-plane of straight lines (lines of force for single particles) in the w-plane, from which it follows that theorems in the w-plane concerning lines and zeros of $P'(w)$ yield theorems in the z-plane concerning equilateral hyperbolas and zeros of $p'(z)$. It is no coincidence that the two properties 1) and 2) yield the same category of curves:

THEOREM 3. *If $\varphi(z)$ is a polynomial in z and $\Phi(w)$ a polynomial in w, then the lines of force in the z-plane due to particles at the zeros of the polynomial $\Phi[\varphi(z)]$ are the images in the z-plane under the transformation $w = \varphi(z)$ of the lines of force in the w-plane due to particles at the zeros of the polynomial $\Phi(w)$.*

The lines of force in the z-plane are (§1.6.2) the curves $\arg \Phi[\varphi(z)] = \text{const}$, and those in the w-plane are the curves $\arg [\Phi(w)] = \text{const}$, so the former are the images of the latter. In §3.6.1 we have been concerned with the case $\varphi(z) \equiv z^2$, $\Phi(w) = w - w_0$, but Theorem 3 applies in more general situations. The W-curves under the hypothesis of Theorem 3 are of course arcs of the lines of force in the two planes, images of each other under the transformation $w = \varphi(z)$.

In the study of p-fold symmetry in O, the transformation $w = z^p$ is of central importance. We use the term *p-hyperbola (with center O) to denote the image in the z-plane of a straight line in the w-plane.* Thus if we set $w = u + iv$, $z = re^{i\theta}$, the equation of a p-hyperbola with center O is in the typical case $u = u_0 > 0$ the curve $r^p \cos p\theta = u_0$; the nearest points of the curve to the origin $z = 0$ are $r = u_0^{1/p}$, $\theta = 2k\pi/p$, $k = 0, 1, 2, \cdots, p - 1$; the asymptotic directions are $\theta = (2k + 1)\pi/2p$, $k = 0, 1, \cdots, 2p - 1$. In the case $u_0 = 0$, the p-hyperbola degenerates to the p lines $\theta = (2k + 1)\pi/2p$, $k = 0, 1, \cdots, 2p - 1$. A p-hyperbola is a complete algebraic curve of degree p. We shall consider the point set $r^p \cos p\theta > u_0$ as the *interior* of the curve, and the set $r^p \cos p\theta < u_0$ as the *exterior* of the curve. When $p = 2$, a p-hyperbola is the usual equilateral hyperbola.

The analog of Theorem 1 is

THEOREM 4. *Let $p(z)$ be a polynomial whose zeros exhibit p-fold symmetry in the origin O.*

1). *If all the zeros of $p(z)$ lie in p alternate closed sectors bounded by p lines which successively intersect at the angle π/p at O, then all zeros of $p'(z)$ lie in those sectors.*

2). *If all the zeros of $p(z)$ lie in the closed interior of a non-degenerate p-hyperbola with center O, then all zeros of $p'(z)$ lie in that closed interior, except for a zero at O of multiplicity $p - 1$.*

3). *If all the zeros of $p(z)$ lie in the closed exterior of a non-degenerate p-hyperbola with center O, then all zeros of $p'(z)$ lie in that closed exterior.*

4). *If all the zeros of $p(z)$ lie on a p-hyperbola with center O, then all zeros of $p'(z)$ also lie on that curve except for a $(p - 1)$-fold zero at O. Any open finite arc of the curve bounded by two zeros of $p(z)$ and containing neither such a zero nor O in its interior contains in its interior precisely one (a simple) zero of $p'(z)$.*

We can write $p(z)$ in the form

$$p(z) \equiv z^k \prod_1^n (z^p - \alpha_j^p), \qquad \alpha_j \neq 0,$$

so the function $[p(z)]^p$ is a polynomial in z^p of the same degree as $p(z)$:

$$P(w) \equiv [p(z)]^p, \qquad w = z^p. \tag{3}$$

Thus under the transformation $w = z^p$ the totality of zeros of $p(z)$ corresponds to the totality of zeros of $P(w)$ and reciprocally; the totality of zeros of $P'(w)$ corresponds to the totality of zeros of $p(z)$ and $p'(z)$, and reciprocally except that $z = 0$ is always a zero of $p'(z)$ unless $z = 0$ is a simple zero of $p(z)$:

$$d[P(w)]/dw = [p(z)]^{p-1}\cdot p'(z)/z^{p-1}.$$

The proof of Theorem 4 is analogous to that of Theorem 1 and is omitted. In case 4), the p-hyperbola may degenerate; then any simple open segment of one of the lines of that curve bounded by zeros of $p(z)$ but containing no such zero and not containing O, contains a single zero of $p'(z)$; if $z = 0$ is not a simple zero of $p(z)$ then $z = 0$ is a zero of $p'(z)$ and reciprocally. In cases 2) and 3) no zero of $p'(z)$ lies on the curve unless it is a multiple zero of $p(z)$ or unless all zeros of $p(z)$ lie on the curve. In case 1) no zero of $p'(z)$ other than O lies on the given lines unless it is a multiple zero of $p(z)$ or unless all zeros of $p(z)$ lie on those lines.

Theorem 4 can be applied repeatedly: *the smallest point set which is convex with respect to p-hyperbolas with center O and which contains all zeros of $p(z)$ also contains all zeros of $p'(z)$ except perhaps O, which is always a zero of $p'(z)$ unless it is a simple zero of $p(z)$.*

The analog of Theorem 2 is

THEOREM 5. *Let all zeros of a polynomial $p(z)$ of degree n have p-fold symmetry in O, and let O be a k-fold zero while all other zeros of $p(z)$ lie in the closed interior of a p-hyperbola H with center O. Then all zeros of $p'(z)$ other than O lie in the closed interior of the p-hyperbola H′ found by shrinking H toward O in the ratio $(n:k)^{1/p}$.*

§3.7. Circular regions and symmetry. Walsh's Theorem (§1.5) can be applied to polynomials which possess symmetry, and under certain conditions leads to interesting results:

THEOREM 1. *Let the zeros of the polynomial $p(z)$ be symmetric in O and lie in the two closed circular regions $C_1: |z - \alpha| \leqq r$, and $C_2: |z + \alpha| \leqq r$, where we have $|\alpha| \geqq 2r$. Then all zeros of $p'(z)$ other than a simple zero at O lie in the closed regions C_1 and C_2.*

We first assume $|\alpha| > 2r$; if the degree of $p(z)$ is n, each of the given regions contains $n/2$ zeros of $p(z)$, so it follows from §1.5 that the zeros of $p'(z)$ lie in C_1, C_2, and $C: |z| \leqq r$; these three regions are mutually exterior, so the region C contains precisely one zero of $p'(z)$. However, it is known (§3.6.1) that $z = 0$ is a zero of $p'(z)$, so all zeros of $p'(z)$ other than O lie in the closed regions C_1 and C_2.

If we have $|\alpha| = 2r$, we approximate $p(z)$ by a variable polynomial whose zeros are symmetric in O and near the respective zeros of $p(z)$, and for which we have variable circles of the theorem satisfying the strong inequality $|\alpha| > 2r$; these circles contain the zeros other than O of the derivative of the variable polynomial. When this approximating polynomial is allowed to approach the original polynomial and the variable circles approach the original circles, we obtain Theorem 1 as stated.

THEOREM 2. *Let the zeros of the real polynomial* $p(z)$ *lie in the two closed circular regions* $C_1: | z - \alpha | \leqq r$ *and* $C_2 : | z - \bar{\alpha} | \leqq r$, *where we have* $| \alpha - \bar{\alpha} | \geqq 4r$. *Then all non-real zeros of* $p'(z)$ *lie in* C_1 *and* C_2; *the polynomial* $p'(z)$ *has precisely one real zero, which lies in the interval* $-r \leqq z - (\alpha + \bar{\alpha})/2 \leqq r$.

In the case $| \alpha - \bar{\alpha} | > 4r$, we denote by C the closed region $| z - (\alpha + \bar{\alpha})/2 | \leqq r$; the three regions C_1, C_2, C are mutually exterior, so it follows from §1.5 that those three regions contain all the zeros of $p'(z)$, and that C contains precisely one such zero. The real polynomial $p'(z)$ of odd degree has at least one

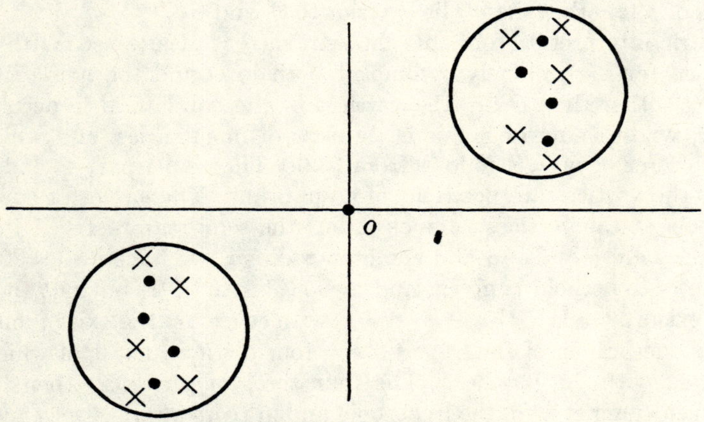

Fig. 7 illustrates §3.7 Theorem 1

real zero, and by Lucas's Theorem all real zeros lie in the interval mentioned, in the closed interior of C; the conclusion follows in this case. If we have $| \alpha - \bar{\alpha} | = 4r$, we complete the proof of Theorem 2 by the use of a variable auxiliary polynomial as in the proof of Theorem 1.

Theorems 1 and 2 are in a sense best possible theorems:

THEOREM 3. *In Theorem* 1 *the number* 2 *in the inequality* $| \alpha | \geqq 2r$ *can be replaced by no smaller number. More explicitly, if the number* q *is arbitrary,* $\frac{1}{2} < q < 1$, *there exist regions* $C_1: | z - \alpha | \leqq r$ *and* $C_2: | z + \alpha | \leqq r$ *with* $| \alpha | = 2qr$ *and a polynomial* $p(z)$ *whose zeros are symmetric in* O *and lie in* C_1 *and* C_2 *such that* $p'(z)$ *has zeros different from* O *not in* C_1 *or* C_2.

We use the fact (§1.4.1) that if a polynomial $p(z)$ of degree four has its zeros at the vertices of a rectangle, then the zeros of the derivative lie at the center of the rectangle and at the intersection of the two circles constructed on the longer sides as diameters. The vertices of the rectangle are chosen as $a \pm ib$, $-a \pm ib$, $b > a$, and the zeros of $p'(z)$ other than O are $\pm i(b^2 - a^2)^{1/2}$. The circles C_1 and C_2 are chosen as $| z - i\beta | = r$ and $| z + i\beta | = r$, with $r^2 = a^2 + (\beta - b)^2$, $\beta > b$, so that C_1 and C_2 pass through pairs of zeros of $p(z)$; we choose $\beta = 2qr$. We keep b fixed and allow a to approach zero; as a approaches zero the equation $r^2 = a^2 + (2qr - b)^2$ defines r so that r approaches $r_0 = b/(2q - 1)$, and we have $r_0 > b$; thus for a sufficiently small, r and β are uniquely defined by the equations given, with $r > b$, $\beta > b$. For a near zero we find by computation $dr/da < 0$. Thus for suitably small a we have $r_0 > r$, $r - qr > qr - b > 0$, whence

$$(r - qr)^2 > (qr - b)^2,$$

$$(2q - 1)^2 r^2 > b^2 - r^2 + (2rq - b)^2,$$

$$(\beta - r)^2 > b^2 - a^2,$$

so the zeros of $p'(z)$ other than O lie exterior to C_1 and C_2.

A geometric interpretation of this choice of $p(z)$ is of interest. With the symmetry chosen for $p(z)$, namely symmetry both in O and the axis of reals, the zeros of $p'(z)$ other than O are the vertices of the equilateral hyperbola whose center is O, whose transverse axis is the axis of imaginaries, and which passes through the zeros of $p(z)$; it follows from §3.6.1 Theorem 1 part 4) and the symmetry that the vertices are positions of equilibrium. The circles of curvature of the hyperbola at the vertices have as radius the semi-major-axis; if the centers of these circles are α and $-\alpha$ and common radius r, we have $| \alpha | = 2r$. If now the equilateral hyperbola is given, and also q, $0 < q < 1$, but not the zeros of $p(z)$, there exists a circle $C_1 : | z - \alpha_1 | = r_1$ with center on the axis of imaginaries, and cutting one branch of the hyperbola in four distinct points of which two lie near a vertex with $| \alpha_1 | = 2qr_1$. The four zeros of $p(z)$ may then be chosen with the given symmetry on the hyperbola and in pairs interior to C_1 and interior to the reflection C_2 of C_1 in O; the zeros of $p'(z)$ other than O are the vertices of the hyperbola, which are exterior to C_1 and C_2.

A different proof of Theorem 3 is indicated in §5.5.3.

The example used in proving Theorem 3 is that of a real polynomial $p(z)$, so we have

THEOREM 4. *In Theorem 2 the number 4 in the inequality* $| \alpha - \bar{\alpha} | \geqq 4r$ *can be replaced by no smaller number. More explicitly, if the number q is arbitrary,* $\frac{1}{2} < q < 1$, *there exist regions* $C_1 : | z - \alpha | \leqq r$ *and* $C_2 : | z - \bar{\alpha} | \leqq r$ *with* $| \alpha - \bar{\alpha} | = 4qr$ *and a real polynomial $p(z)$ whose zeros lie in the regions C_1 and C_2 such that $p'(z)$ has non-real zeros not in C_1 and C_2.*

There are still other occasions of symmetry when a single zero of $p'(z)$ can be approximately located:

THEOREM 5. *Let $p(z)$ be a real polynomial of degree $m_1 + m_2$ with m_k zeros in the region* $C_k : | z - \alpha_k | \leqq r$, $k = 1, 2$, *where α_1 and α_2 are real,* $\alpha_1 < \alpha_2 - 4r$;

let the point $\alpha = (m_2\alpha_1 + m_1\alpha_2)/(m_1 + m_2)$ *lie in the interval* $\alpha_1 + 2r < z < \alpha_2 - 2r$. *Then* $p'(z)$ *has precisely* $m_k - 1$ *zeros in* C_k, $k = 1, 2$, *and has precisely one zero in the interval* $\alpha - r \leqq z \leqq \alpha + r$.

We omit the proof of Theorem 5; the theorem extends in slightly modified form to the cases $\alpha = \alpha_1 + 2r$, $\alpha = \alpha_2 - 2r$, and to the case where C_1 and C_2 are not equal.

Theorems 1, 2, and 5 represent improvements of the theorem of §1.5 in certain cases of symmetry; similar improvements of §3.3 Theorem 1 and its Corollary in cases of symmetry lie at hand. Thus assume the hypothesis of §3.3 Theorem 1; 1) if the points α_j (including the multiplicities m_j) and the zeros of $p(z)$ are symmetric in O, if a circle C_j' contains O in its open interior, is not a circle C_j, and cuts no other circle C_m', then C_j' contains in its closed interior no critical point of $p(z)$ other than O; 2) if the α_j lie on the axis of imaginaries, if the α_j (including the multiplicities m_j) and the zeros of $p(z)$ are symmetric in the axis

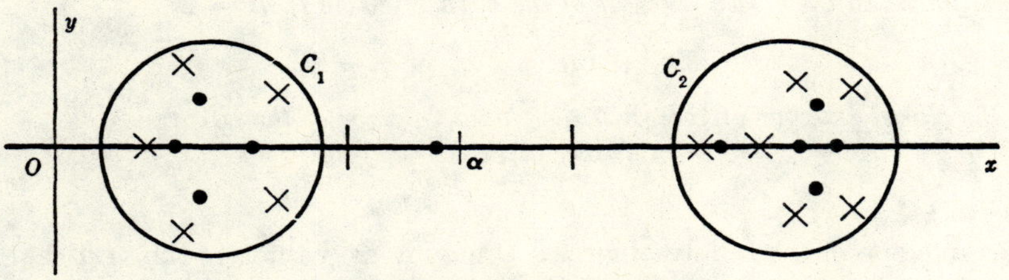

Fig. 8 illustrates §3.7 Theorem 5

of reals, and if the circle C_j' contains O in its open interior, is not a circle C_j, and cuts no other circle C_m', then C_j' contains in its closed interior precisely one critical point of $p(z)$, and that on the axis of reals; 3) if the polynomial $p(z)$ and the points α_j are real, then any circle C_j' which is not a circle C_j and which cuts no other circle C_m', contains in its closed interior precisely one critical point of $p(z)$, and that real; in every case 1)–3) this conclusion holds in slightly modified form if we replace the requirement that C_j' shall cut no other circle C_m' by the requirement that C_j' shall cut no other circle C_m' in more than one point (additional critical points may lie in the closed interior of C_j' at the intersections of C_j' with other circles C_m', but not elsewhere).

§3.8. Higher derivatives. It is not in accord with the plan of the present work to study in detail the zeros of higher derivatives of polynomials. Nevertheless some results on this topic lie near the surface and may well be rapidly considered.

Iteration of Lucas's Theorem yields

THEOREM 1. *If* $p(z)$ *is a polynomial of degree* n *whose zeros lie in a convex point set* Π, *then* Π *contains all zeros of the k-th derivative* $p^{(k)}(z)$, $k = 1, 2, \cdots, n - 1$.

Unless all zeros of $p(z)$ *are collinear, no zero of* $p^{(k)}(z)$ *lies on the boundary of* Π *except a zero of* $p(z)$ *of multiplicity at least* $k + 1$.

It is likewise true that the center of gravity of the zeros of each of these derivatives coincides with the center of gravity of the zeros of $p(z)$.

Jensen [1913] stated the natural extension to higher derivatives of his theorem (§1.4):

THEOREM 2. *If* $p(z)$ *is a real polynomial of degree* n, *construct on each segment* S *joining a pair of conjugate imaginary zeros of* $p(z)$ *an ellipse of eccentricity* $[(k - 1)/k]^{1/2}$ *with* S *as minor axis. Then all non-real zeros of* $p^{(k)}(z)$, $1 \leqq k \leqq n - 2$, *lie in the closed interiors of these Jensen ellipses.*

Here we have occasion to use the

LEMMA. *Let a circle* C *have its center on the axis of reals and its vertical diameter a line segment in the closed interior of the ellipse (circle if* $m = 2$*):*

$$x^2 + (m - 1)y^2 = m - 1, \qquad m > 1. \tag{1}$$

Then the closed interior of C *lies in the closed interior of the ellipse*

$$x^2 + my^2 = m. \tag{2}$$

The envelope of the family of circles C having as diameters the vertical chords of the ellipse (1) is precisely the ellipse (2).

If $(x_1, 0)$ is the center and y_1 the radius of an arbitrary circle C, we have by (1)

$$x_1^2 + (m - 1)y_1^2 \leqq m - 1.$$

The coordinates of an arbitrary point (x, y) on or interior to the circle C satisfy the inequalities

$$(x - x_1)^2 + y^2 \leqq y_1^2 \leqq 1 - x_1^2/(m - 1), \tag{3}$$

$$x^2 - 2x_1x + mx_1^2/(m - 1) + y^2 \leqq 1. \tag{4}$$

We obviously have

$$-\frac{m-1}{m}\left[x^2 - \frac{2mx_1x}{m-1} + \frac{m^2x_1^2}{(m-1)^2}\right] \leqq 0,$$

which by addition with (4) yields

$$x^2/m + y^2 \leqq 1,$$

and the proof is complete.

Theorem 2 now follows by Jensen's Theorem. The non-real zeros of $p'(z)$ lie in the closed interiors of the Jensen circles for the zeros of $p(z)$; the Jensen circles for the zeros of $p'(z)$ are constructed on the segments joining pairs of conjugate imaginary zeros of $p'(z)$ as diameters, and by the Lemma the closed interiors of these circles lie in the closed Jensen ellipses for the zeros of $p(z)$ with eccentricity $1/2^{1/2}$. The non-real zeros of $p''(z)$ lie in the closed interiors of the Jensen circles for the zeros of $p'(z)$, hence in the closed interiors of those ellipses. Continuation of this reasoning establishes the theorem in its entirety. Of course the $(n-1)$-th derivative of $p(z)$ has no non-real zeros.

It is by no means obvious that Theorem 2 is the best possible result in the sense (§1.4.2) that Jensen's Theorem on the critical points of a polynomial is the best possible, for Theorem 2 involved the iteration of Jensen's Theorem. By way of justification of Theorem 2 we establish

THEOREM 3. *If k is given, $1 \leqq k \leqq n$, and if α takes all real values, the non-real zeros of the k-th derivative $p^{(k)}(z)$ of the polynomial*

$$p(z) \equiv (z^2+1)(z-\alpha)^n, \qquad \alpha \text{ real}, \tag{5}$$

take all positions on a suitably chosen ellipse, which as n becomes infinite approaches the ellipse

$$x^2/k + y^2 = 1.$$

In the case $k = 1$, both ellipses are circles; compare §2.2.1 Theorem 2. In equation (5) we set $p_1(z) \equiv (z-\alpha)^n$, $p_2(z) \equiv z^2+1$, so the general formula for the k-th derivative of the product $p_1(z)p_2(z)$ yields

$$\begin{aligned} p^{(k)}(z) &= p_1^{(k)}p_2 + kp_1^{(k-1)}p_2' + k(k-1)p_1^{(k-2)}p_2''/2 + \cdots + p_2^{(k)} \\ &= n(n-1)\cdots(n-k+1)(z-\alpha)^{n-k}(z^2+1) \\ &\quad + 2kn(n-1)\cdots(n-k+2)z(z-\alpha)^{n-k+1} \\ &\quad + k(k-1)n(n-1)\cdots(n-k+3)(z-\alpha)^{n-k+2}; \end{aligned} \tag{6}$$

we write this last equation in the form

$$(z-\alpha)^{-n+k}p^{(k)}(z) = A(z-\alpha)^2 + 2Bz(z-\alpha) + C(z^2+1), \tag{7}$$

where the coefficients depend on k and n and are positive. Separation into real and pure imaginary parts with the elimination of α gives us for the zeros of $p^{(k)}(z)$ the equation $y = 0$ and

$$(A+2B+C)(B^2-AC)x^2 + (A+2B+C)(A+B)^2y^2 = C(A+B)^2. \tag{8}$$

When α is large and positive, the zeros of $p'(z)$ are known (§2.2.1) to be real, so those of $p^{(k)}(z)$ are real. When α is zero, the zeros of $p^{(k)}(z)$ other than α are known to be non-real. These zeros depend analytically on α, so as α decreases continuously from a large positive value the two zeros of the second member of (7) are first real and distinct, then real and coincident, and then become conjugate imaginary on the ellipse (8). When α is zero these two zeros are pure imaginary, so they trace out the entire ellipse (8) as α varies from $+\infty$ to $-\infty$.

Comparison of (6) and (7) shows that as n becomes infinite while k remains constant, the ellipse (8) approaches the ellipse of the theorem.

The study of the zeros of the second derivative of a polynomial $p(z)$ is more difficult than that of the zeros of the first derivative, primarily because the logarithmic derivative enables us to express the zeros of $p'(z)$ as the zeros of a function of which each term contains but one zero of $p(z)$, and that linearly. No such simple expression is known for higher derivatives.

Certain results for higher derivatives are available for polynomials symmetric in O:

THEOREM 4. *If the zeros of a polynomial $p(z)$ of degree n are symmetric in O and lie in the closed exterior of a hyperbola with center O, then the zeros of $p^{(k)}(z)$, $k = 1, 2, \cdots, n - 1$, also lie in that closed exterior.*

Theorem 4 follows at once from §3.6.1 Theorem 1. Likewise if the zeros of $p(z)$ lie in a closed double sector with vertex O, the zeros of the derivative $p^{(k)}(z)$ lie in that closed double sector.

If the zeros of a polynomial $p(z)$ are symmetric in O and lie in the closed interior of an equilateral hyperbola H with center O all zeros of $p''(z)$ need not lie in the closed interior of H. It is sufficient here to choose H as $x^2 - y^2 = 1$, $p(z) \equiv (z^2 - 1)^2$, whence $p''(z) = 4(3z^2 - 1)$, whose zeros do not lie in the closed interior of H.

If the zeros of a polynomial $p(z)$ of degree n are symmetric in O and lie in the closed interior of an equilateral hyperbola H with center O, all zeros of $p''(z)$ lie in the closed interior of the equilateral hyperbola found by shrinking H toward O in the ratio $[(n - 1):1]^{1/2}$. This theorem follows by application of §3.6.1 Theorem 1 part 2, and §3.6.1 Theorem 2; alternate higher derivatives have simple zeros at O, and alternate higher derivatives do not vanish at O. Continued application of these two theorems yields results on the higher derivatives of $p(z)$, but those results can be improved by more delicate reasoning. We state without proof [Walsh, 1948e]:

THEOREM 5. *Let $p(z)$ be a polynomial of degree n whose zeros are symmetric in O, let O be an l-fold zero, and let the remaining zeros lie in the closed interior of a hyperbola H whose center is O. Let $p_0(z)$ be the polynomial of degree n whose zeros are symmetric in O, which has an l-fold zero at O, and whose remaining zeros lie at the vertices of H. Let z_k, $1 \leqq k \leqq n - 2$, be one of the zeros of $p_0^{(k)}(z)$ of smallest positive modulus, and let H_k be the equilateral hyperbola with center O which has z_k as a vertex. Then all zeros of $p^{(k)}(z)$ except perhaps O lie in the closed interior of H_k.*

Theorem 5 extends without change to the case of a polynomial whose zeros possess p-fold symmetry in O.

We state without proof the analog [Walsh, 1921a] for higher derivatives of the theorem of §1.5 due to the present writer:

THEOREM 6. *Let the zeros of the k-th derivative,* $k = 1, 2, \cdots$, *or* $m_1 + m_2 - 1$, *of the polynomial* $q(z) \equiv (z - \alpha_1)^{m_1}(z - \alpha_2)^{m_2}$ *be denoted by* $\beta_1, \beta_2, \cdots, \beta_m$. *Let the circles* C_1 *and* C_2 *whose respective centers are* α_1 *and* α_2 *have the point* α_0 *as external center of similitude in the sense that a stretching of the plane without rotation with* α_0 *as center carries* C_1 *into* C_2; *the case* $\alpha_0 = \infty$ *is not excluded and means that* C_1 *and* C_2 *have the same radius. Let the circles* $\Gamma_1, \Gamma_2, \cdots, \Gamma_m$ *have centers* $\beta_1, \beta_2, \cdots, \beta_m$ *respectively, and let* α_0 *be an external center of similitude for each pair of the circles* C_1, C_2, Γ_j. *If the closed interiors of the circles* C_1 *and* C_2 *are the respective loci of* m_1 *and* m_2 *zeros of the variable polynomial* $p(z)$ *of degree* $m_1 + m_2$, *the locus of the zeros of* $p^{(k)}(z)$ *consists of the closed interiors of the circles* Γ_j. *Any circle* Γ_j *whose closed interior is disjoint from the closed interior of every other circle* Γ_l *contains a number of zeros of* $p^{(k)}(z)$ *equal to the multiplicity of* β_j *as a zero of* $q^{(k)}(z)$; *any set of closed interiors of circles* Γ_j *disjoint from the closed interiors of all other circles* Γ_l *contains a number of zeros of* $p^{(k)}(z)$ *equal to the sum of the multiplicities of the centers* β_j *as zeros of* $q^{(k)}(z)$.

§3.9. Further results. It is not our plan in the present work to set forth all the principal results in the literature on geometrical relations concerning zeros of polynomials, or even concerning critical points of polynomials. We have chosen to emphasize a) results which determine point sets of the plane which are free from critical points (or which contain all critical points) and b) results on critical points of polynomials and rational functions which extend to more general analytic functions and to harmonic functions. Thus we omit c) the enumeration of critical points in a given region and d) the determination of special regions in which a specified number (say at least one, or at most m) of critical points lie.

The critical points of polynomials and of more general rational functions are the foci of certain higher plane curves which are tangent in specified points to lines joining the various zeros and poles of the given function; this relation has been studied by Siebeck [1865] and numerous later writers. For instance the critical points of a polynomial $p(z)$ of degree three whose zeros are not collinear are the foci of the maximum ellipse which can be inscribed in the triangle whose vertices are the zeros of $p(z)$; this ellipse is tangent to the sides of the triangle at their mid-points; this property of the critical points can be used [Walsh, 1920a] to determine a ruler-and-compass construction for the critical points. However, the relation of critical points to the foci of higher plane curves has not yet been exploited in obtaining simple regions which are known to contain or not to contain critical points.

Grace [1902] and Heawood [1907] undertook to extend Rolle's Theorem to the case of polynomials in the complex variable: if the polynomial $p(z)$ of degree n vanishes in the points $z = +1$ and -1, what is the smallest circle (or other region) that will surely contain a zero of $p'(z)$? Grace shows that the circle $|z| = \cot(\pi/n)$ contains in its closed interior at least one zero of $p'(z)$, and it follows by considering the polynomial $p(z)$ whose zeros are the vertices of a regular polygon of n sides that no smaller circle whose radius is a function of n

has this property. The methods developed by Grace and Heawood are of great importance in the more general study of the geometry of the zeros of related polynomials, including the study of the zeros of the higher derivatives of a given polynomial.

As an outgrowth of the Grace-Heawood problem, the following may be suggested [compare Kakeya, 1917]: Let a circle C contain k zeros of a polynomial of degree n; to determine a concentric circle which contains at least m critical points. The Grace-Heawood Theorem enables one to solve this problem in the case $k = 2$, $m = 1$; the results of §1.5 enable one [see Walsh, 1922] to solve the problem in the case $k = n - 1$, $m = n - 2$; the problem in its generality is still unsolved.

For complete references to the literature and a report on the present state of these various problems, the reader may refer to an admirable recent book by Marden [1949], the first full-scale treatment of the general field of the geometry of the zeros of polynomials.

The problem of the determination of the number of critical points in a given region is fundamentally solved by the Principle of Argument (§1.1.2). We have made a number of applications of this principle—indeed it can be used to prove Lucas's Theorem—and on occasion have also determined the number of critical points without that principle. This general problem has recently been investigated by Morse and Heins [1947], to which the reader is referred for an exposition of the results.

The present chapter completes our separate study of the critical points of polynomials, although frequently in the sequel we obtain further results on polynomials related to results on more general rational functions. All the theory of the critical points of polynomials can be applied in, and indeed is a special case of, the study of the critical points of Green's function (Chap. VII). The theory of the critical points of polynomials is also a special case of, and serves as a pattern for, the study of the critical points of more general rational functions.

CHAPTER IV

RATIONAL FUNCTIONS

§4.1. Field of force. In a classical paper, which indeed provides the foundation for our study of the critical points of rational functions, Bôcher [1904] modified the field of force set up in Gauss's Theorem (§1.2) by admitting both positive and negative masses for particles. Each particle repels with a force equal to its mass divided by the distance, so a negative particle attracts instead of repelling. If the algebraic sum of all the masses in the plane is zero, the positions of equilibrium are determined as the zeros of certain algebraic invariants of the ground forms, namely binary forms whose zeros determine the location of the original particles. In the simplest case, the invariant is the jacobian of the two given forms. Bôcher thus made two notable advances: 1) in giving an algebraic interpretation for the positions of equilibrium in the new field of force and 2) in emphasizing the invariant nature of the positions of equilibrium.

§4.1.1. Fundamental theorem. For our present purposes, it is not necessary to introduce the algebraic invariants themselves [compare Porter, 1916, who introduces the logarithmic derivative of $R(z)$ but not its conjugate nor the particles; Walsh, 1918]:

THEOREM 1. *Let $R(z)$ be a rational function whose finite zeros are $\alpha_1, \alpha_2, \cdots, \alpha_m$ and whose finite poles are $\beta_1, \beta_2, \cdots, \beta_n$, each point being enumerated according to its multiplicity as a zero or pole. Then the finite zeros of $R'(z)$ are the finite multiple zeros of $R(z)$ and the finite positions of equilibrium in the field of force due to particles of unit positive mass in the α_k and particles of unit negative mass in the β_k. In addition to the finite zeros of $R'(z)$, that function vanishes at infinity unless infinity is a pole of $R(z)$.*

We may write

$$R(z) \equiv \frac{(z - \alpha_1)(z - \alpha_2) \cdots (z - \alpha_m)}{(z - \beta_1)(z - \beta_2) \cdots (z - \beta_n)}; \tag{1}$$

the totality of finite and infinite zeros and poles of $R(z)$ are displayed in the linear factors of (1), except that when infinity is a zero or pole, the corresponding factor is simply omitted, a number of times corresponding to the multiplicity. We have from (1) by taking the logarithmic derivative

$$\frac{R'(z)}{R(z)} = \sum_1^m \frac{1}{z - \alpha_k} - \sum_1^n \frac{1}{z - \beta_k}. \tag{2}$$

The field of force is represented by the conjugate of the second member of (2), so the finite zeros of $R'(z)$ are precisely the finite multiple zeros of $R(z)$ and the finite positions of equilibrium. Gauss's Theorem (§1.2) is the case $n = 0$.

To study the behavior at infinity, we use the Laurent development in the neighborhood of that point:

$$R(z) = \frac{1}{z^{n-m}} + \frac{A_1}{z^{n-m+1}} + \cdots .$$

In the case $m > n$, $R(z)$ has a pole at infinity and $R'(z)$ does not vanish at infinity. In the case $m = n$, the function $R(z)$ is analytic and different from zero at infinity; the function $R'(z)$ has at least a two-fold zero at infinity. In the case $m < n$, the function $R(z)$ has a zero at infinity of order $n - m$, and $R'(z)$ has a zero there of order $n - m + 1$.

Theorem 1 is now established, but it is appropriate to record the total number of zeros of the derivative and of finite critical points under that theorem with the convention of §1.1.2. The *degree* of $R(z)$ is the greater of the two numbers m and n, or their common value if m and n are equal; thus the degree is the number of zeros or poles in the extended plane. We write

$$R'(z) = \frac{(m - n)z^{m+n-1} + \cdots}{(z^n + \cdots)^2}. \tag{3}$$

Let $R(z)$ have precisely q distinct poles, of respective orders $n_1, n_2, \cdots, n_q$; if these are all finite we have $n_1 + n_2 + \cdots + n_q = n$. These points if finite are respectively poles of $R'(z)$ of orders $n_1 + 1, n_2 + 1, \cdots, n_q + 1$; if the point at infinity is a pole of $R(z)$ of order n_k, it is a pole of $R'(z)$ of order $n_k - 1$ provided we have $n_k > 1$, and otherwise $R'(z)$ is analytic at infinity. In the case $m > n$, the function $R(z)$ is of degree m and has a pole at infinity of order $n_q = m - n$; thus $R'(z)$ in (3) is formally of degree $m + n - 1$, and has no zero at infinity; the number of finite zeros of $R'(z)$ is the formal degree of $R'(z)$ less the sum of the degrees of the common factors of numerator and denominator, namely $m + n - 1 - [(n_1 - 1) + (n_2 - 1) + \cdots + (n_{q-1} - 1)] = m + q - 2$. In the case $m < n$, the function $R(z)$ is of degree n, and has at infinity a zero of order $n - m$; thus $R'(z)$ in (3) is formally of degree $2n$, and has at infinity a zero of order $n - m + 1$; the number of finite zeros of $R'(z)$ is $2n - (n - m + 1) - [(n_1 - 1) + (n_2 - 1) + \cdots + (n_q - 1)] = m + q - 1$. In the case $m = n$, infinity is neither a zero nor a pole of $R(z)$, but is a zero of $R'(z)$ of order at least two; the number of finite zeros of $R'(z)$ is at most $2n - 2 - (n - q) = m + q - 2$. In this case it is not possible to give an unambiguous formula for the number of finite zeros of $R'(z)$, as is illustrated by the example

$$R(z) = \frac{(z - \alpha)^2}{z^2 - 1}, \qquad R'(z) = \frac{2(z - \alpha)(\alpha z - 1)}{(z^2 - 1)^2};$$

here $R'(z)$ has one or two finite zeros according as $\alpha = 0$ or $\alpha \neq 0$, provided $\alpha^2 \neq 1$. We have the

COROLLARY. *In Theorem* 1, *let q denote the number of distinct poles of $R(z)$.*

Then $R'(z)$ has precisely $m + q - 2$ finite zeros in the case $m > n$, $m + q - 1$ finite zeros in the case $m < n$, and has at most $m + q - 2$ finite zeros in the case $m = n$; the total number of critical points of $R(z)$ is the degree of $R(z)$ plus q less two.

The latter part of the Corollary follows from the enumeration already given and from the definition (§1.1.2) of critical point, if we have $m < n$, the point at infinity is a zero of $R(z)$ of order $n - m$, a zero of $R'(z)$ of order $n - m + 1$, and a critical point of $R(z)$ of order $n - m - 1$, so the totality of critical points numbers $n + q - 2$, if we have $m = n$, $R'(z)$ is of degree $2n - [(n_1 - 1) + \cdots + (n_q - 1)] = n + q$, and the point at infinity is a zero of $R'(z)$ of order two more than its multiplicity as a critical point of $R(z)$.

We have made (§1.1.2) the convention that the point at infinity is a critical point of an analytic function $f(z)$ when and only when it is a zero of $f'(z)$ of order greater than two; there are here two great advantages: i) a critical point of an analytic function remains a critical point and of the same order under one-to-one conformal transformation, and ii) the behavior of level loci as to multiple points is typical at all critical points of given order whether finite or infinite; there is nevertheless one great disadvantage: iii) the usual methods (e.g. Principle of Argument) for the enumeration of the zeros of the analytic function $f'(z)$ in an infinite region do not give the number of critical points of $f(z)$ without some adjustment for the point at infinity. Nevertheless, property i) is indispensable in formulating theorems with simplicity and elegance, and its absence necessitates long and involved circumlocutions if all (i.e. finite and infinite) cases are included together; this is our primary justification for the convention made. It is clear that the total number of zeros of $f'(z)$ in a region R is readily found from the number of critical points of $f(z)$ in R, and reciprocally; these two numbers are identical if R is finite or if R is infinite and $f(z)$ has a pole at infinity; the former number exceeds the latter by two if R is infinite and $f(z)$ is analytic at infinity.

Corresponding to the convention made for critical points, we make a similar convention concerning positions of equilibrium. For instance, under the conditions of Theorem 1 the point at infinity is a position of zero force unless it is a zero or pole of $R(z)$; we introduce the convention that *the point at infinity is a position of equilibrium when and only when it is a critical point not a multiple zero of $R(z)$*. This convention is justified on the ground of simplicity, that every critical point is a position of equilibrium and conversely except that every multiple zero of $R(z)$ is a critical point. Moreover the convention is justified by the fact (§4.1.3) that in the field of force on the sphere the behavior of the point at infinity as a position of equilibrium is now no longer exceptional.

When suitably modified, the treatment of §1.6.1 and §1.6.2 applies here. In particular, the lines of force in the field of Theorem 1 are the loci $\arg [R(z)] =$ const, which are orthogonal to the loci $|R(z)| =$ const except at the multiple points of both loci (positions of equilibrium in the field of force), and except at the points α_j and β_k.

§4.1.2. Transformations of the plane. In the field of force described in Theorem 1, it is immaterial whether or not we consider particles to be placed at infinity when infinity is a zero or pole of $R(z)$; the force at a finite point due to a particle at infinity is defined to be zero. But if the configuration is transformed by a non-singular linear transformation of the complex variable

$$w = (\alpha z + \beta)/(\gamma z + \delta), \qquad \alpha\delta - \beta\gamma \neq 0, \tag{4}$$

so that $z = \infty$ corresponds to a finite point $w = \kappa$, then in the w-plane a suitable particle is to be placed at this finite point if $z = \infty$ is a zero or pole of $R(z)$. The behavior of the critical points of $R(z)$ under the transformation (4) depends on the easily determined behavior of the given poles and zeros of $R(z)$, and also on the behavior of the positions of equilibrium.

THEOREM 2. *Under the non-singular transformation* (4) *of the variable* z, *the positions of equilibrium in the field of force of Theorem* 1 *are invariant, with the understanding that in the* w*-plane a suitable particle is to be placed at the image* $w = \kappa$ *of the point* $z = \infty$ *if the latter is a zero or pole of* $R(z)$.

Under the transformation (4), the given rational function $R(z)$ corresponds to a rational function of w whose zeros and poles in the w-plane are the images of the zeros and poles of $R(z)$ in the z-plane, and respectively of the same multiplicities. The critical points in the two planes are studied by considering the multiple zeros of $R(z)$ and its transform, the points at infinity, and the positions of equilibrium in the two planes.

To prove Theorem 2 we merely set

$$\frac{dR}{dw} = \frac{dR}{dz} \cdot \frac{dz}{dw} = \frac{(\gamma z + \delta)^2}{\alpha\delta - \beta\gamma} \cdot \frac{dR}{dz},$$

and scrutinize the first factor in the last member. Even positions of equilibrium at infinity remain positions of equilibrium under the transformation (4).

The positions of equilibrium are invariant under reflection in a line, and also under the transformation $w - \alpha = A/(\bar{z} - \bar{\alpha})$, $A > 0$; any inversion in a circle can be considered as the succession of two such transformations, so we have

COROLLARY 1. *Under inversion in a circle, the positions of equilibrium in the field of force of Theorem* 1 *are invariant.*

Of particular interest in Theorems 1 and 2 is the case that only three distinct points of the extended plane are zeros or poles of $R(z)$, for three given distinct points can be transformed into three arbitrary distinct points of the extended plane by a transformation of type (4). We formulate

COROLLARY 2. *Let the rational function $R(z)$ of degree $m_1 + m_2$ have an m_1-fold zero [or pole] at z_1, an m_2-fold zero [or pole] at z_2, and an $(m_1 + m_2)$-fold pole [or zero] at z_3; the only critical point of $R(z)$ other than z_1, z_2, and z_3 is z_4 as determined by the equation**

$$(5) \qquad (z_1, z_2, z_3, z_4) = (m_1 + m_2)/m_1 .$$

If we choose z_3 at infinity this result follows (compare §1.2) by setting $R(z) \equiv (z - z_1)^{m_1}(z - z_2)^{m_2}$ [or the reciprocal]. Indeed, equation (5) can be written in the equivalent form

$$\frac{m_1}{z_4 - z_1} + \frac{m_2}{z_4 - z_2} = \frac{m_1 + m_2}{z_4 - z_3},$$

from which Corollary 2 is obvious. Of course the point z_4 defined from z_1, z_2, and z_3 by (5) may lie at infinity if none of the points z_1, z_2, z_3 is at infinity, and is a critical point whether finite or infinite.

Although the positions of equilibrium in the field of force of Theorem 1 are invariant, the field of force itself is not necessarily invariant under linear transformation, as we show by means of an example. In Theorem 1 we set $\alpha_1 = 1$, $m = 1$, $\beta_1 = -1$, $n = 1$, so the force $\mathfrak{F}$ at any point z is given by the equation

$$\mathfrak{F} = \frac{1}{\bar{z} - 1} - \frac{1}{\bar{z} + 1} = \frac{2}{\bar{z}^2 - 1};$$

the loci $|\mathfrak{F}| = \text{const}$ are lemniscates $|z^2 - 1| = \text{const}$ with poles in the points ± 1. If we make the substitution $w = (z - 1)/(z + 1)$, the new field of force is due to a single particle at $w = 0$ of mass unity; the loci $|\mathfrak{F}| = \text{const}$ are circles with the common center $w = 0$; these loci are not the images of the former loci.

Even though the fields of force are not invariant under linear transformation, their directions are:

THEOREM 3. *If $R(z)$ is a rational function, and is subjected to a non-singular linear transformation, then the direction (including sense) of the force in the field defined in Theorem 1 is invariant. Otherwise expressed, lines of force are invariant under linear transformation.*

The lines of force in the z-plane are the loci $\arg [R(z)] = \text{const}$, and the force is sensed in the direction of increasing $|R(z)|$; this fact has been proved (§4.1.1) for finite points not zeros or poles of $R(z)$ or $R'(z)$, and is hereby introduced as a convention for the point at infinity; these properties are obviously unchanged by linear transformation. The loci $|R(z)| = \text{const}$ are also unchanged by linear transformation, but (as already indicated by an example) not necessarily the loci $|R'(z)/R(z)| = \text{const}$.

* Cross-ratio is defined in §3.2.

For the sake of reference we formulate the

COROLLARY. *In the field of force due to a positive unit particle at the point* $z = +1$ *and a negative unit particle at the point* $z = -1$, *the lines of force are the circles through those two points; the force is directed on each arc* $(+1, -1)$ *of such a circle from the point* $z = +1$ *toward the point* $z = -1$.

If we transform by setting $w = (z - 1)/(z + 1)$, the lines of force in the w-plane are obviously the straight lines through $w = 0$; their images are the circles of the Corollary, and the sense on each half-line from $w = 0$ to $w = \infty$ corresponds to the sense from $z = +1$ to $z = -1$. There is no position of equilibrium.

A direct proof of the Corollary is readily given. Invert the z-plane in the circle whose center is an arbitrary point $z(\neq \pm 1)$ and whose radius is unity. If α and β denote the inverses of $+1$ and -1, the total force at z due to the given particles is the sum of the vectors αz and $z\beta$, namely the vector $\alpha\beta$. The line $\alpha\beta$ is the inverse of the circle through z, $+1$, and -1, and the direction $\alpha\beta$ is the direction of the tangent to that circle at z, sensed from $+1$ to -1.

Theorem 3 refers to a linear transformation, in contrast to §3.6.2 Theorem 3, but the proof is valid under more general conditions and establishes

THEOREM 4. *If* $\varphi(z)$ *is a rational function of* z *and* $\Phi(w)$ *a rational function of* w, *then the lines of force in the* z*-plane due to particles at the zeros and poles of the rational function* $\Phi[\varphi(z)]$ *are the images in the* z*-plane under the transformation* $w = \varphi(z)$ *of the lines of force in the* w*-plane due to particles at the zeros of the rational function* $\Phi(w)$.

§4.1.3. Stereographic projection. In the study of the critical points of rational functions, two methods are possible: 1) direct study of rational functions as such, and 2) study of the invariants of binary forms and obtaining as an application results on rational functions; the great advantage of 1) is that rational functions seem more fundamental than the invariants, and it is this method that we adopt; the great advantages of 2) are the invariant nature of the results under linear transformation, the treatment of infinity as an unexceptional point (at least on the sphere), and the avoidance of the lack of symmetry between zeros and poles in 1) in that a multiple zero of a rational function is a critical point while a multiple pole is not. One other difference between 1) and 2) is also noteworthy: in 2) the binary forms may be chosen arbitrarily, perhaps with common zeros, so that a point $z = \alpha$ may be the location of both repelling and attracting particles, and then is necessarily a zero of the invariant, while in 1) a single point cannot be both a zero and a pole of a rational function, and under Theorem 1 cannot be the location of more than one kind of particle; of course it is possible to rephrase Theorem 1 to allow for *formally* coincident zeros and poles at a point, but such an artificial point is not necessarily a critical point.

In connection with the invariant formulation of this entire problem, Bôcher [1904] projects the particles and field of force in the plane stereographically onto the sphere, with suitable allowance for particles at infinity:

THEOREM 5. *In Theorem 1 let the particles at infinity be so chosen that the total mass of the fixed particles is zero, and let all the particles in the plane be projected stereographically onto the sphere, retaining the property of repelling with a force equal to the mass divided by the distance. Then at any point P' of the sphere the direction of the force due to the particles on the sphere is tangent to the sphere at P', and coincides at P' with the stereographic projection of the direction of the field of force in the plane; in particular the positions of equilibrium on the sphere (including the possible position at the center of projection) are the stereographic projections of those in the plane.*

We study the stereographic projection of the given plane Π onto a sphere Σ having the unit circle in Π as a great circle of Σ, projecting by means of half-lines terminating in the north pole N, one of the two points of Σ farthest from Π.

To the particles placed on Σ by stereographic projection, including possible particles at N, we artificially add particles at N equal to the negatives of all the other particles; these new particles at N have the total mass zero and hence do not affect the total field of force. Let P and Q be arbitrary finite points of Π and P' and Q' their images on Σ. The force at P' due to a unit positive particle at Q' and a unit negative particle at N on Σ is (§4.1.2 Corollary to Theorem 3) in direction along the circle $Q'P'N$, sensed from Q' to P' to N, and lies in the plane tangent to Σ at P'. The direction in the plane Π of the force at P due to the corresponding particles at Q and infinity is that of the line QP, sensed from Q to P; these two directions and senses on Σ and Π correspond to each other under stereographic projection. The force f at P due to another particle say at a finite point R of Π corresponds to a force at P' on Σ whose direction and sense are the stereographic projection of the direction and sense of f, so by the invariance of angle under stereographic projection the two forces in Π intersect at the same angle as do the two forces on Σ. But it is not yet clear that the two forces in Π have the same relative magnitudes as those on Σ, and thus it is not clear that the directions of the total fields of force correspond under stereographic projection.

We invert Π in the unit circle whose center is P, so the force at P due to the pair of particles at Q and infinity is represented by the vector Q_1P or Q_1N_1, where Q_1 is the inverse of Q and N_1 is P, the inverse of the point at infinity. We similarly study the field of force on Σ by an inversion in space, but first we rotate Σ rigidly about a diameter so that P' takes the position formerly occupied by N, and then invert Σ in the sphere whose center is the new point P' and radius $2^{1/2}$; this inversion is stereographic projection with P' as center of projection, and transforms Σ again into Π. Under this projection let Q' and N on Σ correspond to Q_1' and N_1' in Π. The force at P' due to the pair of particles at Q' and N is half the sum of the vectors in space $Q_1' \, P'$ and $P' \, N_1'$, or half the vector $Q_1'N_1'$, which lies in Π.

The transformation of Π from points Q to points Q_1' consists in succession of stereographic projection onto Σ, a rigid rotation of Σ about the diameter through the poles of the great circle through P' and N, and stereographic projection of Σ onto Π. The rotation of Σ about the diameter is equivalent to successive reflection of Σ in two planes through the diameter; these planes may be chosen respectively to bisect the great circle arc $P'N$, and to pass through N; the first reflection carries P' into N and the second leaves the new P' unchanged. This transformation of Π from points Q to points Q_1' may be interpreted in Π alone, and consists of successive reflection of Π in two "circles", the stereographic projections in Π of the great circles whose planes are the planes in which Σ is reflected. Of these "circles" the first is a circle with P as center (the transformation carries P to infinity) and the second is a straight line L.

In comparing the two fields of force in Π and on Σ, we thus compare the set of vectors Q_1N_1 obtained by inverting Π in the unit circle whose center is P, with the halves of the set of vectors $Q_1'N_1'$ obtained by inverting Π in a suitable circle with center P followed by reflection in a straight line. These two sets of vectors are similar in that one set is obtained from the other merely by a suitable stretching of the plane followed by reflection in a line, and Theorem 5 follows, so far as concerns points different from N.

Of course the size of the circle in which this second reflection is made is readily computed in terms of P, for the circle has P as center and passes through two diametrically opposite points of the unit circle in Π, but that computation is unnecessary for our present purpose. It is clear geometrically that the two sets of vectors Q_1N_1 and $Q_1'N_1'$ must differ qualitatively by a reflection in a line, for in constructing Q_1' we rotate Σ so that P' takes the position of the original N before projecting back onto Π.

Under the hypothesis of Theorem 5, the direction (if any) of the force on Σ at N is that of the direction of the force at infinity in the plane, and is found by continuity considerations from the direction of the force at finite points; a line of force (§4.1.2) passes through the point at infinity. We have already introduced the convention that the point at infinity (assumed not a multiple zero) is considered a position of equilibrium in the plane if and only if it is a critical point. We show that *the point at infinity* (assumed not a multiple zero, and where now no particles are placed at infinity unless it is a zero or pole of $R(z)$) *is a critical point of $R(z)$ if and only if it is a position of equilibrium on the sphere.* The point at infinity is a critical point of $R(z)$ if and only if (§1.1.2) $w = 0$ is a critical point of $R(1/w)$; the transformation $w = 1/z$ is a rigid rotation of Σ about the diameter through $+1$ and -1, which leaves the field of force and positions of equilibrium unchanged on the sphere; the origin $w = 0$ is a critical point of $R(1/w)$ if and only if it is a position of equilibrium or multiple zero. This is our justification for the conventions made concerning infinity as a critical point (§1.1.2), as a position of equilibrium (§4.1.1), and (§4.1.2) as a point in the field of force. Theorem 5 is established.

The great advantage of using the field of force on the sphere rather than in the plane is that on the sphere the point at infinity is not exceptional. This advantage is of relatively minor importance when, as in the present work, we emphasize rational functions rather than algebraic invariants, for with rational functions the point at infinity is inevitably exceptional in its behavior as a critical point.

The methods of the present section are useful in studying the spherical field of force itself; for instance they easily yield the precise analog of §1.5.1 Lemma 1.

§4.2. Bôcher's Theorem. Bôcher did not study the critical points of an arbitrary rational function as such, and the following theorem was first explicitly

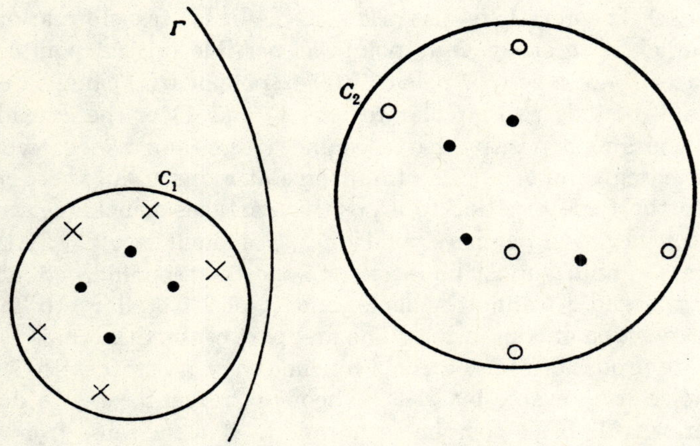

Fig. 9 illustrates §4.2 Bôcher's Theorem

formulated by the present writer [1918]. Nevertheless this theorem is the expression for the derivative of a rational function of a theorem due to Bôcher [1904] concerning the zeros of the jacobian of two binary forms, so we shall use his name.

BÔCHER'S THEOREM. *Let $R(z)$ be a rational function of z. If a circle Γ separates all the zeros of $R(z)$ not on Γ from all the poles not on Γ, but if at least one zero or pole does not lie on Γ, then no point of Γ is a critical point of $R(z)$ unless it is a multiple zero of $R(z)$.*

If two disjoint circular regions C_1 and C_2 contain respectively all the zeros and poles of $R(z)$, then C_1 and C_2 contain all critical points of $R(z)$. Under these conditions, the zeros and poles of $R(z)$ lie respectively in two closed regions T_1 and T_2 bounded by arcs of circles, each circle separating the interior points of T_1 from the interior points of T_2. All the critical points of $R(z)$ lie in T_1 and T_2. If $R(z)$

is of degree n, and has precisely q distinct poles, then the region T_1 *contains precisely* $n - 1$ *critical points and the region* T_2 *contains precisely* $q - 1$ *critical points.*

Of course we use here and in later discussion of rational functions the extended plane or the sphere; possible zeros and poles of $R(z)$ and $R'(z)$ may lie at infinity. The regions T_1 and T_2 are closed regions (which may degenerate) of the extended plane, and the term *circle* is understood to include straight line.

§4.2.1. Proof. It is to be noted that under the conditions of §1.5.1 Lemma 1, the force at a point P of the boundary of the given circular region R due to m particles in R all distinct from P is equivalent to the force at P due to m coincident particles in R, and actually interior to R if at least one of the given particles lies interior to R. To prove Bôcher's Theorem we use the field of force defined in §4.1.1 Theorem 1; by use of a suitable linear transformation if necessary, it is sufficient to study *finite* points as possible critical points. Let z be a finite point of Γ not a zero or pole of $R(z)$ (a pole of $R(z)$ cannot be a critical point). Then Γ bounds two circular regions Γ_1 and Γ_2 of the extended plane, of which Γ_1 contains n positive particles and Γ_2 contains n negative particles, where n is the degree of $R(z)$. The total force at z due to all these particles is equivalent to the force at z due to n positive particles coinciding at a point z_1 of Γ_1 and to n negative particles coinciding at a point z_2 of Γ_2. At least one of these points z_1 and z_2 must lie *interior* to the corresponding circular region. Consequently z_1 and z_2 cannot coincide, and (§4.1.2 Corollary to Theorem 3) z cannot be a position of equilibrium. The first part of the theorem is established.

An alternate proof here can be given by choosing z as finite with Γ a straight line; this choice is allowable by §4.1.2 Theorem 2. The force at z due to each particle interior to Γ_1 has a component normal to Γ in the sense from Γ_1 toward Γ_2, as has the force at z due to each particle interior to Γ_2; at least one such particle exists interior to Γ_1 or Γ_2; the force at z due to each particle on Γ acts along Γ, so z is not a position of equilibrium.

If the circular regions C_1 and C_2 of the theorem exist, through any point z exterior to them can be passed a circle Γ separating C_1 and C_2, so z is not a critical point of $R(z)$. If C_1 and C_2 exist, these regions can be chosen in an infinite number of ways, and the regions T_1 and T_2 can be chosen as regions common to an infinite number of regions C_1 and C_2 respectively.

To determine the number of critical points of $R(z)$ in T_1, we use Bôcher's method of continuity. We assume (compare §4.1.2 Theorem 2) that T_1 is a bounded region. Let all the poles of $R(z)$ be fixed, and allow the zeros of $R(z)$ to vary, to move continuously, and to coalesce at the finite point P_1 of T_1. For this final rational function consider T_1 to shrink to P_1 while T_2 remains fixed; circular regions C_1 and C_2 containing the original T_1 and T_2 respectively also contain P_1 and T_2; it follows that P_1 as a degenerate region and T_2 satisfy the conditions of the theorem. During the variation, the critical points of $R(z)$

vary continuously; none enters or leaves T_1; at the end of the process T_1 (that is, the point P_1) contains precisely one n-fold zero of $R(z)$ and no other zeros of $R(z)$, so T_1 contains $n - 1$ critical points of $R(z)$ both at the end of the process and originally.

To determine the number of critical points of $R(z)$ in T_2, we assume T_2 bounded, as we may do, and consider the critical points of the function $R_1(z) \equiv 1/R(z)$, which number precisely $n - 1$ in T_2. A critical point of $R_1(z)$ in T_2 is a critical point of $R(z)$ if and only if it is not a pole of $R(z)$. If the poles of $R(z)$ in T_2 have the respective multiplicities $m_1, m_2, \cdots, m_q$, with $\sum m_k = n$, these are also the multiplicities of the zeros of $R_1(z)$ in T_2, the multiplicities of these same points as critical points of $R_1(z)$ are $m_1 - 1, m_2 - 1, \cdots, m_q - 1$, whose sum is $n - q$. The remaining critical points of $R_1(z)$ in T_2 have total multiplicity $(n - 1) - (n - q) = q - 1$, which is then the total multiplicity of the critical points of $R(z)$ in T_2.* Bôcher's Theorem is established.

The special case of Bôcher's Theorem in which all poles of $R(z)$ lie at infinity is Lucas's Theorem. The special case that $R(z)$ has but a single pole in the extended plane is included in Bôcher's Theorem and is an immediate consequence of Lucas's Theorem [compare Ford, 1915].

Two particular configurations occur so frequently here that they are worth mentioning explicitly for convenience in reference:

COROLLARY 1. *If the circular regions* $|z| \leqq r_1$ *and* $|z| \geqq r_2(>r_1)$ *contain respectively the zeros and poles of a rational function* $R(z)$ *of degree n, those regions contain all the critical points of* $R(z)$*, and the former region contains precisely* $n - 1$ *critical points.*

COROLLARY 2. *If the two closed sectors* $\alpha \leqq \arg z \leqq \beta(<\alpha + \pi)$ *and* $\alpha + \pi \leqq \arg z \leqq \beta + \pi$ *contain respectively the zeros and poles of a rational function* $R(z)$*, then those sectors contain all critical points of* $R(z)$.

Under the hypothesis of Corollary 2, the origin and point at infinity may be zeros or poles of $R(z)$. Nevertheless, if P is an arbitrary point not in either of the given sectors, the line OP separates all zeros of $R(z)$ not on OP from all poles of $R(z)$ not on OP; except in the trivial case that $R(z)$ is a positive or negative power of z, $R(z)$ has other zeros and poles than those on OP, so in any case P is not a critical point of $R(z)$.

§4.2.2. Locus of critical points. In §1.5 and elsewhere we have determined the actual *locus* of critical points of a variable polynomial defined by prescribed loci

* This formula may also be proved from the fact that when two distinct zeros of $R_1(z)$ of arbitrary multiplicities coalesce, an additional simple critical point coalesces with them; thus when two distinct poles of $R(z)$ of arbitrary multiplicities coalesce, a simple critical point coalesces with them, and when q distinct poles of arbitrary multiplicities coalesce, critical points of total multiplicity $q - 1$ coalesce with them.

for sets of its zeros. It is desirable to study the corresponding facts for the present configuration:

THEOREM 1. *Let the disjoint circular regions C_1 and C_2 be the respective loci of the zeros and poles of a variable rational function of degree greater than unity. Then the locus of the critical points of $R(z)$ consists of C_1 and the interior points of C_2.*

It follows from Bôcher's Theorem that all critical points lie in C_1 plus C_2. An arbitrary point of C_1 may be a multiple zero and thus a critical point of $R(z)$. We now prove the by no means obvious fact that an arbitrary interior point of C_2 may be a critical point of $R(z)$.* It is sufficient, by replacing the given rational function by its reciprocal, to prove as we proceed to do that *an arbitrary interior point of C_1 can be a critical point of $R(z)$ not a multiple zero of $R(z)$*; this result has some intrinsic interest. Let z_0 be an arbitrary finite interior point of C_1, let one zero of $R(z)$ be fixed at z_0, and let another zero $z = \alpha$ be variable in the neighborhood of z_0, while all the remaining zeros and all the poles are fixed at points in their loci remote from z_0. We may write

$$R(z) \equiv (z - z_0)(z - \alpha)\Phi(z), \qquad \Phi(z_0) \neq 0.$$

The critical points of $R(z)$ are the zeros in z of

$$(1) \qquad R'(z) \equiv (z - z_0)(z - \alpha)\Phi'(z) + (2z - z_0 - \alpha)\Phi(z) = 0,$$

an equation which is satisfied by $z = z_0$ when $\alpha = z_0$. Equation (1) can be solved for α:

$$(2) \qquad \alpha - z_0 = \frac{(z - z_0)^2\Phi'(z) + 2(z - z_0)\Phi(z)}{(z - z_0)\Phi'(z) + \Phi(z)}, \qquad \left.\frac{d\alpha}{dz}\right|_{z=z_0} = 2,$$

so α is defined as an analytic function of z. The second member of (2) is a function of z analytic throughout the neighborhood of the point $z = z_0$, so (2) also defines the critical point z as an analytic function of α, and we have $z = z_0$ when $\alpha = z_0$. By the implicit function theorem, as α moves over a certain neighborhood of z_0, the critical point z entirely covers a circular neighborhood $N(z_0)$ of z_0. For small changes of z_0, and for suitably restricted z near z_0, the second member of (2) is uniformly bounded, so by the implicit function theorem the radius of $N(z_0)$ for small changes in z_0 has a positive lower bound independent of z_0. It follows from (1) that z coincides neither with α nor z_0 unless we have $z = \alpha = z_0$. We have shown that for suitably restricted but variable z_0, a neighborhood $N(z_0)$ of fixed size is completely covered except for z_0 itself by critical points of $R(z)$ which are not zeros of $R(z)$. Consequently an arbitrary

* It is possible to revise our definition of critical point so as to admit a multiple pole as a critical point, and if that is done Theorem 1 (but not the later italicized statement) becomes a triviality The enumeration of critical points is simpler under the suggested revised definition, which we consider less discriminating and do not adopt, but under either definition is readily converted to agree with the other.

finite interior point of C_1 can be a critical point of $R(z)$ not a multiple zero of $R(z)$; a linear transformation enables us to include the case of the point at infinity.

A second proof is readily given of the fact that an arbitrary point z interior to C_1 can be a critical point of $R(z)$ not a multiple zero of $R(z)$, by choosing a point z_3 in C_2, and then by choosing z_1 and z_2 in C_1 near z in such a way that we have $(z_1, z_2, z_3, z) = n$, where n is the degree of $R(z)$; it follows from §4.1.2 Corollary 2 to Theorem 2 that z is a critical point of the function $R(z)$ having zeros of multiplicities unity and $n - 1$ in the points z_1 and z_2, and a pole of multiplicity n in z_3.

No boundary point z_0 of C_2 can be a critical point of $R(z)$. For either z_0 is a pole of a fixed $R(z)$ and hence not a critical point, or a new circular region C_2' neighboring to C_2 is disjoint from C_1 and contains all the poles of $R(z)$ but does not contain z_0; then C_1 and C_2' contain all critical points of $R(z)$.

§4.2.3. Specializations. In Bôcher's Theorem the closed regions T_1 and T_2 may degenerate to arcs of a single circle:

THEOREM 2. *Let all the zeros of a rational function $R(z)$ lie on a closed arc T_1 of a circle C, and all poles of $R(z)$ lie on a closed arc T_2 of C disjoint from T_1. Then all critical points of $R(z)$ lie on T_1 and T_2. On any open arc of C bounded by two zeros or by two poles of $R(z)$ and containing no zero or pole of $R(z)$ lies precisely one critical point of $R(z)$.*

At any point of C the force (§4.1.2 Corollary to Theorem 3) is directed* along C. On an open arc of the kind defined, the force reverses sense as z moves from one end to the other, so the arc contains at least one critical point. Such critical points, assuming but one in each designated interval, together with the multiple zeros of $R(z)$ account for all (§4.1.1 Corollary to Theorem 1, or Bôcher's Theorem) critical points of $R(z)$. The proof is complete.

The conclusion that any open arc of C bounded by two zeros or by two poles of $R(z)$ and containing no zero or pole of $R(z)$ contains at least one critical point is valid if all zeros and poles of $R(z)$ merely lie on C without lying on disjoint closed arcs T_1 and T_2; however, it is not true that such an arc always contains precisely one critical point, as is shown by the illustration $R(z) = (z^2 - 1)/(z^2 - 4)^5$, which has three critical points $z = -\frac{1}{2}, 0, \frac{1}{2}$, in the interval $-1 < z < +1$.

Since Bôcher's Theorem refers to the extended plane, all zeros and poles finite and infinite must be taken into account. A somewhat less general theorem with a less restricted hypothesis [Porter, 1916, Marden, 1930a; Walsh, 1939] involves only the finite zeros and poles:

* Thanks to a convention already made (§4.1.2) the point at infinity is not exceptional here. If desired, however, the configuration can be interpreted on the sphere, or can be transformed in the plane if necessary so that C becomes a proper circle. A similar comment applies frequently in the sequel, but will not be repeated explicitly

THEOREM 3. *Let $R(z)$ be a rational function with the finite zeros $\alpha_1, \alpha_2, \cdots, \alpha_m$ and the finite poles $\beta_1, \beta_2, \cdots, \beta_n$. If a line L separates all α_j from all β_k, then no finite critical point of $R(z)$ lies on L. Consequently, if such lines exist, then all finite critical points of $R(z)$ lie in two unbounded closed convex point sets S_1 and S_2 which are separated by every L and which contain respectively the points α_j and the points β_k.*

A line L' which separates all zeros of $R(z)$ not on L' from all poles of $R(z)$ not on L', where at least one such zero or pole exists, passes through no finite critical point of $R(z)$ other than a multiple zero of $R(z)$.

Theorem 3 is an immediate consequence of Bôcher's Theorem. The convexity of S_1 (and similarly of S_2) follows from the fact that S_1 is defined as the set common to the closed half-planes, each of which contains all the α_j and is bounded by a line L.

Let K_1 and K_2 denote the smallest convex sets containing respectively the α_j and the β_k, and let K denote the complement of the set $S_1 + S_2$. At each point P of the plane not in K_1 or K_2, the set K_j subtends a closed sector A_j less than π. The set S_1 consists of K_1 plus all points P such that the sectors A_1 and A_2 intersect, with points of K_1 lying between P and points of K_2 on half-lines terminating in P. The set S_2 consists of K_2 plus all points P such that the sectors A_1 and A_2 intersect, with points of K_2 lying between P and points of K_1 on half-lines terminating in P. The set K consists of all points P such that A_1 and A_2 are disjoint except for the vertex P. If B_1 and B_2 are points of S_1 and S_2 respectively, every line L separates B_1 and B_2 and hence cuts the line B_1B_2 on the finite segment B_1B_2. Disjoint infinite segments of the line B_1B_2 contain B_1 and B_2 and belong to S_1 and S_2 respectively.

The set K may be called projective-convex, in the sense that *if P_1 and P_2 are arbitrary points of K, then (as we shall prove) either the finite line segment P_1P_2 or the infinite line segment $P_1 \infty P_2$ belongs to K.* If a point say of S_1 lies on P_1P_2, then no point of S_1 lies on either $P_1 \infty$ or $P_2 \infty$ (i.e. the infinite segments of the line P_1P_2 not containing P_2 or P_1 respectively), by the convexity of S_1; no point of S_2 lies on $P_1 \infty$ [$P_2 \infty$], for P_2 [P_1] lies in K; thus $P_1 \infty P_2$ contains no point of S_1 or S_2.

In Theorem 3, if a line L exists, there can be determined disjoint circular regions C_1 and C_2 containing respectively all the zeros and all the poles of $R(z)$; if the point at infinity is a zero [pole] of $R(z)$, we choose C_1 [C_2] as a half-plane and C_2 [C_1] as the closed interior of a circle; if the point at infinity is not a zero or pole of $R(z)$, we choose both C_1 and C_2 as the closed interiors of circles. It is to be noted that the region C_1 contains no point of S_2, and that C_2 contains no point of S_1. All the finite critical points of $R(z)$ in C_1 lie in S_1 and all those in C_2 lie in S_2. Mere enumeration of the finite critical points, commencing with a finite region C_1 or C_2 with separate treatments of the cases $m > n$, $m = n$, $m < n$, by use of Bôcher's Theorem and §4.1.1 Corollary to Theorem 1, then yields

COROLLARY 1. *Under the conditions of Theorem 3, the set S_1 contains precisely $m - 1$ or m finite zeros of $R'(z)$ according as $m \geqq n$ or $m < n$. If q denotes the number of distinct poles, finite or infinite, the set S_2 contains precisely $q - 1$ finite zeros of $R'(z)$.*

We add the remark that both Theorem 3 and Corollary 1 can be proved by studying the variation of

$$\arg\left[\sum_{k=1}^{m}\frac{1}{\bar{z}-\bar{\alpha}_k}-\sum_{k=1}^{n}\frac{1}{\bar{z}-\bar{\beta}_k}\right]$$

as z traces the boundaries of S_1 and S_2 (or nearby curves) cut off by circular arcs of large radius. Here §1.5 Lemma 1 is useful.

As a special case of both Theorem 2 and Theorem 3 we have

COROLLARY 2. *Let all the zeros and poles of a rational function $R(z)$ lie on disjoint closed segments S_1 and S_2 respectively of a line; these segments may have infinity as end-points but not as interior points. Then all critical points of $R(z)$ lie on S_1 and S_2. Any open finite or infinite subsegment of S_1 or S_2 bounded by two zeros or two poles of $R(z)$ and containing no zero or pole in its interior contains precisely one critical point of $R(z)$. If $R(z)$ has precisely m finite zeros, n finite poles, and q distinct poles, the set S_1 contains precisely $m - 1$ or m finite zeros of $R'(z)$ according as $m \geqq n$ or $m < n$; the set S_2 contains precisely $q - 1$ finite zeros of $R'(z)$. If S_1 and S_2 are finite, they contain all critical points of $R(z)$.*

§4.2.4. Circular regions as loci. Theorem 3 is concerned with the sets S_1 and S_2 which extend to infinity in the two parts of a double infinite sector, and suggests the desirability of a further result involving point sets that are bounded. We prove an analogue [Walsh, 1921] of the results of §1.5:

LEMMA. *Let the closed interiors of $C_1: |z - \alpha_1| = r_1$ and $C_2: |z - \alpha_2| = r_2$ be the respective loci of the points z_1 and z_2. If we have $m_1 > 0 > m_2$, $m_1 + m_2 \gtrless 0$, then the locus of the point*

$$z = (m_2 z_1 + m_1 z_2)/(m_1 + m_2) \tag{3}$$

is the closed interior of the circle

$$C: \left| z - \frac{m_2\alpha_1 + m_1\alpha_2}{m_1 + m_2} \right| = \frac{m_1 r_2 - m_2 r_1}{|m_1 + m_2|} = r.$$

If z_1 and z_2 are given, we have

$$z - \frac{m_2\alpha_1 + m_1\alpha_2}{m_1 + m_2} = \frac{m_2(z_1 - \alpha_1)}{m_1 + m_2} + \frac{m_1(z_2 - \alpha_2)}{m_1 + m_2},$$

and this second member is not greater than r. If z is given in the closed interior of C, we need merely set

$$\frac{m_2(z_1 - \alpha_1)}{m_1 + m_2} = \left(z - \frac{m_2\alpha_1 + m_1\alpha_2}{m_1 + m_2}\right)(-m_2r_1) \Big/ |m_1 + m_2| r,$$

$$\frac{m_1(z_2 - \alpha_2)}{m_1 + m_2} = \left(z - \frac{m_2\alpha_1 + m_1\alpha_2}{m_1 + m_2}\right) m_1r_2 \Big/ |m_1 + m_2| r,$$

in order to define z_1 and z_2 in their proper loci and related to z by equation (3).

The point $(r_2\alpha_1 + r_1\alpha_2)/(r_1 + r_2)$ is an internal center of similitude for the pair of circles (C_1, C_2), and is an external center of similitude for the pair (C_1, C) or (C_2, C) according as we have $m_1 + m_2 < 0$ or $m_1 + m_2 > 0$.

THEOREM 4. *Let $R(z)$ be a rational function with precisely m finite zeros, which lie in the closed interior of C_1:$|z - \alpha_1| = r_1$, and with precisely n finite poles, which lie in the closed interior of C_2:$|z - \alpha_2| = r_2$. Then if we have $m \gtrless n$, all finite zeros of $R'(z)$ lie in the closed interiors of the circles C_1, C_2, and*

$$C: \left| z - \frac{m\alpha_2 - n\alpha_1}{m - n} \right| = \frac{mr_2 + nr_1}{|m - n|}.$$

If z is a finite zero of $R'(z)$ exterior to C_1 and C_2, we have (§1.5.1 Lemma 1) for suitably chosen α and β in the closed interiors of C_1 and C_2 respectively

$$(4) \qquad \frac{m}{z - \alpha} - \frac{n}{z - \beta} = 0, \qquad z = \frac{m\beta - n\alpha}{m - n},$$

so by the Lemma z lies in the closed interior of C. If the circles C_1, C_2, and C are mutually exterior, and if $R(z)$ has precisely q distinct finite poles, it follows by the method of continuity that those circles contain respectively $m - 1$, $q - 1$, and 1 critical points of $R(z)$.

If the closed interiors of C_1 and C_2 are the respective *loci* of the given finite zeros and poles, then (compare Theorem 1) the locus of the critical points of $R(z)$ consists of the closed interior of C_1 if $m > 1$, the interior of C_2 if $q > 1$, and the closed interior of C, provided the assigned loci are disjoint; compare the discussion in §4.4.3.

Let us modify the hypothesis of Theorem 4 so that we now have $m = n$. If the circles C_1 and C_2 are mutually exterior, the closed interiors of those two circles contain all critical points, by Bôcher's Theorem. If the circles C_1 and C_2 are externally tangent, the proof of the first of equations (4) is valid for z exterior to C_1 and C_2, and the points α and β cannot both coincide with the point of tangency;* thus z cannot be a position of equilibrium, and the closed interiors of C_1 and C_2 contain all critical points. If we have $m = n = 1$, the function

* We note an important difference between rational functions and the jacobian of two binary forms, for in the latter case the forms may be identical here, if the closed interiors of C_1 and C_2 are the respective loci of the zeros of the two forms, and then the jacobian vanishes identically

$R(z)$ has no critical points, no matter what C_1 and C_2 may be. If we have $m = n > 1$, and if the circles C_1 and C_2 have interior points in common, the locus of critical points consists of the entire plane; for instance let z be arbitrary, let β be an interior point of both C_1 and C_2 distinct from z, and let α' and α'' be chosen near β interior to C_1 so that we have

$$\frac{1}{z - \alpha'} + \frac{m - 1}{z - \alpha''} - \frac{m}{z - \beta} = 0;$$

the possibility of this choice becomes clear if we make the transformation $W = 1/(z - Z)$, for the image of $Z = \beta$ need merely divide the segment joining the images of $Z = \alpha'$ and $Z = \alpha''$ in the ratio $(m - 1)$: 1. Then z is a critical point of the corresponding rational function of degree m, for which the simple zero α', the $(m - 1)$-fold zero α'', and the m-fold pole β lie in their proper loci.

§4.2.5. Converse. The fundamental nature of Bôcher's Theorem may be judged by the following converse with reference to the nature of the regions involved:

THEOREM 5. *Let T_1 and T_2 be mutually disjoint closed regions, with the property that whenever the zeros and poles of a rational function $R(z)$ lie in T_1 and T_2 respectively, all critical points of $R(z)$ lie in $T_1 + T_2$. Then through every point z not in $T_1 + T_2$ can be drawn a circle separating T_1 and T_2.*

Suppose a point z_0 not in $T_1 + T_2$ lies on no such circle; we shall reach a contradiction. Transform z_0 to infinity by a linear transformation, then no straight line separates the images (closed and bounded) T_1' and T_2' of T_1 and T_2, and no straight line separates the convex hulls of T_1' and T_2'. If two arbitrary bounded closed convex sets are disjoint, there exists a line (for instance the perpendicular bisector of the shortest segment joining them) that separates the sets. Consequently the convex hulls of T_1' and T_2' are not disjoint, and since T_1' and T_2' are disjoint it follows that either a chord of T_1' contains a point of T_2' or a chord of T_2' contains a point of T_1'; for definiteness we assume the former; the subsequent reasoning requires only obvious modification in the latter case. Let α_1' and α_2' be the end-points of the chord of T_1', hence in T_1', and β_1' a point of T_2' on that chord. We introduce the definition $(\alpha_1', \alpha_2', \beta_1', \infty) = \lambda$, and we have $\lambda > 1$. If $\alpha_1, \alpha_2, \beta_1$ denote the images of $\alpha_1', \alpha_2', \beta_1'$ under the inverse of the former linear transformation, the points α_1 and α_2 are in T_1 and β_1 is in T_2; we have $(\alpha_1, \alpha_2, \beta_1, z_0) = \lambda$. Choose sequences of positive integers k_j and n_j, $j = 1, 2, \cdots$, such that the quotients n_j/k_j are all distinct and approach λ. The rational function $R_j(z)$ of degree n_j with k_j zeros at α_1, $n_j - k_j$ zeros at α_2, and n_j poles at β_1, has (§4.1.2 Corollary 2 to Theorem 2) a critical point at z_j, where we have $(\alpha_1, \alpha_2, \beta_1, z_j) = n_j/k_j$; the z_j are all distinct and z_j approaches z_0 as j becomes infinite. From the assumed properties of T_1 and T_2 it follows that each z_j belongs to $T_1 + T_2$, and hence their limit z_0 belongs to $T_1 + T_2$, contrary to hypothesis. This contradiction completes the proof.

The property of T_1 and T_2 required in Theorem 5 implies that T_1 and T_2 are the precise point sets separated from each other by a certain family of circles. Indeed, let $\{C\}$ be the (non-vacuous) family of all circles separating T_1 and T_2. If a point P say not in T_1 and obviously not in T_2 is separated from T_2 by each circle C, we shall reach a contradiction. By Theorem 5 there exists a circle through P which separates T_1 and T_2; a suitable neighboring circle separates $P + T_2$ from T_1; this circle belongs to the family $\{C\}$ and contradicts our hypothesis on P. Thus *T_1 and T_2 are precisely the point sets separated by a certain family of circles.* Conversely, it follows from Bôcher's Theorem that any two such point sets T_1 and T_2 satisfy the hypothesis of Theorem 5.

It is essential both in Bôcher's Theorem and in Theorem 5 that T_1 and T_2 have no interior point in common, for if β is an interior point both of T_1 and T_2 and if z is an arbitrary point of the plane distinct from β there exist the points α_1 and α_2 in T_1 near β such that $(\alpha_1, \alpha_2, \beta, z) = 2$; this is easily proved as in the latter part of §4.2.4 if z is transformed to infinity. The rational function $R(z)$ of degree 2 with simple zeros in α_1 and α_2 and a double pole in β has a critical point in z.

§4.3. Concentric circular regions as loci. If we consider the field of force defined in §4.1.1 Theorem 1, it is intuitively obvious that there can be no position of equilibrium very near any of the fixed particles. More generally, there can be no position of equilibrium very near and outside of a circle containing a number of fixed particles, all repelling or all attracting, if the other particles are sufficiently remote. Refinement of this intuitive suggestion will yield [Walsh, 1918]:

THEOREM 1. *Let $R(z)$ be a rational function of degree n, let k zeros of $R(z)$ lie in the circular region $C_1: |z| \leqq a$, let the remaining $n - k$ zeros of $R(z)$ lie in the circular region $C_2: |z| \geqq b\ (>a)$, and let the poles of $R(z)$ lie in the circular region $C_3: |z| \geqq c\ (>a)$. Denote by C_0 the circle*

$$(1) \qquad |z| = r_0 = \frac{kbc - nab - (n-k)ac}{(n-k)b + nc - ka},$$

under the assumption $r_0 > 0$.

If we have $a < r_0 \leqq c$, the annulus $a < |z| < r_0$ contains no critical point of $R(z)$. If we have $r_0 > c$, then the annulus $a < |z| < c$ contains no critical point of $R(z)$. In either of these cases, the region C_1 contains precisely $k - 1$ zeros of $R'(z)$.

We have stated Theorem 1 for the case that C_1 contains k *zeros* of $R(z)$; the roles of zeros and poles can be interchanged, except that now the region C_0 contains a number of zeros of $R'(z)$ equal to the number of distinct poles of $R(z)$ in C_0, diminished by unity. *In numerous later results of the present work on critical points of rational functions, the roles of zeros and poles can be interchanged without altering the regions proved to contain all (or no) critical points;* we mention the *fact here without repeating it on each appropriate occasion; the number of critical points in the various regions under the modified hypothesis can be obtained by methods already developed.*

Two lemmas are convenient for reference:

LEMMA 1. *Let the point Q lie exterior to the circle C whose center is O, and let a variable unit positive particle P be required to lie in the closed interior of C. The maximum force both in magnitude and in component along OQ is exerted at Q by P when P is nearest Q, and the minimum both in magnitude and component along OQ is exerted when P is farthest from Q.*

LEMMA 2. *Let the point Q lie interior to the circle C with center O, and let a variable unit positive particle P be required to lie in the closed exterior of C. The maximum force both in magnitude and in component in the sensed direction QO is exerted at Q by P when P is on C nearest Q, and the maximum component in the sensed direction OQ is exerted at Q by P when P is on C farthest from Q.*

Lemmas 1 and 2, like §1.5 Lemma 1, are proved by inversion in the point Q, noting that the force at Q due to the particle P is in magnitude, direction, and sense the vector $P'Q$ where P' is the inverse of P. We proceed to the proof of Theorem 1, using the field of force of §4.1.1 Theorem 1.

Consider the force at a point Q between C_1 and C_2, and between C_1 and C_3, and denote OQ by r. The component in the direction and sense OQ of the force due to the k particles in C_1 is not less than $k/(a + r)$. The component in the direction and sense QO of the force due to the $n - k$ particles in C_2 is not greater than $(n - k)/(b - r)$, and that of the force due to the n particles in C_3 is not greater than $n/(c + r)$. If Q is a position of equilibrium we must have

$$\frac{k}{a+r} \leqq \frac{n-k}{b-r} + \frac{n}{c+r}. \tag{2}$$

$$r_0 \leqq r. \tag{3}$$

The number r_0 defined by (1) need not be positive, but in any case is not greater than b, and is equal to b when and only when $k = n$, in which case the region C_2 serves no purpose and Theorem 1 is §4.2.1 Corollary 1 to Bôcher's Theorem.

We have thus shown that whenever the circle C_0 is exterior to the circle C_1, the annular region $a < |r| < r_1 = \min(r_0, c)$ contains no critical points of $R(z)$. The method of continuity then shows that C_1 contains precisely $k - 1$ critical points of $R(z)$, for if the zeros of $R(z)$ are varied and those in C_1 are caused to coincide at $z = 0$, the method of derivation of (2) shows that no critical point lies interior to C_1 except perhaps at $z = 0$. Theorem 1 is established.

It may be noted that in the case $k > 1$ Theorem 1 cannot be improved as to regions for critical points. For any point of the region C_1 can be a multiple zero of $R(z)$ and hence a critical point. An arbitrary interior point of the region C_3 can be a critical point; compare the proof of §4.2.2 Theorem 1. Inequalities (2) and (3) can become *equalities* for suitable choice of all the particles in their assigned regions; when we temporarily allow first c and then b to become infinite we note that r_0 also becomes infinite, so an arbitrary point of the original annulus $r_0 \leqq |z| < c$ can be a critical point. Thus the regions specified in Theorem 1 cannot be improved. However, in the case $r_0 > c$ it follows from a slight extension of Lemma 2 and from (2) and (3) that no point of the circle C_3 can be a position of equilibrium; no such point can be a multiple zero of $R(z)$, for we have $b \geqq r_0 > c$; thus no point of the circle C_3 can be a critical point. In the case $r_0 > a$, which is necessary for the theorem to be effective, no point of the circle C_1 is a critical point unless it is a multiple zero of $R(z)$.

In the case $k = 1$, and where we assume of course $r_0 > a$, we have shown that the region C_1 contains no critical point, so the region $|z| < r_1$ contains no critical point of $R(z)$.

Theorem 1 can be widely extended, by considering an arbitrary number of circular regions bounded by circles with center O to be the respective loci of prescribed numbers of zeros and poles of $R(z)$. The same method applies, using Lemmas 1 and 2, but involves for the radius (or radii) r_0 equations of higher degree than the first. However, we have here a method for the approximate determination of critical points of rational functions of arbitrary degree by means of the solution of equations of lower degree with real zeros. This method was applied by Montel [1923] in the study of the moduli of the zeros of polynomials, making essential use of Theorem 1.

Theorem 1 is related to several of our previous results. If we set $c = \infty$, we have the case of §3.1.1 Theorem 1 for concentric circles. If we set $a = 0$ and make the substitution $z' = 1/z$, we have the case of §4.2.4 Theorem 4 for concentric circles. The case $a = 0, b = \infty$ is of interest:

COROLLARY 1. *Let $R(z)$ be a rational function of degree n, let the origin O be a zero of order at least $n/2$, let no other finite point be a zero, and let all poles lie in the closed exterior of the circle C whose center is O. Then all critical points of $R(z)$ other than O lie in the closed exterior of C.*

Two applications of Corollary 1 give

COROLLARY 2. *Let $R(z)$ be a rational function of degree $2n$, let $z = 0$ and $z = \infty$ be zeros of order n, and let a closed annulus bounded by circles whose center is the origin contain all poles of $R(z)$. Then this annulus contains all critical points of $R(z)$ except those at $z = 0$ and $z = \infty$.*

The special case $a = 0, b = c$ is also of interest:

COROLLARY 3. *Let $R(z)$ be a rational function of degree n, let $z = 0$ be a zero of order k, and let all other zeros and all the poles of $R(z)$ lie on or exterior to the circle $|z| = b$. Then no critical points of $R(z)$ lie in the region*

$$0 < |z| < kb/(2n - k).$$

Here for an arbitrary zero of $R(z)$ we have a deleted neighborhood which is free from critical points.

Theorem 1, when interpreted in form invariant under linear transformation, deals with three given circular regions C_1, C_2, C_3 bounded by coaxal circles which contain respectively prescribed numbers of zeros or poles of a given rational function. A fourth circle C_0 of the same coaxal family is then determined, which enables us in certain cases to assert that all critical points of the given rational function lie in C_1, C_2, C_3, and a circular region bounded by C_0. We do not now formulate this result in invariant form, for we shall extend (§4.4) the entire theorem to the case that C_1, C_2, and C_3 need not be coaxal. One remark is significant in this connection, however.

It turns out that when (2) is an equality with $r = r_0$ it can be written in the form

$$(-a, b, -c, r_0) = n/k; \tag{4}$$

the points $-a, b, -c, r_0$ lie on the respective circles C_1, C_2, C_3, C_0 in such a way that they lie on a "circle" (in Theorem 1 a line) L through the null circles O and O' of the coaxal family and the points lie alternately on one and the other of the two arcs of L both bounded by O and O'. When L rotates about O, the cross-ratio of suitably chosen intersections with C_1, C_2, C_3, C_0 is unchanged. Thus the circle C_0 can be readily defined in terms of the cross-ratio (4).

§4.4. The cross-ratio theorem. A number of our previous results (§§1.5, 3.1, 3.3, 4.2.4, 4.3), in conjunction with §4.1.2 Corollary 2 to Theorem 2, suggest the use of three given circular regions as loci of respective points representing zeros or poles of a given variable rational function, and the determination of a fourth region which is the locus of a fourth point defined by a real constant cross-ratio with those three points. We shall thus have [Walsh, 1921] not merely a unification of various other results, but a theorem of intrinsic interest.

§4.4.1. Geometric locus. The specific geometric loci just referred to are qualitatively included in

THEOREM 1. *If the given circular regions C_1, C_2, C_3 are the respective loci of the points z_1, z_2, z_3, then the locus C_4 of the point z_4 defined by the real constant cross-ratio*

$$(z_1, z_2, z_3, z_4) = \lambda \tag{1}$$

is also a circular region.

The degenerate cases $\lambda = 0, 1, \infty$ are included merely for completeness. For instance if we have $\lambda = 0$, then by the equation

$$\frac{(z_1 - z_2)(z_3 - z_4)}{(z_2 - z_3)(z_4 - z_1)} = 0 \tag{2}$$

it follows that either the regions C_1 and C_2 have a common point, in which case we may choose $z_1 = z_2$, equation (2) is satisfied for arbitrary z_4, and the locus of z_4 is the entire plane, or C_1 and C_2 have no common point, equation (2) is equivalent to the equation $z_3 - z_4 = 0$, and the locus of z_4 is the region C_3. The cases $\lambda = 1$ and $\lambda = \infty$ are similarly treated; henceforth we suppose λ to be different from 0, 1, and ∞. It follows that no pair of the points z_j coincide unless a third z_k coincides with them.

We introduce the convention that when three of the points z_1, z_2, z_3, z_4 coincide, the cross-ratio of the four points (which is properly not defined) may be considered to have any value, and in particular the value λ. Consequently, if the three given regions have a common point, the points z_1, z_2, z_3 can be allowed to coincide there, and (1) is considered to be satisfied whatever z_4 may be; the locus of z_4 is the entire plane. Whenever one of the given loci, say C_1, is the entire plane, either C_2 and C_3 have a common point and thus C_1, C_2, and C_3 have a common point, or we may choose z_2 and z_3 distinct from each other and fixed, then equation (1) defines a map of the entire z_1-plane onto the entire z_4-plane, so the locus of z_4 is the entire plane. Henceforth we assume that the given regions C_1, C_2, C_3 do not all have a common point, and that none of those regions consists of the entire plane. In any non-trivial case, at least one of the given regions, say C_1, does not degenerate to a point, and the locus of z_4 contains the image of the region C_1 under the transformation (1), where z_2 and z_3 are fixed in their proper loci and z_4 is considered a function of z_1. The locus of z_4 may include the entire plane, as indicated for instance in §3.1.1; on the other hand, if the given regions are relatively small, it is clear by the continuity of the first member of (1) that the locus of z_4 is small. That the locus C_4 of z_4 in Theorem 1 is a closed set follows from the compactness and closure of the regions C_1, C_2, C_3. If C_4 is the entire plane, the conclusion of Theorem 1 is established; in any other case the set C_4 has boundary points, and boundary points of the various loci have a special relation to each other:

LEMMA. *If z_4 is on the boundary of C_4 but does not coincide with two of the three points z_1, z_2, z_3 satisfying* (1), *then z_1, z_2, z_3 lie on the circles C_1, C_2, C_3; the circle C through z_1, z_2, z_3, z_4 cuts the circles C_1, C_2, C_3 at those points at angles equal in magnitude; more explicitly, equation* (1) *with z_1 and z_4 fixed maps the circular region C_2 onto a circular region C_3' having no interior point in common with the region C_3, the boundaries of C_3 and C_3' are either coincident, or tangent at z_3; a similar remark holds for the other pairs (C_1, C_2) and (C_1, C_3).*

In the statement and proof we assume none of the regions C_1, C_2, C_3 degenerate; the modifications for degenerate regions are obvious. Any point common to two of the given regions is a possible point at which two of the points z_1, z_2, z_3 coincide, and hence by our convention is a possible position for z_4. Let z_4^0 lie on the boundary of C_4, suppose the points z_1^0, z_2^0, z_3^0 lie in their proper loci with $(z_1^0, z_2^0, z_3^0, z_4^0) = \lambda$, and suppose z_4^0 does not coincide with two of the three points z_1^0, z_2^0, z_3^0; the three latter points do not all coincide, for C_1, C_2, C_3 have no common point. Then z_1^0, z_2^0, z_3^0 are all distinct; they all lie on the boundaries of their proper loci. For the relation $(z_1, z_2^0, z_3^0, z_4) = \lambda$ defines a non-degenerate linear relation between z_1 and z_4; this relation maps a neighborhood of z_1^0 onto a neighborhood of z_4^0, so if z_1^0 is not on the boundary of the closed region C_1, z_4^0 does not lie on the boundary of the locus C_4; this method shows not merely that z_1^0 lies on the boundary of the locus C_1, but that z_2^0 and z_3^0 lie on the boundaries of the loci C_2 and C_3 respectively. Set now $z_1 = z_1^0$, $z_4 = z_4^0$, so that (1) defines a non-degenerate linear transformation between z_2 and z_3, which maps the circle C_2 into a circle C_3' and maps the circular region C_2 onto a circular region C_3'; the two circular regions C_3 and C_3' can have only the single point z_3^0 and perhaps other boundary points in common; otherwise the transformation between z_2 and z_3 maps the locus C_2 onto a circular region C_3' which has interior points in common with C_3, and we have the non-trivial relation

$$(z_1^0, z_2', z_3', z_4^0) = \lambda$$

where z_1^0, z_2', z_3' are all distinct and lie in their proper loci with z_2' and z_3' *interior points* of their loci, which is known to be impossible.

Equation (1) with two of the variables z_j fixed defines a linear relation between the other two of those variables, and for given λ the latter two always move in the same sense or always move in opposite senses on C. For instance if in (1) we choose z_4 at infinity and z_1 constant we have $(\lambda - 1)\, dz_2 = \lambda\, dz_3$. Thus the relation* $(z_1^0, z_2, z_3, z_4^0) = \lambda$ maps an arc of C interior to the region C_2 terminating in z_2^0 onto an arc of C exterior to the region C_3 terminating in z_3^0, since z_2 and z_3 cannot be simultaneously interior points of their loci, so C cuts the directed circles C_2 and C_3 at z_2^0 and z_3^0 at equal or supplementary angles depending on whether the regions are interior or exterior to their bounding circles; here we define the angle between two circles as the angle between their half-tangents drawn at a point of intersection in the counter-clockwise sense of description. *When C is transformed into a straight line, the tangents to the images of C_1, C_2, C_3 at the images of z_1^0, z_2^0, z_3^0 are all parallel.* The Lemma is established. Incidentally, a slight extension of this proof shows that except at a conceivable vertex of the boundary of C_4, the circle C cuts all the curves C_1, C_2, C_3, C_4 at z_1, z_2, z_3, z_4 at the same or supplementary angles.

The circle C cuts each of the circles C_1, C_2, C_3 in two points, which for the moment we choose as distinct, and if z_1^0 is chosen on C and C_1 there are apparently four distinct ways in which z_2^0 and z_3^0 can be chosen on C_2 and C_3 so that C cuts

---

* When $z_4^0 = \infty$, this relation is a real similarity transformation between z_2 and z_3, with z_1 as fixed point and with z_2^0 and z_3^0 as corresponding points. Under this transformation the image of the region C_2 is a circular region C_3' such that z_1^0 is an internal or external center of similitude for the circles C_2 and C_3'. The points z_1^0, z_2^0, z_3^0 are collinear, and the tangents to the circles C_2 and C_3' at z_2^0 and z_3^0 are parallel.

C_1, C_2, C_3 at the same or supplementary angles. However, the requirement in the Lemma concerning the images of the given regions under transformation (1), with z_4 and one of the other z_j fixed, shows that choice of z_1^0 on C and C_1 determines z_2^0 and z_3^0 on their circles and on C uniquely, even if C is orthogonal to C_1, C_2, and C_3.

If z_4^0 is on the boundary of C_4 but coincides say with z_1^0 and z_3^0, a case not included in the Lemma, then the relation $(z_1^0, z_2, z_3^0, z_4^0) = \lambda$ does not effectively contain z_2, and z_2 may be chosen arbitrarily so that that equation remains satisfied. But if z_2 is chosen fixed and different from z_4^0, then the relation $(z_1, z_2, z_3, z_4^0) = \lambda$ contains effectively both z_1 and z_3, which vary as linear functions of each other, and the conclusion of the Lemma applies to those points. The circles C_1 and C_3 are identical or are tangent at $z_1^0 = z_3^0$; the point z_2 can be chosen on the circle C_2 so that the circle C is orthogonal to C_1, C_2, C_3.

We are now in a position to determine the boundary of C_4, and shall make essential use of the theorem [see for instance Coolidge, 1916]: *If three distinct circles are not all tangent at one point, the circles cutting them at equal angles* [*or cutting a particular one at angles supplementary to the angles cut on the other two*] *belong to a coaxal family.* We consider in order: Cases I, II, III, in which the circles C of the Lemma belong to a (hyperbolic, elliptic, parabolic) coaxal family consisting of the circles respectively through two distinct points, orthogonal to the circles through two distinct points, tangent to a given circle at a given point, as Case IV we shall discuss the situation in which C_1, C_2, C_3 are all tangent at a single point.

In Case I, transform to infinity one of the two points through which pass all circles of the coaxal family, and let the other point be transformed into O. The coaxal family consists of all lines through O, and except in limiting cases the circles C_1, C_2, C_3 are so situated that O is a center of similitude for any pair of them, the point O may or may not lie interior to the circles C_j. When the points z_1, z_2, z_3 are chosen on their respective circles to satisfy the conditions of the Lemma, including the condition that the lines tangent to C_1, C_2, C_3 at z_1, z_2, z_3 are all parallel, it is seen that as C continuously rotates about O those points vary under a continuous similarity transformation with O as center, and that z_4 as defined by (1) thus traces a circle C_4 (which may degenerate to O or the point at infinity) such that O is a center of similitude for any pair of the circles C_1, C_2, C_3, C_4. This situation is not essentially changed, as the reader may determine, if one of the given circles passes through O or the point at infinity (each of the given circles must then pass through either O or the point at infinity) or if one of the given circles is a null circle, necessarily O or the point at infinity.

In Case II, the coaxal family consists of all circles orthogonal to the circles through two given points; we transform one of these points to infinity and one to the origin O, so that the coaxal family consists of the circles with the common center O The circles C_1, C_2, C_3 are then equal circles whose centers are equidistant from O, and these circles may or may not contain O in their interiors. The points z_1, z_2, z_3 satisfying the conditions of the Lemma lie on their circles C_1, C_2, C_3 so that as C varies continuously these points move under a continuous similarity transformation with O as center. The point z_4 defined by (1) then traces a circle C_4 equal to C_1, C_2, C_3, and whose center is equidistant with their centers from O. By virtue of our requirement that the three given regions shall have no point in common, the given circles C_1, C_2, C_3 cannot all pass through O or the point at infinity.

In Case III, the circles C belong to a coaxal family of circles all tangent to each other at a single point, which we choose as the point at infinity; the circles C are then a family of parallel lines. The given circles C_1, C_2, C_3 are all equal with collinear centers, except as one of those circles is a null circle at infinity. As z_1, z_2, z_3 vary continuously on their circles and on C satisfying the conditions of the Lemma, they simply move under a continuous translation of the plane; the point z_4 traces either a circle equal to the proper circles of the set C_1, C_2, C_3 whose center is collinear with their centers, or z_4 lies constantly at infinity. The case that the non-degenerate circles among C_1, C_2, C_3 should all be lines is again excluded by our convention that the three given regions should have no common point.

There are two special non-trivial situations under Theorem 1 and the Lemma which should be noted, and which are not formally included under Cases I–IV. Even when the locus C_4 is not the entire plane, two of the given *regions* C_j may coincide, as in the situation of §1.5.1 Lemma 2; and two of the *circles* C_j may coincide, as in §3.1.1 Lemma 1, even when the corresponding regions do not coincide. In either of these situations, whether or not z_4 on the boundary of C_4 coincides with two of the points z_1, z_2, z_3, and thus whether or not *all* the points z_1, z_2, z_3 are uniquely determined by z_4, it follows from the discussion of the Lemma as already given that z_1, z_2, z_3 may be considered to lie on C_1, C_2, C_3 respectively in such manner that the circle C through the four points z_j cuts C_k at z_k orthogonally ($k = 1, 2, 3$). The circles C belong to a coaxal family under Case I (Cases II and III are excluded by the requirement that the three given regions should have no common point); the discussion essentially as given shows that whenever z_4 lies on the boundary of its locus z_4 lies on a particular circle C_4.

§4.4.2. Discussion of locus. Theorem 1 is now established, for we have shown that whenever the locus C_4 is not the entire plane, C_4 is bounded by points of and hence the whole of a single circle, traced by z_4 when the points z_1, z_2, z_3 trace their circles C_1, C_2, C_3 while satisfying the conditions of the Lemma. It is of interest to have a test as to when the locus C_4 is the entire plane:

COROLLARY. *If the locus C_4 of the point z_4 in Theorem 1 is not the entire plane, then there exists a circle S orthogonal to the circles C_1, C_2, C_3. Let A_1, A_2, A_3 be the closed arcs of S in the closed regions C_1, C_2, C_3 respectively, and let A_4 be the locus of z_4 as determined by* (1) *when the arcs A_1, A_2, A_3 are the loci of the points z_1, z_2, z_3. Then A_4 is the arc of S in the closed region C_4.*

This corollary furnishes a convenient test as to whether C_4 is the entire plane, and also a construction for the region C_4. We construct the (or a) circle S, and then merely determine the locus of z_4 for the points z_1, z_2, z_3 on S and in their proper loci; from the fact that in (1) the point z_4 moves monotonically on S with each of the points z_1, z_2, z_3, it follows that the locus of z_4 on S is an arc A_4 of S. Here the Lemma is again significant in determining the end-points of the arc A_4. The region C_4 is symmetric in S, and its intersection with S is the arc A_4.

To prove the corollary we return to the proof of Theorem 1 and note that in Cases I, II, III the circle S exists as a circle of the coaxal family to which the circles C belong, except perhaps in Case II when in the simplified figure O lies interior to all the circles C_k; we postpone the discussion of this possibility. If the points z_1, z_2, z_3 on S bear the relation to each other prescribed in the Lemma, so also do the points at the opposite intersections of S with the circles C_1, C_2, C_3, as do no other sets of points z_1, z_2, z_3 of S unless a z_4 on S coincides with two of the points z_1, z_2, z_3; this situation is only formally exceptional. The loci C_1, C_2, C_3 are symmetric in S, so the locus C_4 is symmetric in S, and if C_4 is not the entire plane its boundary cuts S in two distinct points z_4. These two points z_4 must correspond then to the two sets of points z_1, z_2, z_3 determined on S by the Lemma, and when z_1, z_2, z_3 move continuously on A_1, A_2, A_3 from one set to the other, the point z_4 remains in its locus (of the Corollary and of the theorem) and moves on S from one intersection of the boundary of C_4 with S to the other intersection. The Corollary is established, with the exception mentioned.

We investigate the situation in still further detail: *Case* II *cannot arise* if C_4 is not the entire plane. We choose for definiteness the notation of the given points and regions so that we have $\lambda > 1$; then on the circle C the points z_1 and z_2 are separated by z_3 and z_4; when z_3 and z_4 are fixed in (1), the points z_1 and z_2 move in opposite senses on C; when z_1 and z_2 are fixed in (1), the points z_3 and z_4 move in opposite senses on C; it follows by the reasoning of the Lemma that in Case II the regions C_1 and C_2 are both the closed interiors or both the closed exteriors of their bounding circles, and that if the region C_4 is not the entire plane, the regions C_3 and C_4 are one the closed interior and the other the closed exterior of their bounding circles. We exclude the case (included in Case I) already treated that C_1, C_2, and C_3 are not all distinct, so it follows by inspection of a circle C tangent to C_1, C_2, C_3 that the circle C_4 is distinct from C_1, C_2, C_3. The regions C_1, C_2, C_3 cannot all be the closed interiors of their bounding circles, for in this case it is clear by simultaneous transformation of the four points z, that the interior of the circle C_4 belongs to the locus C_4, so the locus C_4 is the entire plane. The regions C_1 and C_2 cannot be the closed interiors of their bounding circles with the region C_3 the closed exterior of its bounding circle, for then the region C_4 is the closed interior of its bounding circle and cannot contain the regions common to C_1 and C_3, and common to C_2 and C_3. The regions C_1 and C_2 cannot be the closed exteriors of their bounding circles with C_3 the closed interior of its bounding circle, for then the regions C_1, C_2, C_3 have a point (on C_3) in common, nor can the regions C_1, C_2, C_3 all be the closed exteriors of their bounding circles. The reasoning just given is valid whether the point O lies interior or exterior to the circles C_j, and shows that Case II cannot effectively arise.

The proof of the Corollary is now complete.

The same reasoning shows that *Case* III *cannot effectively arise if the circles* C_1, C_2, C_3 *are all proper circles.* But Case III may arise, for instance in §1.5.1 Lemma 2 with equal circles, if one of the given circles degenerates.

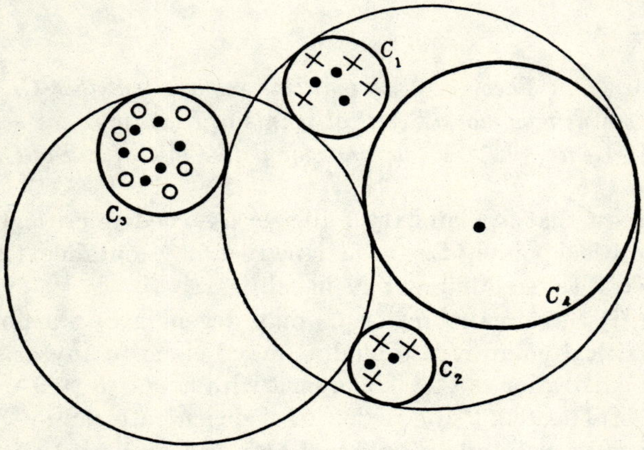

Fig. 10 illustrates §4.4.3 Theorem 2

§4.4.3. Locus of critical points. We now apply Theorem 1 to the study of critical points of rational functions:

THEOREM 2. *Let given circular regions* C_1, C_2, C_3 *be the respective loci of* k *zeros of a rational function* $R(z)$ *of degree* n, *the remaining* $n - k$ *zeros of* $R(z)$, *and all the poles of* $R(z)$, *where we suppose* C_3 *disjoint from* $C_1 + C_2$. *Denote by* C_4 *the locus of the point* z_4 *when* C_1, C_2, C_3 *are the respective loci of points* z_1, z_2, z_3, *with*

$$(z_1, z_2, z_3, z_4) = n/k. \tag{3}$$

Then the locus of critical points of $R(z)$ *consists of* C_1 *(if* $k > 1$*),* C_2 *(if* $n - k > 1$*), the interior points of* C_3, *and of* C_4. *If any of these closed regions* C_j *is disjoint from the other regions, it contains* $k - 1$, $n - k - 1$, $q - 1$, *or* 1 *critical points according as* $j = 1, 2, 3,$ *or* 4, *where* q *is the number of distinct poles of* $R(z)$. *If several of these regions are disjoint from the other regions, they contain together a corresponding total number of critical points.*

Theorem 2 is established by the use of §1.5.1 Lemma 1, §4.1.2 Corollary 2 to Theorem 2, and then by the method of continuity; details are omitted.

Bôcher's Theorem is the special case in which C_1 and C_2 coincide. Theorem 2 does not explicitly use the conclusion of Theorem 1, that C_4 is a circular region.

We make the restriction in Theorem 2 that C_3 shall be disjoint from $C_1 + C_2$, for the zeros and poles of $R(z)$ are assumed to vary independently throughout their loci, and strictly no zero can coincide with a pole. We can interpret the hypothesis broadly, however, admitting the possibility of common points of C_3 and $C_1 + C_2$, requiring that the regions C_1, C_2, C_3 be loci as prescribed, except that no point common to C_3 and $C_1 + C_2$ shall be simultaneously a zero and a pole of $R(z)$:

Corollary 1. *If in Theorem 2 we omit the requirement that C_3 and $C_1 + C_2$ be disjoint, it remains true that all critical points of $R(z)$ lie in the set consisting of C_1 (if $k > 1$), C_2 (if $n - k > 1$), the interior points of C_3, and of C_4.*

We do not assert that in Corollary 1 the set assigned to critical points is the precise *locus* of critical points,* for the assigned set may contain extraneous points as a consequence of several different eventualities.

For instance (i) the circular region C_3 may degenerate to a point, and that point is not a critical point even though it may belong to any or all of the loci C_1, C_2, C_4; an illustration is §4.3 Theorem 1 with $a = 0, b > 0, c = \infty$. Under the hypothesis of Theorem 2 the region C_3 if degenerate cannot belong to C_4, for here we have essentially the situation of §1.5. In Corollary 1 effective deletion of a single *interior* point of C_1 or C_2 by the presence of a degenerate $C_3: z = z_3^0$ interior to C_1 or C_2 does not modify the locus C_4 of Theorem 1 as used in Corollary 1 except by the deletion of that one point from C_4 in Corollary 1, for in Theorem 1 whenever z_1 or z_2 coincides with z_3, so also does z_4 unless the three points z_1, z_2, z_3 all coincide; in the latter case the locus C_4 in Theorem 1 is the entire plane.

Included in (i) is also the possibility that a degenerate C_1 (not in C_2) with $k = 1$ or C_2 (not in C_1) with $n - k = 1$ should lie interior to C_3; such a point cannot be a critical point.

Another difficulty in the determination of the locus of critical points in Corollary 1 is that (ii) there may be certain boundary points z_4 of the region C_4 corresponding to points z_1, z_2, z_3 in Theorem 1 which are not all distinct, a situation which when interpreted in terms of Corollary 1 would require a zero and a pole of $R(z)$ to coincide. We give an illustration of this possibility:

---

* If we assume in Corollary 1 that $R(z)$ is only *formally* of degree n, and has formally k zeros in C_1, $n - k$ zeros in C_2, n poles in C_3, then the point set consisting of C_1 (if $k > 1$), C_2 (if $n - k > 1$), C_3, and C_4 is precisely the combined locus of critical points, formal multiple zeros, and formal multiple poles of $R(z)$. In particular we note that if the three regions C_1, C_2, C_3 have a common point $R(z)$ may be identically constant, and then the derivative $R'(z)$ vanishes identically

The problem involving formal degree and formal zeros and poles is simpler and more elegant but seems less fundamental than the more difficult problem of precise degree and actual zeros and poles.

COROLLARY 2. *Let $R(z)$ be a rational function of degree n, let one zero of $R(z)$ have as its locus the region $C_1: |z| \leqq 1$, let the remaining $n - 1$ zeros lie in the point $C_2: z = 0$, and let the n poles be coincident and have as their loci the region $C_3: |z| \geqq 1$, with the understanding that the n-fold pole shall not coincide with a zero of $R(z)$. Then the locus of critical points consists of the (open) region $C_4: |z| < 1$.*

When z_1 and z_3 are real, in their proper loci, and approach $z = 1$, the point z_4 defined by (3) with $k = 1$ is also real and approaches $z = 1$; when $z_1 = 0$ and $z_3 = 1$, the point z_4 defined by (3) is $z_4 = 0$. Thus the interval $0 \leqq z < 1$ belongs to the locus of critical points, and by the symmetry about O it follows that the entire open region C_4 belongs to the locus. When z is a point of the closed region C_3 not a zero or pole of $R(z)$, the force at z due to the positive particles is equivalent to the force at z due to n positive particles coinciding at an *interior* point of C_1, and the force at z due to the negative particles is equivalent to the force at z due to n negative particles coinciding at some point of the closed region C_3, so z is not a position of equilibrium nor can z be a critical point; this reasoning after suitable transformation of the plane applies also to the original point at infinity. The proof is complete. The regions C_1, C_3, and C_4 all have the same boundary, no point of which can be a critical point. It follows from the reasoning given that *if we do not require in Corollary 2 that the poles of $R(z)$ be coincident, the locus of critical points is the entire plane* (compare §4.2.2 Theorem 1) *with the exception of the circumference $|z| = 1$.*

Another type of phenomenon in (ii), related to Corollary 2, is illustrated in Corollary 3, whose proof we omit:

COROLLARY 3. *Let $R(z)$ be a rational function of degree greater than unity whose zeros are concentrated at $z = 1$ and whose poles have the region $|z| \leqq 1$ with the exception of the point $z = 1$ as their locus. Then the locus of critical points is the closed region $|z| \leqq 1$.*

In Corollaries 2 and 3 we have illustrated the circumstance described in (ii), but still another difficulty (iii) with exceptional points exists. In the proof of Theorem 1 we made the conventions that equation (1) shall be considered satisfied if three of the points z_k coincide, and hence that if two of those points coincide a third must coincide with them. In Corollary 1 if all the zeros of $R(z)$ coincide (not at a formal pole), that point of coincidence is a critical point, but zeros and poles of $R(z)$ cannot coincide at a single point in an admissible situation. Otherwise expressed, there is a lack of symmetry among z_1, z_2, z_3 in Corollary 1 interpreted in the light of Theorem 1, for in Corollary 1 the point z_4 may coincide with z_1 and z_2, but not with z_3 and z_1 or z_2. In the proof of Theorem 1 we introduced *Convention* A: *if the three given regions C_j have a point in common, then the locus C_4 shall be considered the entire plane.* This convention is not entirely justified when Theorem 1 is used to determine the locus of critical points in Corollary 1, as is illustrated by §3.1.2 Theorem 3, where all points of an open half-plane belong to the locus C_4 of Theorem 1 but not to the locus of critical points in Corollary 1. We proceed to investigate (iii) the influence of Convention A in both Theorem 1 and Corollary 1.

Convention A is effective only if the boundaries of the three given regions are all tangent at a single point, including the possibility of identity of two or three bounding circles and including the possibility that one of those regions degenerates to a point but excluding the case (treated in Corollary 3) that two such regions degenerate. In any other case of a point common to all three regions, where the boundaries cut but not all at a zero angle there, transform that point to infinity; through an arbitrary finite point z of the plane can be passed a line which has two disjoint infinite segments respectively in two of the given regions, and also a line which has a single infinite segment common to those two regions; thus whatever λ may be z belongs to the locus C_4 in Theorem 1 and to the locus of critical points in Corollary 1, without reference to Convention A. This reasoning shows also that in Theorem 1 without Convention A the locus C_4 is likewise the entire plane if two of the regions C_1, C_2, C_3 have in common an interior point of the third region; we henceforth exclude this case. The discussion about to be given applies in essence but not in phraseology if the number of distinct circles C_1, C_2, C_3 is less than three; this situation is by no means trivial, for if we choose $C_1{:}\,|z| \leqq 1$, $C_2{:}\,|z| \leqq 1$, $C_3{:}\,|z| \geqq 1$, $\lambda > 1$, then in Theorem 1 without Convention A the locus C_4 is the circular region $|z| \leqq 1$; compare Corollary 2. In Theorem 1 we retain the convention that if any two of the points z_1, z_2, z_3 coincide, then z_4 coincides with them; there is no corresponding convention relating to Corollary 1, for a multiple zero of $R(z)$ is a critical point, and a zero and pole cannot coincide.

Suppose now (Case IV of Theorem 1) the given circles C_1, C_2, C_3 are all tangent at a single point Q. We establish the conclusion of Theorem 1 without Convention A, that the locus Γ_4 of z_4 is a circular region. The set of circles C cutting C_1, C_2, C_3 all at the same angle or a particular one of those circles at an angle supplementary to the angles cut on the other two consists of all circles through Q. For definiteness suppose the region C_3 not degenerate; we replace C_3 by an auxiliary variable closed region C_3' interior to C_3. Theorem 1 as already established applies to the regions C_1, C_2, C_3', and the corresponding locus of z_4 is a circular region C_4' which increases monotonically as C_3' increases. As C_3' approaches C_3 the limit of C_4' is a (closed) circular region C_4^0, which we prove to be identical with Γ_4. Any point z_4^0 in Γ_4 is a limit point of a sequence of points $z_4^{(j)}$ of the C_4', for we have $(z_1^0, z_2^0, z_3^0, z_4^0) = \lambda$, for some choice of z_1^0, z_2^0, and z_3^0, whence by a small displacement of z_3^0 and correspondingly of z_4^0 (although we may have $z_4^{(j)} = z_4^0$) we have $(z_1^0, z_2^0, z_3^{(j)}, z_4^{(j)}) = \lambda$, where $z_3^{(j)}$ is an *interior* point of C_3, and where the sequence $z_4^{(j)}$ approaches z_4^0; then every $z_4^{(j)}$ belongs to some C_4', so z_4^0 belongs to C_4^0. Reciprocally, every point of C_4^0 is a point of Γ_4, as we now show. Choose the variable C_3' symmetric in a fixed circle S orthogonal to C_1, C_2, C_3 at Q. The Corollary to Theorem 1 applies to the loci C_1,

C_2, C_3'; denote by A_4' the intersection of S with the region C_4'. The limit of A_4' is then an arc A_4 of S, which may or may not be the whole circumference. The relations between the end-points (if any) of A_4 as of A_4' correspond to the conditions of the Lemma. The arc A_4 is closed by virtue of the closure of the arcs A_1, A_2, A_3, the monotonic variation of z_4 in (1) under monotonic variation of each of the points z_1, z_2, z_3, and by virtue of our convention that (1) is considered satisfied if z_4 coincides with two of the points z_1, z_2, z_3. The given circles C_1, C_2, C_3 are symmetric in *every* circle S orthogonal to them at Q, and if Q lies at infinity, on such circles S the configuration of the arcs A_1, A_2, A_3, A_4 is independent of the particular circle S chosen. The closure of the arcs A_4 implies the closure of the sum C_4^0 of those arcs, shows that every point of C_4^0 is a point of Γ_4, and completes the proof of Theorem 1 without Convention A. We have also proved the Corollary to Theorem 1 under the present conditions; the method provides both a construction for the locus Γ_4 and a test as to whether that locus is the entire plane. It is worth noticing that an arbitrary point of Γ_4 is a limit of points $z_4^{(j)}$, where we have $(z_1^{(j)}, z_2^{(j)}, z_3^{(j)}, z_4^{(j)}) = \lambda$, and where $z_i^{(j)}$ $(j = 1, 2, 3)$ is an *interior* point of C_i insofar as interior points exist; for this conclusion is true even when the $z_i^{(j)}$ are restricted also to a circle S orthogonal to C_1, C_2, C_3.

Incidentally, we have here an example indicating that in Theorem 1 without Convention A the locus Γ_4 need not vary continuously with the given C_j. In the notation already used, let the variable region C_3' be symmetric in a circle S, where the configuration is so chosen (as in §3.1.2 Lemma 3) that the locus C_4' does not become the entire plane even when C_3' increases monotonically and becomes a circular region whose boundary contains Q. But if C_3' now increases still further and Q becomes an interior point of C_3', the locus C_4' suddenly becomes the entire plane.

We have determined the locus Γ_4 of Theorem 1 primarily for application in connection with Corollary 1 to Theorem 2. Under the hypothesis of Corollary 1 to Theorem 2 the *locus* of critical points consists of C_1 (if $k > 1$), C_2 (if $n - k > 1$), the interior points of C_3, and Γ_4, with the deletion (i.e. deletion from the point set, not merely from this enumeration) of (i) the point C_3 if C_3 is degenerate and in C_1, C_2, or C_4; the point C_1 if C_1 is degenerate with $k = 1$ and in C_3 but not in C_2; and the point C_2 if C_2 is degenerate with $n - k = 1$ and in C_3 but not in C_1, and (ii) each boundary point z_4^0 of Γ_4 which belongs to Γ_4 only because of a relation $(z_1^0, z_2^0, z_3^0, z_4^0) = n/k$ where z_3^0 coincides with z_1^0 or z_2^0. We note that a point z common to the interiors of C_3 and C_1 or C_2 is surely an interior point of Γ_4 in Theorem 1 because there z_3 and z_1 or z_2 may coincide; in Corollary 1 a pole and zero of $R(z)$ cannot coincide at z, but z belongs to the locus of critical points as an interior point of C_3.

In various cases under Theorem 2 and Corollary 1 where C_4 is disjoint from the other regions C_k and hence is known to contain precisely one critical point, that critical point may be located more exactly. For instance if the zeros and poles of $R(z)$ are known to be symmetric in O, it may be possible to conclude, as in §3.7 Theorem 1, that O is the only critical point in C_4; if $R(z)$ is real, since C_4 contains but one critical point that must be real, as in §3.7 Theorems 2 and 5. Compare §§5.1.2 and 8.3 for details and further examples.

§4.4.4. Generalizations. As in §3.3 we generalized the results of §1.5, so Theorem 2 can be extended:

THEOREM 3. *Let $R_0(z)$ be a rational function of z whose zeros $\alpha_1, \alpha_2, \cdots, \alpha_m$ are non-negative with multiplicities $\mu_1, \mu_2, \cdots, \mu_m$, and whose poles $\beta_1, \beta_2, \cdots, \beta_n$ are non-positive with multiplicities $\nu_1, \nu_2, \cdots, \nu_n$; a zero or infinite value of α_k or β_k is not excluded. Let S_k be a circle whose center is α_k and radius $\rho\alpha_k$, $0 < \rho < 1$, and let T_k be a circle whose center is β_k and radius $-\rho\beta_k$; if α_k or β_k is zero or infinite, the corresponding circle is taken as a null circle. Let $R(z)$ be a variable rational function of which μ_k zeros have the closed interior of S_k as their locus and ν_k poles have the closed interior of T_k as their locus, and which has no other zeros or poles. Then the locus of the finite critical points of $R(z)$ consists of the closed interiors of the circles S_j' whose centers are the zeros α_j' of $R_0'(z)$ and whose radii are $\rho \mid \alpha_j' \mid$, and of the interiors of the circles T_j' whose centers are the multiple poles β_j' of $R_0'(z)$ and whose radii are $\rho \mid \beta_j' \mid$. Any circle S_j' or T_j' which is exterior to the other circles S_k' and T_k' contains in its closed interior the number r_j or $q_j - 1$ of critical points, where r_j is the multiplicity of α_j' as a zero of $R_0'(z)$, and q_j is the number of distinct poles of $R(z)$ in the closed interior of T_j'. Any number of such circles S_j' and T_j' exterior to the other circles S_k' and T_k' contain in their closed interiors a number of critical points of $R(z)$ equal to the sum of the corresponding numbers r_j and $(q_j - 1)$.*

It is a consequence of §4.2.3 Theorem 3 that no critical point of $R(z)$ lies in the open double sector $\theta = \sin^{-1} \rho < \arg z < \pi - \theta, \pi + \theta < \arg z < 2\pi - \theta$, which is bounded by the common tangents from O to the circles S_k and T_k. Theorem 3 can now be proved precisely as was §3.3 Theorem 1. In considering the force at a point z in the right-hand half-plane on the boundary of the locus of critical points we replace the negative particles in the closed interior of a circle T_j by positive particles of the same numerical mass in the closed interior of a circle $T_j'(z)$ found by reflecting T_j in z; it will be noticed that the line Oz cuts the circles S_j, T_j, and $T_j'(z)$ all at the same angle. Details of the proof are left to the reader.

In Theorem 3 it is not essential to require $\rho < 1$, provided as before O is an external center of similitude for every pair of circles S_k, and for every pair of circles T_k, and is an internal center of similitude for every pair (S_j, T_k). But if we have $\rho > 1$, the locus of the critical points of $R(z)$ may be the entire plane unless the point at infinity is a circle S_j or T_j, and in any non-trivial case the totality of circles S_j' and T_k' cannot be mutually exterior. Moreover the comments on Theorem 2 concerning exceptional points of the apparent locus of critical points apply in essence here.

Numerous other geometric configurations analogous to that of Theorem 3 lead to analogous results. (a) If the circles S_k and T_k of Theorem 3 are considered, and the closed interiors of the S_k are the loci of prescribed zeros of a polynomial $p(z)$ and the closed exteriors of the T_k are the loci of other prescribed zeros of $p(z)$ which has no further zeros, then the locus of critical points consists (in addition to those of the given regions which are loci of more than one point, but with the possible exception of particular boundary points of all regions involved) of the closed exteriors of various circles whose centers lie on the negative half of the axis of reals, circles for any pair of which together with the T_k the point O is an external center of similitude. (b) Let O be on or interior to each of the circles S_j and T_k, null circles at O and infinity not excluded, and let O be an external center of similitude for any pair of those circles; let the closed interiors of the S_j be assigned as loci of prescribed zeros and the closed exteriors of the T_k be assigned as loci of prescribed poles of a rational function which has no other zeros or poles. Then the locus of critical points consists of the closed interior of a circle S_1' and the closed exterior of a circle T_1', with the possible exception of particular boundary points; the point O is an external center of similitude for any pair of the circles S_j, T_k, S_1', T_1'.

§4.5. Marden's Theorem. In numerous situations (§§1.5, 3.1, 3.3, 3.4, 4.2, 4.2.4, 4.3, 4.4) we have considered various circular regions as the respective loci of prescribed numbers of zeros and poles of a variable rational function, and have determined then the locus of the critical points. The problem of determining the latter locus in complete generality for arbitrarily assigned circular regions was solved by Marden [1930, 1936], who shows that the locus if not the entire plane is ordinarily bounded by the whole or part of a *circular curve* (generalization of circle and of bicircular quartic), namely a curve of degree $2s$ in whose equation the terms of highest degree can be written $(x^2 + y^2)^s$.

MARDEN'S THEOREM. *For $j = 1, 2, \cdots, p$, let σ_j be ± 1 and let $f_j(z)$ denote a polynomial of degree n_j having all its zeros in the fixed circular region C_j defined by*

$$\sigma_j C_j \equiv \sigma_j[|z - \alpha_j|^2 - r_j^2] \leqq 0. \tag{1}$$

Then every finite zero of the derivative of the function $f(z) \equiv f_1(z) \cdot f_2(z) \cdots f_q(z)/f_{q+1}(z) \cdot f_{q+2}(z) \cdots f_p(z)$ not satisfying at least one of the p inequalities (1) *satisfies the inequality*

$$\sum_{j=1}^{p} \frac{n m_j}{C_j(z)} - \sum_{j=1,\ k=j+1}^{j=p,\ k=p} \frac{m_j m_k \tau_{jk}}{C_j(z) C_k(z)} \leqq 0, \tag{2}$$

where $m_j = n_j$ or $-n_j$ according as we have $j \leqq q$ or $j > q$, and where

$$n = \sum_1^p m_j, \qquad \tau_{jk} = |\alpha_j - \alpha_k|^2 - (\mu_j r_j - \mu_k r_k)^2, \qquad \mu_j = \sigma_j \operatorname{sg} m_j.$$

It is also true that conversely any finite z which satisfies (2), with possible exceptional points such as those discussed in §4.4.3, is a zero of $f'(z)$ for suitable choice of the zeros of the $f_j(z)$ in their assigned regions. The locus of zeros of $f'(z)$ is a set of regions (with possible exceptional points) each either a region C_j or a region bounded by the whole or part of the curve whose equation is (2) with the inequality replaced by an equality. Marden shows that this theorem contains numerous theorems that we have established in the present work, as well as various others. Doubtless other interesting properties of the curve bounding (2) both in general and in various special cases remain still to be found.

Marden's Theorem may be proved by the use of algebraic inequalities. A second proof can be given as follows. In considering the critical points of $f(z)$, it is sufficient to treat the special case that the zeros of $f_j(z)$ in C_j coincide at some point z_j in C_j Moreover, a critical point z cannot lie on the boundary of its locus (if the locus has a boundary) unless all the z_j on which z effectively depends lie on the boundaries of their respective loci. Thus we may keep $z_2, z_3, \cdots, z_p$ fixed on the circles $C_2, C_3, \cdots, C_p$, and find the locus of z when z_1 traces C_1. Then we may keep $z_3, \cdots, z_p$ fixed and find the envelope as z_2 traces C_2 of the curve just determined. Next we may keep $z_4, \cdots, z_p$ fixed and find the envelope as z_3 traces C_3 of the envelope. Continuation of this method and study of the topological properties of the envelopes then yields the theorem. For details of the proof and a discussion of the theorem, the reader is referred to the original papers and to Marden [1949].

In a somewhat different order of ideas, inequalities on the zeros alone of a rational function may yield information on the zeros of the derivative. Here we refer the reader to Kakeya [1917], Biernacki [1928], and Dieudonné [1938], in addition to Marden [1949].

CHAPTER V

RATIONAL FUNCTIONS WITH SYMMETRY

In the present chapter we undertake a deeper study of the location of critical points of rational functions which possess symmetry. We consider in part real rational functions, which exhibit symmetry of zeros and poles in the axis of reals, but also consider numerous other kinds of symmetry, for the possibilities are much more diverse than in the case of polynomials (Chapters II and III).

§5.1. Real rational functions. For real rational functions the methods of Chapter II apply in part [Walsh, 1922].

§5.1.1. Real zeros and poles. As a generalization of §2.1 Theorem 1 and in elaborating §4.2.3 Theorem 2 we have

THEOREM 1. *Denote by I_j the interval $(0 <)\alpha_j' \leqq z \leqq \alpha_j''$, $j = 1, 2, \cdots, m$, and by J_k the interval $\beta_k' \leqq z \leqq \beta_k''$ (<0), $k = 1, 2, \cdots, m$. Denote by γ_j and δ_j the combined sets of critical points (counted according to multiplicities) and multiple poles (each counted one less than its order as a pole) of the two functions*

$$R_1(z) = \prod_{j=1}^{m} \frac{(z - \alpha_j')}{(z - \beta_j'')}, \qquad R_2(z) = \prod_{j=1}^{m} \frac{(z - \alpha_j'')}{(z - \beta_j')},$$

respectively, with $\gamma_j \leqq \gamma_{j+1}$, $\delta_j \leqq \delta_{j+1}$. Then if the intervals I_j are the respective loci of the points α_j, $j = 1, 2, \cdots, m$, and the intervals J_k are the respective loci of the points β_k, $k = 1, 2, \cdots, m$, the locus of the critical points of $R(z) = \prod_1^m[(z - \alpha_j)/(z - \beta_j)]$ consists of the intervals $L_j: \gamma_j \leqq z \leqq \delta_j$ if γ_j and δ_j are positive and $\delta_j \leqq z \leqq \gamma_j$ if γ_j and δ_j are negative, $j = 1, 2, \cdots, 2m - 2$, except that end-points are to be omitted from the intervals L_j bounded by multiple poles of $R_1(z)$ and $R_2(z)$, and a fixed pole of $R(z)$ interior to an interval L_j is to be omitted. Any interval L_j which is disjoint from the other distinct intervals L_k contains a number of critical points equal to the multiplicity of γ_j as a critical point of $R_1(z)$ if γ_j is positive.

Theorem 1 may be proved by the methods of §2.1; if the critical points and multiple poles of $R(z)$ are ordered according to algebraic magnitude, the k-th of these points varies continuously and monotonically with α_j and β_j, and if positive increases with α_j and decreases as β_j increases. In fact (§4.2.3 Theorem 2) the critical points of $R(z)$ are all finite; precisely one critical point lies in an open interval bounded by two successive α_j or two successive β_j. Any critical point z of $R'(z)$ not a multiple zero of $R(z)$ satisfies the equation

(1)
$$\frac{R'(z)}{R(z)} = \sum_1^m \frac{1}{z - \alpha_k} - \sum_1^m \frac{1}{z - \beta_k} = 0,$$

from which we find

(2)
$$\frac{\partial z}{\partial \alpha_j} = \frac{1}{(z - \alpha_j)^2} \div \left[\sum_1^m \frac{1}{(z - \alpha_k)^2} - \sum_1^m \frac{1}{(z - \beta_k)^2} \right].$$

If z is positive, it lies in an interval $\alpha_n < z < \alpha_{n+1}$, where we assume $\alpha_1 \leqq \alpha_2 \leqq \cdots \leqq \alpha_m$, and we write from (1)

$$\sum_1^n \frac{1}{z - \alpha_k} + \sum_{n+1}^m \frac{1}{z - \alpha_k} = \sum_1^m \frac{1}{z - \beta_k}, \qquad \sum_1^n \frac{1}{z - \alpha_k} > \sum_1^m \frac{1}{z - \beta_k},$$

$$\sum_1^m \frac{1}{(z - \alpha_k)^2} > \sum_1^n \frac{1}{(z - \alpha_k)^2} \geqq \frac{1}{z - \alpha_1} \sum_1^n \frac{1}{z - \alpha_k}$$
$$> \frac{1}{z - \alpha_1} \sum_1^m \frac{1}{z - \beta_k} > \sum_1^m \frac{1}{(z - \beta_k)^2},$$

whence by (2)

(3)
$$\partial z / \partial \alpha_j > 0.$$

A second proof of (3) may be given as follows. Let all the α_k and β_k except α_n be fixed; when α_n approaches α_{n+1} from the left, so also does z; the relation $\partial z / \partial \alpha_n = 0$ can occur only if two or more critical points of $R(z)$ coincide with z, hence only at a multiple zero of $R(z)$; it is thus clear that we have $\partial z / \partial \alpha_n > 0$, whence (3) follows from (2) for $j = n$ and from (2) for arbitrary j.

A similar proof establishes the inequality

(4)
$$\partial z / \partial \beta_j < 0.$$

It is clear from (3) and (4) that the n-th positive critical point of $R(z)$, where the critical points are arrayed in order of increasing magnitude, varies continuously and monotonically with α_j and β_j, even though several critical points may coincide; the n-th positive critical point is least when the α_j, in their proper intervals I_j, lie in the points α_j' and when the β_j, in their proper intervals J_j, lie in the points β_j''. The locus of the n-th positive critical point is then the corresponding interval L_j.

For a negative critical point z we find as in the proof of (3) and (4) (or by symmetry) the relation $\partial z / \partial \alpha_j < 0$, $\partial z / \partial \beta_j > 0$. However, as β_j varies the points β_j and z may become confounded with a multiple pole of $R(z)$, in which case z of course no longer remains a critical point. But a negative point z, allowed to be either a critical point or a multiple pole, varies continuously and monotonically with α_j and β_j. The conclusion of Theorem 1 now follows at once, the last part by the method of continuity.

Theorem 1 includes §2.1 Theorem 1. The special case of Theorem 1 in which a single transformation $w = \rho z$ $(\rho > 1)$ carries each α_j' into α_j'' and each β_k'' into

β'_k is especially interesting, for this same transformation carries each positive or negative γ_j into δ_j. This special case is a consequence also of §4.4.4 Theorem 3.

The special case in which there are precisely two intervals I_j and one interval J_j is closely related to the cross-ratio theorem (§4.4). Let I_1 be the locus of k zeros, I_2 the locus of $m - k$ zeros, and J_1 the locus of m poles of $R(z)$. The locus of critical points, other than the given intervals, is an interval whose end-points are γ and δ defined by $(\alpha_1', \alpha_2', \beta_1'', \gamma) = m/k$, $(\alpha_1'', \alpha_2'', \beta_1', \delta) = m/k$.

Of course Theorem 1 applies in effect to an arbitrary rational function $R(z)$ whose poles and zeros have as loci closed arcs of a given circle C, provided all the arcs assigned to zeros of $R(z)$ lie on a closed arc A_1 of C and all the arcs assigned to poles of $R(z)$ lie on a closed arc A_2 of C, where A_1 and A_2 are disjoint.

§5.1.2. Non-real zeros. If we permit both non-real zeros and non-real poles, the situation may become relatively complicated. One fairly obvious result is

THEOREM 2. *Let finite intervals I_j and J_k (which may degenerate) of the axis of reals be the respective loci of m_j zeros and n_k poles of the rational function $R_1(z)$ of degree $\sum m_j = \sum n_k$, and let intervals L_j of the axis of reals be the corresponding locus of the real critical points of $R_1(z)$. Let $R(z)$ be a real rational function of degree $\sum m_j$ whose zeros occur in (symmetric) groups of m_j zeros which lie in the closed interior of the circle whose diameter is I_j, and whose poles occur in groups of n_k poles which lie in the closed interior of the circle whose diameter is J_k. Then all real critical points of $R(z)$ not in the I_j or J_k lie in the intervals L_j.*

We note that if z is real, the force at z due to a pair of unit particles at points α and $\bar{\alpha}$ is equal to the force at z due to a double particle at the harmonic conjugate of z with respect to α and $\bar{\alpha}$, a point on the axis of reals which is interior to any circle containing α and $\bar{\alpha}$ if z is exterior to this circle. Consequently if z is a real critical point of $R(z)$, not on an interval I_j or J_k, it is also a position of equilibrium in the field of force corresponding to a suitably chosen function $R_1(z)$ satisfying the prescribed conditions, so z lies in some L_j, as we were to prove.

Jensen circles continue to have significance for real rational functions:

THEOREM 3. *Let $R(z)$ be a real rational function with no finite real poles. Then no non-real critical point of $R(z)$ lies both exterior to all Jensen circles if any for the zeros of $R(z)$, and interior to all Jensen circles for the poles of $R(z)$.*

At a non-real point exterior to all Jensen circles for the zeros of $R(z)$, the force due to particles situated at all such zeros, both real and in pairs of conjugates, has (§1.4.1 Lemma) a non-zero component directed away from the axis of reals; at a point interior to all Jensen circles for the poles of $R(z)$, the total force due to the particles at those poles also has a non-zero component directed away from the axis of reals; the theorem follows.

The cross-ratio theorem (§4.4.3) continues to have some significance for real rational functions:

THEOREM 4. *Let circular regions C_1, C_2, C_3 each symmetric in the axis of reals contain respectively k zeros of a real rational function $R(z)$ of degree n, the remaining $n - k$ zeros of $R(z)$, and all poles of $R(z)$. Denote by C_4 the locus of z_4 as defined by the equation*

$$(z_1, z_2, z_3, z_4) = n/k, \tag{5}$$

when C_1, C_2, C_3 are the respective loci of z_1, z_2, z_3. If C_4 is exterior to the given circular regions, it contains precisely one critical point of $R(z)$, and that real; the remaining critical points lie in C_1, C_2, C_3.

Since C_4 is exterior to the given regions, $C_1 + C_2$ and C_3 are disjoint. It follows from §4.4.3 Theorem 2 that C_4 contains but one critical point of $R(z)$; since C_4 is also symmetric in the axis of reals, that critical point is necessarily real. It follows from §4.4.2 Corollary that the segment of the axis of reals interior to C_4 is the locus of z_4 defined by (5) when the segments of that axis interior to C_1, C_2, C_3 are the respective loci of z_1, z_2, z_3; compare Theorem 2, which gives no information on the non-real critical points.

The same method of proof used for Theorem 4 yields

THEOREM 5. *Let circular regions C_1 and C_2 be mutually symmetric in the axis of reals, and let the circular region C_3 be symmetric in that axis. Let all zeros of a real rational function $R(z)$ lie in C_1 and C_2, and all poles in C_3. Denote by C_4 the locus of z_4 as defined by the equation (5) with $n = 2k$, when C_1, C_2, C_3 are the respective loci of z_1, z_2, z_3. If C_4 is exterior to the given circular regions, it contains precisely one critical point of $R(z)$ and that real; the remaining critical points lie in C_1, C_2, C_3.*

Theorem 5 contains §3.7 Theorem 2.

In both Theorems 4 and 5, the entire segment of the axis of reals in C_4 belongs to the *locus* of the critical points of $R(z)$ when the given circular regions are the loci of the zeros and poles of $R(z)$ proper to the regions.

Theorems 1–5 are not expressed in a form invariant under linear transformation, and new more general results are obtained by such a transformation. As a matter of policy in the present work we ordinarily leave such generalizations to the reader; thus all the results of Chapters I–III concerning polynomials can be interpreted as results on rational functions having but a single pole. However, some results deserve special mention in a transformed situation; thus a consequence of §4.3 Corollary 2 to Theorem 1 is

THEOREM 6. *Let $R(z)$ be a rational function whose only poles, of equal orders, lie in the points $+i$ and $-i$, and let all zeros of $R(z)$ lie in the closed region A*

bounded by two circles of the coaxal family determined by $+i$ *and* $-i$ *as null circles. Then all critical points of* $R(z)$ *lie in* A.

A new proof of Theorem 6 can be given by considering the force at a point P not in A nor at $\pm i$. We invert the given configuration in the unit circle whose center is P, the image of A lies in the closed interior of a circle C_1, the inverse of one of the bounding circles of A, and the images of $+i$ and $-i$ are mutually inverse in C_1. If n denotes the degree of $R(z)$, the force at P due to the particles at the zeros of $R(z)$ is an n-fold vector to P from the center of gravity G_1 of the images of the zeros of $R(z)$, since these images lie in the closed interior of C_1, so also does G_1. The force at P due to the particles at $+i$ and $-i$ is an n-fold vector from P to the center of gravity G_2 of the images of $+i$ and $-i$; the point G_2 lies exterior to C_1, so G_1 and G_2 cannot be identical and P cannot be a position of equilibrium nor a multiple zero of $R(z)$. This completes the proof. We note too that no boundary point z of A other than a multiple zero of $R(z)$ can be a critical point, unless all zeros of $R(z)$ lie on the circle through z bounding A.

In Theorem 6 the function $R(z)$ need not be real, and the conclusion can be sharpened for real rational functions:

THEOREM 7. *Let* $R(z)$ *be a real rational function whose only poles lie in the points* $+i$ *and* $-i$, *and let a J-circle be constructed corresponding to each pair* α_k, $\bar{\alpha}_k$ *of conjugate imaginary zeros of* $R(z)$, *namely the circle orthogonal at* α_k *and* $\bar{\alpha}_k$ *to the circle through* $+i$, $-i$, α_k, $\bar{\alpha}_k$. *If the "interior" of such a J-circle denotes the circular region bounded by it not containing the points* $+i$ *and* $-i$, *then all non-real critical points of* $R(z)$ *lie in the closed "interiors" of those J-circles.*

If we denote by K *any configuration consisting of a closed interval* I *of the axis of reals whose end-points are not in the closed "interiors" of any J-circles and are neither zeros nor critical points of* $R(z)$, *plus the closed "interiors" of all J-circles intersecting* I, *and if* K *contains precisely* k *zeros of* $R(z)$, *then* K *contains precisely* $k - 1$, k, *or* $k + 1$ *critical points of* $R(z)$.

The concept of *J-circle* here is analogous to but by no means identical with that of Jensen circle. In Theorem 7 it is convenient to use the field of force on the sphere Σ (§4.1.3), the stereographic projection of the z-plane Π. It will be noted that $+i$ and $-i$ are diametrically opposite points of Σ, and that the axis of reals projects into a great circle of Σ whose poles are $+i$ and $-i$ and whose plane Π_1 is orthogonal to Π. Each pair of zeros α_k and $\bar{\alpha}_k$ projects into a pair of points α_k' and $\bar{\alpha}_k'$ of Σ which are symmetric in the plane Π_1, and the corresponding J-circle on the sphere, being orthogonal to the circle through $+i$, $-i$, α_k', and $\bar{\alpha}_k'$, is the circle of Σ whose diameter is the line segment $\alpha_k'\bar{\alpha}_k'$.

In three-dimensional space, the field of force due to two unit particles A and B is symmetric about the axis AB, and if Π_1 is the plane bisecting the segment AB and Σ_1 is the sphere constructed on that segment as diameter, it follows from the Lemma of §1.4.1 that at a point P of Π_1 the force lies in Π_1, at a point

P on Σ_1 not in Π_1 the force is parallel to Π_1, and at a point P not on Σ_1 nor in Π_1 the force has a non-zero component toward Π_1 or away from Π_1 according as P is interior or exterior to Σ_1. Thus at a point P of Σ not in Π_1 the force due to the particles at $+i$ and $-i$ is parallel to Π_1; if P lies on Σ but does not lie on Π_1 or in the closed "interior" of the J-circle for the points α_k' and $\bar{\alpha}_k'$, it lies exterior to the sphere whose diameter is the segment joining those two points, and the corresponding force at P has a component directed away from Π_1; the force at P on Σ not on Π_1 due to a single particle on Σ and Π_1 has a component directed away from Π_1. Thus at a non-real point P of Σ not in the closed "interior" of any J-circle the total force has a component directed away from Π_1, so P is not a position of equilibrium.

The case of a single J-circle and no real zeros of $R(z)$ is trivial; all critical points are real. In any other case, there must be either more than one J-circle, or real zeros of $R(z)$. Thus we notice that if z is a non-real critical point not a multiple zero of $R(z)$ nor interior to any J-circle, then z must lie on all J-circles, and there are no real zeros of $R(z)$. The latter part of Theorem 7 follows by the method of §2.3, interpreted either in the plane or (§4.1.3 Theorem 5) on the sphere. The numbers $k-1$, k, $k+1$ are chosen as in §2.3 Theorem 1.

Interpreted on the sphere Σ, Theorem 7 is essentially a limiting case of a more general result. If α and $\bar{\alpha}$ are conjugate imaginary, we call the *J-circle* on Σ the circle which has the line segment $\alpha\bar{\alpha}$ as diameter (in three dimensions) and by its *interior* we mean the part of Σ interior to the sphere whose diameter is $\alpha\bar{\alpha}$. It then follows, as in the proof of Theorem 7 (and in the proof of Theorem 3), that if $R(z)$ is a real rational function with no real poles, then no non-real critical point of $R(z)$ lies exterior to all J-circles for the zeros of $R(z)$ and interior to all the J-circles for the poles of $R(z)$. The pair of points $(+i, -i)$ has properly no J-circle on Σ; but Theorem 7 is a limiting case of the proposition just stated if we choose $R(z)$ as a variable real rational function with precisely two distinct poles which are allowed to approach $+i$ and $-i$.

One special case of Theorems 6 and 7 is worth explicit mention:

COROLLARY. *Let the poles of the rational function $R(z)$ of degree two be the points $+i$ and $-i$, and the zeros the real points α_1 and α_2. The two critical points of $R(z)$ are real and lie on the circle Γ through $+i$ and $-i$ with respect to which α_1 and α_2 are mutually inverse.*

The entire configuration is symmetric in both Γ and the axis of reals, so the set of critical points has that same symmetry and hence is as described. This configuration is particularly simple on the sphere.

§5.1.3. Regions as loci. Theorem 6 suggests the use of circular or other regions as loci of zeros and poles of a real rational function, as indeed do Theorems 4 and 5. We prove

THEOREM 8. *Let the circular regions* $C_1: |z - bi| \leqq r$ *and* $C_2: |z + bi| \leqq r$ *contain all zeros of a real rational function* $R(z)$ *whose poles are all real. The critical points of* $R(z)$ *lie on the set consisting of the axis of reals plus* (i) *the closed interiors of* C_1 *and* C_2 *if we have* $b \geqq 2r$; (ii) *the closed interiors of the circles of the coaxal family determined by* C_1 *and* C_2 *passing through the points*

$$z = \pm i[(b^2 - 2r^2)/2]^{1/2}$$

if we have $2r > b > 2^{1/2}r$.

If all possible rational functions $R(z)$ *of unrestricted degrees are considered, if the locus of their zeros is the closed interiors of* C_1 *and* C_2 *and the locus of their poles is the axis of reals, then the point set assigned in* (i) *and* (ii) *is the locus of critical points; if we have* $b \leqq 2^{1/2}r$ *the locus is the entire plane.*

If the degree of $R(z)$ is two, it follows from Theorem 6 that the locus of critical points is simply the axis of reals. We proceed to the proof of Theorem 8.

An arbitrary point in the closed interior of C_1 is $(\rho \cos \theta, b + \rho \sin \theta)$, with $0 \leqq \rho \leqq r$, and the Jensen circle for this point and its conjugate is

$$(x - \rho \cos \theta)^2 + y^2 - (b + \rho \sin \theta)^2 = 0. \tag{6}$$

In case (i) *the point* $z = i(b - r)$ *is on or interior to this Jensen circle;* a necessary and sufficient condition that this point lie on or within the circle (6) is that the first member of (6) for $x = 0$, $y = b - r$ be negative or zero:

$$-(b + 2\rho \sin \theta)^2/2 + (b - 2r)^2/2 + (\rho^2 - r^2) \leqq 0,$$

and this condition is satisfied, for we have $b + 2\rho \sin \theta \geqq b - 2r \geqq 0$. It follows that the interval $x = 0$, $|y| < b - r$ is interior to all Jensen circles for $R(z)$, so at a non-real point z of that interval the force due to the particles at the zeros of $R(z)$ has a non-zero vertical component directed toward the axis of reals; at such a point z the force due to the particles at the poles of $R(z)$ has no non-zero vertical component directed away from the axis of reals, so z is not a position of equilibrium nor a critical point. The geometric configuration is invariant under any linear transformation which leaves C_1, C_2, and the axis of reals invariant, and a suitable transformation of this group leaves $R(z)$ real and carries an arbitrary non-real preassigned point z' exterior to C_1 and C_2 into a non-real point of the interval $x = 0$, $|y| < b - r$, so z' cannot be a critical point. On the other hand any two points of the axis of reals can be transformed into each other by a transformation of the group just mentioned, so in the study of loci any point of the axis of reals is a point of the locus of the critical points. Any point of the closed interior of C_1 or C_2 belongs to the locus, for it may be a multiple zero of $R(z)$. This completes the proof in case (i).

In case (ii) the point $z_1 = i[(b^2 - 2r^2)/2]^{1/2}$ lies on or interior to each circle (6); a necessary and sufficient condition for this is

$$-(b + 2\rho \sin \theta)^2/2 + (\rho^2 - r^2) \leqq 0, \tag{7}$$

and by the inequality $\rho \leqq r$ this condition is satisfied. It follows that the open interval $(z_1, -z_1)$ is interior to all Jensen circles for $R(z)$, so no non-real critical point lies on that interval. On the other hand, for suitable choice of ρ and θ inequality (7) becomes an equality; if $R(z)$ is of degree $4n$, we can assign n zeros of $R(z)$ to the point of C_1 corresponding to this choice of ρ and θ, and n zeros to the points symmetric in the axes and the origin, while the poles of $R(z)$ are placed at infinity; it follows that z_1 is a critical point. If now the zeros of $R(z)$ are continuously moved on C_1 and C_2, retaining their symmetry in the axes, into the positions $z = \pm i(b - r)$, the critical point at z_1 moves continuously on the axis of imaginaries and takes the position $z = i(b - r)$, so every point of the interval $(z_1, i(b - r))$ belongs to the locus. Every point of the closed interiors of C_1 and C_2 and of the axis of reals is a possible critical point, so the proof in case (ii) can be completed as was the proof in case (i).

If we have $b \leqq 2^{1/2}r$, a necessary and sufficient condition that the origin be on the circle (6) is

$$-(b + 2\rho \sin \theta)^2/2 + \rho^2 - b^2/2 = 0,$$

and this condition is satisfied if we make the allowable choice $\rho = b/2^{1/2}$, $\sin \theta = -1/2^{1/2}$. Consequently the origin is a critical point of a suitably chosen $R(z)$ of degree $4n$ with all poles at infinity; it follows by continuous variation of the zeros of $R(z)$ that all points of the interval $(0, i(b - r))$ belong to the locus, and hence that the locus consists of the entire plane.

It is striking that the limits of cases (i) and (ii) correspond to angles $\pi/3$ and $\pi/2$ subtended at O by C_1. Theorem 8 can be improved for rational functions $R(z)$ of given degree $4m + 2$, $m > 0$, in which case the locus of critical points depends on m, but cannot be improved for functions $R(z)$ of degree $4m$.

In Theorem 8 with zeros and poles interchanged a limiting case is also a limiting case of Theorem 6; a direct proof is easy by counting real critical points:

COROLLARY 1. *If $R(z)$ is a real rational function with precisely two distinct poles, and those conjugate imaginary, and with all zeros real, then all critical points of $R(z)$ are real.*

A result related to Theorem 8 does not primarily involve circular regions as loci:

COROLLARY 2. *Let $R(z)$ be a real rational function whose zeros are all real, and let a circle Γ whose center is on the axis of reals A contain in its closed interior all zeros of $R(z)$ and have in its closed exterior all poles of $R(z)$. Then no non-real critical point of $R(z)$ lies interior to Γ.*

We use the general method of proof of Theorem 6. Denote by Γ' the inverse of Γ and by A' the inverse of the axis of reals in the unit circle whose center is an arbitrary non-real point P interior to Γ, and denote by K the common chord of A' and Γ'. The inverses of all poles of $R(z)$ lie in the closed interior of Γ' and either lie on A' as individuals or occur in pairs of points mutually inverse with respect to A'; the center of gravity of each pair lies exterior to A', hence interior to the half-plane K' bounded by K containing the center of Γ', as does each individual point not at an intersection of A' and Γ', so the center of gravity of all inverses of poles lies in K', and indeed interior to K' unless all poles of $R(z)$ lie on both A and Γ. The inverses of all zeros of $R(z)$ not at the intersections of A' and Γ' lie on the arc of A' exterior to K', so the center of gravity of all inverses of zeros lies in the closed exterior of K' and indeed exterior to K' unless all the zeros of $R(z)$ lie on both A' and Γ'. In any case the two centers of gravity cannot coincide, so P is not a position of equilibrium nor a critical point.

If we modify the hypothesis of Corollary 2 to admit non-real zeros of $R(z)$ interior to Γ but require that P shall lie interior to Γ and exterior to a circle through the intersections of Γ with the axis of reals and containing in its closed interior all zeros of $R(z)$, it is still true that P cannot be a critical point.

On the other hand, Corollary 2 is false if we omit the requirement that Γ shall contain in its closed interior all zeros of $R(z)$.

§5.1.4. Regions as loci, continued. We proceed to extend Theorem 8 to the case where the poles of $R(z)$ have an annular region as their assigned locus. For this purpose it is convenient to choose the loci of zeros and poles as bounded by concentric circles, and here several lemmas are useful.

LEMMA 1. *Let the point* $z = x + iy$ *lie on the circle* $\gamma: |z - a| = r$ *interior to* $C: |z| = 1$ *with* $a \geqq 0$; *we study the function* $F(x) = \Re(z + 1/\bar{z})$. Case I: $r > a$; $\max F(x)$ *occurs for* $x = a + r$, $\min F(x)$ *occurs for* $x = a - r$. Case II: $r < a$, $a - r \geqq (a + r)^3$; $\max F(x) = [(a - r)^2 + 1]/(a - r)$ *occurs for* $x = a - r$; $\min F(x) = [(a + r)^2 + 1]/(a + r)$ *occurs for* $x = a + r$. Case III: $r < a$, $a - r < (a + r)^3$; $\max F(x) = [(a - r)^2 + 1]/(a - r)$ *occurs for* $x = a - r$; $\min F(x) = [(a^2 - r^2)^{1/2} + 1]^2/2a$ *occurs for* $x = [a^2 - r^2 + (a^2 - r^2)^{1/2}]/2a$.

The function $F(x)$ represents the real part of the vector $z + 1/\bar{z}$, which is of some importance in itself, and also represents the horizontal component of the total force at O due to unit attracting particles at z and $1/\bar{z}$. Of course we have $\arg (z + 1/\bar{z}) = \arg z$. We set $\zeta = z + 1/\bar{z}$.

Indeed we have $\zeta\bar{\zeta} = (1 + z\bar{z})^2/z\bar{z} = 4 + (1 - z\bar{z})^2/z\bar{z} \geqq 4$, whence $|\zeta| > 2$ unless $|z| = 1$, in which case we have $|\zeta| = 2$. It follows that when z lies interior to C, the function $|\zeta|$ decreases monotonically as $|z|$ increases, so in Case I we determine, in Lemma 1, $\max F(x)$ and $\min F(x)$ not merely for z on γ, but also for z in the annular region bounded by γ and the inverse of γ in C; in Cases II and III we determine $\max F(x)$ and $\min F(x)$ not merely for z on γ but for z in the closed interior of γ.

For z on γ we have $x^2 + y^2 = r^2 - a^2 + 2ax$, so we may write

$$(8)\qquad \begin{aligned} F(x) &= \Re\left[z\,\frac{z\bar{z}+1}{z\bar{z}}\right] = x\,\frac{r^2 - a^2 + 2ax + 1}{r^2 - a^2 + 2ax}, \\ F'(x) &= \frac{(r^2 - a^2 + 2ax)^2 + r^2 - a^2}{(r^2 - a^2 + 2ax)^2}. \end{aligned}$$

It follows from (8) that in Case I we always have $F'(x) \geqq 0$, from which the conclusion follows in Case I. The points at which $F'(x) = 0$ are $(r < a)$ $x = [a^2 - r^2 \pm (a^2 - r^2)^{1/2}]/2a$, of which the latter is less than $a - r$, and of which the former is greater than $a - r$, and is not less than $a + r$ in Case II but is less than $a + r$ in Case III. Thus in Case II we have $F'(x) \leqq 0$ throughout the interval $a - r \leqq x \leqq a + r$, and the conclusion follows. In Case III we have $F'(a - r) < 0, F'(a + r) > 0$; by direct comparison we find $F(a - r) > F(a + r)$, and this completes the proof of Lemma 1.

LEMMA 2. *Let $R(z)$ be a real rational function of degree four whose zeros are the non-real points α and $\bar{\alpha}$ interior to C and their inverses $1/\bar{\alpha}$ and $1/\alpha$, and whose poles are the real points b and $1/b$, each a double pole, with $0 < b < 1$. If $R(z)$ has a critical point z_0 on the interval $-1 < z_0 < b$, then at a point z with $z_0 < z < b$ the force due to the negative particles is numerically greater than that due to the positive particles; at a point z with $-1 < z < z_0$, the force due to the positive particles is numerically greater than that due to the negative particles. If $R(z)$ has no critical point on the interval $-1 < z < b$, then at every point of that interval the force due to the negative particles is numerically greater than that due to the positive particles.*

Both the zeros and the poles of $R(z)$ are symmetric in C and in the axis of reals, so two of the four critical points are $+1$ and -1, the other two as a pair are inverse in C and in the axis of reals, and if distinct from each other and from the points $+1$ and -1 are simple critical points, and lie on C or on the axis of reals depending on the relation between b and α. However, at a real point z near and to the left of b, the force due to the negative particles is greater than that due to the positive particles; when z moves continuously to the left, this relation can change only at a critical point of $R(z)$, and the relation must reverse as z passes through a simple critical point. Lemma 2 follows.

The situation of Lemma 2 is to be compared with that for the rational function $R_1(z)$ of degree two whose zeros are the points c and $1/c$, and whose poles are the points b and $1/b$, with $0 < c < b < 1$. The two critical points are at $z = +1$ and -1, and throughout the interval $-1 < x < c$ the total force due to the positive particles is numerically greater than that due to the negative particles.

We are now in a position to prove our main result:

THEOREM 9. *Let the zeros and also the poles of a rational function $R(z)$ be symmetric in $C: |z| = 1$, let the zeros lie in the closed regions $|z| \leqq c < 1$ and $|z| \geqq$*

$1/c$, *and the poles in the closed region* $b \leqq |z| \leqq 1/b$, $0 < c < b < 1$. *Then* (i) *if the inequality*

$$(9) \qquad \frac{(1-c^2)(1-3\xi^2)+2(\xi^2-c^2)^{1/2}(1-c^2\xi^2)^{1/2}}{2(1-c^2)\xi(1-\xi^2)} \geqq \frac{1}{\xi+b}+\frac{1}{\xi+1/b}$$

is satisfied for $\xi = c$, *no critical points of* $R(z)$ *lie in the region* $c < |z| < b$; (ii) *if inequality* (9) *is satisfied for some* $\xi = \xi_0$, $c \leqq \xi_0 < b$, *and if we have for* $\xi = \xi_0$

$$(10) \qquad \eta^2 + (3d - d^3)\eta + 1 > 0,$$

with $\eta = (1 - c\xi)/(\xi - c)$, $d = (1 + c^2)/2c$, *then no critical points lie in the region* $\xi_0 < |z| < b$; (iii) *if inequality* (10) *fails to be satisfied for some* ξ_0, $c < \xi_0 < b$, *where* η *and* d *have the same significance as before, then no critical points of* $R(z)$ *lie in the region* $\xi_0 < |z| < b$.

To study the field of force in the z-plane, we consider an arbitrary real value $-\xi$, $c < \xi < b$, and invert the given configuration with $-\xi$ as center of inversion. If the new figure is in the w-plane, with $w = 0$ in the role of $z = -\xi$, the images of the circles $|z| = 1$, $|z| = 1/c$; $|z| = 1/b$ cut the axis of reals in the points $w = 1/(\xi + 1)$, $1/(\xi - 1)$; $1/(\xi + 1/c)$, $1/(\xi - 1/c)$; $1/(\xi + 1/b)$, $1/(\xi - 1/b)$. The image Γ of the circle $|z| = 1$ has the center $w = -\xi/(1 - \xi^2)$ and the radius $1/(1 - \xi^2)$. To study the force at the point $z = -\xi$, we shall make use of Lemma 1; the formulas of Lemma 1 must be modified to allow for the radius $1/(1 - \xi^2)$ of the circle Γ, and to allow for the fact that the center of that circle is not the origin from which we consider the vectors representing forces acting at the point $z = -\xi$. However, we note that the inverse of $z = -\xi$ in the unit circle C is exterior to all regions assigned to the zeros and poles of $R(z)$, so in the w-plane the center of Γ is exterior to all regions assigned to the images of poles and zeros. In Case III of Lemma 1, the minimum horizontal component of a force at $z = -\xi$ directed to the left due to a positive particle in $|z| \leqq c$ and to its inverse in C is numerically the first member of (9).

In the proof of Theorem 9 we restrict ourselves to the study of points $z = -\xi$, $c < \xi < b$, as possible critical points, which is obviously allowable. Moreover, we study only a rational function $R(z)$ whose degree is a multiple of four; this is possible by replacing if necessary a given $R(z)$ by its fourth power.

The center of Γ lies interior to the circles bounding the annular region in the w-plane to which the images of the poles of $R(z)$ are assigned. Here we have Case I of Lemma 1, and it follows that the greatest horizontal force at $z = -\xi$ acting toward the right due to the negative particles occurs if those particles are symmetrically located at the points b and $1/b$; for two particles this force is

$$(11) \qquad \frac{1}{\xi+b}+\frac{1}{\xi+1/b}.$$

The condition $a - r < (a + r)^3$ of Lemma 1 for the images in the w-plane of the positive particles here takes the form

$$\frac{(1-c^2)\xi}{1-c^2\xi^2} - \frac{c(1-\xi^2)}{1-c^2\xi^2} < \left[\frac{(1-c^2)\xi}{1-c^2\xi^2} + \frac{c(1-\xi^2)}{1-c^2\xi^2}\right]^3,$$

$$\frac{(1+c\xi)^3}{(\xi+c)^3} < \frac{1-c\xi}{\xi-c},$$

which by suppression of the positive factor $\eta^2 - 1$ reduces to (10), or to

$$(12) \qquad (\eta-1)^2 + (3d-d^3+2)(\eta-1) + (3d-d^3+2) > 0.$$

Inequalities (10) and (12) are of interest only for $c < \xi < 1$, that is, for $\eta > 1$, and it is apparent that the first member of (12) vanishes at most once in the interval $\eta > 1$, by Descartes's rule. When ξ increases, η decreases, so if (12) is satisfied for a value of ξ, $c < \xi < 1$, then (12) is satisfied for all smaller values of ξ; if (12) is false for a value of ξ, then (12) is false for all larger values of ξ. In terms of Lemma 1, if we have Case III for a value of ξ, then we have Case III for all smaller values of ξ, if we have Case II for a value of ξ, we have Case II for all larger values of ξ. For $\xi = 1$ we have $\eta = 1$, and (12) may or may not be satisfied; for $\xi \to c$ we have $\eta \to \infty$, and (12) is satisfied for sufficiently large η.

We note that the first member of (12) vanishes at some point in the interval $\eta > 1$ when and only when we have $3d - d^3 + 2 < 0$, or $d > 2$; thus we have Case III for all ξ, $c < \xi < 1$, if we have $d \leqq 2$ or $c \geqq 2 - 3^{1/2}$.

In part (iii) of Theorem 9, with $b > \xi > \xi_0$, since we have Case II for $z = -\xi_0$ for the positive particles we also have Case II for $z = -\xi$. Consider the force at $-\xi$ due to a pair of positive particles and a pair of negative particles in their assigned regions. The former force acting toward the left is numerically at least $1/(\xi + c) + 1/(\xi + 1/c)$, and the latter force acting toward the right is at most $1/(\xi + b) + 1/(\xi + 1/b)$; it follows from the discussion of $R_1(z)$ subsequent to Lemma 2 that the former of these expressions is greater than the latter. Hence $z = -\xi$ cannot be a position of equilibrium. The proof as just given is valid not merely in part (iii) but also in parts (i) and (ii) so far as concerns values $z = -\xi$ for which Case II occurs.

In the discussion of part (ii) we may assume $R(z)$ real, for it is obvious from the field of force that if $z = -\xi$ is a critical point of $R(z)$ it is also a critical point of the rational function $R(z)\cdot\bar{R}(z)$, where $\bar{R}(z)$ indicates the rational function whose zeros and poles are the conjugates of those of $R(z)$. We may also assume that $R(z)$ has at most four distinct zeros, symmetric both in the axis of reals and in C; the total force at z due to possible such groups of four distinct zeros in their assigned regions is horizontal and has a certain maximum and a minimum, and by the connectedness of the assigned regions this total force takes on all intermediate values; consequently the total force at z due to several groups each of four zeros is equal to the total force at z due to the same number of groups all coinciding with some group whose points lie in the assigned regions. Similarly we may assume that $R(z)$ has at most four distinct poles, symmetric both in the axis of reals and in C and situated in the assigned regions. We assume then that $z = -\xi$, $\xi > \xi_0$, is a critical point of such a function $R(z)$, and shall reach a contradiction. The force at z due to all positive particles is horizontal and is equal and opposite to the force at z due to all negative particles. Then the maximum force at $z = -\xi$ acting horizontally to the right due to negative particles in their assigned regions is numerically greater than or equal to the minimum horizontal force at $-\xi$ to the left due to positive particles in their assigned regions. It follows from Lemma 2 that also at the point $-\xi_0$ the maximum horizontal force due to the negative particles (necessarily at b and $1/b$) is greater than the horizontal force at $-\xi_0$ to the left due to the positive particles in their previously considered positions, and hence it follows that at the point $-\xi_0$ the maximum horizontal force due to the negative particles in their assigned regions is greater than the minimum force at $-\xi_0$ due to the positive horizontal particles in their assigned regions, and this contradicts our hypothesis (9).

Lemma 1 does not formally apply in the case $z = -c$ (notation of Theorem 9), for here the image in the w-plane of the circle $|z| = c$ passes through the center of Γ. Nevertheless the method of proof of Lemma 1 is valid; the greatest lower bound of the horizontal component of the force at $z = -c$ acting to the left due to two positive particles in their assigned regions is approached when the particles lie on the circle $|z| = c$ and approach $z = -c$, and has as its value the limit as ξ approaches c of the first member of (9). The method of proof already given for part (ii) now holds in essence with $\xi_0 = c$, and this completes the proof of part (i) and of Theorem 9.

The limiting case $c = 0$ of Theorem 9 is precisely Theorem 6 with the roles of zeros and poles interchanged, and with $R(z)$ real. The limiting case $b = 1$ is Theorem 8. It is clear from the proof of Theorem 9 that the regions determined for the critical points cannot be improved without restricting the degree of $R(z)$, but that for odd degrees those regions can be replaced by smaller ones.

The method of continuity shows that if $R(z)$ is of degree $2m$, then precisely $m - 1$ critical points lie in each of the regions $|z| \leqq c$, $|z| \geqq 1/c$ in (i) and in each of the regions $|z| \leqq \xi_0$, $|z| \geqq 1/\xi_0$ in (ii) and (iii).

§5.1.5. W-curves. The curves introduced and studied in §2.6 have their analogs of significance in the study of rational functions. An analog of §2.5 Theorem 1 is

THEOREM 10. *Let $R(z)$ be a real rational function with no finite real poles, with precisely one pair $(\alpha, \bar{\alpha})$ of finite conjugate imaginary non-real poles, and with no non-real zeros. Let z_1 be the algebraically least and z_2 the algebraically greatest of the zeros of $R(z)$. Let A_k $(k = 1, 2)$ be the circular arc bounded by α and $\bar{\alpha}$ of angular measure greater than π which is tangent to the lines αz_k and $\bar{\alpha} z_k$; for $z_k = (\alpha + \bar{\alpha})/2$, A_k consists of two infinite line segments, respectively bounded by α and $\bar{\alpha}$. Then all non-real critical points of $R(z)$ lie in the closed pseudo-lens-shaped region R bounded by A_1 and A_2. No non-real point of A_1 or A_2 can be a critical point unless we have $z_1 = z_2$.*

Theorem 10 is of significance only if $R(z)$ has a pole at infinity; in any other case it follows from Corollary 1 to Theorem 8 that all critical points of $R(z)$

are real. It follows too from Theorem 3 that all non-real critical points lie in the closed exterior of the Jensen circle for α and $\bar{\alpha}$, and it is obvious that no critical point lies on that circle.

Let z_0 be a non-real critical point of $R(z)$ not in R. The line of action L of the force at z_0 due to the particles at α and $\bar{\alpha}$ cuts the axis of reals at a point exterior to the interval $z_1 \leqq z \leqq z_2$, by §2.2.1 Theorem 1. All zeros of $R(z)$ lie on one side of L, so the total force at z_0 cannot be zero and z_0 cannot be a critical point. Likewise if z_0 lies on the boundary of R, then z_0 cannot be a critical point unless all zeros of $R(z)$ lie on L, that is to say, unless we have $z_1 = z_2$, Theorem 10 remains valid in the case $z_1 = z_2$, and R degenerates into the arc $A_1 = A_2$.

Theorem 10 does not explicitly involve the degree of $R(z)$; Theorem 2 does involve the degree and various multiplicities, and can be materially sharpened in the present case.

Theorem 10 represents the use of W-curves for rational functions in a simple but typical situation. However, both in Theorem 10 and in our use of W-curves in Chapter II the point at infinity plays a distinctive role. We now turn to the general situation, for rational functions not necessarily real.

Let $R(z)$ be a rational function with symmetry whose zeros and poles are given geometrically. We wish to determine a point set depending on those zeros and poles but not on their multiplicities, which contains all critical points of $R(z)$. For arbitrary symmetry, a *group* of points is a minimal (i.e. contains no smaller) set of points which possess the prescribed symmetry. For real functions, symmetry is symmetry in the axis of reals, and a group of points is either a single real point or a pair of conjugate imaginary points. Let $g_1, g_2, \cdots, g_\mu$ be the groups of zeros of $R(z)$, and $h_1, h_2, \cdots, h_\nu$ the groups of poles. There are now two alternate methods of procedure. Method I consists in forming each pair (g_i, h_j), where g_i or h_j may be multiply counted, to form a rational function $R_{ij}(z)$ of smallest degree with equal zeros in the points of g_i and equal poles in the points of h_j; this rational function has the prescribed symmetry, and has no other zeros or poles. We then consider the W-curves for each pair $R_{ij}(z)$ and $R_{kl}(z)$, namely the closure of the locus of critical points of the products of all positive powers (not necessarily integral) of these two functions. The function $R(z)$ is the product of a number of factors $R_{ij}(z)$, each to a positive power (not necessarily integral). If there exists a point z_0 and a sector S with vertex z_0 and angle less than π, such that the force at z_0 due to every pair of groups of particles (g_i, h_j) is represented by a vector whose initial point lies in S and whose terminal point is z_0, then there exists a point set Π bounded by W-curves containing all the zeros and poles and all the critical points of $R(z)$ but not containing z_0; such a set Π is precisely the closure of the point set separated from z_0 by all the W-curves. Indeed, as z varies continuously from z_0 without crossing a W-curve, the sector S varies continuously with z; the sector S (chosen as small as possible) can become equal to π only when z meets a W-curve. In the field of force for $R(z)$ no point z exterior to Π can be a critical point of $R(z)$.

An alternate to Method I is Method II, where we assume that the point at infinity satisfies the given symmetry (is a group) and where we now pair each finite group g_i with the point at infinity, obtaining a rational function $R_i(z)$ of smallest degree with zeros in the points of g_i and all poles at infinity; we also pair the point at infinity with each finite group h_j, obtaining a rational function $S_j(z)$ of smallest degree with all zeros at infinity and poles in the points of h_j; these rational functions have the prescribed symmetry, and have no zeros or poles other than those mentioned. We then consider the W-curves of all pairs (R_i, R_j), (R_i, S_j), and (S_i, S_j). The function $R(z)$ is the product of a number of factors $R_i(z)$ and $S_j(z)$, each to a suitable positive integral power. If there exists a point z_0 and a sector S with vertex z_0 and angle less than π, such that the force at z_0 due to every group g_i and the force at z_0 due to every group h_j is represented by a vector whose initial point lies in S and whose terminal point is z_0, then as before there exists a point set Π not containing z_0 bounded by W-curves containing all the zeros, all the poles, and all the critical points of $R(z)$; whenever a finite point z_0 (not a zero or pole of $R(z)$) is a position of equilibrium for no choice of multiplicities (integral or not) of zeros and poles in the points g_i and h_j, such a sector S must exist.

We return to the particular case of real rational functions. For Method II we have already (§2.6) determined the W-curves for pairs (R_i, R_j) and (S_i, S_j). The W-curves for pairs (R_i, S_j) may be determined by precisely those same formulas and methods; the former W-curves are of the form

$$\arg [R_i'(z)/R_i(z)] = \arg [R_j'(z)/R_j(z)],$$

and the latter are of the form

$$\arg [R_i'(z)/R_i(z)] = \arg [S_j'(z)/S_j(z)] + \pi.$$

Thus the W-curve for a pair (g_i, h_j), that is for the pair (R_i, S_j), where g_i and h_j consist each of a single point, is the axis of reals with the finite segment joining those two points omitted. The W-curve for a pair (g_i, h_j), where g_i consists of two non-real points α and $\bar{\alpha}$ and h_j consists of the real point β, is the arc $\alpha\bar{\alpha}$ greater than π of the circle tangent to the line $\alpha\beta$, plus the two disjoint infinite segments of the axis of reals bounded by the points β and $(\alpha + \bar{\alpha})/2$; if we have $\beta = (\alpha + \bar{\alpha})/2$, as may occur for instance in Theorem 10, the circular arc becomes the infinite segments bounded by α and $\bar{\alpha}$ of the line $\alpha\bar{\alpha}$, and the segments of the axis of reals become that entire axis. The W-curve for a pair (g_i, h_j), where g_i and h_j consist respectively of non-real points α and $\bar{\alpha}$, and β and $\bar{\beta}$, consists of the arcs not between the vertical lines $\alpha\bar{\alpha}$ and $\beta\bar{\beta}$, of a bicircular quartic (studied in §2.6.2) plus the segment of the axis of reals not between those lines; if all the points α, $\bar{\alpha}$, β, $\bar{\beta}$ lie on a vertical line, the W-curve consists of that vertical line except for the two finite segments each bounded by a point of each pair and containing no point of either pair, plus the axis of reals.

Method II is similar to Method I, except that in Method II the point at infinity is adjoined artificially as both a zero and a pole, so Method II may lead to less precise results. In Method II the results depend essentially on the position of the point at infinity, and are not invariant under linear transformation; in Method I the results are invariant under linear transformation, but have the relative disadvantage that the W-curves may be algebraically more complicated.

Theorem 10 is an illustration of our general remarks (Method II) on W-curves; the W-curves for pairs of zeros of $R(z)$ are finite segments of the axis of reals, and lie on the segment $z_1 \leqq z \leqq z_2$; the W-curve for the poles $(\alpha, \bar{\alpha})$ and a zero β consists of an arc $\alpha\bar{\alpha}$ in R of the circle tangent to the line $\alpha\beta$ at α, plus the disjoint infinite segments of the axis of reals bounded by the points β and $(\alpha + \bar{\alpha})/2$. At a non-real point z_0 interior to the Jensen circle for α and $\bar{\alpha}$, the force due to the two negative particles has a non-zero component vertically upward, as has the force due to each positive particle, so all vectors representing such forces lie in a sector less than π with vertex z_0.

There is one particular case of a real rational function $R(z)$ in which points z_0 are readily determined, namely where all zeros of $R(z)$ lie in the half-plane $x > 0$ and all poles lie in the half-plane $x < 0$. At any point z_0 of the axis of imaginaries, the force due to each pair (g_i, h_j) in Method I has a non-zero component directed toward the left, and the force due to each pair (R_i, R_j), (R_i, S_j), (S_i, S_j) in Method II likewise has such a component. The vectors representing these forces all lie in a sector with vertex z_0 of angle less than π. Every critical point of $R(z)$ either lies on a W-curve or is separated from the axis of imaginaries by a W-curve.

We remark too that if the prescribed symmetry admits two pairs each of equal particles all on a circle C, the two pairs being of opposite kinds, then the corresponding W-curve consists of arcs of C plus a possible circle C'. The equation of C' can be easily derived as in the proof of §2.2.1 Theorem 1, with one of the given particles at infinity. Compare §§5.6.2 and 9.3.1.

§5.2. Zeros and poles concyclic. We undertake here a deeper study than that of §5.1, now for the case of a real rational function whose zeros and poles are all real, or (the equivalent) for the case of a rational function whose zeros and poles all lie on a circle. This situation is especially significant because it is equivalent to the problem (§§8.7 and 9.3) of the critical points of harmonic measures of arcs of a Jordan region.

In §2.1 we have considered a special rational function whose zeros and poles are concyclic, and in §4.2.3 we have considered more generally a rational function whose zeros lie on an arc of a given circle and whose poles lie on a disjoint arc of that circle; in these cases all critical points likewise lie on that circle—a conclusion to be contrasted with that of Theorem 1 below.

By a non-euclidean (i.e. hyperbolic) line for the interior of a circle C we mean simply the arc interior to C of a circle orthogonal to C. Through any two points in the closed interior of C passes a unique non-euclidean (NE) line. Each NE line separates the interior of C into two open NE half-planes. A NE half-plane is *convex* in the sense that if two points of that set are chosen, the segment of the NE line joining them belongs to the set. Thus the set common to a finite or infinite number of NE half-planes is also convex.

§5.2.1. Zeros and poles interlaced. If $R(z)$ is a rational function, we say that its zeros and poles are *interlaced* on a circle C if those zeros and poles are all simple, and alternate on C. With this definition we have [Walsh, 1947]:

THEOREM 1. *Let the zeros and poles of a rational function $R(z)$ be interlaced on a circle C. No critical points of $R(z)$ lie on C. Let Π be the NE convex polygon whose sides are arcs of the NE lines for the interior of C which join consecutive zeros and consecutive poles of $R(z)$. Then Π contains all critical points of $R(z)$ interior to C. No such critical point lies on the boundary of Π unless $R(z)$ is of degree two, in which case Π degenerates to a point, the unique critical point interior to C.*

Thus a point z interior to C cannot be a critical point of $R(z)$ if a NE line separates z and a zero or pole of $R(z)$ from all the other zeros and poles of $R(z)$.

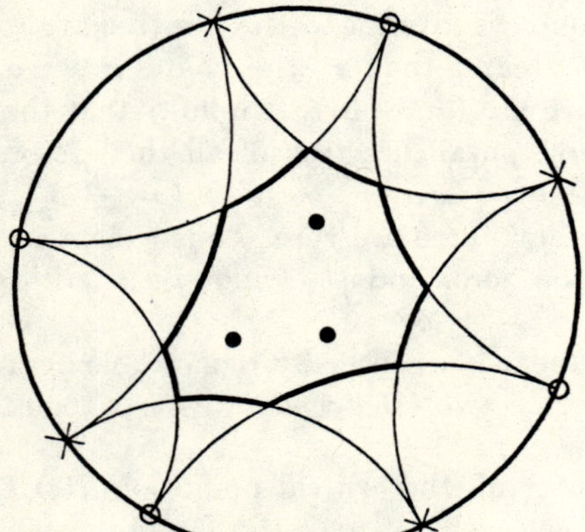

Fig. 11 illustrates §5.2.1 Theorem 1

Theorem 1 is without content for a rational function of degree unity, for such a function has no critical point. There is no non-trivial analog of Theorem 1 for polynomials.

It is readily shown that Π is NE convex, defined as the set common to $2n$ NE half-planes, and is bounded by segments of $2n$ NE lines, where n is the degree of $R(z)$.

Let P be a point of C not a zero or pole of $R(z)$, and let each zero of $R(z)$ be paired in order with an adjacent pole on C, commencing with the zero or pole next following P in the positive (counterclockwise) sense, the zero and pole of a pair not to be separated by P. Each pair of particles in the field of force exerts a force at P which is different from zero and acts along C in a sense independent of the pair, by §4.1.2 Corollary to Theorem 3, so the total force at P is not zero and P is not a critical point of $R(z)$.

In order to study an arbitrary point P interior to C as a possible critical point, we suppose that C is the unit circle and P is the origin O; this supposition involves no loss of generality. Let the zeros of $R(z)$ be the points α_k and the poles the points β_k, so numbered that they lie on C in the positive order α_1, β_1, α_2,

$\beta_2, \cdots, \alpha_n, \beta_n, \alpha_{n+1} = \alpha_1, \beta_{n+1} = \beta_1$. The situation of Theorem 1 is essentially symmetric with respect to zeros and poles. For definiteness let the specific NE line through α_k and α_{k+1} either pass through O or separate O and β_k from all the other points α_j and β_j; for definiteness choose $\beta_k = 1$. The positive arc (α_k, α_{k+1}) has angular measure at least π, and has angular measure π if and only if the NE line passes through O. Either α_k lies on the positive arc $(\beta_k, -1)$, or α_{k+1} lies on the closed positive arc $(-1, \alpha_k)$, or the arc $(\alpha_{k+1}, -1)$ is positive and not greater than the positive arc $(\alpha_k, 1)$, depending on the separation of the points α_k and α_{k+1} by the point -1.

The force at O due to a particle at α_j or β_j is represented by the vector $\alpha_j O$ or $O\beta_j$, and the force due to the pair of particles is represented by the vector $\alpha_j\beta_j$. The point O is a position of equilibrium if and only if the algebraic sum of all the vectors $\alpha_j\beta_j$ vanishes. The vector $\alpha_k\beta_k$ has a positive horizontal component; the point α_{k+1} cannot lie interior to the positive arc $(\beta_k, -\alpha_k)$ unless the positive arc (α_k, β_k) is greater than π; the points $\beta_{k+1}, \alpha_{k+2}, \cdots, \beta_{k-1}$ lie in succession on the positive arc (α_{k+1}, α_k). It follows that the sum of the components in the positive horizontal direction of all the vectors $\alpha_j\beta_j$ is positive, except in the special case $n = 2$, $\alpha_1 = -\alpha_2$, $\beta_1 = -\beta_2$, when the total horizontal component and total force are zero. Thus except in this special case the total force at O cannot be zero, and O cannot be a critical point of $R(z)$; the theorem is established.

The situation of Theorem 1 is especially remarkable because in the non-trivial case that $R(z)$ is of degree two, Π degenerates to a point, the unique critical point interior to C.

Theorem 1 accounts for all the critical points of $R(z)$, since the set of critical points is symmetric with respect to C; the closed interior of Π contains $n - 1$ critical points, where n is the degree of $R(z)$, as does the inverse of Π in C. Theorem 1 has been phrased for the case that C is a proper circle; the essential content of the theorem is true also if C is a line L, in which case the polygon Π is bounded by arcs of circles orthogonal to L, that is to say, is bounded by arcs of NE lines for the half-plane; a polygon Π lies in each of the half-planes bounded by L and contains all the critical points of $R(z)$ in that closed half-plane; arcs of the same circles form the polygons in the two half-planes.

If the zeros and poles of a rational function $R_1(z)$ are all of degree p and alternate on a circle C, Theorem 1 applies to the function $[R_1(z)]^{1/p}$; except for the multiple zeros of $R_1(z)$, all critical points of $R_1(z)$ in the closed interior of C lie in Π.

If R is an arbitrary region whose boundary B consists of a finite number of Jordan arcs, the *harmonic measure* $\omega(z, A, R)$ in the point z of R of a subset A of B consisting of a finite number of Jordan arcs is defined as the unique function harmonic and bounded in R, continuous on $R + B$ except perhaps in a finite number of points, equal to unity on A and to zero on $B - A$. If R is the interior of a circle, this function is represented by Poisson's integral, and at the center of the circle has as value the sum of the angles subtended by the set A divided by

2π; if A is a single arc of the circle B, the NE line joining the end-points of A is the locus $\omega(z, A, R) = \frac{1}{2}$ interior to C. The main conclusion of Theorem 1 can thus be expressed by saying that if z_0 is a critical point of $R(z)$, then for every arc A of C joining two successive zeros or poles of $R(z)$ we have $\omega(z_0, A, R) \leqq \frac{1}{2}$, where R is the interior of C.

If R is the upper half-plane and A the half-line $\arg z = \pi$, it follows by inspection that we have $\omega(z', A, R) = (\arg z')/\pi$. Consequently if R is the interior of a circle C and A an arbitrary arc of C, the locus $\omega(z, A, R) = \lambda$ in R is a circular arc through the end-points of A making angles of $\lambda\pi$ with the complement of A.

§5.2.2. Extensions. Various results analogous to Theorem 1 deserve consideration.

THEOREM 2. *On a circle C let the points a_1', a_1'', b_1', b_1'', $\cdots$, a_n', a_n'', b_n', b_n'', $a_{n+1}' = a_1'$, $a_{n+1}'' = a_1''$, lie in the positive sense in the order indicated. Let all zeros and poles of a rational function $R(z)$ lie on C, precisely m zeros on each arc (a_k', a_k'') and precisely m poles on each arc (b_k', b_k''), where m is independent of k. Let Π_k' denote the region interior to C bounded by the arc (a_k'', b_k') and the NE line $a_k''b_k'$, and let Π_k'' denote the region bounded by the arc (b_k'', a_{k+1}') and the NE line $b_k''a_{k+1}'$. Then Π_k' and Π_k'' contain no critical points of $R(z)$.*

Here and below we use (a, b) to denote the positive arc of C from a to b. Interchange of zeros and poles of $R(z)$—that is, replacing of $R(z)$ by $1/R(z)$—merely interchanges the sets of regions Π_k' and Π_k'', so for definiteness we study a point of Π_k', chosen as the origin O with C the unit circle; we choose the line $a_k''b_k'$ vertical, with a_k'' below b_k'; the origin lies to the right of the line $a_k''b_k'$. If the zeros and poles of $R(z)$ are respectively the points α_j and β_j, the total force at O due to all the particles is the sum of all the vectors $\alpha_j\beta_j$, where the zeros and poles may be paired in any convenient way. We choose to pair the m zeros on each arc (a_j', a_j'') with the m poles on the arc (b_j', b_j''). Then the total force at O is equal to m vectors joining points of (a_k', a_k'') to points of (b_k', b_k''), each of which vectors has an upward vertical component at least equal to the vertical projection of the segment $a_k'b_k''$, plus a number of other vectors, which occur in $n - 1$ sets each of m vectors, namely from points of arcs (a_j', a_j'') to points of arcs (b_j', b_j''), $j \neq k$. These latter vectors lie on or to the left of the line a_kb_k'', and if we choose one vector from each of the $n - 1$ sets, the sum of their vertical components (each directed downward) is less than the vertical projection of the segment $a_k'b_k''$. Consequently the total force at O has a non-zero component directed vertically upward, so O is not a position of equilibrium nor a critical point of $R(z)$. The proof is complete.

Under the conditions of Theorem 2, no interior point of a NE line $a_k''b_k'$ or $b_k''a_{k+1}'$ can be a critical point, as follows by the method just used, nor can an interior point of an arc (a_k'', b_k') or (b_k'', a_{k+1}'), as follows by the method used in Theorem 1.

Theorem 2 is relatively simple to establish, and can be somewhat generalized by a more refined use of the same method. Here a geometric lemma is convenient:

LEMMA 1. *Let A_1 and A_2 be closed disjoint arcs of the unit circle C, and let the circles C_1 and C_2 orthogonal to C pass through the end-points of A_1 and A_2 respectively. Let Γ be the circle of the coaxal family F determined by C_1 and C_2 in which C_1 and C_2 are mutually inverse; then A_1 and A_2 are mutually inverse in Γ. If Γ' is a circle of F which lies between C_1 and Γ, then the inverse A_1' of A_1 in Γ' is an arc of C which contains A_2 in its interior.*

The set of circles orthogonal to C_1 and C_2 (and to every circle of F) is precisely the set of circles through the two null circles of F, so those null circles lie on C. An inversion with one of those null circles as center renders the conclusion obvious.

In the terminology of harmonic measure the arc of Γ interior to C is the locus $\omega(z, A_1, |z| < 1) = \omega(z, A_2, |z| < 1)$. In fact, if an arbitrary point of Γ interior to C is transformed into the origin while C remains the unit circle, it is clear that these two harmonic measures are equal in that point, for $2\pi\omega(O, A_1, |z| < 1)$ is the angle subtended by A_1 at the origin. Moreover the function $\omega(z, A_1, |z| < 1) - \omega(z, A_2, |z| < 1)$ cannot vanish interior to C except on Γ, for in the region interior to C bounded by an arc of Γ and an arc of C containing A_1, this function is harmonic and bounded, non-negative at every boundary point and unity on A_1, hence is positive.

We are now in a position to extend Theorem 2:

COROLLARY 1. *Under the hypothesis of Theorem 2, let Γ denote the circle orthogonal to C in which the arcs (a_k'', b_k') and (a_{k+1}', b_{k-1}'') of C are mutually inverse. Let Π_k denote the region interior to C bounded by an arc of Γ and an arc of C containing (a_k'', b_k'). Then Π_k contains in its interior no critical points of $R(z)$.*

To prove Corollary 1, let z be a point interior to C between Γ and the region Π_k' and let Γ' denote a circle through z of the coaxal family of Lemma 1 for the arcs (a_k'', b_k') and (a_{k+1}', b_{k-1}''). Transform the interior of C into itself so that z is transformed into the origin and Γ' into a vertical diameter, with a_k'' below b_k'. The proof then goes through as before, by means of Lemma 1.

Corollary 1 remains true if the arcs (a_k'', b_k') and (a_{k+1}', b_{k-1}'') are replaced respectively by (b_{k-1}'', a_k') and (b_k', a_{k-1}''). Moreover no point of Γ itself is a critical point of $R(z)$ except in the case $n = 2$, with m zeros of $R(z)$ at a_k'' and at a_{k+1}' and m poles of $R(z)$ at b_k' and at b_{k-1}''.

Under the conditions of Theorem 2, with $a_k' = a_k''$ and $b_k' = b_k''$ for all k, Corollary 1 enables us to construct a subregion of the interior of C which contains all critical points of $R(z)$ interior to C, but except in the special case $n = 2$, Theorem 1 is more powerful here than Corollary 1, as is seen by transforming Γ into a diameter. In Corollary 1 unrestricted a circle Γ_1 joins a point of the arc (b''_{k-1}, a''_k) to a point of the arc (b'_k, a'_{k+1}), and a similar circle Γ_2 determined under Corollary 1 with the roles of zeros and poles interchanged joins a point of the arc (a''_{k-1}, b''_{k-1}) to a point of the arc (a'_k, b'_k). The circles Γ_1 and Γ_2, however, need not intersect, and do not intersect if b''_{k-1} and a'_k are sufficiently close together while the other points involved are distinct.

Under the conditions of Corollary 1, each open arc A of C bounded by two zeros or two poles of $R(z)$ and containing no zero or pole, and belonging to the boundary of one or more regions Π_k (or analogous regions obtained by interchanging the roles of zeros and poles in Corollary 1) contains precisely one critical point. This follows by the method of continuity, by allowing the zeros or poles of $R(z)$ bounding A to move continuously and monotonically in A and to coalesce, while all other zeros and poles of $R(z)$ remain fixed. During this process the critical points of $R(z)$ vary continuously except that when two poles of $R(z)$ coincide the critical point between them (and no other critical point) disappears; when two zeros of $R(z)$ coincide the critical point between them (and no other critical point) coincides with them; no critical point can leave A and enter the interior or exterior of C, and no critical point can leave the interior or exterior of C and enter A. At the end of this process we have precisely one critical point of A accounted for.

Still another result, less easy to apply than Theorem 2 and Corollary 1, is more closely analogous to Theorem 1, of which it is essentially a generalization:

COROLLARY 2. *Under the hypothesis of Theorem 2, let the two angles subtended between NE lines at the point z_0 interior to C by the arcs (a''_k, b'_k) and (b''_k, a'_{k+1}) total more than π. Then z_0 is not a critical point of $R(z)$.*

If C is the unit circle, this condition on z_0 can be written $\omega(z_0, a''_k b'_k, |z| < 1) + \omega(z_0, b''_k a'_{k+1}, |z| < 1) > \frac{1}{2}$.

We shall need a lemma, whose proof is indeed quite easy:

LEMMA 2. *An arc A of fixed length slides rigidly along the unit circle C so that every interior point of A remains on the upper semicircle of C. Then the least horizontal projection of A occurs when one end of A lies on the horizontal diameter.*

Denote by α the length of the arc A, and by β the usual angular coordinate of the initial point of A. The inequality to be proved is

$$\cos \beta - \cos(\alpha + \beta) \geqq 1 - \cos \alpha,$$

$$\sin \alpha \sin \beta \geqq (1 - \cos \alpha)(1 - \cos \beta),$$

$$1 \geqq \tan \frac{\alpha}{2} \tan \frac{\beta}{2} \cdot$$

However, we are assuming $\alpha + \beta \leqq \pi$, whence $\beta/2 \leqq \pi/2 - \alpha/2$,

$$\tan \frac{\beta}{2} \leqq \cot \frac{\alpha}{2} = 1/\tan \frac{\alpha}{2},$$

and the lemma is established.

In continuing the proof of Corollary 2, we assume that C is the unit circle, and that z_0 lies at the origin with $b_k' = 1$. Here we separate the proof into several cases.

Case I: $(a_k'', b_k') \geqq \pi$. Here the conclusion is contained in Theorem 2.

Case II: $(a_k'', b_k') < \pi$ and b_k'' lies on or to the right of the vertical line through a_k''. Each vector from a point of the arc (a_k', a_k'') to a point of the arc (b_k', b_k'') has a horizontal component directed to the right of magnitude not less than the horizontal projection of the vector $a_k''b_k''$, which is the horizontal projection of the vector $\bar{a}_k''b_k''$. However, if a_{k+1}' lies in the upper semicircle we have the following relations among arc lengths considered numerically, making use of the special hypothesis of Corollary 2,

$$(a_{k+1}', -1) = \pi - (b_k'', a_{k+1}') - (b_k', b_k'')$$
$$< (a_k'', b_k') - (b_k', b_k'') = (b_k', \bar{a}_k'') - (b_k', b_k'') = (b_k'', \bar{a}_k'').$$

From Lemma 2 it then follows (whether a_{k+1}' lies in the upper semicircle or not; and if $b_k'' = \bar{a}_k''$, a_{k+1}' must lie in the closed lower semicircle) that no sum of vectors one from a point of each arc (a_j', a_j'') to a point of the arc (b_j', b_j''), $j = 1, 2, \cdots, k-1, k+1, \cdots, n$, can have a horizontal component directed to the left numerically as great as the horizontal component of any vector from a point of (a_k', a_k'') to a point of (b_k', b_k''). The corresponding relation considered for each of the m sets of vectors involved shows that z_0 is not a position of equilibrium.

Case III: $(a_k'', b_k') < \pi$ and b_k'' lies to the left of the vertical line through a_k''. Here b_k'' lies in the upper semicircle and a_{k+1}' lies in the lower semicircle, for by the hypothesis of Corollary 2 we have $(b_k', a_{k+1}') = (b_k', \bar{a}_k'') + (\bar{a}_k'', b_k'') + (b_k'', a_{k+1}') = (a_k'', b_k') + (\bar{a}_k'', b_k'') + (b_k'', a_{k+1}') > \pi + (\bar{a}_k'', b_k'')$. We also have $(b_k'', a_{k+1}') > \pi - (a_k'', b_k') = (-1, a_k'') > (a_{k+1}', a_k'')$. Each vector from an arc (a_j', a_j'') to an arc (b_j', b_j'') whether $j = k$ or $j \neq k$ has a non-vanishing component in the direction and sense a_{k+1}' to $(-a_{k+1}')$, so z_0 is not a position of equilibrium and Corollary 2 is established.

If we modify Corollary 2 so as to require that the sum of angles subtended by the given arcs be not less than π, it is still true that z_0 is not a position of equilibrium unless we have $n = 2$ and the arcs (a_j', a_j'') and (b_j', b_j'') all degenerate to points. Theorem 1 is a consequence of Corollary 2 and of the remark just made, including interchange of the roles of zeros and poles.

We add one further result in this same order of ideas, a result closely related to Corollary 1:

COROLLARY 3. *Let all zeros and poles of the rational function* $R(z)$ *of degree* n *lie on a circle* C, *and let the points* B_1, B_2, B_3, B_4 *lie on* C *in that positive order. Let the arcs* B_1B_2 *and* B_3B_4 *be mutually inverse in the circle* Γ, *and let* Π *denote the subregion of the interior of* C *bounded by an arc of* Γ *and an arc of* C *containing the arc* B_1B_2. *If precisely* k ($k \geqq n/2$) *zeros of* $R(z)$ *lie on the closed arc* B_4B_1, *no zeros or poles lie on the open arc* B_1B_2, *precisely* k *poles of* $R(z)$ *lie on the arc* B_2B_3, *and all the remaining* $n - k$ *zeros and* $n - k$ *poles lie on the open arc* B_3B_4, *then no critical points of* $R(z)$ *lie in* Π.

Let z_0 be a point of Π, and let Γ' be the circle through z_0 of the coaxal family determined by Γ and the circle through B_1 and B_2 orthogonal to C. We choose C as the unit circle, with z_0 the origin, Γ' a vertical diameter, and with B_2 above B_1. It follows from Lemma 1 if z_0 lies between Γ and the NE line B_1B_2, and is otherwise obvious, that the vertical projection of B_1B_2 contains in its interior that of B_4B_3. The total force at O is the sum of at least $n/2$ vectors each from a point of B_4B_1 to a point of B_2B_3, each of these vectors having an upward vertical component at least equal to that of the vector B_4B_3, plus no more than $n/2$ vectors each joining two points of the open arc B_3B_4. Consequently the total force at z_0 has a non-zero component directed vertically upward, and z_0 is not a position of equilibrium nor a critical point.

No point of Γ interior to C can be a critical point of $R(z)$, as follows from the reasoning as given. No point of the arc B_1B_2 of C can be a critical point, as follows by observing the configuration obtained by inverting C in a circle whose center is an arbitrary point of B_1B_2.

§5.2.3. Critical points near a given zero. From the field of force it is obvious that no critical point of a rational function $R(z)$ can lie near a given zero of $R(z)$, provided all the other zeros and all the poles of $R(z)$ are suitably distant. In §4.3 we have made this observation precise, in terms of the multiplicity of the given zero, the degree of $R(z)$, and the minimum distance from the given zero to the other zeros and the poles of $R(z)$. The results of §4.3 remain valid but can be somewhat improved if the zeros and poles of $R(z)$ are concyclic, so we proceed to develop various improvements.

THEOREM 3. *Let* $R(z)$ *be a rational function of degree* n *whose zeros and poles lie on a circle* C. *Let* $z = \alpha_1$ *be a zero of* $R(z)$ *of order* k, *and let positive open arcs* A *and* B *of* C *respectively terminating and commencing in* α_1 *contain no pole or zero of* $R(z)$. *Let the arcs* A *and* B *each subtend by NE lines angles greater than* $\cos^{-1}[(2k - n)/n]$ *at the point* z_0 *interior to* C. *Then* z_0 *is not a critical point of* $R(z)$.

We suppose C to be the unit circle, with z_0 at the origin and $\alpha_1 = 1$. Denote by D the arc of C complementary to the sum of A, B, and their common end-point. If the zeros of $R(z)$ are denoted by $\alpha_1, \alpha_2, \cdots, \alpha_n$ and the poles by $\beta_1, \beta_2, \cdots, \beta_n$, the total force at z_0 due to the corresponding particles is the sum of the vectors $\alpha_j z_0$ and $z_0\beta_j$ or the sum of the vectors $\alpha_j\beta_j$. Then the total force at z_0 is the sum of k vectors from α_1 to a point of D, having a total horizontal component directed to the left greater than

$$k\left[1 - \frac{2k - n}{n}\right] = \frac{2k(n - k)}{n},$$

and of $n - k$ vectors joining points of D to points of D, having a total horizontal component directed to the right less than

$$(n - k)\left[1 + \frac{2k - n}{n}\right] = \frac{2k(n - k)}{n};$$

consequently z_0 is not a position of equilibrium nor a critical point of $R(z)$. Theorem 3 is established.

The condition on z_0, when C is the unit circle, may be written

$$2\pi\omega(z_0, A, |z| < 1) > \cos^{-1}[(2k - n)/n] = \theta_0, \; 2\pi\omega(z_0, B, |z| < 1) > \theta_0.$$

In Theorem 3 we assume that each of the arcs A and B subtends at least a specific angle at z_0. The same method can be extended, however, to include the study of points near A or B. Here it is convenient to phrase the hypothesis in different form, choosing C as a line:

THEOREM 4. *Let $R(z)$ be a rational function of degree n whose zeros and poles all lie on the axis of imaginaries, which has a k-fold zero at the origin, and whose other zeros and poles lie in the intervals $|y| \geqq b$. Then all critical points of $R(z)$ not on the axis of imaginaries lie in the closed exterior of the curve*

$$(1) \qquad n(n - k)(x^2 + y^2)^2 - kb^2[nx^2 + (n - k)y^2] = 0;$$

that is to say, at every critical point of $R(z)$ not on the axis of imaginaries the first member of (1) *is positive or zero.*

Consequently no critical point not on the axis of imaginaries lies interior to the circle $x^2 + y^2 = kb^2/n$, except perhaps at the origin.

No critical point of $R(z)$ except perhaps at the origin lies on the axis of imaginaries in the interval $|y| \leqq kb/(2n - k)$.

The last part of Theorem 4 follows from §4.3 Theorem 1 Corollary 3, and the limit $kb/(2n - k)$ cannot be improved.

Let the point P: (x, y) not lie on the axis of imaginaries; for definiteness let P lie in the right half-plane; we denote the points ib and $-ib$ by E and F respectively, and denote the angles from the positive horizontal to OP, EP, and FP by θ_0, θ_1, θ_2. Invert the entire configuration in the unit circle whose center is P; the image of the axis of imaginaries is a circle C' whose center is denoted by O_1, and P lies at the right-hand end of the horizontal diameter. If we denote the images of O, E, F by O', E', F', the rays O_1O', O_1E', and O_1F' are respectively $\theta = 2\theta_0 + \pi$, $2\theta_1 + \pi$, $2\theta_2 + \pi$. The rays bisecting the arcs $F'PE'$ and $E'O'F'$ are respectively $\theta = \theta_1 + \theta_2$, $\theta_1 + \theta_2 + \pi$.

In the field of force the total force at P is the sum of k vectors $O'P$ plus $n - k$ vectors from points of arc $F'PE'$ to P, plus n vectors from P to points of arc $F'PE'$, hence is equal to the sum of k vectors from O' to points of arc $F'PE'$ plus $n - k$ vectors between pairs of points of arc $F'PE'$. We consider the components of this latter set of vectors along the line through O_1 bisecting the arcs $E'F'$. The point P cannot be a position of equilibrium if the components of the k vectors in the direction $\theta = \theta_1 + \theta_2$ are greater than the components of the remaining $n - k$ vectors in the opposite direction; that is to say, equilibrium at P is impossible if we have

$$(2) \qquad k[\cos (2\theta_0 - \theta_1 - \theta_2) - \cos (\theta_1 - \theta_2)] > (n - k)[1 + \cos (\theta_1 - \theta_2)].$$

If inequality (2) is expressed in terms of b, x, and y, and if a factor x^2 is suppressed, the inequality can be written

$$(3) \qquad n(n - k)(x^2 + y^2)^2 < kb^2[nx^2 + (n - k)y^2].$$

Equation (1) represents a bicircular quartic symmetric in both coordinate axes, of which the origin is an isolated point. The curve has no multiple point corresponding to real branches, for one such point would imply a second one symmetric in the origin to the first, and the line through those multiple points would have at least six intersections with the curve. Indeed, a line through O can meet the curve in at most two points other than O; in the neighborhood of O inequality (3) is clearly satisfied whereas it is clearly not satisfied for large $x^2 + y^2$; thus the curve consists, in addition to the isolated point O, of a Jordan curve in whose interior O lies. Inequality (3) defines (except for the point O) the interior of this curve, and the reverse inequality defines the exterior. We obviously have

$$(n - k)(x^2 + y^2) \leqq nx^2 + (n - k)y^2 \leqq n(x^2 + y^2),$$

so the curve lies in the closed interior of the circle $x^2 + y^2 = kb^2/(n - k)$, and the closed interior of (1) contains the closed interior of the circle $x^2 + y^2 = kb^2/n$. These circles are tangent to the curve on the axis of reals and axis of imaginaries respectively. Theorem 4 is established.

It may seem paradoxical that the limitation $|y| < b_1 = kb/(2n - k)$ does not correspond to the intersection of the curve (1) with the axis of imaginaries. Algebraically the justification for the discrepancy is the suppression of the factor x^2 in deriving (3) from (2). On the other hand, it will be noticed that the limit $|y| = b_1$ is assumed for a rational function whose zeros and poles lie on the axis of imaginaries, and each limit $y = \pm b_1$ corresponds to a *simple* critical point. The critical points of $R(z)$ are symmetric in the axis of imaginaries, and if under such conditions a critical point varies continuously as a function of the zeros and poles of $R(z)$, the critical point can leave the axis only if it is a multiple critical point; the limitation $|y| < b_1$ is derived for simple rather than multiple critical points.

Theorem 3 is contained in Theorem 4, and for points on the axis of reals in Theorem 4, Theorem 3 is no less precise.

§5.2.4. Continuation. In Theorem 4 we have in reality solved a more general problem than that indicated, namely the location of the zeros of the rational function ($\mu_j > 0, \nu_j > 0$)

$$(4) \qquad \frac{k}{z} + \sum \frac{\mu_j}{z - \alpha_j} - \sum \frac{\nu_j}{z - \beta_j}, \qquad \sum \mu_j = n - k, \qquad \sum \nu_j = n,$$

where the points α_j and β_j lie on the segments $|y| \geqq b$ of the axis of imaginaries. We replace the field of force used in the proof of Theorem 4 by the conjugate of (4). The point set obtained in Theorem 4 free from critical points, namely the interior of (1) except points on the axis of imaginaries, cannot be improved in the more general problem, for at an arbitrary point P of (1) not on the axis of imaginaries a suitable choice of $\alpha_j, \beta_j, \mu_j, \nu_j$ not merely replaces (2) by the corresponding *equality* but also implies vanishing of the sum of the vector components in the direction $\theta = \theta_1 + \theta_2 + \pi/2$, and implies that P is a position of equilibrium, a zero of (4).

However, Theorems 3 and 4 admit of improvement if we retain the original hypotheses of those theorems; the improvements express the fact that replacing of (2) by an equality, where k and n are given *integers*, implies that the zeros and poles of $R(z)$ can be so disposed that the component of the total force in the direction $\theta = \theta_1 + \theta_2$ vanishes, but does not automatically imply that the component in the perpendicular direction vanishes, hence does not automatically imply that P is a position of equilibrium. Further investigation of this problem involves the partitions of the numbers k and n, as we proceed to indicate. We restrict ourselves to the situation of Theorem 3, for it is our purpose to illustrate a method rather than to develop an elaborate theory.

Let θ_0 be arbitrary, $0 < \theta_0 < \pi$, and let Δ denote the arc $\theta_0 \leqq \theta \leqq 2\pi - \theta_0$ of the unit circle C, namely the positive arc $\beta_1\beta_2$. In the elaboration of the reasoning already used, two problems present themselves: Problem 1: To determine a single vector equivalent to the sum of k arbitrary vectors whose common initial point is $\alpha_1 = 1$ and whose terminal points lie on Δ, Problem 2: To determine a single vector equivalent to the sum of m arbitrary vectors whose initial and terminal points lie on Δ. We treat these problems in order.

In the case of Problem 1 with $k = 1$, the class of equivalent vectors is precisely the class of vectors from α_1 to points of Δ. The sum of any two such vectors is equivalent to twice the vector from α_1 to the center of gravity of their terminal points. If two points have Δ as their locus, their center of gravity has a readily determined locus Δ_2, which may be found as the locus of points of a variable arc Δ' defined by shrinking Δ toward the point z_0 in the ratio 2:1, and allowing z_0 to trace the arc Δ. The arc Δ' is an arc with vertical chord of a circle of radius $\frac{1}{2}$ which passes through the origin. An end-point of Δ' can be interpreted as the image of an end-point β_j of Δ found by shrinking C toward z_0 in the ratio 2:1, and can also be interpreted as the image of z_0 found by shrinking C and Δ toward β_j in the ratio 2:1. Thus when z_0 varies, the arc Δ' moves so that its end-points lie on two arcs Δ_j' ($j = 1, 2$) found by shrinking Δ toward β_j in the ratio 2:1.

The two arcs Δ_j' have the same size and orientation as Δ', they have a common end-point on the axis of reals, and they are tangent to C in the respective points β_j. The locus Λ_2 is bounded by the arcs Δ, Δ_1', Δ_2', and consists of one closed region if we have $\theta_0 \geqq \pi/2$, and consists of two closed regions with the single point O in common if we have $\theta_0 < \pi/2$. The totality of vectors equivalent to the sum of two vectors each from α_1 to a point of Δ is the class of vectors each double the vector from α_1 to a point of Λ_2.

If now the number k of vectors each from α_1 to a point of Δ is increased to three, the sum is studied by considering the locus Λ_2 just determined as the locus of a point z_1, the arc Δ as the locus of a point z_2, and determining the locus Λ_3 of their weighted center of gravity $(2z_1 + z_2)/3$. The locus Λ_3 is bounded by Δ and the three circular arcs found by shrinking Δ toward β_1, $(\beta_1 + \beta_2)/2$,

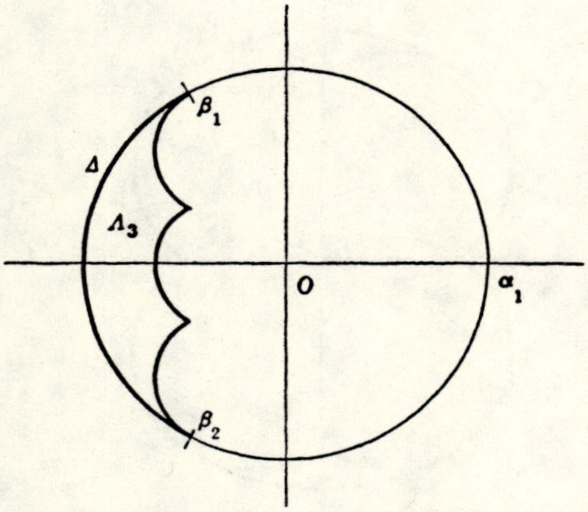

Fig. 12 illustrates §5.2.4 ($\theta_0 > \pi/2$)

and β_2 in the ratio 3:1; the locus Λ_3 consists of one closed region if we have $\theta_0 \geqq \pi/2$, and of three closed regions if we have $\theta_0 < \pi/2$.

In the general case of k vectors each from α_1 to a point of the arc Δ, the locus Λ_k of centers of gravity of their terminal points is bounded by Δ and by k arcs each $(1/k)$-th the size of Δ, oriented like Δ, the terminal point of one arc being the initial point of another; these initial and terminal points of the $k + 1$ arcs are the points β_1, β_2, and the points dividing the segment $\beta_1\beta_2$ into k equal parts. The locus Λ_k consists of one closed region in the case $\theta_0 \geqq \pi/2$, and of k closed regions in the case $\theta_0 < \pi/2$. The sum of k vectors each from α_1 to a point of Δ is equivalent to k times a vector from α_1 to a point of the locus Λ_k, and conversely.

We turn now to Problem 2. Let Q be a fixed point of Δ; we move the arc Δ rigidly without changing its size or orientation so that it always passes through Q; the point-set Φ_1 swept out by Δ represents the terminal point of a vector whose initial point is Q, the vector being equivalent to an arbitrary vector from one point of Δ to another. Denote by C' the circle whose center is Q and radius two. As Δ varies, Δ remains an arc of a variable circle of radius unity constantly tangent to C' at some variable point z_0' of C'. When Δ is magnified with z_0' as center in the ratio 1:2, the end-points of Δ have as respective images two points β_1' and β_2' of C' which are independent of z_0'. The positive arc $\beta_1'\beta_2'$ of C' has the same angular measure and the same orientation on C' as does Δ on C. The end-points of Δ can be considered as the images of z_0' when C' is shrunk toward β_j' ($j = 1, 2$) in the ratio 2:1, so the end-points of Δ trace respectively arcs δ_1 and δ_2 of the circles of radius unity internally tangent to C' at β_1' and β_2', circles found by shrinking C' toward β_j' in the ratio 2:1. The arcs δ_1 and δ_2 have the same angular measure as Δ, are oriented like Δ rotated through the angle π, and have their initial and terminal points on the vertical line through Q.

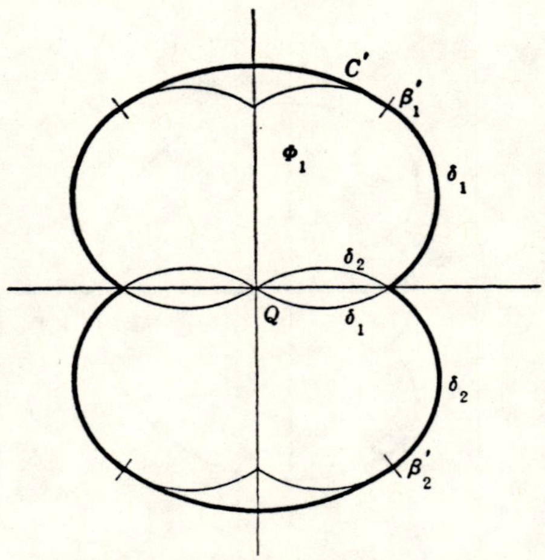

Fig. 13 illustrates §5.2.4 ($\theta_0 < \pi/2$)

The locus Φ_1 is always symmetric in Q, and in the horizontal and vertical lines through Q. In the case $\theta_0 < \pi/2$, this locus is a region bounded by parts of the two arcs δ_1 and δ_2, by parts of their respective reflections δ_1' and δ_2' in Q, and by two arcs of C' each intersecting the vertical line through Q and each terminated by its own points of tangency with δ_1 and δ_2' or δ_2 and δ_1'. In the case $\theta_0 = \pi/2$, the locus Φ_1 consists of the closed interiors of two circles each of radius unity externally tangent at Q and whose centers lie on the vertical line through Q. In the case $\theta_0 > \pi/2$, the locus Φ_1 consists of the two closed regions bounded respectively by δ_1 and δ_2', and by δ_2 and δ_1'; these closed regions have Q as common point. A second method of determining the locus Φ_1 is to study directly all possible vectors in each given direction from a point of Δ to another; here it is convenient to emphasize vectors whose initial or terminal points are end-points of Δ, and of course the same locus Φ_1 is obtained. Any vector from one point of Δ to another is equivalent to a vector from Q to some point of Φ_1, and conversely.

To study the sum of two vectors each having its initial and terminal points on Δ, we study two vectors having their initial points in Q and their terminal points on Φ_1 ; the center of gravity of these terminal points each having Φ_1 as their locus is a new locus Φ_2 , and the sum of the two given vectors is twice the vector from Q to a point of Φ_2 . The locus Φ_2 is conveniently studied by the methods used in §1.5.1 Lemma 2; the boundary of Φ_2 consists of parts of the arcs δ_1 , δ_2 , δ_1' , δ_2' , and if $\theta_0 < \pi/2$ of arcs of the circle C', in every case together with parts of the arcs found from δ_1 and δ_2' by translating them vertically downward a distance $\sin \theta_0$ so that their new centers lie on the horizontal line through Q.

In a similar way, we determine the locus Φ_m , the locus of the center of gravity of m points of Φ_1 ; the locus Φ_m is bounded by parts of the arcs δ_1 , δ_2 , δ_1' , δ_2' , and if $\theta_0 < \pi/2$ of arcs of the circle whose center is Q and radius two, together with parts of the $2(m - 1)$ arcs found from δ_1 and δ_2' by translating them vertically downward through distances $(2j/m) \sin \theta_0$, $j = 1, 2, \cdots, m - 1$. The sum of m vectors each from a point of Δ to a point of Δ is equivalent to m times a vector from Q to a point of Φ_m , and conversely.

We are now in a position to resume the situation of Theorem 3, where k and n are given; we wish to determine θ_0 in such a way that it is the smallest value which enables us to conclude that z_0 is not a critical point of $R(z)$ if the arcs A and B subtend angles greater than θ_0 at z_0 . Let k zeros of $R(z)$ lie at $\alpha_1 = 1$ and the remaining zeros and all the poles lie on the arc Δ. Let us denote by $\rho \cdot \Lambda_k$ the locus found by stretching Λ_k from α_1 in the ratio $1:\rho$, and denote by $\rho \cdot \Phi_{n-k}$ the locus found by choosing Q as α_1 and stretching Φ_{n-k} from α_1 in the ratio $1:\rho$. The point $z_0 = 0$ is a possible critical point of $R(z)$ if and only if the sum of k vectors $\alpha_1 z_0$ plus $n - k$ vectors from points of Δ to O plus n vectors from O to points of Δ vanishes, or if and only if the sum of k vectors from α_1 to points of Δ plus $n - k$ vectors from points of Δ to points of Δ vanishes, or if and only if the two loci $k \cdot \Lambda_k$ and $(n - k) \cdot \Phi_{n-k}$ have points in common. The limiting value of θ_0 is that for which these loci have boundary points but not interior points in common.

The locus $k \cdot \Lambda_k$ is bounded by $k + 1$ circular arcs, of one circle of radius k and k circles of radius unity, provided we have $k > 1$. Each point of the locus lies on or to the left of the vertical line joining the points of tangency of the large circle with two of the smaller circles; the points of the locus on this line (i.e. the "peaks") have the ordinates $\sin \theta_0$ multiplied by $-k, -k + 2, -k + 4, \cdots, k$, respectively. The locus $(n - k) \cdot \Phi_{n-k}$ is bounded in part by two arcs of circles of radius $2(n - k)$ with center α_1 if $\theta_0 < \pi/2$, and whatever θ_0 may be the remaining part of the boundary to the left of the vertical line through α_1 consists of arcs of $n - k + 1$ circles each of radius $n - k$; the centers of these circles have the ordinates $\sin \theta_0$ multiplied by $-n + k, -n + k + 2, -n + k + 4, \cdots, n - k$ respectively.

If k and $n - k$ are both odd or both even—that is to say, if n is even—the "peaks" of $k \cdot \Lambda_k$ lie on the same horizontal lines as the centers of the circles whose arcs form the significant part of the boundary of $(n - k) \cdot \Phi_{n-k}$. For the limiting value of θ_0 we have $k(1 - \cos \theta_0) = (n - k)(1 + \cos \theta_0)$, so θ_0 is as indicated in

Theorem 3; the limiting loci $k \cdot \Lambda_k$ and $(n - k) \cdot \Phi_{n-k}$ have their only common points on the vertical line through the extreme right-hand points of the former, which is the vertical line through the extreme left-hand points of the latter.

If k and $n - k$ are not both odd or both even—that is to say, if n is odd—the "peaks" of $k \cdot \Lambda_k$ lie halfway between the horizontal lines on which lie the centers of the circles whose arcs form the significant part of the boundary of $(n - k) \cdot \Phi_{n-k}$; that is to say, the "peaks" of the former lie opposite the "valleys" of the latter. The radii of the significant circles of the two loci are unity and $n - k$ respectively; the interior angles of intersection of the circular arcs at a peak of $k \cdot \Lambda_k$ (if $k > 1$) are $\pi - 2\theta_0$ or $2\theta_0 - \pi$ according as we have $\theta_0 < \pi/2$ or $\theta_0 \geqq \pi/2$, and the exterior angles at a valley of $(n - k) \cdot \Phi_{n-k}$ are not less than this number. Thus the limiting value of θ_0 occurs when a peak meets a valley; the corresponding equation for θ_0 is

$$k(1 - \cos \theta_0) = (n - k) \left[\left(1 - \frac{\sin^2 \theta_0}{(n - k)^2} \right)^{1/2} + \cos \theta_0 \right],$$

$$\cos \theta_0 = (1 + 2kn - n^2)/(n^2 - 1).$$

We have thus established

THEOREM 5. *If n is even, Theorem 3 cannot be improved by replacing the number $(2k - n)/n$ by a larger number, but if n is odd that number may be replaced by $(1 + 2kn - n^2)/(n^2 - 1)$.*

The methods just developed enable one easily to determine all special rational functions $R(z)$ for which z_0 is a critical point, if the arcs A and B of Theorem 3 subtend angles at z_0 of precisely θ_0 .

§5.2.5. Zeros and poles on prescribed arcs. The methods already used yield various other results, of which we present several illustrations.

THEOREM 6. *Let k zeros of the rational function $R(z)$ of degree n lie on an arc A of a circle C, and let the remaining $n - k$ zeros and all poles of $R(z)$ lie on an arc B of C disjoint from A. If z_0 is a point interior to C at which opposite angles less than $2\theta_0$ are subtended between NE lines by arcs of C containing A and B respectively, then z_0 is not a critical point of $R(z)$ if we have*

$$\cos \theta_0 = (n - k)/(n + k), \quad n + k \text{ even},$$

$$\cos^2 \theta_0 = [(n - k)^2 - 1]/[(n + k)^2 - 1], \quad n + k \text{ odd}.$$

We suppose that C is the unit circle, z_0 the origin, A an arc whose center is $z = 1$, and B an arc whose center is $z = -1$; in the proof we identify A and B with the arcs containing them. We choose A and B symmetric with respect to O, each subtending the angle 2θ there; it remains to find the limiting value θ_0 of θ that O may be a position of equilibrium.

The usual field of force consists of k positive particles on A, $n - k$ positive particles on B, and n negative particles on B. So far as concerns the force at O, these negative particles on B are equivalent to respective positive particles in the diametrically opposite points, which lie on A. Thus we study A as the locus of $n + k$ positive particles, B as the locus of $n - k$ positive particles, and consider O as a possible position of equilibrium.

The locus Λ_m of the center of gravity of m positive particles on A consists of A if $m = 1$ and otherwise of a closed region bounded by A and by m arcs each in size and orientation a shrinking of A in the ratio $m:1$; the end-points of these m arcs divide into m equal parts the segment joining the end-points of A. If m is even, the point of Λ_m nearest to O lies on the axis of reals and is at the distance $d_m = \cos\theta$; if m is odd, the point of Λ_m nearest to O is at the distance from O

$$d_m = \left[\frac{1}{m^2} + \frac{m^2 - 1}{m^2}\cos^2\theta\right]^{1/2}.$$

We denote by $m \cdot \Lambda_m$ the locus found by stretching Λ_m from O in the ratio $1:m$, so the force at O due to m positive particles on A is always a vector to O from a point of $m \cdot \Lambda_m$, and conversely. Likewise we define the similar locus Φ_m of the center of gravity of m particles on the arc B, and by $(-m)\cdot\Phi_m$ the locus found by stretching Φ_m from O in the ratio $1:m$ and then rotating about O through the angle π. The force at O due to m positive particles on B is always a vector from O to a point of $(-m)\cdot\Phi_m$, and conversely.

If θ is small, the total force at O due to the $n + k$ positive particles on A is numerically greater than that due to the $n - k$ positive particles on B; in other words, the locus $(n + k)\cdot\Lambda_{n+k}$ lies to the right of and is disjoint from the locus $(-n + k)\cdot\Phi_{n-k}$. A necessary and sufficient condition that O be a position of equilibrium for suitable choice of $R(z)$ is that these two loci intersect. If k and n are fixed but θ is allowed to increase, there is a limiting value θ_0 of θ; for $\theta < \theta_0$ the loci do not intersect, and for $\theta \geqq \theta_0$ the loci do intersect. The left-hand boundary of $(n + k)\cdot\Lambda_{n+k}$ consists of $n + k$ arcs of circles each of radius unity, while the right-hand boundary of $(-n + k)\cdot\Phi_{n-k}$ consists of a single circle of center O and radius $n - k$. Thus the limiting value of θ occurs when we have $(n + k)d_{n+k} = n - k$, and Theorem 6 follows.

In Theorem 6, if the opposite angles at z_0 subtended by A and B are precisely $2\theta_0$, then z_0 is a critical point of $R(z)$ only for certain special configurations which are readily discussed by the present methods. We remark that the case $n + k$ odd, $n - k = 1$, $\theta_0 = \pi/2$ is of particular interest, for here the right-hand boundary of $(-n + k)\cdot\Phi_{n-k}$ is a semicircle of radius unity which also belongs to the boundary of $(n + k)\cdot\Lambda_{n+k}$; of course this limiting case is itself not properly included in Theorem 6, but if $n = 2$ the origin is a critical point no matter in what point the k zeros of $R(z)$ are chosen on the right-hand semicircle, if $n - k$ zeros are chosen diametrically opposite, with k poles at $+i$ and $n - k$ poles at $-i$.

In our proof of Theorem 6 we have in reality solved also another problem, namely the question as to when the origin can be a critical point of a polynomial of degree $2n$, when $n + k$ zeros lie on an arc A subtending at O an angle 2θ and when $n - k$ zeros lie on the diametrically opposite arc B. The limiting values θ_0 of θ are as given, and the special limiting configurations all retain their full significance. For instance if $n + k$ is odd, $n - k = 1$, $\theta_0 = \pi/2$, then the special polynomial has $(n + k - 1)/2$ zeros at each of the points $\pm i$, one zero at an arbitrary point on the right-hand semicircle, and another zero at the diametrically opposite point; the origin is a critical point.

If $R(z)$ is given, the locus of points z_0 that satisfy the conditions of Theorem 6 is readily constructed. Let disjoint closed arcs A and B of C contain respectively k zeros of $R(z)$ and all the remaining zeros and poles of $R(z)$. Denote by D_1 and D_2 the arcs of C adjacent to and separating A and B. Let θ_0 be computed from the formulas given. We construct four loci interior to C: the loci of points z_0 from which arcs A and B respectively subtend angles less than $2\theta_0$ (these loci are bounded by arcs of C complementary to A and B and by circular arcs interior to C joining their end-points), and the loci of points z_0 from which arcs D_1 and D_2 respectively subtend angles greater than $\pi - 2\theta_0$ (these loci are bounded by the arcs of C concerned and by circular arcs interior to C joining their end-points). *The set of points z_0 common to these four loci is the locus of points z_0 satisfying the hypothesis of the theorem.*

We note first that if z_0 satisfies the hypothesis of the theorem, it belongs to each of these four loci. Conversely, let z_0 assumed at the center of C belong to each of these four loci. Denote by A' the minimal arc of C which contains A, contains no point of B, but contains all points of C diametrically opposite to points of B; denote by B' the minimal arc of C which contains B, contains no point of A', but contains all points of C diametrically opposite to points of A. The arcs A and B are each of measure less than π and are separated by a suitable diameter of C, so the arcs A' and B' exist, are diametrically opposite, and each of measure less than π. Denote by D_1' and D_2' the two diametrically opposite arcs complementary to $A' + B'$, the order $A'D_1'B'D_2'$ on C being that of AD_1BD_2. If A' coincides with A or B' with B, the arcs A' and B' contain A and B respectively, are of measure less than $2\theta_0$, and are diametrically opposite; in any other case, either D_1' coincides with D_1 or D_2' coincides with D_2, so D_1' and D_2' are of measure greater than $\pi - 2\theta_0$, and A' and B' are of measure less than $2\theta_0$ and are diametrically opposite; in any case the hypothesis of Theorem 6 is satisfied.

We mention a result of more elementary nature than Theorem 6:

THEOREM 7. *Let all zeros and poles of a rational function $R(z)$ of degree n lie on a circle C, let an arc A of C contain at least k zeros of $R(z)$ and an arc B of C contain at least k poles of $R(z)$. If z_0 is a point interior to C at which opposite angles less than $2\theta_0$ are subtended between NE lines by arcs of C containing A and B respectively, then z_0 is not a critical point of $R(z)$ if we have $\cos\theta_0 = (n - k)/k$.*

We choose z_0 as the origin, with A the arc $(-\theta_1, \theta_1)$, $0 < \theta_1 < \theta_0$, and B the arc $(\pi - \theta_1, \pi + \theta_1)$. The force at z_0 due to k positive particles in A has a horizontal component toward the left greater than $k \cos \theta_0$, as has the force at z_0 due to k negative particles in B. The force at z_0 due to all the remaining $2(n - k)$ positive and negative particles is in magnitude not greater than $2(n - k)$, so equilibrium is impossible if we have $2k \cos \theta_0 \geqq 2(n - k)$; this inequality is valid, and Theorem 7 is established.

A further result is of this same general character:

THEOREM 8. *Let all zeros and poles of a rational function $R(z)$ of degree n lie on a circle C, let a closed arc A of C contain k zeros of $R(z)$ and a closed arc B of C disjoint from A contain the remaining $n - k$ zeros and the n poles of $R(z)$. If z_0 is a point interior to C at which two arcs separating A and B subtend by NE lines each an angle greater than φ_0, where $\cos \varphi_0 = (2k - n)/n$, then z_0 is not a critical point of $R(z)$.*

As a preliminary remark, we study the function

$$f(\theta_0) = \frac{1 + \cos(\varphi + \theta_0)}{1 + \cos \theta_0}, \qquad 0 \leqq \theta_0 \leqq \pi - \varphi,$$

where φ is fixed, $0 < \varphi < \pi$. We have $f(0) \geqq 0$, $f'(0) < 0$; the condition for the vanishing of $f'(\theta_0)$ can be written

$$\tan \frac{\theta_0}{2} = \cot \frac{\varphi}{2}, \text{ or } \theta_0 = \pi - \varphi.$$

Thus the function $f(\theta_0)$ decreases monotonically in the interval $0 < \theta_0 < \pi - \varphi$, and we have established the inequalities

$$(5) \qquad 0 \leqq f(\theta_0) \leqq (1 + \cos \varphi)/2, \qquad 0 \leqq \theta_0 \leqq \pi - \varphi.$$

We choose C as the unit circle, z_0 at the origin, with A symmetric in the axis of reals, and $z = 1$ the mid-point of A. Two arcs each of length $\varphi > \varphi_0$ separate and adjoin A and B. Thus the arc A: $(-\theta_0, \theta_0)$ contains precisely k zeros of $R(z)$, the arcs $(\theta_0, \theta_0 + \varphi)$ and $(-\theta_0 - \varphi, -\theta_0)$ contain in their interiors no zeros or poles, and all the remaining zeros and poles lie in B, which is contained in the arc $(\theta_0 + \varphi, 2\pi - \theta_0 - \varphi)$.

The force at O due to the k positive particles on A and the n negative particles on B has a horizontal component to the left of magnitude at least $k \cos \theta_0 - n \cos(\theta_0 + \varphi)$; the $n - k$ positive particles on B exert a force at O having a horizontal component to the right not greater than $n - k$, so O is not a position of equilibrium if we have

$$k \cos \theta_0 - n \cos(\theta_0 + \varphi) > n - k,$$

$$\frac{k}{n} > \frac{1 + \cos(\theta_0 + \varphi)}{1 + \cos \theta_0} = f(\theta_0).$$

It follows from (5) that O is not a position of equilibrium if we have

$$\frac{k}{n} > \frac{1 + \cos \varphi}{2},$$

and this condition is a consequence of our assumption $\varphi > \varphi_0$. Theorem 8 is established.

Theorem 8 differs from Theorem 6, for in Theorem 8 the arcs A and B are not assumed to subtend at z_0 *opposite* angles between NE lines. Theorem 8 is particularly easy to apply. If $R(z)$ is given, the arcs A and B may be chosen and φ_0 computed; denote by D_1 and D_2 the two arcs of C separating A and B. The locus of points interior to C from which D_k subtends an angle greater than φ_0 is the locus $\omega(z, D_k, |z| < 1) > \varphi_0/2\pi$, and is bounded by D_k and by an easily constructed circular arc interior to C whose end-points are those of D_k. The point set if any common to these two loci, $k = 1, 2$, consists of all points z_0 satisfying the conditions of the theorem.

If k and n are both even, the limit φ_0 given in Theorem 8 cannot be improved; all special configurations for which D_1 and D_2 subtend the angle φ_0 at z_0 for which z_0 is a critical point of $R(z)$ can easily be enumerated. It is not difficult to study in detail the case in which k and n are not both even.

By way of contrast with our results on rational functions with concyclic zeros and poles, we add a further remark, a consequence of Bôcher's Theorem: *Let all the finite zeros and poles of a rational function $R(z)$ lie on a proper circle C, and let the point at infinity be a pole or zero of $R(z)$. Then no critical point of $R(z)$ other than a multiple zero lies on C.*

The general study of the location of critical points of a rational function whose zeros and poles lie on a circle is intimately connected with the study of the critical points of harmonic measure and more generally of an arbitrary function harmonic in a circle; each of these problems has applications to the other; compare §9.3, especially §9.3.5.

§5.3. Hyperbolic plane. We proceed to consider a rational function whose zeros lie interior to the unit circle C and are symmetric to its poles with respect to C; such a function is of constant modulus on C. This class of functions is of considerable interest, for as we shall prove (Radó) the most general function $R(z)$ which defines a 1-to-m conformal transformation of the interior of $C: |z| = 1$ into itself is a function of this class of the form:

$$(1) \qquad R(z) = \lambda \prod_{k=1}^{m} \frac{z - \alpha_k}{1 - \bar{\alpha}_k z}, \qquad |\alpha_k| < 1, \qquad |\lambda| = 1.$$

It follows at once by Rouché's Theorem (§1.1.2) that $R(z)$ as given by (1) takes on an arbitrary value w_0 with $|w_0| < 1$ precisely m times interior to C, for we have on $C: |w_0/R(z)| < 1$. Conversely, if $f(z)$ defines a 1-to-m map of $|z| < 1$ onto itself and transforms the m points α_k into the origin, then the quotient $f(z)/R(z)$, when suitably defined in the points α_k, is analytic and different from zero interior to C, and its modulus approaches unity as z interior to C approaches C, so the quotient is a constant of modulus unity.

The results which we establish [Walsh, 1939] on functions of type (1) have immediate application to the critical points of an arbitrary analytic function

(§6.2.1), and to the critical points of a Blaschke product (§6.2.2), the limit of a sequence of functions of type (1). Our results are expressed in terms of NE (hyperbolic) geometry for the interior of C.

§5.3.1. Analog of Lucas's Theorem. Bôcher's Theorem (§4.2) obviously applies at once to $R(z)$; the circle C or any circle interior to C which contains in its interior all the points α_k separates the zeros of $R(z)$ from the poles of $R(z)$, and hence passes through no critical point of $R(z)$. Such a circle contains in its interior precisely $m - 1$ critical points of $R(z)$. But we can be more precise:

THEOREM 1. *Let $R(z)$ be defined by* (1). *A NE line Γ for the interior of C not passing through all the points α_k and not separating any pair of points α_k can pass*

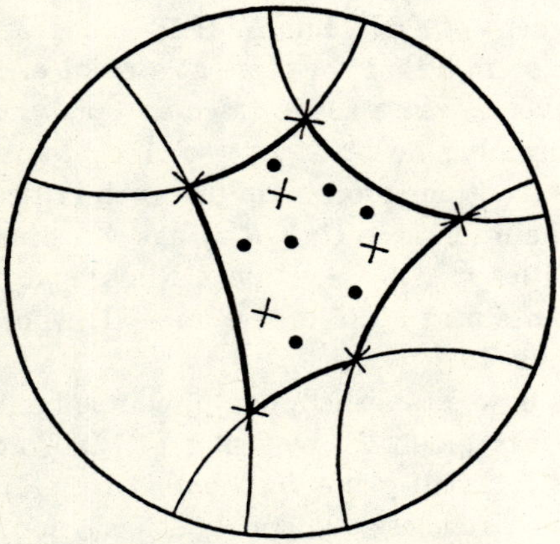

Fig. 14 illustrates §5.3.1 Theorem 1

through no critical point of $R(z)$ interior to C other than a multiple zero of $R(z)$. Consequently the smallest closed NE convex polygon Π containing all the points α_k also contains all the critical points of $R(z)$ interior to C. No such critical point other than a multiple zero of $R(z)$ lies on the boundary of Π unless Π degenerates to a segment of a NE line Δ. In the latter case, any open arc of Δ bounded by two points α_k and containing no point α, contains precisely one critical point of $R(z)$.

When the α_k are given, each NE line Γ (§5.2) of Theorem 1 bounds two closed NE half-planes, one containing all the α_k and the other containing in its interior no α_k. The set Π is the set common to all the former NE half-planes, and is obviously NE convex in the sense that whenever two points belong to Π, so does the segment of the NE line joining them.

We consider the field of force of §4.1.1 Theorem 1, and in order to study whether a particular point P interior to C is a critical point, we transform P into the origin O by a linear transformation which carries the interior of C into itself. This transformation does not alter the form of $R(z)$, nor does it affect any of the essential elements in Theorem 1. Let Γ satisfy the conditions of Theorem 1; we consider Γ as the segment $-1 < z < 1$, with the points α_k not on Γ lying in the upper half-plane, and we study the force at O due to the positive particles at the points α_k and to the negative particles at their inverses $\beta_k = 1/\bar{\alpha}_k$. The force at O (assumed not a point α_j) due to the particle at α_k is the vector $\beta_k O$, and the force at O due to the particle at β_k is the vector $O\alpha_k$, so the force at O due to the pair of particles is the vector $\beta_k \alpha_k$. At least one such vector has a non-vanishing component vertically downward; no such vector has a non-vanishing component vertically upward, so O is not a position of equilibrium nor a critical point of $R(z)$.

The polygon Π contains all positions of equilibrium interior to C and contains all points α_k, hence contains all critical points of $R(z)$ interior to C. No such critical point other than a multiple zero of $R(z)$ lies on the boundary of Π unless Π degenerates to a segment of a NE line Δ. If Π so degenerates, it follows from §4.1.2 Corollary to Theorem 3 that the force at a point of Δ is directed along Δ; this force obviously reverses sense as we trace an open arc A of Δ bounded by two points α_k and containing no point α_j, so A contains at least one critical point of $R(z)$; no arc A contains more than one such critical point, for precisely $m - 1$ critical points lie interior to C; if $R(z)$ has p distinct zeros interior to C, of respective multiplicities $m_1, m_2, \cdots, m_p$, whose sum is m, those zeros are respectively critical points if at all of orders $m_k - 1$, whose sum is $m - p$, and there are $p - 1$ intervals A.

Theorem 1, which is now established, can of course be proved without transforming to O an arbitrary point P, by using §4.1.2 Corollary to Theorem 3. If α_k lies interior to a NE half-plane H bounded by a NE line Γ through P, the force at P due to the particles at α_k and $1/\bar{\alpha}_k$ has a non-vanishing component orthogonal to Γ in the sense exterior to H. Theorem 1 can obviously be applied to the study of the critical points of $R(z)$ exterior to C; all such critical points lie in the inverse of Π with respect to C, which is indeed the smallest closed NE convex polygon for the exterior of C containing the poles of $R(z)$. Of course positions of equilibrium in the field of force are invariant under interchange of zeros and poles of $R(z)$, and under inversion in a circle C (§4.1.2 Corollary 1 to Theorem 2); these two operations taken together yield again the original field of force, and invert in C the positions of equilibrium.

Theorem 1 is the precise NE analog of Lucas's Theorem (§1.3.1). A polynomial $p(z)$ of degree m maps the euclidean plane $|z| < \infty$ conformally onto itself in a 1-to-m manner, and is the most general function defining such a map, just as $R(z)$ maps the NE plane conformally onto itself in a 1-to-m manner, and is the most general function defining such a map. The functions $p(z)$ and $R(z)$ have precisely m zeros, no poles, and $m - 1$ critical points in the regions involved. The polygons specified in Lucas's Theorem and in Theorem 1 are the smallest convex sets containing the zeros of the given functions, in the respective senses of euclidean and non-euclidean geometry. Each theorem is invariant under an arbitrary one-to-one conformal map onto itself of the euclidean or non-euclidean plane.

If the fixed points $\alpha_1, \alpha_2, \cdots, \alpha_m$ are given interior to $|z| = r$, the most general function analytic in $|z| < r$, vanishing precisely in the points α_k, whose modulus is continuous in $|z| \leqq r$ and equal to r^m on $|z| = r$, is given by

$$f(z) \equiv \lambda \prod_{k=1}^{m} \frac{z - \alpha_k}{1 - \bar{\alpha}_k z/r^2}, \qquad |\lambda| = 1;$$

the function $R(z)$ defined by (1) is the special case $r = 1$. When r becomes infinite, $f(z)$ approaches uniformly in any fixed circle C_0 the polynomial $p(z) \equiv \lambda\prod(z - \alpha_k)$; the derivative $f'(z)$ approaches $p'(z)$ uniformly in C_0 and the critical points of $f(z)$ in C_0 approach those of $p(z)$ in C_0. A variable circle Γ_0 through two fixed points in C_0 which is orthogonal to the circle $|z| = r$ has a radius that becomes infinite with r, for the center of Γ_0 lies exterior to the circle $|z| = r$. The NE polygon Π for $f(z)$ for the region $|z| < r$ approaches the Lucas polygon for $p(z)$, so Lucas's Theorem is a limiting case of and can be proved from Theorem 1.

Theorem 1 has the property of utilizing the positions but not the multiplicities of the points α_k as zeros of $R(z)$, and among theorems with that same property is a "best possible" theorem. There exists a precise NE analog of §1.3.2 Theorem 1, proved by the same method.

§5.3.2. Extensions. Theorem 1 can be extended to include the case of poles of $R(z)$ interior to C:

THEOREM 2. *Let $R(z)$ be a rational function of z with the zeros $\alpha_1, \alpha_2, \cdots, \alpha_m$ and the poles $\beta_1, \beta_2, \cdots, \beta_n$ interior to $C: |z| = 1$, and the poles $1/\bar{\alpha}_k$ and zeros $1/\bar{\beta}_k$ exterior to C, and no other zeros or poles. If L is a NE line for the interior of C which separates each α_j from each β_k, then no critical point of $R(z)$ lies on L. Consequently, if such a line L exists, the critical points of $R(z)$ interior to C lie in two closed NE convex point sets Π_1 and Π_2 which respectively contain all the α_k and all the β_k, and which are separated by every L.*

If a NE line L' for the interior of C separates all the α_j not on L' from all the β_k not on L', and if at least one α_j or β_k does not lie on L', then no point of L' is a critical point of $R(z)$ unless it is a multiple zero of $R(z)$.

Here the function $R(z)$ is the quotient of two functions of the kind defined by (1), not necessarily of the same degree. The sets Π_1 and Π_2 are NE convex because each is the set common to a number of NE half-planes.

We use precisely the method used to prove Theorem 1; it is sufficient to prove the last part of Theorem 2. Let P interior to C be an arbitrary point of L' not a zero or pole of $R(z)$; it is no loss of generality to assume, as we do, that P is the origin O and L' the segment $-1 < z < 1$; suppose too that all the points α_k not on L' are above L' and all the points β_k not on L' are below L'. The force at O

due to a positive particle at α_k and a negative particle at $1/\bar{\alpha}_k$ is the sum of the vector from $1/\bar{\alpha}_k$ to O and the vector $O\alpha_k$, namely the vector from $1/\bar{\alpha}_k$ to α_k, which has a non-negative downward component; the force at O due to a negative particle at β_k and a positive particle at $1/\bar{\beta}_k$ is the vector from β_k to $1/\bar{\beta}_k$, which has a non-negative downward component; for at least one value of k the vector from $1/\bar{\alpha}_k$ to α_k or from β_k to $1/\bar{\beta}_k$ has a non-zero downward component, so O is not a position of equilibrium.

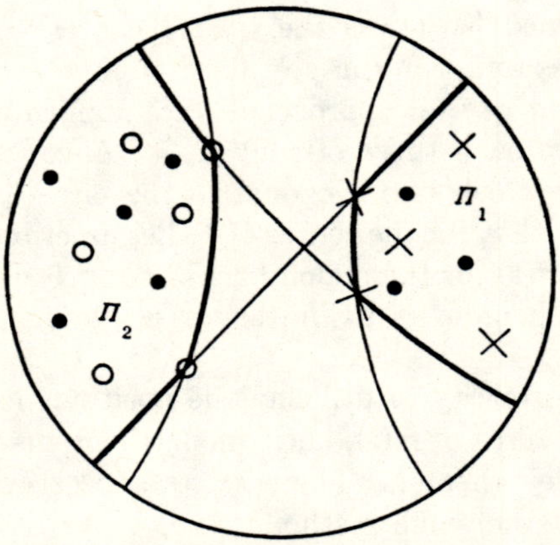

Fig. 15 illustrates §5.3.2 Theorem 2

Theorem 2 is the NE analog of §4.2.3 Theorem 3, and the latter can be proved from Theorem 2 just as Lucas's Theorem can be proved from Theorem 1.

Like the first part of Theorem 1, Theorem 2 can be sharpened in a particular case:

COROLLARY 1. *Let $R(z)$ be a rational function with the zeros $\alpha_1, \alpha_2, \cdots, \alpha_m$ and poles $\beta_1, \beta_2, \cdots, \beta_n$ on the NE line L interior to $C: |z| = 1$, and the poles $1/\bar{\alpha}_k$ and zeros $1/\bar{\beta}_k$ exterior to C, and no other zeros or poles. On L let two disjoint segments L_1 and L_2 terminated by C contain respectively all the points α_k and all the points β_k; then L_1 and L_2 contain all the critical points of $R(z)$ interior to C. On any open segment of L bounded by two zeros or two poles of $R(z)$ and containing no such zero or pole lies a unique critical point of $R(z)$.*

It follows from Theorem 2 that all critical points of $R(z)$ interior to C lie on L; the convex sets Π_1 and Π_2 of Theorem 2 can be chosen as minimal segments L_1 and L_2 respectively. Let A be an open segment of L_1 bounded by the points α_1 and α_2 and containing no α_k in its interior. At every point of A the total force is directed along A itself, near α_1 this force is directed toward α_2 and near α_2 is directed toward α_1, so the force vanishes at some intermediate point of A and A contains at least one critical point. To study the number of critical points on

A we use the "method of continuity." Allow α_1 to vary continuously and monotonically along A and to approach α_2, while the other zeros and poles of $R(z)$ remain fixed, except the pole in $1/\bar{\alpha}_1$. During this process we do not alter the multiplicities of α_1 and α_2 as zeros of $R(z)$, so none of the critical points of $R(z)$ (which vary continuously and interior to C lie on L) can enter or leave the arc A. When α_1 coincides with α_2, precisely one new critical point (i.e. other than those already at α_1 and α_2) coincides with α_2; all critical points originally in A and conceivably others coincide with α_2; no critical point originally at α_1 or α_2 fails to coincide with α_2; thus A must always contain precisely one critical point. Interchanging the roles of zeros and poles now completes the proof.

Corollary 1 is the NE analog of §4.2.3 Corollary 2 to Theorem 3; to complete the analogy we add a further remark:

Corollary 2. *Under the conditions of Corollary 1 with $m = n$, no critical point of $R(z)$ lies on an open arc A of L having one end-point on C and containing no zero or pole of $R(z)$, and no critical point lies at an end-point of L on C. Under the conditions of Corollary 1 with $m > n$, no critical point of $R(z)$ lies on an open arc A of L bounded by a zero of $R(z)$ and a point of C and containing no zero or pole of $R(z)$, and no critical point lies at the end-point of A on C.*

It is sufficient to consider a point P on an open arc A of L bounded by a zero of $R(z)$ or at the end-point of such an arc on C, with $m \geqq n$. On the circle C_0 of which L is an arc the particles α_k and $1/\bar{\alpha}_k$ separate P from the particles β_j and $1/\bar{\beta}_j$; this fact implies by inversion in the unit circle whose center is P that the force exerted at P by the former pair of particles is greater in magnitude than the force at P exerted by the latter pair; the forces exerted at P by all the pairs α_k, $1/\bar{\alpha}_k$ act in the same sense, so P is not a position of equilibrium.

The last part of Corollary 1 applied both to the interior and exterior of C accounts for all but two critical points of $R(z)$; since positions of equilibrium are symmetric in the two circles C and C_0, it follows from Corollary 2 that these two remaining critical points either (i) coincide at an end-point of L on C, namely an end-point of L_1 or L_2 according as we have $m < n$ or $m > n$; or (ii) lie on C in distinct points symmetric with respect to C_0; or (iii) lie on C_0 in distinct points symmetric with respect to C, on an open arc of C_0 bounded by a zero or pole of $R(z)$ interior to C according as we have $m < n$ or $m > n$ and bounded by the inverse of that point with respect to C, but containing no zero or pole of $R(z)$. If we have $m = n$, case (ii) must occur.

In Corollary 1 (and in Theorem 2) the force at each point of C is orthogonal to C; case (ii) is characterized by the fact that the forces at the two end-points of L are directed one interior and the other exterior to C, and case (iii) is characterized by the fact that those two forces are both directed interior or both directed exterior to C, it is then clear simply by noting the direction of the forces that in case (ii) each arc of C bounded by the end-points of L contains at least one (and hence only one) critical point, and that in case (iii) a particular arc of L bounded by a point of C and a zero or pole of $R(z)$ and containing no such zero or pole contains at least one (and hence only one) critical point.

Of course the condition $m = n$ in Theorem 2, with all the α_j coincident and all the β_k coincident, implies that $R(z)$ has no critical point other than perhaps α_1 interior to C.

Under the conditions of Corollary 1 it may occur that critical points of $R(z)$ lie on C but not on the circle L_0 of which the NE line L is an arc, so we now give an example in which they are precisely determined. Let L_0 be a proper circle orthogonal to C and let M denote the diameter of L_0 through the origin. Let the point α_1 lie on L_0 interior to C but not on M, and let β_1 be the inverse of α_1 in M; then $1/\bar{\alpha}_1$ and $1/\bar{\beta}_1$ lie on L_0; of course if α_1 and β_1 are distinct and given arbitrarily interior to C, then both L_0 (a circle through α_1 and β_1 orthogonal to C) and M (a circle orthogonal to both C and L_0, and with respect to which α_1 and β_1 are mutually inverse) exist, and the configuration is equivalent to our original one under a suitable linear transformation. Denote by γ either of the intersections of M and C; the circle through α_1, γ, and $1/\bar{\alpha}_1$, is tangent to M at γ, so the force at γ due to the pair of particles at α_1 and $1/\bar{\alpha}_1$ is directed along M toward the exterior of C. Similarly the force at γ due to the pair of particles at β_1 and $1/\bar{\beta}_1$ is an equal force directed along M toward the interior of C. Thus γ is a position of equilibrium and lies on C but not on L_0.

The proof just given with only obvious modifications yields the first part of

COROLLARY 3. *Under the conditions of Corollary* 1 *with* $m = n$, *if the points* α_k *and* β_k *are mutually inverse in a circle* M *orthogonal to* C *and* L, *then the intersections of* M *with* C *are critical points of* $R(z)$; *these are the only critical points of* $R(z)$ *other than those mentioned in the last part of Corollary* 1.

The second part of Corollary 3 follows merely by counting the total number of critical points.

A second proof of the first part of Corollary 3 can be given by noticing that $|R(z)|$ is constant on both C and M, and hence the intersections of C and M are critical points of the harmonic function $\log |R(z)|$ and of $R(z)$; even if $R(z)$ does not satisfy the conditions of Corollary 1, but if the zeros and poles of $R(z)$ are mutually inverse in both C and M, then the intersections of C and M (which obviously cannot be zeros or poles of $R(z)$) are critical points of $R(z)$. In Corollary 3, the given conditions are symmetric with respect to C and M.

We add a numerical example to show that *under the conditions of Corollary* 1, *the number of critical points of* $R(z)$ *interior to* C *depends not merely on the orders of the zeros and poles of* $R(z)$ *but also on their relative positions*. We choose

$$-1 < \alpha < 1, \alpha \neq 0,$$

and set

$$R(z) = \frac{z(\alpha z - 1)^2}{(z - \alpha)^2}, \qquad R'(z) = \frac{(\alpha z - 1)[\alpha z^2 - (3\alpha^2 - 1)z + \alpha]}{(z - \alpha)^3};$$

the critical points of $R(z)$ other than $z = 1/\alpha$ are

$$\frac{3\alpha^2 - 1}{2\alpha} \pm \frac{1}{2\alpha}(1 - \alpha^2)^{1/2}(1 - 9\alpha^2)^{1/2}.$$

If we have $|\alpha| > 1/3$, both these critical points are non-real and on C, for the critical points are conjugate imaginary and their product is unity; if we have $|\alpha| < 1/3$, precisely one of them is real and interior to C and the other is real and exterior to C; if we have $\alpha = \pm 1/3$, these two critical points coincide at $z = \mp 1$.

§5.3.3. Analog of Jensen's Theorem. Just as Theorem 1 is the NE analog of Lucas's Theorem, so we have [Walsh, 1946a] a NE analog of Jensen's Theorem:

THEOREM 3. *Let $R(z)$ be a real rational function all of whose zeros $\alpha_1, \alpha_2, \cdots, \alpha_m$ lie interior to $C: |z| = 1$ and whose poles are the points $1/\bar{\alpha}_1, 1/\bar{\alpha}_2, \cdots, 1/\bar{\alpha}_m$. Then all non-real critical points of $R(z)$ interior to C lie on or within the NE Jensen circles constructed on the NE segments joining pairs of conjugate imaginary zeros of $R(z)$ as NE diameters. A non-real point on a Jensen circle but not interior to any Jensen circle is not a critical point unless it is a multiple zero of $R(z)$.*

If $z = \alpha$ is interior to C, the *NE vertical* at α is the direction of the NE line through α orthogonal to the axis of reals. If α is non-real the NE segment joining α and $\bar{\alpha}$ is the segment of this NE line bounded by α and $\bar{\alpha}$, and the corresponding NE Jensen circle J is the circle through α and $\bar{\alpha}$ whose NE center lies on this segment. If α lies on the axis of imaginaries, the NE Jensen circle is simply the usual Jensen circle; if α does not lie on the axis of imaginaries but is transformed onto that axis by a linear transformation which carries $|z| < 1$ into itself and also the axis of reals into itself, then J is carried into the ordinary Jensen circle for the images of α and $\bar{\alpha}$.

As a matter of convenience we establish

LEMMA 1. *Let a be arbitrary, $0 < a < 1$, and consider the field of force due to unit positive particles at $+ai$ and $-ai$, and unit negative particles at $+i/a$ and $-i/a$. Denote by J the circle whose center is O and radius a. Then at any non-real point z between C and J the force has a positive component in the direction of the NE vertical directed away from Ox.*

At a non-real point z between C and J, the force due to the pair of particles at $+ai$ and $-ai$ has a non-zero (euclidean) vertical component directed away from the axis of reals Ox, by the Lemma of §1.4.1, and by the same lemma the force at z due to the pair of particles at $+i/a$ and $-i/a$ also has a non-zero vertical component directed away from Ox; thus the total resultant force at z has a positive vertical component directed away from Ox.

At a point z between C and J, the force due to the particles at ai and i/a has the direction of the circle through those three points, in the sense of the arc $(ai, z, i/a)$; the center of this circle lies on a horizontal line which does not cut C; consequently the force at z has a non-vanishing horizontal component directed away from the axis of imaginaries Oy unless z lies on Oy. The same conclusion applies to the force at z due to the particles at $-ai$ and $-i/a$, and hence applies to the force at z due to all four particles.

If z lies on Oy between C and J, the total force at z due to the four particles is directed vertically in the sense away from Ox. If z lies between C and J but not on Ox or Oy, we denote by V the NE vertical direction at z in the sense away from Ox; then V lies in the right angle between the horizontal direction sensed away from Oy and the vertical direction sensed away from Ox. The total force $\mathfrak{F}$ at z is the vector sum of its horizontal and vertical components $\mathfrak{F}_x$ and $\mathfrak{F}_y$, and the component of $\mathfrak{F}$ in the direction V is the sum of the component in the direction V of the horizontal component $\mathfrak{F}_x$ and the component in the direction V of the vertical component $\mathfrak{F}_y$. Both summands are positive, so Lemma 1 is established.

The set of circles symmetric in Ox and orthogonal to C represents the direction V and is invariant under the group of linear transformations which carry $|z| < 1$ into itself and leave Ox invariant, so under this group the NE vertical at a point is transformed into the NE vertical at the image point. It consequently follows that *if α lies interior to C but not on Ox, and if we denote by J the NE Jensen circle for α and $\bar{\alpha}$, then the force at a non-real point z interior to C and exterior to J, due to unit positive particles at α, $\bar{\alpha}$ and unit negative particles at $1/\bar{\alpha}$, $1/\alpha$, has a non-vanishing component in the direction and sense V.* Indeed, this conclusion holds when α lies on Oy, and (§4.1.2 Theorem 3) is invariant under the transformations considered.

A relation similar to Lemma 1 holds for pairs of particles on Ox:

LEMMA 2. *If the point b lies on Ox interior to C, then at any non-real point z interior to C the force due to a unit positive particle at b and a unit negative particle at $1/b$ has a non-vanishing component in the direction and sense V.*

If we have $b = 0$, the force has the direction and sense Oz, and the conclusion is immediate. The lemma follows from the invariance (§4.1.2 Theorem 3) of lines of force under linear transformations which transform the interior of C into itself and leave Ox invariant.

We are now in a position to establish Theorem 3. Let z be a non-real point interior to C and not on or within any NE Jensen circle. In the usual field, the force at z has a non-vanishing component in direction and sense V, whether we consider the force due to a positive particle at a real point interior to C and a negative particle at the inverse point, or the force due to pairs of positive particles at conjugate imaginary points interior to C and to pairs of negative particles at their inverses, or the total force in the field determining the critical points of

$R(z)$. Thus the point z cannot be a position of equilibrium, cannot be a multiple zero of $R(z)$, hence cannot be a critical point of $R(z)$.

The proof of Lemma 1 holds with only minor changes if z is a non-real point on J, $z \neq \pm ai$; the force at z due to the particles at $+ai$ and $-ai$ is horizontal, and the force at z due to the particles at $+i/a$ and $-i/a$ has a non-vanishing vertical component directed away from Ox. The remainder of Theorem 3 follows by the method already used.

We note here a lack of parallelism between the euclidean and NE cases; in the former case the force due to a pair of conjugate imaginary particles is horizontal on their Jensen circle, but in the latter case the force at a point of J due to the positive particles at $+ai$ and $-ai$ and the negative particles at $+i/a$ and $-i/a$ does not have either the direction perpendicular to V or the direction of the euclidean horizontal.*

Just as Lucas's Theorem and Jensen's Theorem are mutually complementary in the sense that neither contains the other, so their NE analogs, Theorems 1 and 3, are mutually complementary; when both theorems apply non-trivially, each gives some information not provided by the other. Jensen's Theorem is a limiting case of, and can be proved from, Theorem 3.

Theorem 3 has been phrased for a *real* rational function, but a suitable modification applies to an arbitrary rational function analytic interior to C whose zeros and poles are mutually inverse with respect to C, and whose zeros are symmetric with respect to an arbitrary NE line.

As a complement to Theorem 3 we have

THEOREM 4. *Under the conditions of Theorem* 3, *suppose we have* $-1 < \alpha < \beta < +1$, *where neither* α *nor* β *is a zero or critical point of* $R(z)$ *nor on or within a NE Jensen circle. Let* K *denote the configuration consisting of the segment* $\alpha\beta$ *plus the closed interiors of all NE Jensen circles which intersect* $\alpha\beta$, *and let* k *denote the number of zeros of* $R(z)$ *in* K. *Then* K *contains precisely* $k - 1$, k, *or* $k + 1$ *critical points of* $R(z)$.

If the forces at α *and* β *are directed away from each other, this number is* $k - 1$; *if those forces are directed both toward the right or both toward the left, this number is* k; *if those forces are directed toward each other, this number is* $k + 1$.

Theorem 4 is the analog of §2.3 Theorem 1, and may be proved by the method of proof of that theorem, using the Principle of Argument; we omit the details.

§5.3.4. Circular regions as loci. We proceed to discuss the critical points of rational functions whose poles are symmetric to the zeros in a given circle, and where various circular regions are assigned to certain numbers of poles and zeros. The methods to be used are quite similar to those of §§5.1.3 and 5.1.4.

---

*The locus on which the force due to this quadruple of particles has the direction perpendicular to V consists of the axis of reals and a certain bicircular quartic; the latter can be used in Theorem 3 instead of the NE Jensen circles.

LEMMA. *Let the point* $z = x + iy$ *lie on the circle* $\gamma: |z - a| = r$ *interior to* $C: |z| = 1$, *with* $a > 0$; *we study the function* $F(x) = \Re(z - 1/\bar{z})$ *in the interval* $a - r \leqq x \leqq a + r$. *Case* I: $r < a$; max $F(x)$ *occurs for* $x = a + r$, min $F(x)$ *occurs for* $x = a - r$. *Case* II: $r > a, r - a \geqq (r + a)^3$; max $F(x) = [(a - r)^2 - 1]/(a - r)$ *occurs for* $x = a - r$, min $F(x) = [(a + r)^2 - 1]/(a + r)$ *occurs for* $x = a + r$. *Case* III: $r > a, r - a < (r + a)^3$; min $F(x) = -[(r^2 - a^2)^{1/2} - 1]^2/2a$ *occurs for* $x = [a^2 - r^2 + (r^2 - a^2)^{1/2}]/2a$, max $F(x) = [(a - r)^2 - 1]/(a - r)$ *occurs for* $x = a - r$.

The function $F(x)$ is the real part of the vector $z - 1/\bar{z}$, which is of intrinsic importance, and is also the horizontal component of the total force at O due to a unit repelling particle at z and a unit attracting particle at the inverse of z in C.

In the Lemma we include also a more general problem, namely that of max $F(x)$ and min $F(x)$ for z in the closed interior of γ, $r < a$. In fact, both maximum and minimum are assumed when z is on the boundary γ, for if z is an interior point of the region we may move z (remaining in the region) along the line Oz in either sense; motion in one sense increases $F(x)$ and motion in the other sense decreases $F(x)$. Similarly, the Lemma includes the determination of max $F(x)$ and min $F(x)$ for z in the closed annular region bounded by C and γ, if we have $r > a$.

For z on γ we have $x^2 + y^2 = r^2 + 2ax - a^2$, whence

$$F(x) = \Re\left[z \frac{z\bar{z} - 1}{z\bar{z}}\right] = x \frac{r^2 + 2ax - a^2 - 1}{r^2 + 2ax - a^2},$$

$$F'(x) = \frac{(r^2 + 2ax - a^2)^2 - r^2 + a^2}{(r^2 + 2ax - a^2)^2}. \tag{2}$$

It follows from (2) that in Case I we always have $F'(x) > 0$, from which the conclusion in Case I follows. Indeed, the vector $z - 1/\bar{z}$ is a vector from a point of γ' (the inverse of γ) to a point of γ; it is obvious geometrically that the greatest and least horizontal projections have the values indicated; in fact these correspond also to the least and greatest values of $|z - 1/\bar{z}|$.

With $r > a$, the points at which $F'(x)$ vanishes are $[a^2 - r^2 \pm (r^2 - a^2)^{1/2}]/2a$, of which the latter x_1 is always less than $a - r$; the former x_2 is always greater than $a - r$, and is greater than, equal to, or less than $a + r$ according as we have $r - a$ greater than, equal to, or less than $(r + a)^3$. Thus in Case II we have $F'(x) < 0$ throughout the interval $a - r \leqq x < a + r$, so max $F(x)$ occurs for $x = a - r$, and min $F(x)$ occurs for $x = a + r$. In Case III we have $F'(x) > 0$ in $x_2 < x \leqq a + r$, $F'(x) < 0$ in $a - r \leqq x < x_2$, so min $F(x)$ occurs for $x = x_2$; by direct comparison we note $F(a - r) > F(a + r)$, so max $F(x)$ occurs for $x = a - r$, and the Lemma is established.

We are now in a position to prove

THEOREM 5. *Let the poles of the rational function* $R(z)$ *be inverse to its zeros in the unit circle* C, *and let* $R(z)$ *have* m *zeros in the closed region* $|z| \leqq c$ (<1), n

zeros in the closed region $(c<)\ b \leqq z \leqq 1$, *and no other zeros or poles interior to* C. *If the equation*

$$(3)\quad \varphi(r) = m(c - 1/c)(b - r)(r - 1/b) + n(b - 1/b)(c + r)(r + 1/c) = 0$$

has a zero $r = r_0$ *satisfying the inequalities* $c < r_0 < b$, *then* $R(z)$ *has no critical points in the annulus* $c < | z | < r_0$, *and has precisely* $m - 1$ *critical points in the region* $| z | \leqq c$, *and precisely* n *critical points in the region* $r_0 \leqq | z | < 1$.

In the proof of the first part of Theorem 5, it is sufficient to consider the point $P: z = -r, c < r < b$, as a position of equilibrium in the field of force. Invert in the unit circle Γ whose center is P, and denote by C' the inverse of C. The force at P due to a positive particle in the region $| z | \leqq c$ and the corresponding negative particle in the region $| z | \geqq 1/c$ is the vector to P from the inverse S in Γ of the former particle, plus the vector from P to the inverse T in Γ of the latter particle, which is equivalent to the vector ST. The region assigned to S is the closed interior of a circle exterior to C' and the region assigned to T is the closed interior of a circle interior to C'; the center of C' is exterior to this latter circle. Except for the size of C', we have precisely the situation of the Lemma, Case I; similarly Case II or Case III applies in the study of positive particles in the region $b \leqq | z | \leqq 1$. The point P cannot be a position of equilibrium if the horizontal component of the force at P due to the m positive particles and their inverses acting to the left is greater in magnitude than the horizontal component of the force at P due to the n positive particles and their inverses acting to the right. That is to say, P cannot be a position of equilibrium if we have

$$(4)\qquad \frac{m}{r + c} - \frac{m}{r + 1/c} > \frac{n}{b - r} - \frac{n}{(1/b) - r},$$

which is equivalent to $\varphi(r) > 0$. Equation (3) is a reciprocal equation (unless $\varphi(r)$ reduces to a constant times r) in the sense that if it is satisfied for a particular value $r = r_1$, it is also satisfied for the reciprocal value $r = 1/r_1$. The inequality $\varphi(b) < 0$ is clear by inspection, *so a necessary and sufficient condition for the existence of* r_0 *is the inequality* $\varphi(c) > 0$, *which may be written*

$$(5)\qquad m(1 - c^2)(b - c)(1 - bc) - 2cn(1 - b^2)(1 + c^2) > 0.$$

If inequality (5) is satisfied, $\varphi(r)$ has at least one zero r_0 in the interval $c < r < b$ and at least one zero in the interval $1/c < r < 1/b$, thus has no other zeros, and (4) is satisfied throughout the interval $c < r < r_0$. Reciprocally, if $\varphi(r)$ has a zero r_0, $c < r_0 < b$, then $\varphi(1/r_0) = 0$, so we have $\varphi(r) > 0$ for $c < r < r_0$ and (4) is satisfied for these values of r. This completes the proof of the first part of Theorem 5. We note that the number r_0 can be replaced by no larger number. Moreover, $R(z)$ has no critical point on the circle $| z | = r_0$ unless the m zeros coincide at some point z_1 on the circle $| z | = c$ and the n zeros coincide at the point $-bz_1/c$.

Also, this method shows that if r_0 exists, $R(z)$ has no critical point on the circle $|z| = c$ except perhaps a multiple zero of $R(z)$.

The method of continuity yields the number of critical points in the region $|z| \leqq c$. We keep b fixed and consider r_0 (assumed to exist) as a function $r_0(c)$. When c is decreased monotonically, there are old configurations excluded for $R(z)$ while no new ones are admitted. Thus, since $r_0(c)$ is a bound that cannot be improved, $r_0(c)$ never decreases. As c is decreased to zero, we also allow the zeros of $R(z)$ in the region $|z| \leqq c$ to move continuously and monotonically in modulus to zero. In the final configuration, there are no critical points in the region $0 < |z| < r_0(0)$, hence precisely $m - 1$ critical points of $R(z)$ in the region $|z| < r_0(0)$. The critical points vary continuously during the process, and none ever lies in the region $c < |z| \leqq r_0(c) \leqq r_0(0)$, where c is variable. Thus originally there were precisely $m - 1$ critical points in the region $|z| \leqq c$. The circle C separates the zeros and poles of $R(z)$, hence passes through no critical point and contains precisely $m + n - 1$ critical points. Consequently the annulus $r_0 \leqq |z| < 1$ contains precisely n such critical points, and Theorem 5 is established.

In Theorem 5 the case $c = 0$ is not excluded; here equation (3) takes the form

$$-m(b - r)(r - 1/b) + nr(b - 1/b) = 0,$$

which is still a reciprocal equation whose first member is negative for $r = b$; condition (5) is always satisfied.

Theorem 5 is obviously not a complete analog of §4.3 Theorem 1, for in Theorem 5 we have allowed but two regions in the NE plane $|z| < 1$ as loci, and those regions loci of zeros but not of poles. It is entirely possible to extend Theorem 5 by the continued use of the Lemma, assigning an arbitrary number of circular regions interior to C as loci of zeros and poles of $R(z)$. These extensions represent improvements in accuracy but some loss in elegance, and are left to the reader.

Theorem 5 is of significance for regions bounded by circles having the origin as common center or, by a suitable linear transformation, having an arbitrary point interior to C as common NE center. We define NE distance from the origin as $\frac{1}{2} \log [(1 + h)/(1 - h)]$, where h (<1) is euclidean distance from the origin, and NE distance is invariant under linear transformation that carries the interior of C into itself.

COROLLARY. *Let $z = \alpha$ be an arbitrary point interior to the unit circle C, C_1 be the NE circle whose NE center is α and NE radius $\frac{1}{2} \log [(1 + c)/(1 - c)]$, and C_2 be the NE circle whose NE center is α and NE radius $\frac{1}{2} \log [(1 + b)/(1 - b)]$, $c < b$. Let the rational function $R(z)$ of degree $m + n$ have its poles symmetric with respect to its zeros in C, have precisely m zeros in the closed interior of C_1 and precisely n zeros interior to C but in the closed exterior of C_2. If equation (3) has a zero $r = r_0$ satisfying the inequalities $c < r_0 < b$, then $R(z)$ has no critical points interior to the annulus bounded by C_1 and the NE circle C_0 whose NE center is α and NE radius $\frac{1}{2} \log [(1 + r_0)/(1 - r_0)]$, and has precisely $m - 1$ critical* *points in the closed interior of* C_1 *and precisely n critical points in the closed annular region bounded by* C *and* C_0.

§5.3.5. Analog of Bôcher's Theorem. Theorem 2 is obviously the NE analog of §4.2.3 Theorem 3 rather than of Bôcher's Theorem, for in Theorem 2 we do not require $m = n$, nor do we obtain closed regions interior to the NE plane within which the critical points must lie. Nor is Theorem 5 an analog of Bôcher's Theorem, for Theorem 5 as stated involves regions containing only zeros of $R(z)$ interior to C, and the suggested extension of Theorem 5 involving both zeros and poles interior to C is by no means as simple as Bôcher's Theorem. One of our main tools in the euclidean plane is §1.5.1 Lemma 1, so we proceed to the study of its NE analog.

Let γ be the closed interior of a NE circle interior to C, and let z_0 be a point interior to C but exterior to γ. If the force at z_0 due to n arbitrary positive particles in γ and negative particles at the points inverse with respect to C is equivalent to the force at z_0 due to n *coincident* positive particles in γ and n negative particles at their inverse with respect to C, for all values of n, we say that z_0 *has Property A with respect to* γ. Thus if a point z_0 has Property A with respect to two such circular regions γ_1 and γ_2, and if γ_1 and γ_2 are disjoint and contain respectively n positive particles and n negative particles, then the force at z_0 due to these $2n$ particles and their inverses in C with reversed masses, is equivalent to the force at z_0 due to an n-fold positive particle at a point z_1 in γ_1, an n-fold negative particle at a point z_2 in γ_2, and their inverses in C with reversed masses. It follows from Corollaries 1 and 2 to Theorem 2 that z_0 is not a position of equilibrium, so z_0 is not a critical point of the corresponding rational function, which is analogous to the conclusion of Bôcher's Theorem. On the other hand, suppose P lies in C but exterior to a circular region γ_1 with respect to which P does not have Property A; for a certain set of n positive particles the force $\mathfrak{F}$ at P due to them and their inverses in C with reversed masses is not equivalent to the force at P due to n coincident positive particles in γ_1 and their inverses in C with reversed masses. It follows from Theorem 1 that $\mathfrak{F}$ is not zero, and it follows by inversion of C in P that an *arbitrary* $\mathfrak{F}_1$ not zero is equivalent to the force at P due to n coincident positive particles at a unique suitably chosen point α_1 interior to C and negative particles at their inverses in C. Thus P is a position of equilibrium in the field due to the original n particles in γ_1 and n negative particles at a suitably chosen point α_1 interior to C but *exterior to* γ_1 and their inverses in C with reversed masses; this situation is analogous to the negation of Bôcher's Theorem.

Our present topic is thus intimately related to Property A, which we proceed to investigate. It follows from inversion in an arbitrary point P exterior to γ, with C' and γ' the images of C and γ, that the force at P due to the set of n particles in γ and their inverses with respect to C is equivalent to the sum of n vectors from points in the closed interior of γ' to the points inverse with respect to C'. A necessary and sufficient condition for Property A is that the sum of n

such vectors of the class always be equivalent to an n-fold vector of the class; if these vectors are translated to a common initial point, this necessary and sufficient condition is that the locus of terminal points be a convex set, a condition which we shall study in some detail. By a slight change of notation, we replace C' by $C:|z| = 1$ and γ' by $\gamma:|z - a| = r$, $r < a - 1$, so that the closed interior of γ is exterior to C. For z in the closed interior of γ we study the vector $w = z - 1/\bar{z}$; as in the proof of the Lemma of §5.3.4 it is sufficient to consider z on γ itself. We set $z = a + re^{i\varphi}$, $w = u + iv$, and straightforward algebraic computation of $du/d\varphi$ and $dv/d\varphi$ yields

$$(6) \qquad \frac{dv}{du} = \frac{-(a^2 + r^2)\cos\varphi - 2ar + \cos\varphi(a^2 + r^2 + 2ar\cos\varphi)^2}{-\sin\varphi[(a^2 - r^2) + (a^2 + r^2 + 2ar\cos\varphi)^2]}.$$

This denominator vanishes only when $\sin\varphi = 0$, for we have $a > r$. Further computation and suppression of a factor $a^2 + r^2 + 2ar\cos\varphi$, which is equal to $z\bar{z}$ and positive, shows that the algebraic sign of $d(dv/du)/d\varphi$ is the same as the algebraic sign of the function

$$(7) \qquad \begin{aligned} F(\varphi) = {} & (a^2 + r^2 + 2ar\cos\varphi)^3 - 2r(r + a\cos\varphi)(a^2 + r^2 + 2ar\cos\varphi) \\ & + 8a^2r\sin^2\varphi(r + a\cos\varphi) - (a^2 - r^2). \end{aligned}$$

The vector w lies on the half-line from O through z, and $|w|$ increases with $|z|$, so the question of the convexity of the locus of the terminal points of the vector w is precisely the convexity of the curve whose parametric equation we are considering, which is the question of the positive character of $F(\varphi)$ for $0 < \varphi < \pi$.

We now choose specifically $r = 1$, $\cos\varphi = -15/8a$, $a = 3\cdot 2^{-1/2}$. Then we have $F(\varphi) = -63/32 < 0$, so the locus of w is not convex and we have established

THEOREM 6. *There exists a rational function $R(z)$ of degree four with two zeros interior to a circle γ interior to $C:|z| = 1$ and a double pole interior to C but not in the closed interior of γ, whose zeros are symmetric with respect to its poles in C. A critical point of $R(z)$ lies interior to C but exterior to γ.*

By inspection of (7) it is clear that $F(\varphi)$ is positive in the interval $0 < \varphi < \pi$ provided r is sufficiently small in comparison with a. This fact deserves further investigation; we shall deduce only some of the more immediate results.

§5.3.6. Continuation. We show that the condition

$$(8) \qquad r + a\cos\varphi \geqq 0$$

implies

$$(9) \qquad F(\varphi) > 0.$$

We make repeated use of the inequalities

$$(10) \qquad a > 1 + r,\ a^2 - r^2 = (a - r)(a + r) > a + r > 1 + 2r.$$

In particular we have $a^2 < (a^2 - r^2)^2$, whence

$$2r(r + a\cos\varphi) + a^2 < (a^2 - r^2)^2 + 2(a^2 - r^2)(r^2 + ar\cos\varphi) + (r^2 + ar\cos\varphi)^2. \tag{11}$$

We also have $a^2 + ar\cos\varphi \geqq a^2 - r^2$ from (8), whence by (11)

$$2r(r + a\cos\varphi) + a^2 < (a^2 + r^2 + 2ar\cos\varphi)^2, \tag{12}$$

and since we have $a^2 + r^2 + 2ar\cos\varphi \geqq (a - r)^2 > 1$, we have by (12): $a^2 < (a^2 + r^2 + 2ar\cos\varphi)^3 - 2r(r + a\cos\varphi)(a^2 + r^2 + 2ar\cos\varphi)$, which implies (9).

In equation (7) we make the substitution $r = \alpha(a - 1)$, so that α is necessarily less than unity; we obtain under the hypothesis $a\cos\varphi + r < 0$, by combining the first term of the second member of (7) with $-a^6$, and by use of the inequalities $(a - r)^2 \leqq a^2 + r^2 + 2ar\cos\varphi = a^2 + 2r(r + a\cos\varphi) - r^2 < a^2 - r^2$, the relation $F(\varphi)/(a - 1) = \alpha[\alpha(a - 1) + 2a\cos\varphi]\{[a^2 + \alpha^2(a - 1)^2 + 2a\alpha(a - 1) \cdot\cos\varphi]^2 + a^2[a^2 + \alpha^2(a - 1)^2 + 2a\alpha(a - 1)\cos\varphi] + a^4\} - 2\alpha[\alpha(a - 1) + a \cdot\cos\varphi][a^2 + \alpha^2(a - 1)^2 + 2a\alpha(a - 1)\cos\varphi] + 8a^2\alpha(1 - \cos^2\varphi)[\alpha(a - 1) + a \cdot\cos\varphi] + (a^2 + a^3 + a^4 + a^5) + \alpha^2(a - 1) > \alpha[\alpha(a - 1) - 2a][3a^4] + 8a^2\alpha[\alpha(a - 1) - a] + (a^2 + a^3 + a^4 + a^5) + \alpha^2(a - 1) = a^5(1 - 6\alpha + 3\alpha^2) + a^4(1 - 3\alpha^2) + a^3(1 - 8\alpha + 8\alpha^2) + a^2(1 - 8\alpha^2) + a\alpha^2 - \alpha^2$. If we now require $\alpha \leqq 1/6$, this last expression is greater than $a^3(3 - 14\alpha + 8\alpha^2) + a^2(1 - 8\alpha^2) > 0$.

The requirement $\alpha = r/(a - 1) \leqq 1/6$ is easily expressed in geometric form. Since $F(\varphi) > 0$ for $0 < \varphi < \pi$ is a sufficient condition for Property A, and since by Corollaries 1 and 2 to Theorem 2 no point interior to C is a position of equilibrium in the field of force due to two distinct equal and opposite particles interior to C and their inverses with reversed masses, we have proved

THEOREM 7. *Let the mutually exterior circles γ_1 and γ_2 lie interior to $C: |z| = 1$, let n zeros of the rational function $R(z)$ lie in the closed interior of γ_1 and n poles lie in the closed interior of γ_2, where $R(z)$ has no other zeros or poles interior to C and where the zeros of $R(z)$ are inverse to the poles in C. Let each of the circles γ_1 and γ_2 subtend a maximum angle at z_0 not greater than $2\sin^{-1}(1/6)$ by means of circles through z_0 intersecting on C, where z_0 lies interior to C but exterior to γ_1 and γ_2. Then z_0 is not a critical point of $R(z)$.*

The geometric condition on z_0 is not difficult to interpret. Choose γ_1 as a circle whose center is O and radius r_1; consider a point z_0, $r_1 < z_0 < 1$ and the angle β subtended at z_0 by γ_1 between circles through the points z_0 and -1 tangent to γ_1; this configuration may also be interpreted by transforming the point $z = -1$ to infinity. The set of points z_0 in the interval $r_1 < z_0 < 1$ at which we have $\beta \leqq 2\sin^{-1}(1/6)$ is either empty or a set $r_2 \leqq z_0 < 1$; so the corresponding set of points z_0 no longer required to lie on the given interval, but at which γ_1 subtends an angle not greater than that prescribed, is precisely the set $r_2 \leqq |z| < 1$; more generally this is an annulus bounded by C and by a circle of the coaxal family determined by γ_1 and C, namely a circle NE concentric with γ_1.

For the circle γ_2 there exists a corresponding annulus, and Theorem 7 asserts that a point z_0 common to these annuli cannot be a critical point of $R(z)$.

Before leaving this general topic of rational functions whose zeros are symmetric with respect to their poles in $C: |z| = 1$, we return briefly to the general situation of Theorem 5, and obtain an analog of Bôcher's Theorem somewhat different from Theorem 7:

THEOREM 8. *Let the m zeros of the rational function $R(z)$ interior to $C: |z| = 1$ lie in the closed interior of the circle γ_1 interior to C, and let the n poles ($n \leqq m$) of $R(z)$ interior to C lie in the annular region bounded by C and a circle γ_2 interior to C. Let the zeros of $R(z)$ be symmetric to the poles in C. If P is a point interior to C and γ_2 but exterior to γ_1, and if the greatest NE distance from P to γ_1 is less than the least NE distance from P to γ_2, then P is not a critical point of $R(z)$.*

In the proof we choose P as the origin, and choose γ_1 as the circle $|z - a| = r$, $0 < a < 1 - r$. From the Lemma of §5.3.4 (or by a direct proof) it follows that the horizontal component to the left of the force at O due to the m particles and their inverses with reversed masses is not less than it would be if the m particles were concentrated at the point $z = a + r$, and a direct proof shows that the horizontal component to the right of the force at O due to the n particles and their inverses with reversed masses is not greater than it would be if these n particles were concentrated at the point $z = b$, where $b(> a + r)$ is the least distance from O to γ_2. Direct comparison of these two forces at O, or an inversion with O as center, then shows that O cannot be a position of equilibrium.

The geometric condition on P can be easily expressed in terms of the NE distances ρ_1 and ρ_2 from P to the centers of γ_1 and γ_2 and the NE radii r_1 and r_2 of γ_1 and γ_2, namely the condition $\rho_1 + r_1 < r_2 - \rho_2$, or $\rho_1 + \rho_2 < r_2 - r_1$. This latter inequality is precisely the condition that P shall lie interior to a certain NE ellipse whose foci are the centers of γ_1 and γ_2; if the circles γ_1 and γ_2 are concentric this ellipse degenerates to a circle concentric with them of NE radius $(r_2 - r_1)/2$. In addition, P shall lie exterior to γ_1.

The proof of Theorem 8 requires only an obvious alteration to yield the

COROLLARY. *Let m zeros of the rational function $R(z)$ lie in the closed interior of the circle γ_1 interior to $C: |z| = 1$, and let zeros of $R(z)$ or poles or both of total multiplicity not greater than m lie in the annular region bounded by C and a circle γ_2 interior to C. Let $R(z)$ have no other zeros or poles interior to C, and let the zeros of $R(z)$ be symmetric to the poles in C. If P is a point interior to γ_2 but exterior to γ_1, and if the greatest NE distance from P to γ_1 is less than the least NE distance from P to γ_2, then P is not a critical point of $R(z)$.*

§5.3.7. NE half-planes as loci. From our discussion of §§5.3.4–5.3.6 it is clear that circular regions in the hyperbolic plane differ essentially in properties from circular regions in the euclidean plane. For the case that the circular regions are NE half-planes, there are phases of agreement as well as of difference, as we proceed to show in more detail. A precise NE analog of §1.5.1 Lemma 1 for half-planes is the

LEMMA. *Let the point P interior to $C:|z| = 1$ lie exterior to the NE half-plane H. The force at P due to k unit positive particles in H and to unit negative particles at their inverses with respect to C is equivalent to the force at P due to k coincident positive particles in H and to negative particles at their inverses with respect to C.*

In the proof it is sufficient to choose the point P as the origin O, for if Q denotes the point in H at which the k equivalent coincident particles lie, the point P (that is, O) is a critical point of the rational function which has zeros in the given k points of H, poles in the inverses of these points, a k-fold pole in Q, a k-fold zero in the inverse of Q, and no other zeros or poles; the Lemma for arbitrary P follows by a linear transformation leaving the interior of C invariant, by §4.1.2 Theorem 2.

We choose, then, P as the origin, and choose H as the point set interior to C in the closed interior of the circle Γ:

$$r^2 - 2ar \cos \theta + 1 = 0, \qquad a > 0, \tag{13}$$

where (r, θ) are polar coordinates. The force at O due to a positive particle at A and a negative particle at the inverse A' of A with respect to C is represented by the sum of the vectors $A'O$ and OA, or by the vector $A'A$. On each half-line through O the totality of such vectors consists of all vectors directed toward O whose lengths are not greater than the segment of the half-line interior to the circle (13). We reverse all these vectors in sense, and lay them off from O as initial points. The locus L of terminal points is then a convex set. Indeed, the boundary of the locus has by (13) the equation

$$r^2 = 4(a^2 \cos^2 \theta - 1), \tag{14}$$

for which the angle ψ from the radius vector to the curve is characterized by

$$\tan \psi = \frac{r\, d\theta}{dr} = \frac{-2(a^2 \cos^2 \theta - 1)}{a^2 \sin 2\theta},$$

whence by differentiation

$$\frac{d\,(\tan \psi)}{d\theta} = \frac{-4[-1 + (2 - a^2) \cos^2 \theta]}{a^2 \sin^2 2\theta},$$

which is non-negative by virtue of the inequalities $a > 1$ and $a \cos \theta \geqq 1$. For $\theta = 0$ we have $\psi = \pi/2$, and for $a \cos \theta = 1$ we have $\psi = \pi$ with $r = 0$; the curve lies in the sector $\cos \theta \geqq 1/a$. Consequently the curve (14) which bounds L is convex, so L is a convex set. If the terminal points of k vectors lie in L, so does the center of gravity of those terminal points, and this is equivalent to the conclusion of the Lemma.

In our application of the Lemma it is also convenient to choose a special configuration, but which is readily generalized by a linear transformation of the region $|z| < 1$ onto itself:

THEOREM 9. *Let $R(z)$ be a rational function of z of degree $k + m$ whose zeros are symmetric to its poles with respect to the unit circle C. Let the origin O be a k-fold zero, and let m zeros lie interior to C in the closed interior of the circle Γ whose equation is* (13). *Then all critical points of $R(z)$ interior to C not at O lie in the closed Jordan region S in the closed interior of C bounded by the arc A_0:$\cos\theta \geqq (1/a)$ of C and an arc A_1 of the curve*

$$r^2 - 2ar\cos\theta - (2mr/k)(a^2\cos^2\theta - 1)^{1/2} + 1 = 0. \tag{15}$$

It will appear in the course of the proof that S contains all points common to the interiors of C and Γ, so by the Lemma it is in addition sufficient to study the critical points of $R(z)$ when the m zeros coincide at a point α of the assigned locus of those zeros. Thus we study the zeros of the function

$$R_0(z) = \frac{k}{z} + \frac{m}{z - \alpha} - \frac{m}{z - 1/\bar{\alpha}}, \qquad |\alpha| < 1,$$

where α lies in the closed interior of Γ. The poles of $R_0(z)$ are collinear and there are precisely two zeros, both collinear with the poles; one zero lies in the interval $(0, \alpha)$ and the other in the interval $(1/\bar{\alpha}, \infty)$ not containing 0. If $\arg\alpha$ is fixed, the sole zero β of $R_0(z)$ interior to C lies on the interval $(0, \alpha)$, and as α varies in its prescribed locus so that $|\alpha|$ varies monotonically, $|\beta|$ likewise varies monotonically in the same sense, by §5.1.1 Theorem 1. When $|\alpha|$ approaches unity, so does $|\beta|$, by Hurwitz's Theorem, for $R_0(z)$ approaches k/z uniformly in the neighborhood of any point $z \neq 0$, $|z| \neq 1$. For fixed $\theta = \arg\alpha$, the limiting value of $|\beta|$ (other than $|\beta| = 1$) is given by α on Γ, whence from (13) this limiting value of $r = |\beta|$ satisfies

$$\frac{k}{r} + \frac{m}{r - a\cos\theta + (a^2\cos^2\theta - 1)^{1/2}} - \frac{m}{r - a\cos\theta - (a^2\cos^2\theta - 1)^{1/2}} = 0,$$

which is equivalent to (15). We note that for fixed θ this limiting value of r is less than the value of r on Γ interior to C, so the proof of Theorem 9 is complete.

The arc A_1 lies in the sector $|\arg z| \leqq \cos^{-1}(1/a)$, and cuts C on the lines bounding this sector, hence is not a circular arc. Clearly Theorem 9 is not a perfect analog of §3.1.2 Theorem 2 with C_1 a null circle.

We add some geometric remarks concerning A_1. If (15) is transformed into rectangular coordinates, it is seen that A_1 is an arc of a bicircular quartic. We have shown that A_1 except for its end-points lies exterior to the circle Γ. From a study of $R_0(z)$ using §5.1.1 Theorem 1 it follows that on each half-line $\theta = \arg\alpha$ cutting Γ the intersection of the half-line with A_1 lies to the right of the intersection nearer O of the half-line with the circle

$$|z - ka/(k + m)| = k(a^2 - 1)^{1/2}/(k + m):$$

compare §1.5.1 Lemma 2. Moreover if (15) is satisfied we obviously have

$$r^2 - 2ar \cos \theta - (2amr/k) \cos \theta + 1 < 0,$$

so the arc A_1 lies interior to the circle

$$| z - (k + m)a/k | = [(k + m)^2 a^2/k^2 - 1]^{1/2},$$

which is orthogonal to C.

In preparing to prove the analog of Theorem 9 where the m zeros of $R(z)$ are replaced by m poles, it is convenient to study the zeros of the function

$$(16) \qquad R_1(z) = \frac{k}{z} - \frac{m}{z - \alpha} + \frac{m}{z - 1/\bar{\alpha}}, \qquad |\alpha| < 1,$$

where for the present we choose α as real and positive. The equation $R_1(z) = 0$ is equivalent to a reciprocal equation, hence if z_0 is a zero so also is $1/z_0$. If the zeros are not both real they both lie on C. Interpretation of the second member of (16) as a field of force shows that no zero of $R_1(z)$ lies in the interval $0 < z < \alpha$. The vanishing of $R_1(z)$ in the interval $-1 < z < 0$ is equivalent to the vanishing of $f(z) = kz^2 - (k\alpha + k/\alpha + m\alpha - m/\alpha)z + k$ in that interval for which (since we have $f(0) > 0$) a necessary and sufficient condition is

$$(17) \qquad f(-1) = 2k + k\alpha + k/\alpha + m\alpha - m/\alpha < 0.$$

Condition (17) can be written $(0<)\alpha < (m - k)/(m + k)$. If this latter condition is satisfied, the function $f(z)$ has precisely one zero in the interval $-1 < z < 0$; direct differentiation of $f(z)$ shows that this zero considered as a function of α moves on the axis of reals in the sense opposite to that of α. The Lemma, including its limiting case for P on the boundary of H, now yields the analog of Theorem 9:

THEOREM 10. *Let $R(z)$ be a rational function of z of degree $k + m$ whose zeros are symmetric to its poles with respect to the unit circle C. Let the origin be a k-fold zero, and let m poles lie interior to C in the closed interior of the circle Γ whose equation is (13). If we have $k \geqq m$ or if we have $k < m$ with $a - (a^2 - 1)^{1/2} \geqq (m - k)/(m + k)$, then no critical points of $R(z)$ lie in C except at O or interior to Γ. If we have $k < m$ with $a - (a^2 - 1)^{1/2} < (m - k)/(m + k)$, let θ_0 (>0) denote the value of θ for which $r = (m - k)/(m + k)$ on Γ; then all critical points of $R(z)$ interior to C except at O and interior to Γ lie in the sector $\pi - \theta_0 < 0 < \pi + \theta_0$ between C and an arc of the curve*

$$r^2 - 2ar \cos \theta - (2mr/k)(a^2 \cos^2 \theta - 1)^{1/2} + 1 = 0.$$

This last equation is found from (15) by reversing the signs of m, r, and $\cos \theta$. With $k < m$ the inequality $a - (a^2 - 1)^{1/2} < (m - k)/(m + k)$ is equivalent to $a > (m^2 + k^2)/(m^2 - k^2)$.

The point set interior to C assigned by Theorem 10 to the critical points of $R(z)$ is the actual *locus* of critical points interior to C for all possible functions $R(z)$ satisfying the given conditions except in the case $m = 1$, in this latter case not every point interior to both C and Γ can be a critical point. Indeed, a necessary and sufficient condition that $f(z)$ vanish in the interval $(0 < \alpha < 1)$ $\alpha < z < 1$ is

$$f(1) = 2k - k\alpha - k/\alpha - m\alpha + m/\alpha < 0, \tag{18}$$

which can be expressed in the form $(0<)(k - m)/(k + m) > \alpha(<1)$. If this condition is fulfilled, the function $f(z)$ has precisely one zero in the interval $0 < z < 1$, and direct differentiation of $f(z)$ shows that this zero considered as a function of α moves on the axis of reals in the same sense as does α. It follows that *in the case* $k = 1$, $m = 1$, *and also in the case* $m = 1$, $a - (a^2 - 1)^{1/2} \geqq (k - m)/(k + m) > 0$, *no critical point of* $R(z)$ *lies interior to both* C *and* Γ; *in the case* $m = 1$, $a - (a^2 - 1)^{1/2} < (k - m)/(k + m) > 0$, *we denote by* θ_1 (> 0) *the value of* θ *for which we have* $r = (k - m)/(k + m)$ *on* Γ, *and all critical points of* $R(z)$ *interior to both* C *and* Γ *lie in the sector* $-\theta_1 < \theta < \theta_1$ *between* C *and an arc of the curve*

$$r^2 - 2ar \cos \theta + (2mr/k)(a^2 \cos^2 \theta - 1)^{1/2} + 1 = 0.$$

With $k > m$ the inequality $a - (a^2 - 1)^{1/2} < (k - m)/(k + m)$ is equivalent to $a > (k^2 + m^2)/(k^2 - m^2)$.

Of course Theorems 9 and 10 when applicable yield results not contained in §5.3.1 Theorem 1 and §5.3.2 Theorem 2, but in a given situation it may be desirable to apply both Theorem 9 and §5.3.1 Theorem 1 or both Theorem 10 and §5.3.2 Theorem 2.

§5.4. Elliptic plane. As a companion piece to the study (§5.3) of rational functions whose zeros are symmetric with respect to their poles in the unit circle, involving hyperbolic NE geometry, we study [Walsh, 1948] rational functions whose zeros and likewise whose poles possess central symmetry on the Neumann sphere. The symmetry required here is precisely the symmetry required in Klein's classical model of elliptic geometry.

§5.4.1. Analog of Bôcher's Theorem. We prove the analog of Bôcher's Theorem (§4.2) and of §5.3.2 Theorem 2:

THEOREM 1. *Let* $R(z)$ *be a rational function whose zeros and whose poles occur in diametrically opposite points of the sphere. Let* P *be an arbitrary point of the sphere, let* H *be the open hemisphere containing* P *whose pole is* P, *and let a great circle* G *through* P *separate all the zeros of* $R(z)$ *in* H *not on* G *from all the poles of* $R(z)$ *in* H *not on* G, *where at least one zero or pole lies in* H *not on* G. *Then* P *is not a critical point of* $R(z)$ *unless it is a multiple zero.*

As in §4.1.3, we choose the sphere as having the unit circle in the z-plane as a great circle; but we continue to study the field of force in the plane rather than on the sphere. If z_0 is an arbitrary point of the sphere, the diametrically opposite point may be found by successive reflection in three mutually orthogonal great circles of the sphere, or in the plane may be found by successive reflection in the unit circle and the two coordinate axes, so the diametrically opposite point is $-1/\bar{z}_0$.

In the plane choose the image of P as the origin O, the image of H as the interior of the unit circle $C: |z| = 1$, and the image of G the axis of imaginaries. The force at O due to a pair of positive particles at z_0 and $-1/\bar{z}_0$ is

$$(1) \qquad (-1/\bar{z}_0) + z_0 = -z_0 \frac{1 - z_0 \bar{z}_0}{z_0 \bar{z}_0},$$

so if z_0 lies interior to C this force is directed along the line from z_0 to O; if z_0 lies on C this force is zero. Under the conditions of Theorem 1, each pair of zeros of $R(z)$ in the plane not on C has precisely one zero z_0 interior to C, and the total force at O due to the pair of particles is directed from z_0 toward O; this force has a non-vanishing horizontal component unless z_0 lies on the axis of imaginaries. The total force at O due to particles at a pair of poles of $R(z)$ also has a non-vanishing horizontal component in this *same* sense unless the particles lie on the axis of imaginaries. Since $R(z)$ has at least one zero or pole in H not on G, the total force at O is not zero and O is not a position of equilibrium, as we were to prove.

It is desirable in Theorem 1 to choose H as open, for the force at O due to a pair of particles on the boundary of H is zero, and no restriction need be made as to the location of those particles with reference to G; on the other hand, it is essential to the truth of the theorem to require that at least one zero or pole shall lie in H not on G, for if all zeros and poles lie on the boundary of H, the point O is surely a position of equilibrium; if all zeros and poles in H lie on G, then O may be a point of symmetry or otherwise and be a critical point not a multiple zero. However, we have

COROLLARY 1. *Let $R(z)$ be a rational function whose zeros and whose poles occur in diametrically opposite points of the sphere. Let P be an arbitrary point of the sphere, let H be the open hemisphere containing P whose pole is P, and let all zeros and poles of $R(z)$ in H lie on a great circle G through P, where P separates the zeros in H from the poles in H, at least one of these sets being non-vacuous. Then P is not a critical point of $R(z)$.*

If *all* zeros and poles of $R(z)$ lie on G, it follows by consideration of the field of force on the sphere that any open arc of G bounded by two zeros or by two poles of $R(z)$ but containing no zero or pole contains at least one critical point; if four mutually disjoint closed arcs of G contain in pairs all zeros but no poles and all poles but no zeros of $R(z)$, then those four arcs contain all critical points of $R(z)$ on G, and the other critical points lie at the poles of G:

COROLLARY 2. *Let $R(z)$ be a rational function whose zeros and whose poles occur in diametrically opposite points of the sphere, and all lie on the stereographic projection G of the axis of reals. Let the zeros of $R(z)$ lie on the arcs $0 < z < 1$ and $-\infty < z < -1$, and the poles of $R(z)$ lie on the arcs $-1 < z < 0$ and $1 < z < +\infty$. Each open arc of G bounded by two zeros or by two poles of $R(z)$ and containing no zero or pole of $R(z)$ contains precisely one critical point of $R(z)$; the points $+i$ and $-i$ are poles of the great circle G and are simple critical points; there are no other critical points except multiple zeros of $R(z)$.*

We proceed to consider the geometric conditions on P in Theorem 1 when $R(z)$ is given. On the sphere as in the plane we define a circular region as a closed region bounded by a circle, and we frequently use the same notation for a region as for its boundary. Let the zeros of $R(z)$ lie in two diametrically opposite disjoint circular regions C_1 and C_3 of the sphere, and the poles lie in two diametrically opposite disjoint circular regions C_2 and C_4 disjoint from C_1 and C_3. Construct the circular region S_k ($k = 1, 2, 3, 4$) which is the locus of points P such that no point of the region C_k is at a spherical distance from P greater than $\pi/2$. Thus the circular regions C_k and S_k have the same poles, and their spherical radii are complementary. Denote by T the open region (assumed not empty) common to S_1 and S_2. If P is a point of T, the open hemisphere whose pole is P and which contains P contains the regions C_1 and C_2 and contains no point of C_3 or C_4. Taken together with suitably chosen arcs of C_1 and C_2, the great circles Γ_1 and Γ_2 tangent to C_1 and C_2 and separating those circles separate T into three regions R_1, R_2, R, of which R_1 and R_2 may be empty. The region R consists of all points P in T which lie on great circles G separating C_1 and C_2, and R is open. The region R_1 closed with respect to T is disjoint from R, and consists of all points of T common to all hemispheres bounded by circles G and containing points of C_1; thus R_1 consists of all points of T not in R which lie on great circles cutting both C_1 and C_2, but on the maximal arcs of those great circles bounded by points of C_1 and by points of the boundary of T, namely arcs in T containing no points of C_2; the region R_2 is similarly defined by permuting subscripts. The regions R_1 and R_2 may contain the whole or parts of the regions C_1 and C_2 respectively. Both R_1 and R_2 are convex with respect to great circles. It follows from Theorem 1 that all critical points of $R(z)$ in T lie in R_1 and R_2; no critical point of $R(z)$ lies in R.—We have constructed the geometric configuration C_1, C_2, R_1, R_2, R, T for simplicity of exposition; under suitable conditions corresponding regions R_1, R_2, R, T may be used without the introduction of circular regions C_k.

§5.4.2. Circular regions as loci. The relation between Theorem 1 and §5.3.2 Theorem 2 is far closer than mere analogy. In the proof of Theorem 1 we have studied the origin as a possible position of equilibrium in the field of force due to positive unit particles at points α_k and $-1/\bar{\alpha}_k$, and to negative unit particles at the points β_k and $-1/\bar{\beta}_k$. So far as concerns the force at O, this field is equivalent to the field due to positive unit particles at α_k and $1/\bar{\beta}_k$, and to negative unit particles at β_k and $1/\bar{\alpha}_k$; the latter is precisely the field used in the proof of §5.3.2 Theorem 2, with each positive unit particle symmetric to a negative unit particle in C. In the original field of force for Theorem 1, the total numbers of particles α_k and β_k are the same, but pairs of diametrically opposite particles on C itself may be omitted or provided artificially at pleasure. In the light of this remark, it follows that the two fields of force are entirely equivalent; results established for either field apply also to the other. All the results of §5.3 can be interpreted in the present study of elliptic geometry; but for elliptic geometry the geometric configuration depends essentially on the point O at which the force is considered.

Theorem 1 is an exact analog of Bôcher's Theorem insofar as the latter refers to critical points on circles which separate zeros and poles, but not insofar as the theorems relate to circular regions containing zeros and poles. Indeed, it follows from §5.3.5 Theorem 6 that no precise analog exists in this respect:

THEOREM 2. *There exists a rational function $R(z)$ of degree four whose zeros and whose poles occur in pairs in diametrically opposite points of the sphere, with the property that a point P is a critical point of $R(z)$, where the open hemisphere H containing P and whose pole is P contains a circular region C_1 containing precisely two zeros of $R(z)$ and not containing P, and H contains precisely one double pole of $R(z)$, which is exterior to C_1.*

An analog and consequence of §5.3.6 Theorem 8 is

THEOREM 3. *Let $R(z)$ be a rational function whose zeros and whose poles occur in pairs in diametrically opposite points of the sphere, let P be an arbitrary point of the sphere, and let H denote the closed hemisphere containing P whose pole is P. Let all zeros of $R(z)$ in H lie in the circular region γ_1 interior to H not containing P, and let all poles of $R(z)$ in H lie in a closed annular subregion of H not containing P bounded by a circle γ_2 and by the boundary of H. If the greatest spherical distance from P to γ_1 is less than the least spherical distance from P to γ_2, then P is not a critical point of $R(z)$.*

Let r_1 and r_2 denote the spherical radii of γ_1 and γ_2, and ρ_1 and ρ_2 the spherical distances from P to the poles in H of those circles. The condition on P can then be written $\rho_1 + r_1 < r_2 - \rho_2$, or $\rho_1 + \rho_2 < r_2 - r_1$. This latter condition expresses the condition that P should lie interior to a certain spherical ellipse whose foci are the poles of γ_1 and γ_2; if these poles are identical, the ellipse is a circle of spherical radius $(r_2 - r_1)/2$. In addition, P must lie exterior to the circular region γ_1; it is also true that P must lie between the circles γ_1 and γ_2, but no point of the ellipse is on γ_2 or separated by γ_2 from γ_1.

As a simple application of Theorems 1 and 3 we prove

THEOREM 4. *Let $R(z)$ be a rational function whose zeros and whose poles occur in pairs in diametrically opposite points of the sphere. Let all the zeros lie in diametrically opposite circular regions C_1 and C_3, and all the poles lie in diametrically opposite circular regions C_2 and C_4, where the regions C_k are mutually disjoint. Denote by Z_1 and Z_2 the closed zones of the sphere which are the loci of the great circles whose poles lie in C_1 and C_3 and in C_2 and C_4 respectively. Suppose that for the circles C_1 and C_2, and likewise for C_2 and C_3, the spherical length of the common tangent T, chosen as the shorter arc of a great circle separating the corresponding regions, is not less than the sum of the spherical diameters of those circular regions.* Then all critical points of $R(z)$ lie in the regions C_k and Z_1 and Z_2.*

Assume a point P of the sphere not in a region C_k or a zone Z_j to be a critical point of $R(z)$; we shall reach a contradiction. The open hemisphere H containing P whose pole is P must contain in its interior the whole of one of the regions C_1 and C_3, and no point of the other of those regions, for P does not lie in Z_1; similarly for the regions C_2 and C_4. For definiteness suppose H to contain C_1 and C_2.

By Theorem 1 no great circle through P can separate C_1 and C_2. For definiteness suppose the pole of C_1 in H to be nearer P than the pole of C_2 in H. There exists a great circle C through P which cuts C_1 and C_2 at supplementary angles, those circles being oriented in the same sense on the sphere; thus C may be defined as the great circle through P and through the intersection of the common tangents T to C_1 and C_2. Then C cuts C_1 and C_2 in such a way that the points P, A_1, B_1, A_2, B_2 lie in that order on an arc of C in H, where A_k and B_k lie on C_k; the points A_k and B_k do not coincide unless C is tangent to C_k at A_k. The spherical distance A_1B_2 is not less than the length of T, which by hypothesis is not less than the sum of the spherical diameters of C_1 and C_2. The circle Γ whose pole is P and whose spherical radius is PB_2 less the spherical diameter of C_2 cannot cut C_2. Moreover Γ cuts C between A_1 and B_2 at a point whose distance from A_1 is not less than the diameter of C_1, so Γ cannot cut C_1, and hence C_1 is separated by Γ from C_2. It then follows from Theorem 3 that P is not a critical point of $R(z)$; this contradiction completes the proof of Theorem 4.

If in Theorem 4 the function $R(z)$ is of degree $2m$, and if the regions C_k are disjoint from the Z_j, the regions C_1 and C_3 contain each $m - 1$ critical points of $R(z)$, for it follows from the method of proof of Theorem 3 (i.e. of §5.3.6 Theorem 8) that at a point near but exterior to C_k the force has a component directed away from C_k, and the Principle of Argument applies; the region $Z_1 + Z_2$ contains then precisely two critical points, as is obviously the case if $R(z)$ has precisely two distinct zeros and two distinct poles.

---

* This condition is merely a convenient one for use in the proof; it may be replaced by other conditions less restrictive.

We shall not study in further detail critical points of real rational functions whose zeros and whose poles occur in pairs in diametrically opposite points of the sphere, but the results of §§5.1 and 5.2 are of significance here, and §5.1.2 Theorem 7 yields a complete analog of Jensen's Theorem.

Rational functions whose zeros and poles occur in diametrically opposite points of the sphere can be further studied by the methods of §§5.3.5–5.3.7, but we shall not elaborate this remark. All results concerning such functions, when suitably formulated, are invariant under rigid rotation of the sphere, but are not necessarily invariant under arbitrary linear transformation; consequently new results are frequently obtained by such a transformation. The symmetry required in the present theorems is that of diametrically opposite points of the sphere, or otherwise expressed, that to any point z corresponds a paired point z' found by successive reflections in three particular given mutually orthogonal circles; any symmetry in the plane or on the sphere which can be described in the latter terms can be transformed by a linear transformation into the symmetry of diametrically opposite points of the sphere. A special case of both kinds of symmetry is defined in the plane by anti-inversion in the unit circle $C: | z | = 1$, where we have $z' = -1/\bar{z}$; any circle through z and z' cuts C in two diametrically opposite points.

§5.5. Symmetry in the origin. The primary method for the study of rational functions whose zeros and whose poles are symmetric in the origin:

$$(1) \quad R(z) = \frac{z^k(z^2 - \alpha_1^2)(z^2 - \alpha_2^2) \cdots (z^2 - \alpha_m^2)}{(z^2 - \beta_1^2)(z^2 - \beta_2^2) \cdots (z^2 - \beta_n^2)}, \qquad \alpha_i \neq 0, \qquad \beta_j \neq 0,$$

as in the case of polynomials (§3.6) is by the transformation $w = z^2$, setting

$$(2) \qquad F(w) = [R(w^{1/2})]^2.$$

Thus $F(w)$ is a rational function of the same degree as $R(z)$; if α_j is a zero of $R(z)$ of given order, then α_j^2 is a zero of $F(w)$ of twice that order; if β_j is a pole of $R(z)$ of given order, then β_j^2 is a pole of $F(w)$ of twice that order; if $z = 0$ is a zero or pole of $R(z)$, then $w = 0$ is a zero or pole of $F(w)$ of the same order; if $z = \infty$ is a zero or pole of $R(z)$, then $w = \infty$ is a zero or pole of $F(w)$ of the same order. Moreover we have

$$(3) \qquad F'(w) = R(w^{1/2})R'(w^{1/2})/w^{1/2},$$

so every critical point w of $F(w)$ corresponds to zeros or critical points $z = \pm w^{1/2}$ of $R(z)$, and every critical point z of $R(z)$ corresponds to a critical point $w = z^2$ of $F(w)$ except that $z = 0$ and $z = \infty$ are critical points of $R(z)$ unless they are zeros of the first order or poles.

§5.5.1. Regions bounded by concentric circles. As an illustration of this general method we establish the analog of §4.3 Theorem 1:

THEOREM 1. *Let $R(z)$ be a rational function of degree n whose zeros and whose* *poles are symmetric in the origin O. Let k zeros of R(z) lie in the circular region* $C_1:|z| \leqq a$, *let the remaining* $n - k$ *zeros lie in the circular region* $C_2:|z| \geqq b(>a)$, *and let the poles of R(z) lie in the circular region* $C_3:|z| \geqq c(>a)$. *Denote by* C_0 *the circle*

$$(4) \qquad |z| = r_0 = \left[\frac{kb^2c^2 - na^2b^2 - (n-k)a^2c^2}{(n-k)b^2 + nc^2 - ka^2}\right]^{1/2},$$

where we suppose $r_0 > 0$.

If we have $a < r_0 \leqq c$, *then the annulus* $a < |z| < r_0$ *contains no critical point of R(z). If we have* $r_0 > c$, *then the annulus* $a < |z| < c$ *contains no critical point of R(z). In either of these cases, the region* C_1 *contains precisely* $k - 1$ *zeros of* $R'(z)$.

The rational function $F(w)$ defined by (2) is of degree n, has k zeros in the region $|w| \leqq a^2$, the remaining $n - k$ zeros in the region $|w| \geqq b^2(>a^2)$, and all its poles in the region $|w| \geqq c^2(>a^2)$. It follows from §4.3 Theorem 1 that the annulus specified in Theorem 1 if existent contains no critical point of $R(z)$. The method of continuity, where the zeros and poles of $R(z)$ are varied in such a way that the symmetry is preserved, shows that C_1 contains precisely $k - 1$ zeros of $R'(z)$.

The number r_0 defined by (4) depends not on the specific values of k and n but only on the ratio k/n, and if this ratio is given, Theorem 1 cannot be improved; for we may choose both k and n even, so that Theorem 1 is simply a transformation of §4.3 Theorem 1, which cannot be improved. In the proof of §4.3 Theorem 1, the extremal function has (in the present notation) k zeros on the circle $|z| = a^2$, $n - k$ zeros on the circle $|z| = b^2$, and n poles on the circle $|z| = c^2$. Under the conditions of Theorem 1 the functions $R(z)$ and $F(w)$ vanish at the origin if k is odd, so $F(w)$ cannot be the previous extremal function if $a \neq 0$. But whenever k and n are both even, Theorem 1 cannot be improved.

When k is even, so also is n; for if n were odd, the region $|z| \geqq b$ would contain the odd number $n - k$ of zeros of $R(z)$ and the region $|z| \geqq c$ would contain the odd number n of poles of $R(z)$, so the point at infinity would be both a zero and pole of $R(z)$, which is impossible.

If k is odd and n is even, the regions $|z| \leqq a$ and $|z| \geqq b$ both contain odd numbers of zeros, so the origin and point at infinity are both zeros of $R(z)$. The inequality (2) of §4.3 here becomes in the present notation

$$(5) \qquad \frac{k-1}{a^2+r} + \frac{1}{r} \leqq \frac{n-k-1}{b^2-r} + \frac{n}{c^2+r},$$

as a necessary condition for equilibrium at a point w with $|w| = r$. We denote by r_1 the smallest positive zero of the equation

$$r^3 + [ka^2 - (n-k)b^2 - (n-1)c^2]r^2$$
$$+ [kb^2c^2 - (n-k)a^2c^2 - (n-1)a^2b^2]r + a^2b^2c^2 = 0;$$

we find by substitution that r_1 is not greater than b^2. If we have $r < r_1$, inequality (5) is impossible, so Theorem 1 remains valid if we set $r_0 = r_1^{1/2}$.

If k is odd and n is odd, the origin is a zero of $R(z)$ and the point at infinity is a pole; the inequality corresponding to (5) is

$$\frac{k-1}{a^2+r}+\frac{1}{r} \leqq \frac{n-k}{b^2-r}+\frac{n-1}{c^2+r}, \tag{6}$$

as a necessary condition for equilibrium at a point w with $|w| = r$. We denote by r_1 the smallest positive zero of the equation

$$r^3 - [(k-2)a^2 - (n-k-1)b^2 - nc^2]r^2 \\ - [kb^2c^2 - (n-k+1)a^2c^2 - (n-2)a^2b^2]r - a^2b^2c^2 = 0;$$

we find by substitution that r_1 is not greater than b^2. If we have $r < r_1$, inequality (6) is impossible, so Theorem 1 remains valid if we set $r_0 = r_1^{1/2}$.

Of course Theorem 1 can be established without the use of the transformation $w = z^2$, for instance by the use of lemmas analogous to §4.3 Lemmas 1 and 2, §5.1.4 Lemmas 1 and 2, and the Lemma of §5.3.4. Thus let positive particles at the points α and $-\alpha$ lie in the closed interior of the unit circle C; the algebraically greatest and least horizontal components of the total force at a point $z(>1)$ correspond to the values $\alpha = \pm 1$ and $\alpha = \pm i$ respectively; indeed, the force at z due to the two given particles is the conjugate of

$$\frac{1}{z-\alpha}+\frac{1}{z+\alpha}=\frac{2z}{z^2-\alpha^2}, \qquad |\alpha| \leqq 1,$$

which is numerically greatest and is horizontal when $\alpha^2 = 1$, and is numerically least and horizontal when $\alpha^2 = -1$. Let positive particles at the points α and $-\alpha$ lie in the closed exterior of the unit circle C; the algebraically greatest and least horizontal components of the total force at a point z, $0 < z < 1$, correspond to the values $\alpha = \pm i$ and $\alpha = \pm 1$ respectively; an inversion with z^2 as center and interpretation of α^2 as an arbitrary point in the closed exterior of C renders this conclusion apparent.

§5.5.2. Regions bounded by equilateral hyperbolas. Theorem 1 is based entirely on the moduli of the zeros and poles of $R(z)$. A simple result [Walsh, 1947b] depending only on the arguments of those zeros and poles is

THEOREM 2. *Let $R(z)$ be a rational function whose zeros and whose poles are symmetric in the origin O. Let S_1 and S_2 be closed double sectors with vertex O, disjoint except for O, having angular openings less than $\pi/2$, which contain respectively the zeros and poles of $R(z)$. Denote by S_3 and S_4 the maximal double sectors with vertex O with the sides of S_3 perpendicular to those of S_4 with the property that any pair of mutually perpendicular lines in S_3 and S_4 respectively separates the interior of S_1 from the interior of S_2. Then no zero of $R'(z)$ lies in the interior of S_3 or S_4.*

As in the proof of Theorem 1, we set $w = z^2$ and define the rational function $F(w)$ by equation (2). The zeros of $F(w)$ lie in the closed simple sector S_1' whose vertex is $w = 0$ and which is the image in the w-plane of the double sector S_1; the poles of $F(w)$ lie in the closed simple sector S_2' whose vertex is $w = 0$ and which is the image in the w-plane of the double sector S_2; the sectors S_1' and S_2' have no common finite point other than $w = 0$, and each sector is less than π. It follows from Bôcher's Theorem that no line L in the w-plane which separates S_1' and S_2' (except for $w = 0$) passes through a critical point of $F(w)$ other than a possible multiple zero of $F(w)$ at O or infinity; of course zeros and poles of $F(w)$ may lie on L at $w = 0$ and $w = \infty$, but if $F(w)$ has no zeros or poles other than in those two points, all critical points of $F(w)$ lie in those points. The locus of all lines L is an open double sector in the w-plane, whose image in the z-plane consists of the two open double sectors S_3 and S_4; the latter contain in their interiors no zeros of $R(z)$ and hence no critical points of $R(z)$.

The boundaries of the double sectors S_3 and S_4 separate the z-plane into the double sectors S_3 and S_4 and also double sectors S_5 and S_6 containing respectively S_1 and S_2. The numbers of critical points of $R(z)$ in S_5 and S_6 are readily determined by considering the w-plane, and depend on the origin and point at infinity as possible zeros and poles of $R(z)$. For instance if neither origin nor point at infinity is a zero or pole of $R(z)$, then each of the two simple sectors composing S_5 contains in its interior precisely $(n - 2)/2$ critical points of $R(z)$, where n is the degree of $R(z)$.

If $R(z)$ has the symmetry required in Theorems 1 and 2, and if $F(w)$ is defined by (2), it is clear that any result whatever relating to the location of the critical points of $F(w)$ in the w-plane can by the transformation $w = z^2$ be formulated as a result in the z-plane relating to the location of the critical points of $R(z)$; compare §3.6.1. But this method may involve relatively complicated algebraic curves in the z-plane. An alternate method is to consider various curves and other loci directly in the z-plane; we proceed to some illustrations of this latter method.

THEOREM 3. *Let $R(z)$ be a rational function whose zeros and whose poles are symmetric in the origin.*

1). *If an equilateral hyperbola H whose center is O separates all zeros of $R(z)$ not on H from all poles not on H, and if at least one zero or pole not on H exists, then no critical point of $R(z)$ other than the origin or point at infinity or a multiple zero of $R(z)$ lies on H.*

2). *If an equilateral hyperbola H_1 whose center is O contains in its closed interior all zeros of $R(z)$, and if an equilateral hyperbola H_2 whose center is O lies in the exterior of H_1, contains H_1 in its interior, and contains in its closed exterior all poles of $R(z)$, then between H_1 and H_2 lie no critical points of $R(z)$. No finite critical points other than O and multiple zeros of $R(z)$ lie on H_1 or H_2.*

3). *If an equilateral hyperbola H_1 whose center is O contains in its closed interior* *all zeros of* $R(z)$, *and if an equilateral hyperbola* H_2 *whose center is* O *contains in its exterior the closed interior of* H_1 *and contains in its closed interior all poles of* $R(z)$, *then between* H_1 *and* H_2 *lie no critical points of* $R(z)$ *other than* O. *No finite critical points other than* O *and multiple zeros lie on* H_1 *or* H_2.

4). *If an equilateral hyperbola* H *whose center is* O *passes through all zeros and poles of* $R(z)$, *and if on each branch of* H *the zeros and poles respectively lie on two finite or infinite disjoint arcs of* H, *then all critical points of* $R(z)$ *other than* O *lie on* H. *On any open finite arc of* H *bounded by two zeros or two poles of* $R(z)$ *and not containing* O *or a zero or pole of* $R(z)$ *lies a unique (a simple) critical point of* $R(z)$.

Here we expressly admit degenerate equilateral hyperbolas, namely pairs of mutually perpendicular lines. For a degenerate equilateral hyperbola either double sector which it bounds can be considered as interior or exterior; an arc of a degenerate curve passing through O is a broken line with a right angle at O. Two points one in the interior and one in the exterior of an equilateral hyperbola H are considered separated by H, but not two points both in the interior of H.

Theorem 3 is most easily proved by means of the transformation $w = z^2$ and study in the w-plane of the function $F(w)$ defined by (2), making use of Bôcher's Theorem. A proof is readily given also in the z-plane alone. If two unit positive particles are symmetric in O, the corresponding lines of force are (§§1.6.3 and 2.2) equilateral hyperbolas with center O through those particles; on each branch of the hyperbola the force is directed along that branch in the sense away from the particle which lies on that branch. If we identify each point P of the plane with its reflection in O, it is readily shown (§3.6.1) that through two distinct points (i.e. pairs of points) passes a unique equilateral hyperbola with center O. Theorem 3 can thus be proved by a study of the field of force in the z-plane; details are left to the reader.

A special case involving symmetry yields

COROLLARY 1. *Let the zeros and also the poles of a rational function* $R(z)$ *be symmetric in* O, *let the zeros lie on the axis of reals and the poles on the axis of imaginaries. Then all critical points of* $R(z)$ *lie on those axes. Any open finite segment of either axis bounded by two zeros or two poles and containing no zero or pole of* $R(z)$ *contains a unique (a simple) critical point of* $R(z)$.

In Part 4) of Theorem 3 an arc of H is not permitted to contain O, for with such a function as $R(z) = z^4 - 1$ the origin is not a simple critical point; but in Corollary 1 a segment of an axis is permitted to contain O, for if O is not a zero or pole of $R(z)$ the origin $w = 0$ is not a critical point of $F(w)$ and $z = 0$ is a simple critical point of $R(z)$.

A more general situation than Corollary 1 is worthy of mention:

COROLLARY 2. *Let the zeros and also the poles of a rational function $R(z)$ be symmetric in O and all lie on the coordinate axes. Let the poles of $R(z)$ lie on two finite or infinite segments of the axis of imaginaries which are symmetric in O and contain no zeros of $R(z)$. Then all critical points of $R(z)$ lie on the coordinate axes. Any open finite segment of either axis not containing O and bounded by two zeros or two poles and containing no zero or pole of $R(z)$ contains a unique (a simple) critical point of $R(z)$.*

Theorem 3 yields a new result by inversion with O as center. The inverse of an equilateral hyperbola with O as center is a Bernoullian lemniscate with O as center. We phrase merely the analog and immediate consequence of Part 3):

COROLLARY 3. *Let the zeros and poles of the rational function $R(z)$ be symmetric in O. Let all zeros of $R(z)$ lie in the closed interior of a Bernoullian lemniscate L_1 whose center is O, and let all poles of $R(z)$ lie in the closed interior of a Bernoullian lemniscate L_2 whose center is O and whose axis is orthogonal to that of L_1. Then all finite critical points of $R(z)$ lie in the closed interiors of L_1 and L_2. No critical point of $R(z)$ other than O or a multiple zero of $R(z)$ lies on L_1 or L_2.*

§5.5.3. Circular regions as loci of zeros and poles. We continue to use the geometry of equilateral hyperbolas in the z-plane, and shall establish

THEOREM 4. *Let (C_1, C_2) and (C_3, C_4) be two pairs of circles, each pair symmetric in the origin O, let the two lines of centers of pairs be orthogonal, and let all the circles subtend the same angle not greater than $\pi/3$ at O. Let $R(z)$ be a rational function of degree $2m$ whose zeros and whose poles are symmetric in O, and let the closed interiors of C_1 and C_2 be the locus of the zeros of $R(z)$ and the closed interiors of C_3 and C_4 be the locus of the poles of $R(z)$. Then the locus of the critical points of $R(z)$ consists of O and the point at infinity, and (if we have $m > 1$) of the closed interiors of C_1 and C_2 and the open interiors of C_3 and C_4.*

The closed interiors of C_1 and C_2 contain precisely $m - 1$ critical points each, and if $R(z)$ has precisely $2q$ distinct poles the interiors of C_3 and C_4 contain precisely $q - 1$ critical points each.

The origin O is a point at which the force due to each pair of positive particles is zero, as well as the force due to each pair of negative particles, so O is always a position of equilibrium and a critical point of $R(z)$; the same conclusion applies to the point at infinity, for the given conditions are unchanged by the transformation $z' = 1/z$. In the case $m = 1$, there are in any non-degenerate case precisely two critical points, so the locus of critical points consists of O and the point at infinity. In the case $m > 1$ it follows from the discussion of §4.2.2 as applied in the w-plane, $w = z^2$, that any point in the closed interior of C_1 or C_2 and any point in the interior of C_3 or C_4 can be a critical point of $R(z)$. It remains to show that any finite critical point lies at O or interior to a C_j or on C_1 or C_2.

Here a lemma is convenient for reference:

LEMMA. *If the point* $P:(x_1, y_1)$ *lies on the equilateral hyperbola* $x^2 - y^2 = c^2$, *and if* N *is the intersection with the x-axis of the normal to the curve at* P, *then the interior of the circle* Γ *whose center is* N *and radius* NP *lies interior to the curve.*

At P the slope of the curve is x_1/y_1 and of the normal $-y_1/x_1$, so N is the point $(2x_1, 0)$, and Γ is the circle $(x - 2x_1)^2 + y^2 = x_1^2 + y_1^2 = 2x_1^2 - c^2$. At an arbitrary point (x, y) on Γ we have $x^2 - y^2 - c^2 = 2(x - x_1)^2$, which is positive unless we have $x = x_1$. Thus every point of Γ other than P and its reflection in Ox lies interior to the hyperbola, and the lemma follows. The radius of Γ increases monotonically with x_1.

As a limiting case of the Lemma, it follows that the interior of the circle of curvature at the vertex lies interior to the curve; the curvature at the vertex is $1/c$, the limit of

$$\frac{|y''|}{(1 + y'^2)^{3/2}} = \frac{c^2}{(x^2 + y^2)^{3/2}}.$$

Denote by A_k the arc of C_k which is convex toward O, and bounded by the points of tangency of the tangents to C_k from O. We prove *that the arc* A_1 *is the envelope of a family of equilateral hyperbolas* H *with center* O, *each of which separates the interior of* C_1 *from* C_3 *and* C_4. For definiteness choose the center of C_1 on the positive half of the axis of reals, and let $A:x = a$ be the intersection of A_1 with that axis. The equilateral hyperbola H_0 through A whose axes are the coordinate axes is $x^2 - y^2 = a^2$, whose radius of curvature at the point A is a. The radius r of C_1 is not greater than a, by virtue of our hypothesis that the angle subtended at O by C is not greater than $\pi/3$, so every point of C_1 other than A lies interior to H_0, and the interior of C_1 lies interior to H_0. An arbitrary equilateral hyperbola $H_1:x^2 - y^2 = c^2$ $(0 < c < a)$ is now considered, which contains H_0 and therefore C_1 in its interior. The hyperbola H_1 is rotated rigidly in the clockwise sense about O until it becomes a hyperbola H, tangent in the first quadrant to C_1, say at some point P_1. The relation $OP_1 > OA \geqq r$ implies, by the Lemma, that the interior of C_1 lies interior to H. Corresponding to each c, $0 < c < a$, there exists such a hyperbola H, and the points of tangency P_1 (on the upper half of A_1) vary monotonically and continuously with c; if P is an arbitrary point of the upper half of A_1, there exists a hyperbola H tangent to A_1 at P. In a similar manner we obtain a hyperbola H tangent to A_1 at an arbitrary point of the lower half of A_1. Through an arbitrary point on any finite line segment bounded by O and a point of A_1 passes at least one curve H. Each curve H is tangent to C_1, hence tangent to C_2, and separates the interiors of C_1 and C_2 from the interiors of C_3 and C_4. No non-degenerate curve H is tangent to both C_1 and C_3 or C_4, but if we admit the value $c = 0$, the degenerate hyperbolas H are tangent to C_1, C_2, C_3, and C_4.

In a similar manner we obtain a family G of equilateral hyperbolas with common center O all tangent to A_3 and A_4, such that the interiors of C_3 and C_4 lie interior to each hyperbola and are separated by the curve from the interiors of C_1 and C_2; if P is an arbitrary point of A_3, a curve of the family G is tangent to A_3 at P. Through an arbitrary point on any finite line segment bounded by O and a point of A_3 passes at least one curve G.

It follows from Theorem 2 that no critical points of $R(z)$ lie on a line through O not cutting a circle C_k. Denote by R_k the closed region not containing O bounded by A_k and by the infinite segments not containing O of the lines through O tangent to A_k terminated by the points of tangency. Through an arbitrary point P on any finite open line segment bounded by O and a point of A_k passes at least one curve H or G, and it follows from Part 1) of Theorem 3 if this curve is non-degenerate and from Part 1) and Corollary 1 if this curve is degenerate that P is not a critical point. Thus all finite critical points of $R(z)$ other than O lie in the regions R_k. The substitution $z' = 1/z$ and application of the result just proved now shows that all finite critical points other than O lie in the closed interiors of the C_k.

To consider whether a particular point z_0 of a circle C_k can be a critical point, we choose z_0 on A_k and consider the hyperbola H or G tangent to C_k at z_0. Again by Part 1) of Theorem 3 and by Corollary 1 it follows that z_0 is not a critical point of $R(z)$ unless z_0 lies on C_1 or C_2 and is a multiple zero of $R(z)$; no point of the circle C_3 or C_4 can be a critical point. The remainder of Theorem 4 now follows at once by the method of continuity.

In Theorem 4 we do not exclude the limiting case that C_3 and C_4 coincide with the point at infinity. Here we have the situation of §3.7 Theorems 1 and 3, and it follows that Theorem 4 is false if we omit the requirement that the angle subtended at O by the circles C_k be not greater than $\pi/3$. In this connection a further result is of interest:

THEOREM 5. *Let C_1 and C_2 be circles mutually symmetric in the origin O, subtending at O an angle greater than $\pi/3$, and let their closed interiors contain the zeros (assumed symmetric in O) of a polynomial $p(z)$. Then all critical points of $p(z)$ except perhaps O lie in a closed finite region or regions R bounded by an arc of C_1, an arc of C_2, and two arcs of an equilateral hyperbola whose center is O, one of whose axes passes through the centers of C_1 and C_2 and which is tangent to each of the circles C_k in two distinct points, all arcs so chosen that R contains the closed interiors of C_1 and C_2.*

If the circles C_1 and C_2 subtend angles $\pi/2$ at O, the hyperbola degenerates into the tangents to C_1 and C_2 from O, and the regions R are bounded by the finite segments of those lines between O and the circles plus suitable intercepted arcs of C_1 and C_2. If C_1 and C_2 subtend angles at O less than $\pi/2$, the set R is composed of two disjoint closed regions each bounded by an arc of one of the circles C_k and by an arc of a hyperbola tangent to that circle in two distinct points, the transverse axis of the hyperbola being the line of centers of C_1 and

C_2. If the circles C_1 and C_2 subtend angles at O greater than $\pi/2$, including the possibility that O lies on or interior to the circles, the set R is a single region bounded by arcs of C_1 and C_2 and by arcs of a hyperbola whose transverse axis is perpendicular to the line of centers of C_1 and C_2.

We omit the details of the proof of Theorem 5, but that theorem may be proved by precisely the method of proof of Theorem 4, distinguishing the cases just described and considering arcs A_1 and A_2 of C_1 and C_2 as the envelopes of a suitable family of equilateral hyperbolas. Indeed, the set R is the smallest set containing C_1 and C_2 which is convex with respect to the family of equilateral hyperbolas with center O, and hence (§3.6.1) contains all critical points of $p(z)$. Moreover, the set R can be replaced by no smaller closed set without restricting the degree of $p(z)$, as follows from §1.3.2 Theorem 1 used in the plane of $w = z^2$. Compare also Theorem 9 below.

§5.5.4. Multiple symmetry in O. The results already proved for rational functions $R(z)$ whose zeros and whose poles are symmetric in O extend to rational functions $R(z)$ whose zeros and whose poles exhibit multiple symmetry in O. The analog of Theorem 1 presents no difficulty; equation (4) is here replaced by

$$|z| = r_0 = \left[\frac{kb^p c^p - na^p b^p - (n-k)a^p c^p}{(n-k)b^p + nc^p - ka^p}\right]^{1/p}.$$

This limit can be improved in various special cases in which the origin and point at infinity are zeros and poles of orders $1, 2, \cdots, p-1$; we omit the details.

The present topic may be studied either directly in the z-plane, or by transformation onto the plane of $w = z^p$; compare the methods of §3.6.2. Thus a generalization of Theorem 2 is

THEOREM 6. *Let $R(z)$ be a rational function whose zeros and whose poles exhibit p-fold symmetry in O. Let $S_1, S_2, \cdots, S_p, S_{p+1} = S_1$ be closed sectors with common vertex O, each of angular opening less than π/p, with S_{k+1} found from S_k by rotation through the angle $2\pi/p$ about O, and let the S_k contain all zeros of $R(z)$. Let $T_1, T_2, \cdots, T_p, T_{p+1} = T_1$ be closed sectors with common vertex O, disjoint from the S_k except for O, each of angular opening less than π/p, with T_{k+1} found from T_k by rotation through the angle $2\pi/p$ about O and let the T_k contain all poles of $R(z)$. Denote by $\Sigma_1, \Sigma_2, \cdots, \Sigma_p$ the maximal sectors (which exhibit p-fold symmetry about O) which are the locus of a variable frame of p equally spaced half-lines emanating from O which separates the interiors of the S, from the interiors of the T_k. Then no zeros of $R'(z)$ lie in the interior of the Σ_k.*

A generalization of Theorem 3, whose proof is similar to that of Theorem 3, is

THEOREM 7. *Let $R(z)$ be a rational function whose zeros and whose poles possess p-fold symmetry in O.*

1). *If a p-hyperbola H (§3.6.2) whose center is O separates all zeros of $R(z)$*

not on H from all poles of R(z) not on H, and if at least one zero or pole does not lie on H, then no finite critical point other than O or a multiple zero of R(z) lies on H.

2). *If a p-hyperbola* H_1 *whose center is O contains in its closed interior all zeros of R(z), and if a p-hyperbola* H_2 *whose center is O lies in the exterior of* H_1, *contains* H_1 *in its interior, and contains in its closed exterior all poles of R(z), then between* H_1 *and* H_2 *lie no critical points of R(z). No finite critical points other than O and multiple zeros of R(z) lie on* H_1 *or* H_2.

3). *If a p-hyperbola* H_1 *whose center is O contains in its closed interior all zeros of R(z), and if a p-hyperbola* H_2 *whose center is O contains in its exterior the closed interior of* H_1 *and contains in its closed interior all poles of R(z), then between* H_1 *and* H_2 *lie no critical points of R(z) other than O. No finite critical points of R(z) other than O and multiple zeros lie on* H_1 *or* H_2.

4). *If a p-hyperbola H whose center is O passes through all zeros and poles of R(z), and if on each branch of H the zeros and poles respectively lie on finite or infinite disjoint arcs of H, then all finite critical points of R(z) other than O lie on H. On any open finite arc of H bounded by two zeros or by two poles of R(z) and not containing O or a zero or pole of R(z) lies a unique critical point of R(z).*

In Theorem 7 we do not exclude degenerate p-hyperbolas; such a curve is a set of p equally spaced lines through O. For a degenerate p-hyperbola either set of the alternate sectors into which it divides the plane can be considered as interior or exterior. An arc of a degenerate curve passing through O is a broken line of two segments with an angle π/p at O. Two points one in the interior and one in the exterior of a p-hyperbola H are considered separated by H, but not two points both in the interior of H. As a special case under Part 4) we have the

COROLLARY. *Let R(z) be a rational function whose zeros and whose poles possess p-fold symmetry in O, and lie respectively on the two sets each of p equally spaced half-lines emanating from O forming together a degenerate p-hyperbola. Then all critical points of R(z) lie on that degenerate p-hyperbola. Any open finite segment of the curve bounded by two zeros or two poles of R(z), not containing O and containing no zero or pole of R(z), contains a unique critical point of R(z).*

In the Corollary we have excluded an arc of the curve passing through O or the point at infinity, for if O or the point at infinity is not a zero or pole of $R(z)$ it is a critical point of order $p - 1$.

A typical non-degenerate p-hyperbola with center O is the image in the z-plane ($z = re^{i\varphi}$) of the line $u = a(>0)$ in the plane of $w = u + iv$, and has for its equation

$$r^p \cos p\varphi = a. \tag{7}$$

The inverse of (7) in the unit circle has the equation $ar^p = \cos p\varphi$, and is also useful; for instance Corollary 3 to Theorem 3 admits an immediate generalization involving these inverse curves.

§5.5.5. Multiple symmetry in O, continued. In continuing our investigations, we shall need further properties of the curve (7). We shall prove later that the circle of curvature at a vertex of a p-hyperbola with center O subtends at O the angle $2 \sin^{-1}(1/p)$, so we prove a lemma related to that of §5.5.3:

LEMMA. *Let Γ be a circle whose center lies on the positive half of the axis of reals and which subtends an angle at O less than π/p but greater than $2 \sin^{-1}(1/p)$. If a p-hyperbola H symmetric in the coordinate axes is tangent to Γ in two distinct points, then the interior of Γ lies interior to H.*

For variety we study the situation in the plane of $w = z^p$, where Γ and H are considered in the z-plane. Denote by $2\theta_0$ the angle subtended at O by Γ, so that we have $\sin^{-1}(1/p) < \theta_0 < \pi/2p$. Denote by Γ_1 the image in the w-plane of Γ, so that Γ_1 is a Jordan curve in the right-hand half of the w-plane symmetric in the axis of reals; the interior of Γ_1 is the image of the interior of Γ. We need to show that if $w = u + iv$, and if a line $u = \text{const}$ is tangent to Γ_1 in two distinct points, then the interior of Γ_1 lies to the right of that line. Let Γ be the circle $|z - b| = \rho$, so that for z on Γ we have $z = b + \rho e^{i\theta}$, where the angle θ is measured at the center of Γ, whence

$$\text{(8)} \qquad \frac{dz}{d\theta} = i\rho e^{i\theta}, \qquad \frac{dw}{dz} = pz^{p-1}, \qquad \frac{d(u + iv)}{d\theta} = ip(z - b)z^{p-1}.$$

The tangent to Γ_1 is vertical when and only when $du/d\theta = 0$, for which it is necessary and sufficient that $(z - b)z^{p-1}$ be real. For z on the upper half of Γ we have $0 < \arg z \leqq \theta_0 < \pi/2p$. As z commences with the point $z = b + \rho$ and traces Γ in the counterclockwise sense, both $\arg z$ and $\arg (z - b)$ increase monotonically until we have $\arg z = \theta_0$, $\arg (z - b) = \theta_0 + \pi/2$. Thus $\alpha = \arg [(z - b)z^{p-1}] = \arg [dw/d\theta] - \pi/2$ commences with the value zero and increases to the value $p\theta_0 + \pi/2$, which is less than π; it follows from (8) that throughout the interval $0 < \theta < \theta_0 + \pi/2$ we have $du/d\theta < 0$. In order to study the values $\theta > \theta_0 + \pi/2$, we set $\varphi = \arg z$, so the angle Ozb is $\theta - \varphi$, and from the triangle Ozb we note the relations

$$\text{(9)} \qquad \frac{\sin (\theta - \varphi)}{b} = \frac{\sin \varphi}{\rho}, \qquad d\theta - d\varphi = \frac{b \cos \varphi d\varphi}{\rho \cos (\theta - \varphi)}.$$

We are especially concerned with $\alpha = \theta + (p - 1)\varphi$, in the interval $\theta_0 + \pi/2 < \theta < \pi$, so we set $f(\theta) = b \cos \varphi + p\rho \cos (\theta - \varphi)$, whence by (9)

$$\text{(10)} \qquad d\alpha = d\theta + (p - 1)d\varphi = f(\theta)d\varphi/\rho \cos (\theta - \varphi).$$

We have $f(\theta_0 + \pi/2) = b \cos \theta_0 > 0, f(\pi) = b - p\rho < 0$, so $f(\theta)$ vanishes at least once in the interval $\theta_0 + \pi/2 < \theta < \pi$; for these same values of θ we have by (9)

$$\begin{aligned} d[f(\theta)] &= -b \sin \varphi d\varphi - p\rho \sin (\theta - \varphi)(d\theta - d\varphi) \\ &= -b \sin \varphi[pd\theta - (p - 1)d\varphi] < 0, \end{aligned}$$

so $f(\theta)$ vanishes but once. Thus as θ increases from the value $\theta_0 + \pi/2$, α commences with the value $p\theta_0 + \pi/2 (< \pi)$ and by (10) increases monotonically for an interval and then decreases monotonically, to the value π when $\theta = \pi$. In the interval $\theta_0 + \pi/2 < \theta < \pi$ we always have $\alpha < \pi + (p - 1)\theta_0 < 3\pi/2$. Thus when z traces the upper half of Γ in the counterclockwise direction, the function $(z - b)z^{p-1}$ never vanishes, and with increasing argument traces an arc from the positive half of the axis of reals into the first and second quadrants, across the negative half of the axis of reals into the third quadrant, thence with decreasing argument to the negative half of the axis of reals. Consequently $du/d\theta = \Re[ip(z - b)z^{p-1}]$ vanishes but once for $0 < \theta < \pi$, and indeed is negative throughout the interval $0 < \theta < \theta_0 + \pi/2$, negative throughout a certain interval $\theta_0 + \pi/2 \leqq \theta < \theta_1 < \pi$, and positive throughout the interval $\theta_1 < \theta < \pi$.

In the corresponding behavior in the w-plane, the point w tracing Γ_1 commences at $w = (b + \rho)^p$ on the axis of reals with vertical direction, moves above that axis and to the left in the upper half-plane, then reaches a minimum abscissa u_0 (necessarily positive) at which Γ_1 has a vertical tangent, after which w moves to the right and reaches the axis of reals with vertical direction at the point $w = (b - \rho)^p$. Of course Γ_1 is symmetric in the axis of reals, so the interior of Γ_1 lies to the right of the double tangent $u = u_0$, and the Lemma is proved. We have shown also that if Γ is given, with $\sin^{-1}(1/p) < \theta_0 < \pi/2p$, then a unique p-hyperbola H of form (7) exists which is tangent to Γ in two distinct points.

If b is fixed and ρ is allowed to decrease, the circle Γ satisfying the conditions of the Lemma shrinks monotonically, as must also the curve Γ_1. The double tangent $u = u_0$ to Γ_1 moves to the right. The limiting case Γ^0 of Γ under the Lemma corresponds to the value $\rho = b/p$, and from the analysis already given it follows that for Γ^0 we have $f(\pi) = 0$ but $f(\theta) > 0$ throughout the interval $0 < \theta < \pi$, so as w traces the upper half of the image Γ_1^0 of Γ_0, we have $du/d\theta < 0$. As ρ approaches b/p, the limit of the double tangent is the line $u = (b - b/p)^p$, which has contact of order greater than two with Γ_1^0 at the point $w = (b - b/p)^p$, so Γ^0 is the circle of curvature of the corresponding p-hyperbola at the vertex. In the z-plane the corresponding situation is that of a variable circle tangent to a variable p-hyperbola in two points; as the radius of the circle shrinks, the center remaining fixed, there is a limiting case in which the two points of tangency coincide with the vertex; with the aid of a suitable stretching of the z-plane, this situation can be interpreted as that of a variable circle tangent to a fixed p-hyperbola in two distinct points which approach the vertex; the variable circle then approaches the circle of curvature at the vertex, and the interior of the circle of curvature lies interior to the p-hyperbola.

If H is an arbitrary p-hyperbola with a vertex on the positive half of the axis of reals, and if Γ_0 is the circle of curvature at that vertex, then stretching the plane by the transformation $z' = kz$, $k > 1$, but keeping H fixed moves Γ^0 into the interior of H, and the new Γ^0 has no point in common with H; this is a consequence of the fact that an arbitrary half-line from O which cuts the new Γ^0 cuts H between O and Γ^0 and cannot cut H elsewhere. If P is an arbitrary point of H not a vertex but on the branch intersecting the positive half of the axis of reals, and if N denotes the intersection with that axis of the normal to H at P, then the circle Γ whose center is N and radius NP must subtend a larger angle at O than does Γ^0 yet subtends a smaller angle than π/p, the angle between the asymptotes of that branch. Consequently Γ satisfies the conditions of the Lemma, and the interior of Γ lies interior to the curve.

It follows now, by the method used in §5.5.3, that if Γ is any circle which subtends an angle at O not greater than $2 \sin^{-1} (1/p)$, then the arc A_1 of Γ convex toward O between the tangents to Γ from O can be considered as the envelope of a family of p-hyperbolas with center O. The details of the proof of Theorem 4 are a complete pattern for the proof of

THEOREM 8. *Let each of the two sets of circles $C_1, C_2, \cdots, C_p$ and $D_1, D_2, \cdots, D_p$ possess p-fold symmetry in O, where the half-lines from O to the centers of the C_k bisect the successive angles between the half-lines from O to the centers of the D_k. Let all the circles subtend the same angle, not greater than $2 \sin^{-1} (1/p)$, at O. Let $R(z)$ be a rational function of degree mp whose zeros and whose poles possess p-fold symmetry in O, and let the closed interiors of the C_k and the D_k respectively be the loci of the zeros and poles of $R(z)$. Then the locus of the critical points of $R(z)$ consists of O and the point at infinity, and if we have $m > 1$ of the closed interiors of the C_k and the open interiors of the D_k.*

The closed interiors of the C_k contain precisely $m - 1$ critical points each, and if $R(z)$ has precisely pq distinct poles the interiors of the D_k contain precisely $q - 1$ critical points each.

§5.5.6. Polynomials and multiple symmetry in O. To extend Theorem 5 we need an elaboration of the previous Lemma:

LEMMA. *Let Γ be a circle $|z - b| = \rho$, $b > 0$, which subtends an angle at O greater than π/p. Then there exists a p-hyperbola H with center O tangent to Γ in two distinct points such that the interior of Γ lies exterior to H.*

We retain the notation of §5.5.5, and note by (8) that $\arg (dw) = \alpha + \pi/2$ increases monotonically with θ in the interval $0 \leqq \theta \leqq \pi/p$; the corresponding values of w then lie on a convex arc Γ_1' of Γ_1 in the closed upper half of the w-plane. Denote by Γ_1'' the reflection of Γ_1' in the axis of reals, which is the image of the arc of Γ corresponding to the interval $0 \geqq \theta \geqq -\pi/p$. Then the Jordan curve $\Gamma_2 = \Gamma_1' + \Gamma_1''$ consists wholly of points of Γ_1 and contains the origin $w = 0$ in its interior, and all other points of Γ_1 lie interior to Γ_2. Indeed in the interval $-\pi/p \leqq \theta \leqq \pi/p$ the least value of $|z|$ is greater than any value of $|z|$ in the interval $\pi/p < \theta < (2p - 1)\pi/p$, so all the values of $|w|$ in the former interval are greater than any in the latter interval. At the image of the point $z = b + \rho$, the curve Γ_2 has a vertical tangent; and at the image of the first point at which $\arg z = \pi/p$, since except at the origins angles are unchanged by the transformation, the curve Γ_2 has a tangent which makes an acute angle with the positive horizontal. It follows that Γ_2 has a unique vertical double tangent $u = u_0$ and the interior of Γ_2 lies entirely to the right of that double tangent. This conclusion is valid even if O lies interior to Γ, so the Lemma follows. The conclusion remains valid also in the limiting case that Γ subtends an angle π/p at O, in which case the p-hyperbola degenerates; the limiting case $b = 0$ is trivial.

Further considerations of convexity in the w-plane or of convexity in the z-plane with respect to p-hyperbolas now yield

THEOREM 9. *Let the set of circles $C_1, C_2, \cdots, C_p$ possess p-fold symmetry in O, and let their closed interiors contain the zeros (assumed p-fold symmetric in O) of a polynomial $p(z)$.*

1). *If each circle C_k subtends at O an angle not greater than $2 \sin^{-1} (1/p)$, then all critical points of $p(z)$ except O lie in the closed interiors of the C_k.*

2). *If each circle C_k subtends at O an angle less than π/p but greater than $2 \sin^{-1} (1/p)$, there exists a p-hyperbola H with center O each of whose p branches is doubly tangent to one of the circles C_k; all critical points of $p(z)$ except O lie in the closed interiors of p suitably chosen Jordan curves, each consisting of an arc of a C_k and a finite arc of H doubly tangent to the arc of C_k.*

3). *If each circle C_k subtends at O an angle π/p, all critical points of $p(z)$ lie on the set of closed finite line segments joining O to points of the C_k.*

4). *If each circle C_k subtends at O an angle greater than π/p, there exists a p-hyperbola H each of whose branches is tangent to two of the circles C_k, and the interiors of the C_k lie in the exterior of H. All critical points of $p(z)$ lie in a closed Jordan region whose boundary consists of an arc of each of the circles C_k plus an arc of each of the p branches of H, the latter arc tangent to two of the circles C_k.*

Part 3) is of course a limiting case of both 2) and 4). In both 2) and 4) an axis of H coincides with the line joining O to the center of C_k, and another axis coincides with the bisector of the angle between the lines joining O to the center of successive circles C_k. A limiting case of 4) is that all the circles C_k are identical. In each case the set assigned to the critical points contains the interiors of all the C_k.

The set specified as containing the critical points of $p(z)$ can be replaced by no smaller closed set without restricting the degree of $p(z)$.

§5.6. Skew symmetry in O. If the zeros of a rational function are symmetric to its poles in O, some of our previous results may be applicable. For instance if the zeros and poles lie on respective disjoint arcs of a circle, all critical points lie (§4.2.3 Theorem 2) on that circle. In continuing these results, sectors with vertex O are of particular interest.

§5.6.1. Sectors containing zeros and poles. Bôcher's Theorem is itself significant for this kind of symmetry; if two disjoint circular regions contain respectively the zeros and poles of a rational function, they contain all critical points; indeed these circular regions may have a single point in common without altering this conclusion.

Bôcher's Theorem and its complements (§4.2.1 Corollary 2, §4.2.2 Theorem 1) yield [Walsh, 1946a] the first part of

THEOREM 1. *Let the zeros of the rational function $R(z)$ be symmetric to its poles in the origin O. Let the zeros and poles of $R(z)$ lie in the respective halves of a closed double sector S of angular opening less than π and vertex O.*

1). *All critical points of $R(z)$ lie in S. No critical point of $R(z)$ other than a multiple zero lies on a line bounding S unless all zeros and poles of $R(z)$ lie on that line.*

2). *If all zeros and poles of $R(z)$ lie in the closed exterior [interior] of a circle C whose center is O, then no critical points of $R(z)$ lie on the axis of S interior [exterior] to C. No critical point of $R(z)$ other than a multiple zero lies on C and on the axis of S unless all zeros and poles of $R(z)$ lie on C.*

3). *If the angular opening of S is not greater than $\pi/2$, and if the zeros and poles of $R(z)$ lie in the respective closed regions R_1 and R_2 common to S and a closed annulus bounded by circles with center O, then R_1 and R_2 contain all critical points of $R(z)$. No critical point P of $R(z)$ other than a multiple zero lies on the boundary of R_1 or R_2 unless all zeros and poles lie on a boundary line of S containing P or lie on the circle whose center is O and radius OP. If $R(z)$ is of degree m, then $m - 1$ critical points lie in R_1; if $R(z)$ has precisely q distinct poles, then $R(z)$ has precisely $q - 1$ critical points in R_2.*

4). *If the angular opening of S is β, $\pi > \beta \geqq \pi/2$, and if all zeros and poles of $R(z)$ lie exterior [interior] to a circle C whose center is O, then no critical points of $R(z)$ lie interior [exterior] to C in the closed double sector S_1 whose axis is the axis of S and whose angular opening is $\pi - \beta$.*

Of course O can be neither a zero nor a pole nor a critical point of $R(z)$, nor can the point at infinity.

In the proof of Part 2) we use the fact (§4.1.2 Corollary to Theorem 3) that the lines of force due to a positive unit particle at α and a negative unit particle at $-\alpha$ are the circles through those points. Thus at a point P on the axis of S but interior [exterior] to C, and in the half of S in which the zeros of $R(z)$ lie, the force at P due to each pair of particles, one at a zero of $R(z)$ and the other at a pole symmetric to the zero in O, has a non-vanishing component in the direction and sense PO [OP], so P cannot be a position of equilibrium or a critical point of $R(z)$. If P lies on C and on the axis of S, this conclusion is still valid unless all zeros and poles lie on C, or unless P is a multiple zero of $R(z)$; thus 2) is established. The parts of 2) which refer respectively to the interior and exterior of C are of course equivalent to each other by means of the transformation $z' = 1/z$.

If in 2) we allow S to have the angular opening π, it is not necessarily true that no critical points of $R(z)$ lie on the axis of S interior [exterior] to C. We exhibit the counterexample

$$R(z) = \frac{(z-1)(z-a)^2}{(z+1)(z+a)^2}, \qquad a = -\frac{1}{2}(3 + 7^{1/2}),$$

which has the critical points $z = \pm i/2^{1/2}$, and these lie interior to $C: |z| = 0.9$, whereas the zeros and poles of $R(z)$ lie exterior to C; we choose S as the sum of the simple sectors $0 \leqq \arg z \leqq \pi$ and $0 \geqq \arg z \geqq -\pi$ pertaining respectively to the zeros and poles. However, if in 2) we allow S to have the angular opening π, and if we require that at least one pair of zeros and poles of $R(z)$ shall lie interior to S, then no critical points lie on the axis of S interior [exterior] to C.

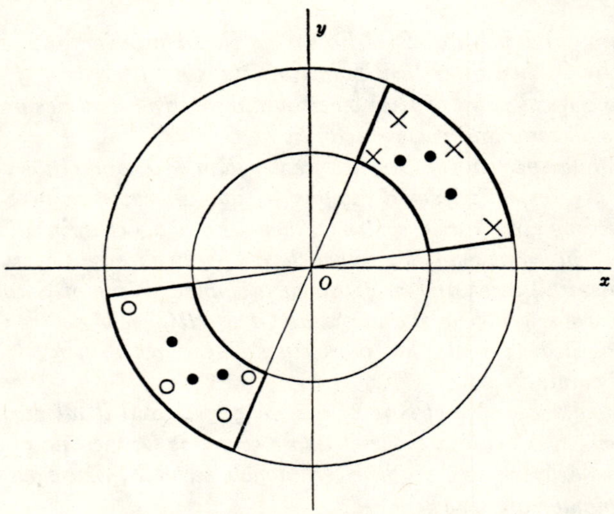

Fig. 16 illustrates §5.6.1 Theorem 1

Under the hypothesis of 3), it follows from 1) that all critical points lie in S and indeed interior to S except for multiple zeros, unless all zeros and poles lie on a boundary line of S; and it follows from 2) by the use of the auxiliary sector found by reflecting S in a line OP through an arbitrary point P interior to S but not in R_1 or R_2 that P cannot be a critical point. The method of proof of 2) shows that no boundary point of S not in the closed annulus can be a critical point, and 2) asserts that no boundary point P of the annulus can be a critical point unless it is a multiple zero or unless all zeros and poles of $R(z)$ lie on the same bounding circle as P. The method of continuity now completes the proof of 3).

Part 4) follows in the case $\beta > \pi/2$ from 2) by a judicious choice of auxiliary sectors and in the case $\beta = \pi/2$ from 3). The counterexample already given shows that the condition $\beta < \pi$ cannot be replaced by $\beta \leqq \pi$.

In a somewhat different order of ideas from Theorem 1, and indeed related to §4.3 Theorem 1, we prove

THEOREM 2. *Let the zeros of a rational function $R(z)$ be symmetric to its poles in O, let a unique zero, of order k, of $R(z)$ lie on the circle $|z| = 1$, and let all the remaining m zeros of $R(z)$ lie in the closed exterior of the circle $|z| = R$. Then all critical points other than multiple zeros of $R(z)$ lie in the closed exterior of the circle $|z| = [(kR^2 - mR)/(k + mR)]^{1/2}$, if existent.*

The force at the point z due to a unit positive particle at α and a unit negative particle at $-\alpha$ is the conjugate of

$$\frac{1}{z - \alpha} - \frac{1}{z + \alpha} = \frac{2\alpha}{z^2 - \alpha^2}.$$

Thus the force at z_0 due to a pair of unit particles on the unit circle is in magnitude at least $2/(1 + |z_0|^2)$, and the force due to a pair of unit particles on or exterior to the circle $|z| = R(>|z_0|)$ is at most $2R/(R^2 - |z_0|^2)$. Consequently equilibrium is impossible if we have

$$\frac{2k}{1 + |z_0|^2} > \frac{2mR}{R^2 - |z_0|^2}, \qquad |z_0|^2 < \frac{kR^2 - mR}{k + mR};$$

this last member is less than R^2, and Theorem 2 follows. The theorem is not incapable of improvement.

§5.6.2. W-curves. In order to sharpen part 4) of Theorem 1, we proceed to the study of W-curves. In rectangular coordinates $z = x + iy$ we consider unit positive particles at the points i and $a + ib$, and unit negative particles at the points $-i$ and $-a - ib$. The line of force for the pair $+i$ and $-i$ through the point (x_1, y_1) is the circle through those three points, namely the circle $(x - c)^2 + y^2 = 1 + c^2$, with $(x_1 - c)^2 + y_1^2 = 1 + c^2$, whose slope at the point (x_1, y_1) is $(-x_1^2 + y_1^2 - 1)/2x_1y_1$. The line of force for the pair $a + ib$ and $-a - ib$ through the point (x_1, y_1) is the circle through those three points, namely the circle $(x - d)^2 + (y + ad/b)^2 = (a - d)^2 + (b + ad/b)^2$, with $(x_1 - d)^2 + (y_1 + ad/b)^2 = (a - d)^2 + (b + ad/b)^2$, whose slope at the point (x_1, y_1) is

$$\frac{b}{a}\,\frac{-x_1^2 + y_1^2 - a^2 - b^2 + 2ax_1y_1/b}{x_1^2 - y_1^2 - a^2 - b^2 + 2bx_1y_1/a}.$$

The equation of the W-curve is found by equating these two slopes, and is

$$(x^2 + y^2)^2 - (a^2 + b^2 - 1)(x^2 - y^2 + 2bxy/a) - (a^2 + b^2) = 0.$$

Any number of points of the curve are readily constructed; if Γ_1 is a circle through $+i$ and $-i$ not separating the points $a + ib$ and $-a - ib$, then any circle Γ_2 through the latter two points tangent to Γ_1 is tangent to Γ_1 in a point of this bicircular quartic.

In polar coordinates (r, θ), where we set $a + ib = Be^{i\alpha}$ and suppose $B > 1$, $-\pi/2 < \alpha < \pi/2$, this bicircular quartic takes the form

$$(1) \qquad r^4 - (B^2 - 1)r^2 \cos(2\theta - \alpha)/\cos\alpha - B^2 = 0.$$

Corresponding to each value of θ there is a unique positive value of r, by Descartes's Rule. The significant part of (1) is that joining the two positive particles: $\alpha \leqq \theta \leqq \pi/2$, and the two negative particles: $\alpha + \pi \leqq \theta \leqq 3\pi/2$. The curve as a whole is symmetric in the origin and in the lines $\theta = \alpha/2$ and $\theta = (\alpha + \pi)/2$.

For the values $\theta = \pi/2$ and $\theta = \alpha + \pi/2$ and their reflections in O the only positive value of r is unity; these are essentially the only values of θ for which r is unity. For the values $\theta = 0$ and $\theta = \alpha$ and their reflections in O the only positive value of r is B; these are essentially the only values of θ for which r is B. The maximum and minimum values of r are found by setting $dr/d\theta = 0$, and correspond to $\theta = \alpha/2 + n\pi$ and $\theta = (\alpha + \pi)/2 + n\pi$. These values of r are respectively given by

$$r^2 = \pm \frac{B^2 - 1}{2 \cos\alpha} + \frac{[(B^2 - 1)^2 + 4B^2 \cos^2 \alpha]^{1/2}}{2 \cos \alpha}.$$

There are no other relative maxima and minima of r, except in the trivial excluded case $B = 1$, in which r is identically unity.

We always have $r_{\max} > B$, $r_{\min} < 1$. As α approaches zero with B fixed, we have $r_{\max} \to B$, $r_{\min} \to 1$. As α approaches $\pi/2$ with B fixed we have $r_{\max} \to \infty$, $r_{\min} \to 0$; these results are related to the counterexample given above. With fixed α as B approaches unity, both $r_{\max}$ and $r_{\min}$ approach unity. With fixed α, as B becomes infinite so also does $r_{\max}$, but $r_{\min}$ approaches the value $(\cos\alpha)^{1/2}$; this last fact will be of significance later. Direct computation of the derivative shows too that as B increases $r_{\max}$ increases and $r_{\min}$ decreases, except in the case $\alpha = 0$, when $r_{\min}$ has the constant value unity.

The W-curve that concerns us is the locus of positions of equilibrium in the field of force due to masses $\pm m_1$ at the points $\pm i$, and to masses $\pm m_2$ at the points $\pm Be^{i\alpha}$, where m_1 and m_2 are positive but need not be rational. Let C_1 denote the circle through $+i$, $-i$, and $Be^{i\alpha}$, and let C_2 denote the circle through $Be^{i\alpha}$, $-Be^{i\alpha}$, and i. Then *the arc of the W-curve joining the points* $Be^{i\alpha}$ *and* i *lies in the finite region* R *bounded by the two arcs of* C_1 *and* C_2 *joining* $Be^{i\alpha}$ *and* i *and not passing through* $-i$ *and* $-Be^{i\alpha}$ *respectively*. If R' denotes the reflection of R in O, each of the circles C_1 and C_2 separates the interior of R from the interior of R'. It follows then by Bôcher's Theorem extended to include the case where m_1 and m_2 are not necessarily rational, that R and R' contain all positions of equilibrium. Consequently the significant arcs of the W-curve lie in the two regions R and R', and in the sectors $\alpha < \arg z < \pi/2$ and $\alpha + \pi < \arg z < 3\pi/2$ respectively.

Whenever the zeros of a rational function $R(z)$ are symmetric to its poles in O, the W-curves may well be of significance; compare the discussion of §5.1.5. Under the conditions of Theorem 1, all critical points lie in S, and neither O nor the point at infinity can be a critical point, so a certain point set Π bounded by W-curves contains all critical points. Moreover, we have an approximate construction for the W-curves, so a suitable set containing Π can be constructed with ease.

Our immediate purpose in studying the W-curves is to sharpen Part 4) of Theorem 1, to which we now return. For definiteness assume the zeros of $R(z)$ to satisfy the relations $1 \leqq |z| \leqq B$, $\alpha \leqq \arg z \leqq \pi/2$. The case $\alpha \geqq 0$ is included in Part 3) of Theorem 1, so we assume $0 > \alpha > -\pi/2$. No critical points lie in the region $|z| < 1$, $0 < \arg z < \alpha + \pi/2$, so we fasten our attention on the critical points in the region R_0: $|z| < 1$, $\alpha + \pi/2 \leqq \arg z \leqq \pi/2$. At such a point z, the characteristic sector which is the locus of terminal points of all vectors with initial points z representing forces each due to a unit positive particle at an admissible zero z_1 and a unit negative particle at $-z_1$ is bounded by the directions at z corresponding to $z_1 = +i$ and $z_1 = Be^{i\alpha}$; if the angle between Oz and Oz_1 is acute, increasing $|z_1|$ with z and $\arg z_1$ constant rotates the direction of the force toward the vector from z to O, but if the angle between Oz and Oz_1 is obtuse, increasing $|z_1|$ with $\arg z_1$ constant rotates the direction of the force in the opposite sense. The two directions bounding this characteristic sector are opposite if and only if z lies on the W-curve. At a point z with $\alpha + \pi/2 \leqq \arg z \leqq \pi/2$ and near O the characteristic sector has approximately the angular opening $-\alpha + \pi/2$, which is less than π. As z recedes from O along the ray Oz, the characteristic sector remains in size less than π until z reaches the W-curve. Thus *the only possible positions of equilibrium in R_0 lie in the closed region T bounded by an arc of the W-curve in R_0 and an arc of $|z| = 1$*; the W-curve cuts the circle $|z| = 1$ on the half-lines $\arg z = \pi/2$ and $\arg z = \alpha + \pi/2$.

As an approximation to T but more easily constructed and containing T in its interior, we may use the region bounded by the arc $|z| = 1$, $\alpha + \pi/2 \leqq \arg z \leqq \pi/2$, by an arc of the circle through i, $Be^{i\alpha}$, $-Be^{i\alpha}$ in the sector $(\alpha + \pi)/2 \leqq \arg z \leqq \pi/2$, and by the reflection of this latter arc in the line $\arg z = (\alpha + \pi)/2$; for the W-curve lies in R and is symmetric in the line $\arg z = (\alpha + \pi)/2$. Another approximation to T can be found by using the circular arcs here for the case $B = \infty$, namely the line segment through i in the direction $\arg z = \alpha$ in the sector $(\alpha + \pi)/2 \leqq \arg z \leqq \pi/2$, and the reflection of that line segment in the line $\arg z = (\alpha + \pi)/2$. Still another approximation to T is found by using the W-curve (1) corresponding to the value $B = \infty$; the curve (1) becomes an equilateral hyperbola, whose minimum distance to O is $(\cos \alpha)^{1/2}$.

By the use of the W-curve (1) already used we can study the possible positions of equilibrium in the region $|z| > B$, $\alpha \leqq \arg z \leqq 0$. A new W-curve, symmetric with respect to the former one in the line $\arg z = \alpha/2 + \pi/4$, is needed to study the positions of equilibrium in the regions $|z| > B$, $\alpha + \pi/2 \leqq \arg z \leqq \pi/2$, and $|z| < 1$, $\alpha \leqq \arg z \leqq 0$. Possible positions of equilibrium are symmetric in O. In sum, we have the following complement to Theorem 1:

THEOREM 3. *Let the zeros of the rational function $R(z)$ be symmetric to its poles in O, and let the zeros lie in the closed region R_1: $\alpha \leqq \arg z \leqq \pi/2$, with $-\pi/2 <$*

$\alpha < 0$, $(0<)B_1 \leqq |z| \leqq B_2$; the poles of $R(z)$ lie in the closed region R_2 found by reflecting R_1 in O. Any critical point of $R(z)$ not in R_1 or R_2 lies in one of the four closed regions T_1 bounded by an arc of $|z| = B_1$ or $|z| = B_2$ which is a bounding arc of R_1 or R_2 and by an arc of the curve

$$(2) \qquad r^4 - (B_2^2 - B_1^2)r^2 \cos(2\theta - \alpha)/\cos\alpha - B_1^2B_2^2 = 0,$$

or in one of the four closed regions T_2 bounded by an arc of $|z| = B_1$ or $|z| = B_2$ which is a bounding arc of R_1 or R_2 and by an arc of the curve

$$(3) \qquad r^4 + (B_2^2 - B_1^2)r^2 \cos(2\theta - \alpha)/\cos\alpha - B_1^2B_2^2 = 0.$$

One arc A_1 of (2) bounding a region T_1 has the end-points $B_1 i$ and $B_1 e^{(\alpha+\pi/2)i}$ and lies in the region $(0<)r_1 \leqq |z| \leqq B_1$, $\alpha + \pi/2 \leqq \arg z \leqq \pi/2$, where we have

$$r_1^2 = \frac{[(B_2^2 - B_1^2)^2 + 4B_1^2B_2^2\cos^2\alpha]^{1/2}}{2\cos\alpha} - \frac{B_2^2 - B_1^2}{2\cos\alpha};$$

another arc A_2 of (2) bounding a region T_1 has the end-points B_2 and $B_2 e^{i\alpha}$ and lies in the region $B_2 \leqq |z| \leqq r_2$, $\alpha \leqq \arg z \leqq 0$, where we have

$$r_2^2 = \frac{[(B_2^2 - B_1^2)^2 + 4B_1^2B_2^2\cos^2\alpha]^{1/2}}{2\cos\alpha} + \frac{B_2^2 - B_1^2}{2\cos\alpha};$$

the other two arcs of (2) bounding regions T_1 are symmetric to A_1 and A_2 in O. The four regions T_2 may be found from the four regions T_1 by rotation through an angle $\pm\pi/2$ or by reflection in the line $\arg z = \alpha/2 + \pi/4$.

If $R(z)$ is of degree n, then R_1 plus the adjacent regions T_1 and T_2 contains precisely $n - 1$ critical points of $R(z)$, and if $R(z)$ has precisely q distinct poles, then R_2 plus the adjacent regions T_1 and T_2 contains precisely $q - 1$ critical points.

It will be noted that the curve (3) may be found from (2) not merely by reflection in the line $\arg z = \alpha/2 + \pi/4$, but alternately by inversion in the circle $|z| = (B_1B_2)^{1/2}$ in which R_1 and R_2 are self-inverse. Consequently our discussion of approximations to the region T is readily applied to the regions T_1 and T_2.

In Theorem 3 no set of smaller regions suffices without restricting the degree of $R(z)$; thus any point of R_1 may be a multiple zero ($n > 1$) and hence a critical point of $R(z)$; moreover W-curves vary continuously with the particles involved, and corresponding to two positive and two negative particles on a circle Γ with center O exhibiting the symmetry of Theorem 3, the W-curves are the two arcs of Γ joining the positive particles and joining the negative particles; when B_2 varies and approaches the fixed value B_1, the curve (2) approaches the circle $r = B_1$; thus for all choices of pairs of particles in Theorem 3 the W-curves completely fill the regions T_1 and T_2. Here we apply also the methods of §4.2.2 for points of R_2.

However, for fixed n the regions defined in Theorem 3 as containing the critical points are unnecessarily large. For instance in the case $n = 2$, if the two zeros of $R(z)$ are α_1 and α_2, $\alpha_1 \neq \alpha_2$, the two critical points are $\pm(\alpha_1\alpha_2)^{1/2}$, which lie in R_1 and R_2; the regions T_1 and T_2 are extraneous.

As an illustration that throws light on the preceding results, we mention

$$R(z) = \frac{(z-1)^{m_1}(z+a)^{m_2}}{(z+1)^{m_1}(z-a)^{m_2}}, \qquad m_1 > 0, \quad m_2 > 0, \quad a > 1.$$

The critical points other than 1 and $-a$ are given by

$$z^2 = \frac{a^2 - am}{1 - am}, \qquad m = m_2/m_1.$$

If m is small and positive, we have approximately $z = \pm a$; when m increases, so does z^2, and z^2 becomes infinite as m approaches $1/a$. For the values $1/a < m < a$, we have $z^2 < 0$, and z^2 takes all values between $-\infty$ and zero, so z takes all pure imaginary values. When $m = a$ we have $z = 0$; for $m > a$, z^2 increases with m and takes all values between zero and unity, so z takes all values $-1 \leqq z \leqq +1$. For the particular value $m = 1$ we have $z^2 = -a$. The W-curve consists of the two coordinate axes with the exception of the two segments $1 \leqq z^2 \leqq a^2$.

The illustration just given is a justification for a further complement to Theorem 1:

THEOREM 4. *Let the zeros of the rational function $R(z)$ be symmetric to its poles in O, and let a closed double sector S of angular opening less than π with vertex O contain all zeros and poles of $R(z)$. Let S' denote the double sector found by rotating S about O through the angle $\pi/2$. If in each half of S the circle Γ with center O separates all zeros of $R(z)$ not on Γ from all poles of $R(z)$ not on Γ, and if not all zeros and poles lie on Γ, then exterior to S' no critical point other than a multiple zero of $R(z)$ lies on Γ. Consequently if an annulus bounded by circles with center O separates all zeros from all poles in each half of S, then that annulus contains in its interior no critical points of $R(z)$ exterior to S'.*

The content of Theorem 4 is essentially different in the cases that the angular opening of S is less than, equal to, or greater than $\pi/2$. In any case, if a point P does not lie in S', there is a closed sector S_P, half of S, such that for all Q in S_P the angle POQ is acute. Let P lie on Γ, and suppose for definiteness that in S_P the zeros of $R(z)$ not on Γ lie exterior to Γ and the poles of $R(z)$ not on Γ lie interior to Γ. The force at P due to a pair of symmetric particles on Γ acts along Γ; the force due to a positive particle in S_P and to its symmetric particle has a non-vanishing component in the direction and sense PO, as has the force due to a negative particle in S_P and to its symmetric particle. Thus P is not a position of equilibrium. The remainder of Theorem 4 follows at once.

As background for the W-curves used in Theorem 3, we mention the following theorem:

The locus of points of tangency of pairs of circles belonging to two given coaxal families is a bicircular quartic.

Of course this quartic may degenerate. Let one coaxal family consist of the circles through the two points α_1 and β_1, and the other consist of the circles through the points α_2 and β_2. The former family consists of the lines of force due to a unit positive particle at α_1 and a unit negative particle at β_1, so the circle $\alpha_1 z \beta_1$ has at z the direction of the conjugate of

$$\frac{1}{z-\alpha_1} - \frac{1}{z-\beta_1} = \frac{\alpha_1 - \beta_1}{(z-\alpha_1)(z-\beta_1)}. \tag{4}$$

The condition for tangency at z of the circles of the two coaxal families can be written

$$\arg \frac{\alpha_1 - \beta_1}{(z-\alpha_1)(z-\beta_1)} = \arg \frac{\alpha_2 - \beta_2}{(z-\alpha_2)(z-\beta_2)} \quad (\text{mod } \pi),$$

$$\arg \frac{(\alpha_1 - \beta_1)(z-\alpha_2)(z-\beta_2)}{(\alpha_2 - \beta_2)(z-\alpha_1)(z-\beta_1)} = 0 \quad (\text{mod } \pi),$$

$$\mathfrak{I}[(\alpha_1 - \beta_1)(z-\alpha_2)(z-\beta_2)(\bar{\alpha}_2 - \bar{\beta}_2)(\bar{z} - \bar{\alpha}_1)(\bar{z} - \bar{\beta}_1)] = 0,$$

which is a degenerate or non-degenerate bicircular quartic.

If the first of the given coaxal families consists of all circles mutually tangent at α_1, each member of (4) is to be replaced by $A/(z-\alpha_1)^2$, where A is a suitably chosen constant. If the first of the given coaxal families consists of all circles orthogonal to the circles through α_1 and β_1, each member of (4) is to be multiplied by i. In any case the conclusion persists.

§5.7. Symmetry in z and $1/z$. In various cases of symmetry, the critical points of a function of z can profitably be studied by transforming the entire z-plane. One illustration is that of symmetry in O (§§3.6 and 5.5), and another is the present situation.

§5.7.1. Polynomials in z and $1/z$. It is appropriate to distinguish between rational functions such as $z + 1/z$ which are unchanged by the substitution $z = 1/z'$ and rational functions the set of whose zeros and the set of whose poles are unchanged by that substitution; the latter condition is satisfied by the function $z - 1/z$, but the former condition is not. If a polynomial in z and $1/z$, namely $P(z) = a_0 z^m + a_1 z^{m-1} + \cdots + a_{m+n} z^{-n}$, with $a_0 \cdot a_{m+n} \neq 0$, has its set of zeros and its set of poles unchanged by the substitution $z = 1/z'$, the order n of $z = 0$ as a pole must be equal to the order m of $z = \infty$ as a pole, whence $m = n$. If $z = \alpha$ is a zero of $P(z)$, so also is $z = 1/\alpha$. Thus the function $z^n P(z)$ is a polynomial in z not vanishing in the origin, with each factor $(z - \alpha_k)$ paired with another factor $(z - 1/\alpha_k)$ unless $\alpha_k = 1/\alpha_k$, namely $\alpha_k = \pm 1$. In any case, even $\alpha_k = \pm 1$, we may write

$$R(z) = [P(z)]^2 = a_0^2 \prod_{k=1}^{2n} (z^2 + 1 - \alpha_k z - z/\alpha_k)/z, \tag{1}$$

so $R(z)$ *is a polynomial in* $(z + 1/z)$. Our primary object is the study of the zeros of $P'(z) = R'(z)/2[R(z)]^{1/2}$. The poles of $R(z)$ are the poles of $P(z)$ and are not zeros of $P'(z)$, so it is sufficient for us to study the zeros of $R'(z)$. That is to say, in specifying a point set S on which all the zeros of $P'(z)$ lie, we ordinarily include in S all possible zeros of $P(z)$, for unless a special restriction is imposed the zeros of $P(z)$ may have multiplicity greater than unity and hence may be zeros of $P'(z)$ and must be included in S; the only other zeros of $P'(z)$ are zeros of $R'(z)$. We shall prove

THEOREM 1. *Let* $P(z)$ *be a polynomial in* z *and* $1/z$ *the set of whose zeros and the set of whose poles are symmetric in* z *and* $1/z$. *If the zeros of* $P(z)$ *lie*

1). *on any two segments* S_1 *of the positive half of the axis of reals, or on any two segments* S_2 *of the negative half of that axis, or on a set* A *consisting of two arcs of the unit circle* C *symmetric in that axis, or on any connected set consisting of* $S_1 + A$ *or* $S_2 + A$ *or* $S_1 + S_2 + C$, *where in every case we assume that the assigned set is unchanged by the transformation* $z = 1/z'$, *then the zeros of* $P'(z)$ *other than* $z = \pm 1$ *lie on that same point set.*

2). *on the axis of imaginaries or on two finite segments of that axis images of each other under the transformation* $z = 1/z'$, *so also do the zeros of* $P'(z)$ *other than* $z = \pm 1$.

3). *in the annulus* $r \leqq |z| \leqq 1/r$, *so do the zeros of* $P'(z)$.

4). *in the sector* $-\alpha \leqq \arg z \leqq \alpha(<\pi/2)$ [*or in the sector* $-\alpha + \pi \leqq \arg z \leqq \alpha + \pi$], *the zeros of* $P'(z)$ *other than* $z = \pm 1$ *lie in that same sector.*

5). *in the right-hand half-plane, in the closed exterior of the unit circle in the first quadrant and in the closed interior of the unit circle in the fourth quadrant, then the zeros of* $P'(z)$ *other than* $z = -1$ *lie in those same closed regions.*

We make use of the well-known transformation $w = \frac{1}{2}(z + 1/z)$, setting $z = x + iy = r(\cos\theta + i \sin\theta)$, $1/z = (\cos\theta - i\sin\theta)/r$, $w = u + iv$, whence

$$u = \tfrac{1}{2}(r + 1/r)\cos\theta, \; v = \tfrac{1}{2}(r - 1/r)\sin\theta. \tag{2}$$

Thus the infinite segments $x > 0$ and $x < 0$ of the x-axis correspond to the infinite segments $u > 1$ and $u < -1$ of the u-axis. The y-axis corresponds to the v-axis. The circle $|z| = 1$ corresponds to the segment $-1 \leqq u \leqq 1$ of the u-axis, and the two circles $|z| = r \neq 1$ and $|z| = 1/r$ correspond to the ellipse (2), where θ is a parameter, with foci $w = +1$ and -1, and semi-axes $(r + 1/r)/2$ and $|r - 1/r|/2$. The two half-lines $\arg z = \pm\theta (0 < \theta < \pi/2)$ correspond to the curve (2) where θ is constant, namely the right-hand branch of the hyperbola $u^2/\cos^2\theta - v^2/\sin^2\theta = 1$, whose foci are $w = +1$ and -1.

We write, under the conditions of Theorem 1,

$$R(z) = [P(z)]^2 = p(w), \qquad R'(z) = p'(w)\, dw/dz = p'(w)(z^2 - 1)/2z^2, \tag{3}$$

where $p(w)$ is a polynomial in w. The points $z = 0$ and ∞ are poles of $P(z)$, hence not critical points. To find a point set in the z-plane containing all zeros of

$P'(z)$, it is sufficient to determine a set containing all zeros of $P(z)$, the images of the points where $p'(w)$ vanishes, and the points $z = +1$ and -1.

In part 1) the image in the w-plane of the given set S is a single segment S' of the u-axis, and the entire image in the z-plane of S' is the set S. Since all zeros of $p(w)$ lie on S', so also do all zeros of $p'(w)$, from which the conclusion follows. It is to be noted that any open subsegment of S_1 or S_2 or any open arc of C bounded by two zeros of $P(z)$ but containing no zero of $P(z)$ contains at least one zero of $P'(z)$; if the subsegment or arc does not contain either of the points $z = \pm 1$, it contains precisely one zero of $P'(z)$.

It is important in part 1) to choose the given set S invariant under the transformation $z = 1/z'$; otherwise we may choose $P(z)$ with the zeros 2, 3, 1/2, 1/3, and S_1 as the sum of the two intervals $0.333 \leqq z \leqq 0.334$, $\frac{1}{2} \leqq z \leqq 3$; here S_1 does not contain the zero of $P'(z)$ in the interval $1/3 < z < 1/2$. A similar remark applies in the other parts of Theorem 1.

It is of course not sufficient in 1) to require merely that the zeros of $P(z)$ be real, nor that they lie on arbitrary arcs of the unit circle; this is illustrated by the example $P(z) = (z - 2 + 1/z)(z + 2 + 1/z) = [(z^2 - 1)/z]^2$, whose derivative vanishes in the points $z = +i$ and $-i$.

It is in order here to state explicitly

COROLLARY 1. *Let the circles C_1 and C_2 be mutually orthogonal. Let $R(z)$ be a rational function with precisely two distinct poles, of equal order, on C_1 and mutually inverse in C_2. Let all the zeros of $R(z)$ lie on C_1 and C_2 in such manner that the entire configuration is symmetric in C_1 and in C_2. Then all critical points of $R(z)$ lie on C_1 and C_2; in particular the intersections of C_1 and C_2 are critical points unless they are simple zeros of $R(z)$.*

We transform the poles of $R(z)$ to zero and infinity so that C_1 is transformed into the axis of reals and C_2 into the unit circle. The conclusion follows from Part 1). A special case is §5.1.2 Corollary to Theorem 7; compare also §5.1.2 Theorem 6 and §5.5.2 Corollary 2 to Theorem 3.

In Part 2) the image S' of the given set S is connected, hence is the whole v-axis or a finite or infinite segment of that axis. The set S' cannot be the v-axis with the exception of a finite segment, for the complete image in the z-plane of the latter set must contain in its interior $z = 0$ and $z = \infty$, and cannot consist of two disjoint finite segments of the y-axis; the set S is the complete image in the z-plane of S'. The set S' contains all zeros of the polynomial $p(w)$, hence contains all zeros of $p'(w)$, and the conclusion follows.

By means of 2) we shall prove

COROLLARY 2. *Let C be the unit circle and let the rational function $R(z)$ have but two poles, of equal order, and in diametrically opposite points of C. Let the zeros of $R(z)$ occur in diametrically opposite points of C. Then all critical points except $z = 0$ and $z = \infty$ lie on C.*

In the usual expression for $R(z)$ as the quotient of two products of linear factors, each factor of the form $z - \alpha$ is multiplied by a corresponding factor $z + \alpha$, so $R(z)$ satisfies the equation $R(z) \equiv R(-z)$. If β and $-\beta$ are the poles of $R(z)$, the function $z' = (z - \beta)/(z + \beta)$ is pure imaginary on C, so that transformation carries C into the axis of imaginaries. We write $R(z) \equiv R_1(z') \equiv R_1[(z - \beta)/(z + \beta)] \equiv R_1[(-z - \beta)/(-z + \beta)] \equiv R_1(1/z')$, so 2) applies. Corollary 2 can also be proved by the methods of §5.5, and is related to the results of §5.2 and §5.4.

To establish 3), we merely notice that the image S' in the w-plane of the given annulus S is the closed interior of the ellipse (2), where θ is a parameter; reciprocally the image in the z-plane of S' is S. The conclusion follows from Lucas's Theorem by the convexity of S', and from the fact that S contains the points $z = 1$ and -1. Part 3) is less general than §5.1.2 Theorem 6.

Part 4) follows from the fact that the image in the w-plane of the sector $-\theta \leqq \arg z \leqq \theta (< \pi/2)$ is a closed convex region, the closed interior of the right-hand branch of the hyperbola (2), where r is a parameter. As a limiting case, $\alpha = \pi/2$ is not excluded.

Part 5) follows from the fact that the closed first quadrant of the w-plane is convex. It is similarly proved that if the zeros of $P(z)$ in the first and second quadrants lie in the closed exterior of the unit circle, and those in the third and fourth quadrants lie in the closed interior of the unit circle, then all zeros of $P'(z)$ lie in those closed regions.

We formulate some other easily proved results:

COROLLARY 3. *Let $P(z)$ be a polynomial in z and $1/z$ the set of whose zeros and the set of whose poles are symmetric in z and $1/z$ and also symmetric in O. If the zeros of $P(z)$ lie in [exterior to] the double sector $-\alpha \leqq \arg z \leqq \alpha(\leqq \pi/4)[\alpha \geqq \pi/4]$, $-\alpha + \pi \leqq \arg z \leqq \alpha + \pi$, so also do all zeros of $P'(z)$ except those at $z = \pm 1$ and $z = \pm i$.*

Under the transformation $w = \frac{1}{2}(z + 1/z)$ the two factors $(z^2 + 1 - \alpha_1 z - z/\alpha_1)/z$ and $(z^2 + 1 + \alpha_1 z + z/\alpha_1)/z$ become $(2w - \alpha_1 - 1/\alpha_1)$ and $(2w + \alpha_1 + 1/\alpha_1)$, so the factors of $R(z) = [P(z)]^2$, which is a polynomial in w, correspond to zeros in w which are symmetric in the origin $w = 0$. The image in the w-plane of the given double sector is the closed interior [exterior] of a hyperbola of eccentricity not greater [not less] than $2^{1/2}$, a point set containing the zeros of $p(w)$, and hence by §3.6.1 Corollary to Theorem 1 containing those of $p'(w)$ other than $w = 0$.

COROLLARY 4. *Let $P(z)$ be a polynomial in z and $1/z$ whose zeros and whose poles are symmetric in z and $1/z$ and also symmetric in O. If the zeros of $P(z)$ in the first and third quadrants lie in the closed exterior of the unit circle, and those in the second and fourth quadrants lie in the closed interior of the unit circle, then all zeros of $P'(z)$ lie in those same closed regions.*

The function $p(w)$ defined by (3) is a polynomial in w whose zeros are symmetric in $w = 0$ and which lie in the closed first and third quadrants of the w-plane. It follows from §3.6.1 Theorem 1 that the zeros of $p'(w)$ lie in those quadrants, and hence that the zeros of $P(z)$ lie in the regions stated. In Corollary 4, there are no requirements on the zeros of $P(z)$ on the axes.

In Theorem 1 we have collected some of the more obvious and simple results. These results may be combined to give further results—for instance 4) combines well with 1), 3), and 5). Still further information can be obtained by transforming other theorems of Chapters I, II, and III from the w-plane to the z-plane.

§5.7.2. General rational functions. Let $R(z)$ be a rational function the set of whose zeros and the set of whose poles are unchanged by the transformation $z = 1/z'$. We write

$$(4) \qquad R(z) = Az^k \frac{(z - \alpha_1) \cdots (z - \alpha_m)}{(z - \beta_1) \cdots (z - \beta_n)}, \qquad A \neq 0,$$

where we assume numerator and denominator to have no non-trivial common factor, and assume each α_j and β_j different from zero. We assume too $k \geqq 0$; in the contrary case it is sufficient to apply the subsequent reasoning to the function $1/R(z)$. The point $z = 0$ is a zero of $R(z)$ of order k, and $z = \infty$ is a zero of order $n - (m + k) = k$, whence $2k = n - m$. If we set $P(z) \equiv (z - \alpha_1) \cdots (z - \alpha_m)$, $Q(z) \equiv (z - \beta_1) \cdots (z - \beta_n)$, the two sets of numbers α_k and β_k are unchanged by the substitution $z = 1/z'$, so by the reasoning previously used (§5.7.1) the functions $[P(z)]^2/z^m$ and $[Q(z)]^2/z^n$ are polynomials in $(z + 1/z)$; their quotient is equal to $[R(z)]^2/A^2$ by virtue of the equation $2k = n - m$, and *is a rational function of* $(z + 1/z)$.

Our primary interest is in the zeros of $R'(z)$, where if we set $S(z) = [R(z)]^2$ we have $R'(z) = S'(z)/2[S(z)]^{1/2}$. The poles of $S(z)$ are the poles of $R(z)$, and are not zeros of $R'(z)$; the multiple zeros of $R(z)$ are also zeros of $R'(z)$, so it will be sufficient for us to study the zeros of $S'(z)$. The analog of Theorem 1 is

THEOREM 2. *Let the set of zeros and the set of poles of the rational function $R(z)$ be unchanged under the transformation $z = 1/z'$.*

1). *Let the zeros of $R(z)$ lie on a set S, namely any two segments S_1 of the positive half of the axis of reals, or any two segments S_2 of the negative half of the axis of reals, or the set A symmetric in the axis of reals and consisting of two arcs of the unit circle C, or any connected set consisting of $S_1 + A$ or $S_2 + A$ or $S_1 + S_2 + C$, assuming in every case that S is unchanged by the substitution $z = 1/z'$. Let the poles of $R(z)$ lie on a set S' of this same character and disjoint from S. Then all critical points except $z = +1$ and -1 lie in $S + S'$.*

2). *Let the zeros of $R(z)$ lie on a set S consisting of two (finite) segments of the axis of imaginaries, the set S unchanged by the transformation $z = 1/z'$, and let the poles of $R(z)$ lie on a second such set S' disjoint from S. Then all critical points of $R(z)$ except $z = +1$ and -1 lie on $S + S'$.*

3). *If the zeros of* $R(z)$ *lie in the sector* $-\alpha \leqq \arg z \leqq \alpha(<\pi/2)$, *and the poles in the sector* $-\alpha + \pi \leqq \arg z \leqq \alpha + \pi$, *then all critical points of* $R(z)$ *lie in those two sectors.*

4). *If the zeros of* $R(z)$ *lie in the closed right-hand half-plane, and in the closed exterior of* $C: |z| = 1$ *in the first quadrant and in the closed interior of* C *in the fourth quadrant, and if the poles of* $R(z)$ *lie in the closed left-hand half-plane, in the closed interior of* C *in the second quadrant and in the closed exterior of* C *in the third quadrant, where a point on the axis of reals may be arbitrarily assigned to either adjacent quadrant and a point on the axis of imaginaries is assigned to the right-hand quadrant if a zero and to the left-hand quadrant if a pole, then the critical points of* $R(z)$ *lie in these same closed regions.*

5). *Let the zeros and the poles of* $R(z)$ *be symmetric in* O, *with all zeros on the set composed of the unit circle plus the axis of reals, and all poles on the axis of imaginaries. Then all critical points of* $R(z)$ *lie on these two sets.*

In every case we write $w = \frac{1}{2}(z + 1/z)$,

$$(5) \qquad S(z) = [R(z)]^2 = R_1(w), \; S'(z) = R_1'(w)(z^2 - 1)/2z^2,$$

where $R_1(w)$ is a rational function of w. To determine a point set in the z-plane containing all critical points of $R(z)$, it is sufficient to determine a set containing all zeros of $S(z)$, containing the images of the critical points of $R_1(w)$, and containing the points $z = +1$ and -1. In each part of Theorem 2 each set S or S' has an image in the w-plane of which the original set is the complete image in the z-plane.

In part 1) the two sets S and S' correspond in the w-plane to two disjoint intervals of the axis of reals, which contain respectively the zeros and poles of $R_1(w)$; the conclusion follows from §4.2.3 Corollary 2 to Theorem 3. In part 2) the given sets correspond to two disjoint intervals of the axis of imaginaries, and the conclusion follows as before.

Part 3) is included merely for comparison, and follows without the conformal map from §4.2.1, Corollary 2 to Bôcher's Theorem or with the conformal map from §4.2.3 Theorem 3.

In part 4) the regions assigned to the zeros of $R(z)$ form the image of the first quadrant of the w-plane, and the regions assigned to the poles of $R(z)$ form the image of the third quadrant. Those two closed quadrants contain respectively the zeros and poles of the function $R_1(w)$. Through any point P exterior to those closed quadrants can be drawn a line OP which separates the zeros of $R_1(w)$ not on OP from the poles of $R_1(w)$ not on OP. If all the zeros and poles of $R_1(w)$ lie on OP, they lie in the two points $z = 0$ and $z = \infty$, and all critical points of $R_1(w)$ lie in that pair of points; in any case, whether other zeros or poles exist or not, the point P is not a critical point of $R_1(w)$, by §4.2.3 Theorem 3, and its image in the z-plane cannot be a critical point of $R(w)$.

In part 5) the function $R(z)$ satisfies one of the functional equations $R(z) \equiv \pm R(-z)$, so the function $S(z)$ defined by (5) satisfies the equation $S(z) \equiv S(-z)$

and the function $R_1(w)$ satisfies the equation $R_1(w) \equiv R_1(-w)$. The conclusion follows from §5.5.2 Corollary 1 to Theorem 3.

The connection between the present results (§5.7) on symmetry in z and $1/z$ and those involving symmetry in O (§5.5) is far closer than mere analogy. The transformation $w = \frac{1}{2}(z + 1/z)$ can be considered as the succession of transformations $z' = (z - 1)/(z + 1)$, $z'' = z'^2$, $w = (1 + z'')/(1 - z'')$, hence as the transform of the transformation $w = z^2$ by the substitution $z' = (z - 1)/(z + 1)$. Nevertheless symmetry in z and $1/z$ is intrinsically important, suggests new geometric configurations, and seems worth independent discussion.

The preceding study of polynomials in z and $1/z$ and rational functions in z whose sets of zeros and poles are unchanged by the substitution $z = 1/z'$ applies also in essence to polynomials in z and $1/z$ and rational functions in z whose sets of zeros and poles are unchanged by the substitution $z = -1/z'$. Indeed, if we set $Z = iz$, $Z' = iz'$, the latter substitution becomes $Z = 1/Z'$, and Theorems 1 and 2 with their corollaries can be applied.

§5.8. Miscellaneous results. We shall include here a number of isolated and scattered results involving a combination of symmetries, less complete and less systematic in nature than those previously developed. Indeed, those now to be given and related fields deserve further investigation.

§5.8.1. Combinations of symmetries. In the hyperbolic plane it is appropriate to study also symmetry in O:

THEOREM 1. *Let the zeros of a rational function $R(z)$ be symmetric to its poles in the unit circle C, and let both zeros and poles be symmetric in the origin O. If all zeros of $R(z)$ lie interior to C in a closed double sector with vertex O and angle less than $\pi/2$, then all critical points interior to C lie in that sector.*

The function $[R(z)]^2$ is a rational function of $w = z^2$, whose zeros are symmetric to its poles in the circle $|w| = 1$; the zeros of this function lie interior to that circle in a sector with vertex $w = 0$ which is the image of the given sector in the z-plane, and whose angle is not greater than π and which is therefore NE convex. Theorem 1 follows from §5.3.1 Theorem 1. The same method yields

THEOREM 2. *Let the zeros of a rational function $R(z)$ be symmetric to its poles in the unit circle C, and let both zeros and poles be symmetric in O. Let the zeros of $R(z)$ interior to C lie in a closed double sector S_1 with vertex O and angle less than $\pi/2$, and the poles of $R(z)$ interior to C lie in a closed double sector S_2 found by rotating S_1 through an angle $\pi/2$ about O; then all critical points interior to C lie in those two double sectors.*

A rational function may be skew-symmetric in two perpendicular lines, and then is symmetric in their intersection:

THEOREM 3. *Let the zeros of a rational function $R(z)$ be symmetric to its poles in each of the coordinate axes, and let the zeros in the closed first quadrant lie in the sector S: $0 < \alpha \leqq \arg z \leqq \beta$, $\pi/4 \leqq \beta \leqq \pi/2$. Let no poles of $R(z)$ lie in the closed first quadrant. Then all critical points of $R(z)$ lie in S and its reflections in O and the coordinate axes.*

In the plane of $w = z^2$, the zeros of the rational function $[R(z)]^2$ of w are symmetric to its poles in the axis of reals and lie in the sector $2\alpha \leqq \arg w \leqq 2\beta$, which in the upper half-plane is NE convex. By §5.3.1 Theorem 1 all critical points of that function in the upper half-plane lie in the sector, so the conclusion follows. We add a further consequence of §5.3.1 Theorem 1:

COROLLARY. *Under the conditions of Theorem 3, a circle with center O which contains no [or all] zeros and poles of $R(z)$ contains no critical points of $R(z)$ other than O [or all critical points other than the point at infinity].*

Theorems 1, 2, 3 all involve symmetry in O, and indicate anew the usefulness of the transformation $w = z^2$; this comment applies also to

THEOREM 4. *Let the zeros of a polynomial $p(z)$ be symmetric in both coordinate axes, and let every zero $z_0 = x_0 + iy_0$ in the open first quadrant satisfy the conditions*

$$x_0^2 + 2x_0y_0 - y_0^2 \leqq a^2 \quad \text{if} \quad 0 < \arg z_0 \leqq \pi/4, \tag{1}$$

$$-x_0^2 + 2x_0y_0 + y_0^2 \leqq a^2 \quad \text{if} \quad \pi/4 \leqq \arg z_0 < \pi/2. \tag{2}$$

Then all critical points except on the coordinate axes lie in the closed interior of the circle $| z | = a$.

We note first that *if $P(w)$ is a real polynomial and if every non-real zero $w_0 = u_0 + iv_0$ satisfies the condition $| u_0 | + | v_0 | \leqq r$, then all non-real critical points lie in the closed interior of the circle $| w | = r$.* All Jensen circles for $P(w)$ have both intersections with the axis of reals in the closed interval $-r \leqq w \leqq r$, so all Jensen circles and all *non-real* critical points lie in the closed region $| w | \leqq r$. No non-real critical point lies on the circle $| w | = r$ except perhaps a multiple zero of $P(w)$ at $w = \pm ir$.

Theorem 4 now follows from the result just stated, by setting $w = z^2$; condition (1) implies that z_0 in the sector $0 < \arg z_0 \leqq \pi/4$ shall lie in the closed exterior of the hyperbola whose image in the w-plane is the line $u + v = a^2$, and condition (2) implies that z_0 in the sector $\pi/4 \leqq \arg z < \pi/2$ shall lie in the closed exterior of the hyperbola whose image in the w-plane is the line $v = u + a^2$. No critical point except on the coordinate axes lies on the circle $| z | = a$ other than a multiple zero on a line $y = \pm x$.

§5.8.2. Circular regions and symmetry in a line. In §§5.3.5 and 5.3.6 we have studied in some detail the problem of replacing two pairs of particles, each pair consisting of a unit positive particle and a unit negative particle mutually symmetric in a circle, by an equivalent double pair exhibiting that same symmetry. Equivalence here indicates equivalence so far as concerns force exerted at a preassigned point. We now proceed to study that same problem of equivalent pairs, but where we have symmetry instead of skew-symmetry.

LEMMA 1. *If the point z traces the circle $\gamma: |z - a| = r$, with $a \geqq 2r + 1$, the point $w = \frac{1}{2}(z + 1/\bar{z})$ traces a convex curve Γ.*

The point w is the midpoint of the segment joining z and its inverse in $C: |z| = 1$. We set $z = a + re^{i\varphi}$, $w = u + iv$, from which straightforward algebraic computation yields

$$(3) \qquad \frac{dv}{du} = \frac{(a^2 + r^2)\cos\varphi + 2ar + \cos\varphi(a^2 + r^2 + 2ar\cos\varphi)^2}{\sin\varphi\,[a^2 - r^2 - (a^2 + r^2 + 2ar\cos\varphi)^2]}.$$

As z traces γ, the point w lies on the segment Oz and traces a curve Γ symmetric in the axis of reals which has vertical tangents at the points corresponding to $\varphi = 0$ and $\varphi = \pi$. The convexity of Γ depends on the positive character of $d(dv/du)/d\varphi$, and this, after suppression of the factor $z\bar{z} = a^2 + r^2 + 2ar\cos\varphi$, depends on the positive character of the function

$$(4) \quad f(\cos\varphi) = (a^2 + r^2 + 2ar\cos\varphi)^3 + 2r(r + a\cos\varphi)(a^2 + r^2 + 2ar\cos\varphi) - 8a^2r\sin^2\varphi(a\cos\varphi + r) - a^2 + r^2.$$

Further algebraic computation shows that the discriminant of the quadratic $f'(\cos\varphi)$ in $\cos\varphi$ has the sign of $-a^2 + r^2 + 1$; the latter function is negative by virtue of our hypothesis $a \geqq 2r + 1$, so $f'(\cos\varphi)$ has no real zero. Consequently the cubic $f(\cos\varphi)$ in $\cos\varphi$ has but one real zero. By virtue of this same hypothesis we find $f(1) > 0$, $f(-1) > 0$, so $f(\cos\varphi)$ and $d(dv/du)/d\varphi$ are positive throughout the interval $0 < \varphi < \pi$, and Lemma 1 follows.

Study of the dependence of w as a function of z for z on a half-line through O shows that if the locus of z is the closed interior of γ, then the locus of w is the closed interior of Γ.

We remark that the number *two* that appears in the inequality $a \geqq 2r + 1$ is the smallest number for which the conclusion can be established, for if we set $a = 1 + (1 + \epsilon)r$ we find $f(-1)/(a - r) = 4(\epsilon - 1)r + \cdots$, where only terms involving higher powers of r are omitted; thus for given $\epsilon < 1$, for suitably chosen $r > 0$ the curve Γ is not convex.

LEMMA 2. *Let the circles C_1 and $\bar{C}_1$ be disjoint and mutually symmetric in the axis of reals. Let C_1 and the point z_0 lie in the upper half-plane, and let C_1 subtend at z_0 a maximum angle not greater than $\pi/3$ between circles tangent to C_1 intersecting on the axis of reals. Then the force at z_0 due to m unit particles in the closed interior of C_1 and m unit particles at their inverses in the axis of reals is equal to the force at z_0 due to an m-fold particle in C_1 and its inverse in the axis of reals.*

After inversion in the unit circle whose center is z_0 and if necessary a change of scale and orientation, we have precisely the situation of Lemma 1; there are given m points in the closed interior of γ; the center of gravity of the m corresponding points w lies in the closed interior (a convex region) of Γ; this closed region is the *locus* of w for z in the closed interior of γ, so Lemma 2 follows.*

We are now in a position to prove

THEOREM 5. *Let C_1 and C_2 be mutually exterior circles in the upper half-plane, and $\bar{C}_1$ and $\bar{C}_2$ their respective reflections in the axis of reals. Let A_k $(k = 1, 2)$ be the closed annular region (if any) bounded by two circles of the coaxal family determined by C_k and $\bar{C}_k$ which is the axis of reals plus the locus of a non-real point z_0 such that the maximum angle subtended at z_0 by the nearer of the two circles C_k and $\bar{C}_k$ between circles tangent to that circle and intersecting on the axis of reals is not greater than $\pi/3$. Let $R(z)$ be a real rational function all of whose zeros lie in the closed interiors of C_1 and $\bar{C}_1$, and all of whose poles lie in the closed interiors of C_2 and $\bar{C}_2$. Then no non-real critical points of $R(z)$ lie in the intersection of A_1 and A_2.*

The annular region A_k has already (§5.3.6) been considered as locus of z_0.

As a preliminary remark, we notice that if $R_1(z)$ is a real rational function with precisely two zeros α and $\bar{\alpha}$, and precisely two poles β and $\bar{\beta}$, then all critical points of $R_1(z)$ not multiple zeros are real. Indeed, all critical points lie on the circle through α, $\bar{\alpha}$, β, and $\bar{\beta}$, and precisely one critical point lies between α and $\bar{\alpha}$ and one between β and $\bar{\beta}$, by §4.2.3 Theorem 2; symmetry in the axis of reals shows that these two critical points are real.

We proceed now to Theorem 5. If z_0 is a non-real point in $A_1 \cdot A_2$, the force at z_0 due to the m pairs of positive particles in the closed interiors of C_1 and $\bar{C}_1$ is by Lemma 2 equivalent to the force at z_0 due to a pair of m-fold positive particles symmetric in the axis of reals lying in the closed interiors of C_1 and $\bar{C}_1$; the force at z_0 due to the m pairs of negative particles in the closed interiors of C_2 and $\bar{C}_2$ is equivalent to the force at z_0 due to a pair of m-fold negative particles symmetric in the axis of reals lying in the closed interiors of C_2 and $\bar{C}_2$. It follows from our preliminary remark that z_0 is not a position of equilibrium; z_0 lies exterior to the given circles and is not a multiple zero, so Theorem 5 is established.

Under the conditions of Theorem 5 there exists a circle whose center is on the axis of reals which separates C_1 and C_2, and hence separates $C_1 + \bar{C}_1$ and $C_2 + \bar{C}_2$; Bôcher's Theorem can be applied, and yields further information on critical points.

Lemma 2 applies also in the study of the critical points of a real polynomial some of whose zeros lie in given circular regions; for instance we may have k zeros

---

* We remark incidentally that the conclusion of Lemma 2 holds for real z_0 with no restriction on C_1.

in the closed interior of each of the circles C_1 and $\bar{C}_1$ symmetric in the axis of reals, while the remaining zeros either are concentrated at a single point, or lie in the closed interior of a circle whose center lies on the axis of reals, or lie in the closed interiors of two circles C_2 and $\bar{C}_2$ symmetric in the axis of reals. These applications are left to the reader.

§5.8.3. Circular regions and real polynomials. We studied in §§5.1.3 and 5.1.4, and in more detail in §5.8.2, pairs of circular regions symmetric to each other in the axis of reals as loci of the zeros or poles of a rational function, but did not emphasize there the possibility of using a single circular region symmetric in the axis of reals as such a locus. The primary reason for this omission is, as we shall later indicate, that no analog of §5.8.2 Lemma 2 exists for a *single* circular region symmetric in the axis of reals. However, this situation is not entirely beyond treatment by the methods already developed. We proceed to establish a result, of minor interest in itself, but which connects some of our previous results, and which especially seems to point the way to a new chapter of this present theory which deserves investigation.

LEMMA 1. *Let the circle Γ be orthogonal to the unit circle. The mid-point of the two intersections with Γ of a line through the origin lies on the circle through the origin, through the center of Γ, and through the intersections of Γ with the unit circle.*

In polar coordinates (r, θ), let the center of Γ be $(a, 0)$, whence the radius of Γ is $(a^2 - 1)^{1/2}$ and the equation of Γ is

$$r^2 - 2ar \cos \theta + 1 = 0.$$

For given θ, the two values of r have the sum $2a \cos \theta$, and the average is $a \cos \theta$; the mid-point of the two intersections lies on the curve $r = a \cos \theta$, as we were to prove.

We note that on any half-line $\theta = \text{const}$ (it is sufficient to choose here $\theta = 0$), the distance from the origin to this mid-point decreases continuously with a.

LEMMA 2. *Let C be the unit circle and P a non-real point exterior to C. Let C_0 be the circle through $+1$, -1, P, and let C_1 be the reflection of C_0 in the axis of reals. Let A_k $(k = 0, 1)$ be the arc of C_k in the closed interior of C, and let R_k be the closed lens-shaped region bounded by A_k and the line segment $(-1, +1)$; we set $R = R_0 + R_1$.*

The force at P due to a pair of unit particles in the closed interior of C and symmetric in the axis of reals is equivalent to the force at P due to a double particle in R_1; conversely, the force at P due to a double particle in R_1 is equivalent to the force at P due to a pair of unit particles in the closed interior of C and symmetric in the axis of reals. The force at P due to m pairs of unit particles in the closed interior of C, each pair symmetric in the axis of reals, is equivalent to the force at P due to a particle of mass $2m$ in R.

Invert the given figure in the unit circle whose center is P; the image of the axis of reals is a circle γ, that of C is a circle C' orthogonal to γ. The inverse of A_0 is a line segment A_0' and that of A_1 is an arc A_1' of a circle through the center of γ; both A_0' and A_1' are terminated by the intersections of C' and γ. Since P lies exterior to C, the interior of C corresponds to the interior of C'. A pair of points in the closed interior of C and symmetric in the axis of reals corresponds to a pair of points in the closed interior of C' and symmetric in γ; by Lemma 1 the mid-point of the latter pair lies in the lens-shaped region interior to C' bounded by A_1' and an arc of γ, which proves the first part of Lemma 2. The second part of Lemma 2 follows similarly from Lemma 1. The last part of Lemma 2 is a consequence of the convexity of the lens-shaped region bounded by A_0' and A_1'. We remark too that no closed region smaller than R suffices here, unless some restriction is made on m; in particular no analog exists of §5.8.2 Lemma 2.

LEMMA 3. *If the variable point α lies on the unit circle and x_1 is real and fixed, every non-trivial non-real critical point of the polynomial $p(z) \equiv (z - \alpha)^m(z - \bar{\alpha})^m (z - x_1)^k$, where we have*

$$(m + k)^2 - m^2x_1^2 > 0, \tag{5}$$

lies on the circle S:

$$(m + k)(2m + k)(x^2 + y^2) - 2m(2m + k)x_1x + 2m^2x_1^2 - k(m + k) = 0. \tag{6}$$

If condition (5) is not satisfied, the polynomial $p(z)$ has no non-real critical point.

The non-trivial critical points satisfy the equation

$$\frac{m}{z - \alpha} + \frac{m}{z - \bar{\alpha}} + \frac{k}{z - x_1} = 0,$$

which by the substitution $\alpha = \cos\theta + i \sin\theta$ may be written

$$2m(x + iy - \cos\theta)(x + iy - x_1) \\ + k(x + iy - \cos\theta - i \sin\theta)(x + iy - \cos\theta + i \sin\theta) = 0.$$

Separation into real and pure imaginary parts and elimination of θ yields the equations $y = 0$ and (6). Equation (6) represents a proper circle, the point $z = mx_1/(m + k)$, or no locus, according as the first member of (5) is positive, zero, or negative; this completes the proof.

If we modify the hypothesis of Lemma 3 by requiring $|\alpha| \leqq 1$ instead of $|\alpha| = 1$, we prove that all non-trivial non-real critical points of $p(z)$ lie on or interior to S. In the case $\alpha = 0$ there are no non-real critical points. If we have $0 < |\alpha| = \rho < 1$, the polynomial

$$p(\rho z)/\rho^{2m+k} = (z - \alpha/\rho)^m(z - \bar{\alpha}/\rho)^m(z - x_1/\rho)^k$$

has (by Lemma 3) its non-trivial non-real critical points (if any) on the circle

$(m + k)(2m + k)(x^2 + y^2) - 2m(2m + k)\, x_1x/\rho + 2m^2x_1^2/\rho^2 - k(m + k) = 0$, so $p(z)$ has its non-trivial non-real critical points (if any) on the circle

$$(7) \quad (m + k)(2m + k)(x^2 + y^2)/\rho^2 - 2(2m + k)\, x_1x/\rho^2 + 2m^2x_1^2/\rho^2 - k(m + k) = 0.$$

If the coordinates (x, y) of any point satisfy this last equation, the first member of (6) is negative, so (x, y) lies interior to S whenever (5) is valid. If (5) is not valid, no non-real point lies on (7).

We are now in a position to prove our result:

THEOREM 6. *Let $p(z)$ be a real polynomial of degree $2m + k$, with $2m$ zeros in the closed unit circle C and a k-fold zero in the real point x_1. If* (5) *is satisfied, all non-real critical points of $p(z)$ not in the closed interior of C lie in the closed interior of S defined by* (6). *If* (5) *is not satisfied, $p(z)$ has no non-real critical points exterior to C. Whether or not* (5) *is satisfied, all real critical points not in the closed interior of C or at x_1 lie in the interval S_0 :*

$$(2mx_1 - k)/(2m + k) \leqq z \leqq (2mx_1 + k)/(2m + k).$$

Let P be a non-real critical point of $p(z)$ not in the closed interior of C; for definiteness choose P in the upper half-plane. By Lemma 2, the force at P due to the $2m$ particles in the closed interior of C is equivalent to the force at P due to $2m$ coincident particles in R, situated say at some point Q. Then P is a position of equilibrium in the field of force due to $2m$ particles at Q and k particles at x_1, so Q lies in the upper half-plane, namely in R_1 (notation of Lemma 2). It follows from Lemma 2 that P is a position of equilibrium in the field due to a k-fold particle at x_1 and a pair of m-fold particles symmetric in the axis of reals and in the closed interior of C. It follows from the remark supplementary to Lemma 3 that P lies in the closed interior of S, provided (5) is satisfied, and it follows that P cannot exist if (5) is not satisfied. The remainder of Theorem 6 follows from Walsh's Theorem.

If condition (5) is valid, the interval S_0 intersects S, so we have no conclusion regarding the number of critical points in S. It is to be noted, however, that the point set assigned by Theorem 6 to the critical points of $p(z)$ cannot be improved; any point of the closed interior of C or S or of S_0 can be a critical point, if we have $m > 1$.

The reasoning used in the proof of Theorem 6 holds with only minor changes if we have $k > -2m > 0$, so that $p(z)$ is now a real rational function of degree k with $-2m$ poles in the closed interior of C, a pole at infinity of order $k + 2m$, and a k-fold zero in the real point x_1. Lemmas 1 and 2 require no change, and Lemma 3 remains valid, as does the discussion relating to (7). Likewise the application of Lemma 2 as in the proof of Theorem 6 remains valid, and we are in a position to apply Lemma 3; here we make essential use of the fact that if the non-real point P is a position of equilibrium in the field of force due to a real particle of mass k and a non-real particle of mass $2m$, then the latter lies in the upper or lower half-plane with P (in the case $k > -2m > 0$). It follows that if (5) is satisfied, all non-real critical points of $p(z)$ lie in the closed interiors of C and S; if (5) is not satisfied, all non-real critical points lie in the closed interior of C. Real critical points can be investigated by §4.2.4 Theorem 4. The case $-2m > k > 0$, $x_1^2 \geqq 1$, is also readily discussed; the real rational function $p(z)$ has $-2m$ poles in the closed interior of C, a k-fold zero at x_1, and a $(-2m - k)$-fold zero at infinity; if z_0 is exterior to C and non-real, the circle $+1, -1, z_0$ separates R (notation of Lemma 2) from $z = x_1$ and $z = \infty$, so by Bôcher's Theorem z_0 is not a critical point; all critical points of $p(z)$ exterior to C are real; compare §5.1.2 Theorem 4. In the case $-2m = k > 0$, $x_1^2 \geqq 1$, the real rational function $p(z)$ has $-2m$ poles in the closed interior of C and a k-fold zero at x_1; it follows from Bôcher's Theorem that all critical points not at $z = x_1$ lie interior to C.

Theorem 6 is set forth primarily as indicative of a general theory, analogous to that of Marden, but where we now restrict ourselves to real polynomials or rational functions. Other similar theorems have already been developed (§§3.7, 5.1.3, 5.1.4, 5.8.2). The theory may presumably be elaborated by continued study of the envelopes of various plane curves involved.

The methods used in Theorem 6 apply also in a somewhat different order of ideas. We prove

THEOREM 7. *Let $p(z)$ be a real polynomial of degree n with $2k$ non-real zeros in the closed interior of the unit circle and no other non-real zeros. Then all non-real critical points of $p(z)$ lie on the point set consisting of the closed interior of the ellipse*

$$(8) \qquad x^2 + (2n - 2k)y^2/(n - 2k) = (2n - 2k)/n,$$

plus (in the case $k > 1$) the closed interior of the unit circle.

For convenience we formulate

LEMMA 4. *If the points α and $\bar{\alpha}$ lie in the closed interior of the unit circle C, then the closed interior of the circle Γ whose center is $(\alpha + \bar{\alpha})/2$ and radius $[(n - 2k)/n]^{1/2} \cdot |\alpha - \bar{\alpha}|/2$ lies in the closed interior of the ellipse* (8).

The circle Γ has its center on the axis of reals and has its vertical diameter in the closed interior of the ellipse $x^2 + ny^2/(n - 2k) = 1$; the conclusion follows from the Lemma of §3.8.

In the case $k = 1$, the conclusion of Theorem 7 follows from Lemma 4 and §2.2.2 Theorem 3. In the case $k > 1$, the closed interior of C clearly contains all non-real multiple zeros of $p(z)$. Let P be a non-real critical point of $p(z)$ exterior to C, so P is a position of equilibrium in the field of force; for definiteness suppose P to lie in the upper half-plane. In the notation of Lemma 2, the force at P due to the k pairs of non-real particles in the closed interior of C is equivalent to the force at P due to $2k$ coincident particles at some point Q in R. But P is a position of equilibrium in the field due to $2k$ particles at Q and to $n - 2k$ *real* particles, so Q lies in the upper half-plane, hence in R_1. Then by Lemma 2, P is a position of equilibrium in the field due to a pair of k-fold particles mutually symmetric in the axis of reals and lying in the closed interior of C, and to $n - 2k$ real particles. The conclusion of Theorem 7 now follows from Lemma 4 and §2.2.2 Theorem 3.

The point set specified in Theorem 7 is the precise *locus* of non-real critical points for all possible polynomials $p(z)$, in the respective cases $k = 1$ and $k > 1$.

CHAPTER VI

ANALYTIC FUNCTIONS

Two general methods obviously apply in the extension of our preceding results on polynomials and more general rational functions to the study of other analytic functions: i) a more general analytic function can often be expressed as the uniform limit of a series or sequence of rational functions, either by expressing an infinite sum or product as the limit of its partial sums or products or by approximating a given integral by suitable Riemann sums, and known results for the functions of the sequence may yield new results for the limit; ii) both specific conformal mappings, such as the use of the exponential function, and more or less arbitrary conformal mappings, such as mapping a given region onto the interior of a circle, may lead from old results to new ones. We shall employ both i) and ii), as well as a combination of those methods.

We shall not infrequently study the critical points of non-uniform analytic functions. It is only a slight modification of our study of polynomials to envisage the more general class of functions

$$\prod_{k=1}^{n} (z - \alpha_k)^{\mu_k}, \ \mu_k > 0,$$

where the exponents need not be integers, to extend Gauss's Theorem and Lucas's Theorem to include this class, and indeed to elaborate a complete new theory; compare §§1.2 and 1.3.2. Similarly we may envisage the class of functions of the form

$$\prod_{j=1}^{m} (z - \alpha_j)^{\mu_j} \Big/ \prod_{j=1}^{n} (z - \beta_j)^{\nu_j}, \qquad \mu_j > 0, \qquad \nu_j > 0,$$

to consider the critical points as positions of equilibrium in the field of force due to positive particles of masses μ_j at the points α_j and negative particles of masses ν_j at the points β_j, to establish the analog of Bôcher's Theorem, and indeed to generalize the entire theory already developed. We refrain from explicitly formulating these more general results, but shall use them on occasion.

§6.1. Entire and meromorphic functions. Let the numbers α_k be different from zero, and let the series $\sum_1^\infty |\alpha_k|^{-1}$ converge; the sequence α_k becomes infinite with k. For k sufficiently large we have $|\alpha_k| > 2R$, where R is given, whence for $|z| \leqq R$

$$2|z - \alpha_k| \geqq 2(|\alpha_k| - |z|) = |\alpha_k| + (|\alpha_k| - 2|z|) \geqq |\alpha_k|,$$

$$2/|\alpha_k| \geqq 1/|z - \alpha_k|;$$

so the series

$$\sum_{1}^{\infty} (z - \alpha_k)^{-1} \tag{1}$$

converges uniformly on any closed set containing none of the points α_k. The conjugate of this last series is precisely the limit of the field of force used in Gauss's Theorem (§1.2) as the number of particles becomes infinite.

The series (1) can be integrated term by term over a path from the origin to z, provided the path passes through no point α_k; this integration yields the multiple-valued analytic function

$$\sum_{1}^{\infty} \log (1 - z/\alpha_k), \tag{2}$$

any two values of which differ by an integral multiple of $2\pi i$. The exponential of (2)

$$f(z) = \prod_{1}^{\infty} (1 - z/\alpha_k) \tag{3}$$

is then analytic and single-valued at every finite point of the plane. Reciprocally, if $f(z)$ is defined by (3), where $\sum |\alpha_k|^{-1}$ converges, it follows by use of §1.6.3 Lemma 1 that series (2) converges uniformly in every closed simply-connected region containing no point α_k, to the value $\log [f(z)]$. Consequently the series (1) converges uniformly in every closed simply-connected region containing no point α_k, to the value $f'(z)/f(z)$. We have thus expressed the entire function $f(z)$ as the uniform limit of a series of polynomials. Results on the critical points of $f(z)$ can be derived (i) by use of Hurwitz's Theorem from results on the critical points of polynomials or (ii) directly by study of the field of force represented by the conjugate of (1) and involving an infinite number of fixed particles; this field of force is (on any closed bounded set containing no point α_k) the uniform limit of the corresponding field for a finite number of particles. Method (i) is ordinarily more suggestive, but method (ii) may be more powerful. Method (ii) here yields as in the proof (§1.3.1) of Lucas's Theorem:

THEOREM 1. *Let the series* $\sum |\alpha_k|^{-1}$ *converge, and let* Π *be the smallest convex set containing all the points* α_k; *then* Π *contains all critical points of the function* $f(z)$ *defined by* (3). *Unless* Π *degenerates to a line or a segment of a line, no critical point of* $f(z)$ *other than a multiple zero lies on the boundary of* Π.

The latter part of Theorem 1 follows easily by method (ii), but follows only with difficulty if at all by method (i). On the other hand, method (i) yields at once a Corollary [Laguerre] which can be proved only with difficulty by method (ii):

COROLLARY. *If* Π *in Theorem* 1 *degenerates to a line or a segment of a line, any open segment of the line bounded by two points* α_k *and containing no point* α_k *contains precisely one critical point of* $f(z)$.

It follows from Rolle's Theorem that such a segment contains at least one critical point, and follows by Hurwitz's Theorem from the fact that Π contains all critical points of the polynomials uniformly approximating to $f(z)$, at most one critical point of each polynomial in the given segment, that the given segment contains at most one critical point of $f(z)$.

In a similar manner may be proved Jensen's Theorem [1913] for entire functions, namely the extension of Jensen's Theorem for polynomials, and also the extension of §2.3 Theorem 1; we require $f(z)$ to be real, that is, the uniform limit of a sequence of *real* polynomials.

THEOREM 2. *Let $f(z)$ defined by (3) be real, with $\sum |\alpha_k|^{-1}$ convergent. Then all non-real critical points of $f(z)$ lie in the closed interiors of the Jensen circles. A non-real point z_0 not a multiple zero of $f(z)$ nor interior to a Jensen circle for $f(z)$ is a critical point if and only if $f(z)$ has no real zeros and all Jensen circles pass through z_0.*

Let the real points α and β lie exterior to all Jensen circles for $f(z)$, and let neither α nor β be a zero nor a critical point. Denote by K the configuration consisting of the segment $\alpha\beta$ together with the closed interiors of all Jensen circles intersecting that segment. If K contains k zeros of $f(z)$, then K contains $k - 1$, k, or $k + 1$ critical points of $f(z)$ according as the forces $f'(\alpha)/f(\alpha)$ and $f'(\beta)/f(\beta)$ are both directed outward, one inward and the other outward, or both directed inward.

These same methods can be used to prove a generalization [Walsh, 1925] of §3.3 Theorem 1:

THEOREM 3. *Let the numbers a_k be real, with $\sum |a_k|^{-1}$ convergent, and let the critical points of the function $F(z) = \prod_1^\infty (1 - z/a_k)$ be $c_1, c_2, \cdots$, necessarily real. If r is constant, and for every k we have $|\alpha_k - a_k| \leqq r$, then all critical points of $f(z)$ defined by (3) lie in the circles $C_k : |z - c_k| = r$.*

Any set S of circles C_k exterior to all the other circles C_j contains a number of critical points of $f(z)$ equal to the sums of the multiplicities of the corresponding points c_k as critical points of $F(z)$.

The inequality $|a_k| \geqq 2r$ implies $2|\alpha_k| \geqq 2(|a_k| - r) = |a_k| + (|a_k| - 2r) \geqq |a_k|$, $2/|a_k| \geqq 1/|\alpha_k|$, so the series $\Sigma |\alpha_k|^{-1}$ converges, and the proof follows readily.

The methods used in Theorems 1–3 obviously apply in other situations; thus the function

$$\sin z = z \prod_1^\infty (1 - z^2/k^2\pi^2)$$

is on any bounded set the uniform limit of the polynomials found by taking here a finite product instead of an infinite product; it follows that all critical points are real, and that precisely one lies on each of the intervals $k\pi < z < (k + 1)\pi$.

The quotient of two entire functions:

$$(4) \qquad f(z) = \prod_1^\infty (1 - z/\alpha_j) \Big/ \prod_1^\infty (1 - z/\beta_k), \qquad \sum |\alpha_j|^{-1} \text{ and } \sum |\beta_k|^{-1} \text{ convergent,}$$

is a meromorphic function, and Bôcher's Theorem is readily applied, particularly in the form of §4.2.3 Theorem 3 and Corollary 2. We have [essentially due to Porter, 1916]

THEOREM 4. *Let $f(z)$ be defined by (4); if a line L separates all the points α_j from all the points β_k, then no critical point of $f(z)$ lies on L. Consequently, if such lines exist, then all finite critical points of $f(z)$ lie in two unbounded closed convex point sets which are separated by every L and which contain respectively the points α_j and the points β_k.*

A line L' which separates all zeros of $f(z)$ not on L' from all poles of $f(z)$ not on L', where at least one such zero or pole exists, passes through no critical point of $f(z)$ other than a multiple zero.

If all the α_j lie on a closed segment S_1 of a line λ and all the β_k lie on a closed segment S_2 of λ disjoint from S_1, then all critical points of $f(z)$ lie on S_1 and S_2; any open segment of λ bounded by two α_j or two β_k and containing no α_j or β_k contains precisely one critical point of $f(z)$.

By way of contrast to Theorem 4, we note that §5.2.1 Theorem 1 has a complete analog here; as an illustration, the function

$$\tan \pi z = \pi z \prod_1^\infty (1 - z^2/j^2) \Big/ \prod_0^\infty [1 - 4z^2/(2k + 1)^2]$$

can be expressed on any closed bounded set as the uniform limit of a sequence of rational functions whose zeros and poles are interlaced on the axis of reals. It follows that no critical points lie interior to the circles $|z - j| = \frac{1}{2}$ and $|z - j - \frac{1}{2}| = \frac{1}{2}, j = \cdots, -1, 0, 1, 2, \cdots$.

Entire functions with exponential factors may also be treated by the present methods; a simple illustration is

THEOREM 5. *Let $p(z)$ be a polynomial of degree n whose zeros lie in the closed region $C: |z - \alpha| \leqq r$, and let a $(\neq 0)$ be constant. Then the critical points of the function $f(z) = e^{az}p(z)$ lie in C and in the region $C': |z - \alpha + n/a| \leqq r$. If C and C' are mutually exterior, they contain respectively $n - 1$ and 1 critical points.*

We have

$$(5) \qquad \frac{f'(z)}{f(z)} = a + \sum_{k=1}^{n} \frac{1}{z - \alpha_k},$$

where the α_k are the zeros of $p(z)$. If z lies exterior to C and is a zero of $f'(z)$, we have (§1.5.1 Lemma 1) for some α_0 in C

$$a + \frac{n}{z - \alpha_0} = 0, \qquad z = \alpha_0 - \frac{n}{a},$$

so z lies in C'. The remainder of the theorem follows by the method of continuity. Lucas's Theorem is the limiting case $a = 0$; as a approaches zero, C' becomes infinite. Theorem 5 extends [Walsh 1922b] to the case where the exponent az is replaced by an arbitrary polynomial, and can be further extended to the case where $p(z)$ is replaced by a product of form (3).

The theory of the critical points of entire and meromorphic functions is an extensive one, begun by Laguerre and continued by numerous writers. It is the plan of the present work to indicate only these few simple but typical illustrations.

§6.2. The hyperbolic plane. A theorem of considerable utility dates from 1898:

MACDONALD'S THEOREM. *Let the function $f(z)$ be analytic in a simply-connected region R whose boundary B consists of more than one point. Let $f(z)$ be of constant modulus not zero on B in the sense that whenever z in R approaches B the modulus $|f(z)|$ approaches a non-zero constant M. Then in R the number of zeros of $f(z)$ exceeds the number of critical points by unity.*

Map R onto the interior of the unit circle C in the w-plane, with $F(w) \equiv f(z)$ in corresponding points. The zeros a_k of $F(w)$ interior to C are finite in number, say m, and the function

$$g(w) = \prod_{k=1}^{m} \frac{w - a_k}{1 - \bar{a}_k w}, \qquad |a_k| < 1, \tag{1}$$

has precisely those same zeros interior to C, and $g(w)$ is of modulus unity on C. Thus the quotient $F(w)/g(w)$ has no zeros or poles interior to C and is of constant modulus M on C, hence is identically constant throughout the interior of C. The circle C separates the zeros of $g(w)$ from the poles, so by Bôcher's Theorem the interior of C contains precisely $m - 1$ critical points of $g(w)$ and of $F(w)$. Critical points of $f(z)$ and of $F(w)$ correspond to each other under the conformal map, so the theorem follows.

It follows also that $f(z)$ takes on each value w, $|w| < M$, precisely m times in R.

§6.2.1. Analytic and meromorphic functions. Macdonald's Theorem is of considerable importance in analysis, and it is clear from the proof just given that the results of §5.3 can be used to make Macdonald's Theorem more precise, by describing in more detail the location of the critical points. Thus we have

THEOREM 1. *Under the conditions of Macdonald's Theorem let the zeros of $f(z)$ in R be $\alpha_1, \alpha_2, \cdots, \alpha_m$, and let R be provided with hyperbolic (NE) geometry by mapping R onto the interior of a circle. A NE line Γ not passing through all the points α_k and not separating any pair of points α_k can pass through no critical point* *of $f(z)$ other than a multiple zero of $f(z)$. Consequently the smallest closed NE convex polygon Π containing all the points α_k also contains all the critical points of $f(z)$ in R. No such critical point other than a multiple zero of $f(z)$ lies on the boundary of Π unless Π degenerates to a segment of a NE line Λ. In the latter case, any open arc of Λ bounded by two points α_k and containing no point α_j contains precisely one critical point of $f(z)$.*

If the points α_k are (NE) symmetric in a NE line Λ, all critical points of $f(z)$ in R not on Λ lie in the closed interiors of the Jensen circles, namely NE circles whose diameters are the segments joining pairs of points symmetric in Λ. A point not on Λ but on a Jensen circle and not interior to a Jensen circle is not a critical point unless it is a multiple zero of $f(z)$. If A and B are points of Λ, neither A nor B a zero nor critical point of $f(z)$ nor on or within a Jensen circle, let K denote the configuration consisting of the NE segment AB plus the closed interiors of all Jensen circles intersecting that segment. Let k denote the number of zeros of $f(z)$ in K; then K contains precisely $k - 1$, k, or $k + 1$ critical points of $f(z)$.

Let $z = \alpha$ be an arbitrary point interior to R, let C_1 be the NE circle whose center is α and NE radius $\frac{1}{2} \log [(1 + c)/(1 - c)]$, and let C_2 be the NE circle whose center is α and NE radius $\frac{1}{2} \log [(1 + b)/(1 - b)]$, $c < b$. Let $f(z)$ have precisely m_1 zeros interior to C_1, m_2 zeros in R exterior to C_2, and no other zeros in R. If the equation $m_1(c - 1/c)(b - r)(r - 1/b) + m_2(b - 1/b)(c + r)(r + 1/c) = 0$ has a zero $r = r_0$ satisfying the inequalities $c < r_0 < b$, then $f(z)$ has no critical points interior to the annulus bounded by C_1 and the NE circle C_0 whose center is α and NE radius $\frac{1}{2} \log [(1 + r_0)/(1 - r_0)]$, and has precisely $m_1 - 1$ critical points in the closed interior of C_1.

The first part of Theorem 1 is due to the present writer [1939], and was proved later by Gontcharoff [1942]; the second part is due to Walsh [1946a].

If we assume $f(z)$ no longer analytic but meromorphic in R, the same general method is applicable. Conformal map of R onto the region $|w| < 1$ transforms $f(z)$ into a function $F(w)$ meromorphic in that region, having there the zeros $a_1, a_2, \cdots, a_m$ and the poles $b_1, b_2, \cdots, b_n$. We set

$$g_1(w) = \prod_1^n \frac{w - b_k}{1 - \bar{b}_k w}, \qquad |b_k| < 1,$$

and the function $f(z)g_1(w)/g(w)$ is analytic in R, of constant modulus M on the boundary, hence identically constant in R. It is no longer possible (§5.3.2) always to state the precise number of critical points in R, but various conclusions can be formulated:

THEOREM 2. *Let the function $f(z)$ be meromorphic in a simply-connected region R whose boundary B consists of more than one point, and let $f(z)$ be of constant non-zero modulus on B. Denote by $\alpha_1, \alpha_2, \cdots, \alpha_m$ the zeros and by $\beta_1, \beta_2, \cdots, \beta_n$ the poles of $f(z)$ in R. If L is a NE line for R which separates each α_j from each β_k, then no critical point of $f(z)$ lies on L. Consequently, if such a line L exists, the critical points of $f(z)$ in R lie in two closed NE convex point sets Π_1 and Π_2 which* *respectively contain all the α_j and all the β_k, and which are separated by every L. If a NE line L′ separates all the α_j not on L′ from all the β_k not on L′, and if at least one α_j or β_k does not lie on L′, then no point of L′ is a critical point of f(z) unless it is a multiple zero of f(z).*

If two disjoint segments λ_1 and λ_2 of a NE line λ each terminating in B contain respectively all the α_j and all the β_k, then λ_1 and λ_2 contain all the critical points of f(z) in R. Any open segment of λ bounded by two points α_j or by two points β_k and containing no point α_j or β_k contains precisely one critical point of f(z). If we have $m = n$, all critical points of f(z) in R lie on such segments of λ.

Other results of §5.3 and of §5.8.1 are readily carried over to the present situations.

§6.2.2. Blaschke products. The formal infinite product

$$(2) \qquad B(z) = \prod_1^\infty \frac{\bar{\alpha}_n(z - \alpha_n)}{|\alpha_n|(\bar{\alpha}_n z - 1)}, \qquad |\alpha_n| < 1,$$

where the factors $\bar{\alpha}_n$ and $|\alpha_n|$ are omitted if we have $\alpha_n = 0$, is called a *Blaschke product*, and we prove the well known fact that (2) *is convergent* or divergent in* $|z| < 1$ *with the product* $\prod |\alpha_n|$. Each factor in the infinite product (2) is in modulus less than unity, so the partial products are uniformly bounded. If an infinity of the numbers α_n are zero, the product $\prod |\alpha_n|$ diverges by definition; and the infinite product in (2) has an infinity of factors of modulus $|z|$, hence approaches zero for every z, $|z| < 1$, and thus diverges. If only a finite number of the α_n are zero, the convergence and divergence of both $B(z)$ and $\prod |\alpha_n|$ is unchanged if we suppress the vanishing α_n, which we do. If the product in (2) as thus modified is convergent, it is convergent for the value $z = 0$, so the product $\prod |\alpha_n|$ is convergent. Conversely, let the latter product be convergent; it remains to show that the product in (2) as modified (i.e. assuming $\alpha_n \neq 0$) is convergent. Each sequence of partial products is uniformly bounded in $|z| < 1$ and forms a normal family in that region, so from every infinite subsequence there can be extracted a new subsequence which converges uniformly in every closed subregion; all such subsequences converge to the value $\prod |\alpha_n|$ $(\neq 0)$ for $z = 0$, no limit function vanishes identically, the α_n have no limit point in $|z| < 1$, so by Hurwitz's Theorem no limit function can vanish in $|z| < 1$ except in the points α_n. If we choose any two convergent subsequences, repeating the terms of one subsequence if necessary so that the zeros of the terms of the latter subsequence always form a proper subset of the zeros of the corresponding terms of the former subsequence, the quotients of corresponding terms are also uniformly bounded in the region $|z| < 1$, are there in modulus less than unity, and in the point $z = 0$ approach the value unity; by Hurwitz's Theorem the quotients approach the value unity uniformly in any closed subregion, so the two subse-

* An infinite product is *convergent* when and only when the series of logarithms of the individual factors, with the possible suppression of a finite number of factors, exists and converges.

quences of partial products converge to the same limit; this limit is then independent of the subsequence chosen, and by Vitali's Theorem the infinite product in (2) converges uniformly, in every closed subregion of $|z| < 1$, as we were to prove.

The critical points of $B(z)$ defined by (2) may be studied (i) by considering $B(z)$ as the uniform limit of the sequence of partial products, and by applying known results for the latter, and also (ii) by taking the logarithmic derivative of that sequence, obtaining a new sequence uniformly convergent in every closed subregion of $|z| < 1$ containing no point α_n, and interpreting the resulting equation

$$\frac{B'(z)}{B(z)} = \sum_1^\infty \left[\frac{1}{z - \alpha_n} - \frac{1}{z - 1/\bar{\alpha}_n}\right]$$

by taking conjugates and considering the second member as defining a field of force. It follows that the conclusion of Theorem 1 except the last part applies to $B(z)$ in the region $R: |z| < 1$.

The same two methods (i) and (ii) apply in essence to the study of the critical points of the quotient of two convergent Blaschke products, defined by (2) and

$$B_1(z) = \prod_1^\infty \frac{\bar{\beta}_n(z - \beta_n)}{|\beta_n|(\bar{\beta}_n z - 1)}, \quad |\beta_n| < 1.$$

The conclusion of Theorem 2, including the last sentence, without the restriction $m = n$, applies to the critical points of $B(z)/B_1(z)$ in the region $R: |z| < 1$.

In a different order of ideas, let the function $f(z)$ be analytic but not identically constant in a region R of finite multiple connectivity, and be of constant modulus $M \neq 0$ on the boundary B of R. If the universal covering surface of R is mapped onto the interior of $C: |w| = 1$ in the w-plane, the function $f(z)$ is transformed into a bounded function $F(w)$ of constant modulus almost everywhere on C for angular approach. It can then be shown (we omit the proof) that $F(w)$ is a constant multiple of a Blaschke product interior to C, and consequently our results on Blaschke products apply to $F(w)$ in the w-plane and to $f(z)$ in the z-plane. We do not elaborate this remark here, for we shall later (§8.9) include the results suggested in the study of harmonic functions.

§6.3. Functions with multiplicative periods. A class of functions studied by Pincherle [1880] and possessing a multiplicative period: $f(z) \equiv f(\rho z)$, $\rho > 1$, is of considerable importance in our later work, so we study it in some detail.

§6.3.1. Uniform functions. Let there be given the real number $\rho(>1)$ and the points $\alpha_1, \alpha_2, \cdots, \alpha_n$ and $\beta_1, \beta_2, \cdots, \beta_n$, all different from zero, with $\rho^k\alpha_i \neq \rho^m\beta_j$, $k = \cdots, -1, 0, 1, 2, \cdots, m = \cdots, -1, 0, 1, 2, \cdots$. We wish to study a function $f(z)$ meromorphic in the entire plane except at $z = 0$ and $z = \infty$ whose zeros are the points $\rho^m\alpha_j$ and whose poles are the points $\rho^m\beta_j$. For convenience we assume none of these points $\rho^m\alpha_j$ or $\rho^m\beta_j$ equal to unity; a suitable rotation or stretching of the plane will ensure this condition.

We write formally

$$\begin{aligned}\Phi(z) \equiv \sum_{-\infty}^{\infty}\Big[\frac{1}{z-\rho^m\alpha_1}+\frac{1}{z-\rho^m\alpha_2}+\cdots+\frac{1}{z-\rho^m\alpha_n} \\ -\frac{1}{z-\rho^m\beta_1}-\frac{1}{z-\rho^m\beta_2}-\cdots-\frac{1}{z-\rho^m\beta_n}\Big].\end{aligned} \tag{1}$$

If we now set

$$\frac{1}{z-\rho^m\alpha_1}-\frac{1}{z-\rho^m\beta_1}=\frac{\rho^m(\alpha_1-\beta_1)}{(z-\rho^m\alpha_1)(z-\rho^m\beta_1)}=\frac{\alpha_1-\beta_1}{\rho^m(\alpha_1-z/\rho^m)(\beta_1-z/\rho^m)},$$

where these last fractions are interpreted for $m\to-\infty$ and $m\to\infty$ respectively, it is clear by the Weierstrass M-test that the series in (1) converges uniformly on every closed bounded set containing none of the points $0, \rho^m\alpha_j, \rho^m\beta_j$. Term-by-term integration of the second member from $z=1$ to $z=z$ over an arbitrary path on which the series converges uniformly yields a function which is not necessarily single-valued, a series of terms of the form

$$\int_1^z\frac{dz}{z-\rho^m\alpha_j}-\int_1^z\frac{dz}{z-\rho^m\beta_j}=\log\frac{(z-\rho^m\alpha_j)(1-\rho^m\beta_j)}{(z-\rho^m\beta_j)(1-\rho^m\alpha_j)},$$

and the exponential of this new function is single-valued:

$$\Psi(z)\equiv\prod_{-\infty}^{\infty}\frac{(z-\rho^m\alpha_1)\cdots(z-\rho^m\alpha_n)(1-\rho^m\beta_1)\cdots(1-\rho^m\beta_n)}{(z-\rho^m\beta_1)\cdots(z-\rho^m\beta_n)(1-\rho^m\alpha_1)\cdots(1-\rho^m\alpha_n)}, \tag{2}$$

where the product converges uniformly on every closed bounded set containing none of the points $0, \rho^m\alpha_j, \rho^m\beta_j$. If S is any closed bounded set not containing the origin, and if the factors which have poles and zeros on S are omitted, the resulting product converges uniformly on S, so $\Psi(z)$ is meromorphic in the entire plane except at the origin and infinity, and possesses precisely the zeros $\rho^m\alpha_j$ and poles $\rho^m\beta_j$.

A functional equation satisfied by $\Psi(z)$ is found by writing

$$\begin{aligned}\Psi(\rho z)&\equiv\prod_{-\infty}^{\infty}\frac{(z-\rho^{m-1}\alpha_1)\cdots(z-\rho^{m-1}\alpha_n)(1-\rho^m\beta_1)\cdots(1-\rho^m\beta_n)}{(z-\rho^{m-1}\beta_1)\cdots(z-\rho^{m-1}\beta_n)(1-\rho^m\alpha_1)\cdots(1-\rho^m\alpha_n)}\\ &\equiv\prod_{-\infty}^{\infty}\frac{(z-\rho^m\alpha_1)\cdots(z-\rho^m\alpha_n)(1-\rho^{m+1}\beta_1)\cdots(1-\rho^{m+1}\beta_n)}{(z-\rho^m\beta_1)\cdots(z-\rho^m\beta_n)(1-\rho^{m+1}\alpha_1)\cdots(1-\rho^{m+1}\alpha_n)},\end{aligned}$$

where the infinite products converge except for $z=0, \rho^m\alpha_j, \rho^m\beta_j$. Then we may write

$$\begin{aligned}\frac{\Psi(\rho z)}{\Psi(z)}&\equiv\prod_{-\infty}^{\infty}\frac{(1-\rho^{m+1}\beta_1)\cdots(1-\rho^{m+1}\beta_n)(1-\rho^m\alpha_1)\cdots(1-\rho^m\alpha_n)}{(1-\rho^{m+1}\alpha_1)\cdots(1-\rho^{m+1}\alpha_n)(1-\rho^m\beta_1)\cdots(1-\rho^m\beta_n)}\\ &=\lim_{m\to\infty}\frac{(1-\rho^m\beta_1)\cdots(1-\rho^m\beta_n)}{(1-\rho^m\alpha_1)\cdots(1-\rho^m\alpha_n)}=\frac{\beta_1\cdots\beta_n}{\alpha_1\cdots\alpha_n}.\end{aligned}$$

We denote this last number by σ, and have established the functional equation

$$\Psi(\rho z) \equiv \sigma\Psi(z), \qquad \sigma = \beta_1 \cdots \beta_n/\alpha_1 \cdots \alpha_n . \tag{3}$$

We have proved all but the last part of

THEOREM 1. *Let the real number* ρ (>1) *and the points* $\alpha_1, \alpha_2, \cdots, \alpha_n$ *and* $\beta_1, \beta_2, \cdots, \beta_n$ *all different from zero be given, with* $\rho^k\alpha_i \neq \rho^m\beta_j$. *Assume all the* $\rho^m\alpha_i$ *and* $\rho^m\beta_j$ *different from unity. Then the function* $\Psi(z)$ *defined by* (2) *has zeros and poles precisely in the points* $\rho^m\alpha_i$ *and* $\rho^m\beta_j$ *respectively, and is otherwise analytic except in the points* 0 *and infinity. The function* $\Psi(z)$ *satisfies the functional equation* (3), *and except for a constant factor is uniquely defined by these properties.*

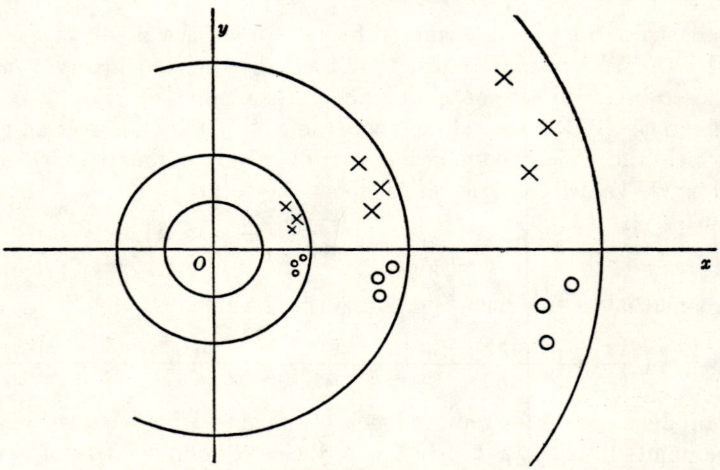

Fig. 17 illustrates §6.3.1 Theorem 1

If a second function $\Psi_1(z)$ possesses these properties, the quotient $\Psi(z)/\Psi_1(z)$ is analytic at every point of the plane other than the origin and point at infinity. The quotient satisfies the equation $\Psi(\rho z)/\Psi_1(\rho z) \equiv \Psi(z)/\Psi_1(z)$, hence in the whole plane other than the origin and point at infinity takes only the values taken in the annulus $1 \leqq |z| < \rho$, hence is uniformly bounded in the entire plane and reduces to a constant.

In Theorem 1 the points α_i and β_j are not uniquely determined by the given geometric set of points, for any α_i or β_j may be replaced by $\rho^k\alpha_i$ or $\rho^k\beta_j$, where k is a positive or negative integer. Such a replacement modifies σ by a positive or negative power of ρ, and naturally modifies $\Psi(z)$. However, we may always assume $1 \leqq |\sigma| < \rho$; as a special case we have the

COROLLARY. *If the function* $\Psi(z)$ *satisfies the conditions of Theorem* 1, *including* (3) *in the special form of the functional equation* $\Psi(\rho z) \equiv \rho^k\Psi(z)$, *where* k *is a positive or negative integer, then by proper change of notation if necessary we may choose* $\beta_1 \cdots \beta_n/\alpha_1 \cdots \alpha_n = 1$.

Each member of (1) has the value $\Psi'(z)/\Psi(z)$, so by §4.2.1 Corollary 2 to Bôcher's Theorem we have

THEOREM 2. *Under the conditions of Theorem* 1, *if the two sectors* $\alpha \leqq \arg z \leqq \beta \; (<\alpha + \pi)$ *and* $\alpha + \pi \leqq \arg z \leqq \beta + \pi$ *contain respectively the points* α_j *and* β_k, *then those sectors contain all critical points of* $\Psi(z)$.

If the α_j *and* β_k *lie respectively on the two halves terminated by O of a line L through O, all critical points of* $\Psi(z)$ *lie on L. Any open segment of L bounded by two points* α_j *or two points* β_k *containing no* α_j *or* β_k *contains precisely one critical point.*

If the zeros and poles of $\Psi(z)$ are interlaced on a line (not necessarily on a half-line), the analog of §5.2.1 Theorem 1 is valid.

If for every j we have $\beta_j = \bar{\alpha}_j$, and if the α_j lie in the upper half-plane, the function $\Psi(z)$ is essentially a Blaschke product for the upper half-plane. Here σ cannot be unity except in the case $n > 1$. By §5.3.1 Theorem 1 we have

THEOREM 3. *Under the conditions of Theorem* 1, *let the equation* $\beta_j = \bar{\alpha}_j$ *hold for every j, so that* $\Psi(z)$ *is unity for real z. If the points* α_j *lie in the sector* $0 < \alpha \leqq \arg z \leqq \beta$, $\alpha \leqq \pi/2 \leqq \beta < \pi$, *so also do all critical points of* $\Psi(z)$ *in the upper half-plane.*

In particular, if here all points α_j *lie on the axis of imaginaries, so do all critical points, and any open interval of that axis bounded by two points* α_j *and containing no* α_j *contains precisely one critical point. The function* $\Psi(z)$ *satisfies the functional equation* $\Psi(\rho z) \equiv (-1)^n \Psi(z)$.

The sector $\alpha \leqq \arg z \leqq \beta$ is NE convex, considered as a point set of the upper half-plane. More generally any set NE convex in the upper half-plane and containing the α_j contains all critical points of $\Psi(z)$ in the upper half-plane.

§6.3.2. Non-uniform functions. In the latter part of Theorem 1 and in its Corollary, we have assumed both parts of (3) satisfied; that is to say, in the assumed functional equation

$$\Psi(\rho z) \equiv \sigma \Psi(z) \tag{4}$$

we have made it part of the hypothesis that σ is related to the given zeros and poles by the equation

$$\sigma = \beta_1 \cdots \beta_n / \alpha_1 \cdots \alpha_n . \tag{5}$$

These assumptions are unnecessarily restrictive:

THEOREM 4. *Let the real number* ρ (>1) *and the points* $\alpha_1, \alpha_2, \cdots, \alpha_n$ *and* $\beta_1, \beta_2, \cdots, \beta_{n'}$, *all different from zero, be given, with* $\rho^k \alpha_i \neq \rho^m \beta_j$. *Let the uniform function* $\Psi_1(z)$ *have precisely the zeros and poles* $\rho^m\alpha_i$ *and* $\rho^m\beta_j$, *all different from unity, and be otherwise analytic except in the points* 0 *and infinity, and satisfy the equation* (4). *Then by a suitable change of notation of the* α_i *and* β_j *if necessary, we have* $n = n'$, *equation* (5) *is satisfied, and* $\Psi_1(z)$ *is a constant multiple of the function* $\Psi(z)$ *of Theorem* 1.

We introduce the transformation $w = u + iv = \log z$, $z = e^w$, so the z-plane is mapped onto each of the strips $v_0 \leqq v < v_0 + 2\pi$. The function $F(w) \equiv \Psi_1(e^w) \equiv \Psi_1(z)$ satisfies the functional equations

$$F(w + 2\pi i) \equiv F(w), \tag{6}$$

$$F(w + \tau) \equiv \sigma F(w), \qquad \tau = \log \rho, \tag{7}$$

and from (7) we also have

$$F'(w + \tau) \equiv \sigma F'(w); \tag{8}$$

of course $F(w)$ is single-valued and meromorphic in the w-plane.

We form the integral

$$\frac{1}{2\pi i}\int \frac{F'(w)\,dw}{F(w)}$$

taken in the counterclockwise sense over a rectangle whose sides are the following:

$$\begin{aligned}
&u_0 \leqq u \leqq u_0 + \tau\,, && v = v_0\,;\\
&v_0 \leqq v \leqq v_0 + 2\pi, && u = u_0 + \tau;\\
&u_0 \leqq u \leqq u_0 + \tau\,, && v = v_0 + 2\pi;\\
&v_0 \leqq v \leqq v_0 + 2\pi, && u = u_0\,;
\end{aligned} \tag{9}$$

we choose this rectangle so that no zero or pole of $F(w)$ lies on a side. The two parts of the integral over the horizontal sides taken together vanish, as do the two parts of the integral over the vertical sides. Thus the integral is zero, and it follows that *the rectangle contains precisely as many zeros as poles of* $F(w)$; let us denote these zeros and poles by $\log \alpha_1$, $\log \alpha_2$, $\cdots$, $\log \alpha_n$, $\log \beta_1$, $\log \beta_2$, $\cdots$, $\log \beta_n$. In the w-plane *any* zero or pole of $F(w)$ differs from one of these special zeros or poles by an integral multiple of $2\pi i$ plus an integral multiple of τ. It follows too that *each annular region* $r < |z| < r\rho$ *having on its boundary no zero or pole of* $\Psi_1(z)$ *contains the same number of zeros as poles of* $\Psi_1(z)$.

We compute also the integral

$$\frac{1}{2\pi i}\int w\,\frac{F'(w)\,dw}{F(w)}$$

over the rectangle (9), which by (6), (7), and (8) reduces to

$$\frac{1}{2\pi i}\left[-2\pi i\int_{u_0+iv_0}^{u_0+\tau+iv_0}\frac{F'(w)\,dw}{F(w)}+\tau\int_{u_0+iv_0}^{u_0+2\pi i+iv_0}\frac{F'(w)\,dw}{F(w)}\right]$$

$$=\frac{1}{2\pi i}\left[-2\pi i\log F(w)\Big|_{u_0+iv_0}^{u_0+\tau+iv_0}+\tau\log F(w)\Big|_{u_0+iv_0}^{u_0+2\pi i+iv_0}\right]$$

$$=-\log\sigma+2\pi Mi+N\tau,$$

where M and N are suitably chosen integers. On the other hand, this integral has the value

$$\sum_{j=1}^{n}(\log\alpha_j-\log\beta_j)=\log\frac{\alpha_1\cdots\alpha_n}{\beta_1\cdots\beta_n}.$$

In these two expressions for the integral, the term $2\pi Mi$ is of no significance here, and the term $N\tau=\log\rho^N$ can be eliminated if in the previous notation we replace the pole β_1 by the pole $\rho^N\beta_1$; this completes the proof of (5) and hence by Theorem 1 of Theorem 4. Replacement of β_1 by $\rho^N\beta_1$ in Theorem 1 obviously replaces σ in (3) by $\rho^N\sigma$. This is of course not serious as a modification, because the function $\psi(z)\equiv z^N$ satisfies the functional equation $\psi(\rho z)\equiv\rho^N\psi(z)$, and (compare the Corollary to Theorem 1) if $\Psi(z)$ satisfies the equation $\Psi(\rho z)\equiv\rho^N\sigma\Psi(z)$, then $\Psi(z)/\psi(z)$ satisfies the corresponding equation without the factor ρ^N.

More generally, if $\Psi(z)$ is an arbitrary function satisfying equation (4), it is clear that a branch of the function $\Psi_1(z)\equiv z^\lambda\Psi(z)$ satisfies the equation $\Psi_1(\rho z)\equiv\sigma\rho^\lambda\Psi_1(z)$, whether λ is real or not; we define z^λ as $e^{\lambda\log z}$. A converse is easily proved:

THEOREM 5. *Let the real number $\rho(>1)$ and the points $\alpha_1,\alpha_2,\cdots,\alpha_n$ and $\beta_1,\beta_2,\cdots,\beta_{n'}$, all different from zero, be given, with $\rho^k\alpha_i\neq\rho^m\beta_j$. Let the function $\Psi_1(z)$ not necessarily uniform have branch points only in $z=0$ and $z=\infty$, and let the branch $\Psi_1(z)$ go into the branch $e^{2\pi\lambda i}\Psi_1(z)$ when z traces in the positive sense a circle whose center is the origin. Let each branch satisfy the equation $\Psi_1(\rho z)\equiv\sigma\rho^\lambda\Psi_1(z)$, where (5) is not assumed. Let the zeros and poles of $\Psi_1(z)$ be $\rho^m\alpha_i$ and $\rho^m\beta_j$ respectively, all different from unity, where each branch of $\Psi_1(z)$ is analytic except in the points β_j, 0, ∞. Then with a slight change of notation of the α_j and β_k if necessary, we have $n=n'$, equation (5) is satisfied, and we can write $\Psi_1(z)\equiv cz^\lambda\Psi(z)$, where $\Psi(z)$ is the function of Theorem 1 and c is a constant.*

The function $\Psi_2(z)\equiv\Psi_1(z)/z^\lambda$ is uniform, has precisely the zeros $\rho^m\alpha_j$ and poles $\rho^m\beta_j$, and satisfies the equation $\Psi_2(\rho z)\equiv\sigma\Psi_2(z)$; thus $\Psi_2(z)$ satisfies the conditions of Theorem 4, and Theorem 5 follows.

The critical points of $\Psi_1(z)$ of Theorem 5 may be studied by writing

$$\Psi_1'(z)/\Psi_1(z)\equiv\Phi(z)+\lambda/z,\qquad\lambda=\lambda_1+i\lambda_2,\tag{10}$$

in the notation (1). Taking conjugates in (10) gives a field of force in which the positions of equilibrium are precisely the critical points of $\Psi_1(z)$ other than the multiple zeros. At the origin we place an ordinary positive or negative particle whose field is represented by $\lambda_1/\bar{z}$, and in addition a *skew particle*, here of mass $-\lambda_2$, whose field is represented by $-i\lambda_2/\bar{z}$. We consider a skew particle of positive or negative mass μ at z_0 to exert a force at z equal to $i\mu/(\bar{z} - \bar{z}_0)$, so the force is in magnitude that of the ordinary (i.e. not skew) particles hitherto considered, and the direction angle of the new force is found from that of the old by adding $\pi/2$. A positive skew particle can be considered a *right* skew particle, meaning that if a right forearm, thumb away from the plane, points from z_0 to z, then the sense of the force exerted at z by a positive skew particle at z_0 is indicated by the fingers at right angles to the forearm. Similarly a negative skew particle can be considered as a *left* skew particle.

For convenience in reference we formulate the

LEMMA. *Let $R(z)$ be a rational function with zeros $\alpha_1', \cdots, \alpha_n'$ and poles $\beta_1', \cdots, \beta_n'$; we consider the function $F(z) \equiv z^\lambda R(z)$, $\lambda = \lambda_1 + i\lambda_2$, with $\lambda_2 \neq 0$. The critical points of $F(z)$ other than multiple zeros of $F(z)$ are the positions of equilibrium in the field of force due to unit positive particles at the points α_j', unit negative particles at the points β_j', a particle of mass λ_1 at 0, and a skew particle of mass $-\lambda_2$ at 0.*

Let the α_j' and β_j' lie in respective closed sectors S_1 and S_2 with vertex O and angles less than π, namely the halves of a double sector, let Σ_1 be the open sector with vertex O following S_1 and preceding S_2 in the positive sense of rotation about O, and let Σ_2 be the open sector with vertex O following S_2 and preceding S_1. If we have $\lambda_2 < 0$, the sector Σ_1 contains no critical points of $F(z)$, and if we have $\lambda_2 > 0$ the sector Σ_2 contains no critical points of $F(z)$. In these respective cases, the boundaries of Σ_1 and Σ_2 pass through no critical points other than multiple zeros of $F(z)$.

The first part of the Lemma follows merely by taking the conjugate of the function $F'(z)/F(z)$.

Let P lie in the sector Σ_1, with $\lambda_2 < 0$. The force at P due to each particle α_j' not at 0 or infinity has a non-zero component orthogonal to OP in the counterclockwise sense of rotation about O, as have the force at P due to each particle β_j' not at 0 or infinity and the force at P due to the skew particle at O; the force at P due to the ordinary particle at O acts along OP; the component of the force due to the skew particle is certainly not zero, so P is not a position of equilibrium nor a critical point of $F(z)$. The remainder of the Lemma follows similarly.

The Lemma obviously applies if the rational function $R(z)$ is replaced by a function satisfying the conditions of Theorem 4, so we have

THEOREM 6. *Under the conditions of Theorem 5, with $\lambda = \lambda_1 + i\lambda_2$, let sectors $\alpha \leqq \arg z \leqq \beta (< \alpha + \pi)$ and $\alpha + \pi \leqq \arg z \leqq \beta + \pi$ contain respectively the points α_j and β_k. If we have $\lambda_2 \leqq 0$, the sector $\beta < \arg z < \alpha + \pi$ contains no critical points of $\Psi_1(z)$. If we have $\lambda_2 \geqq 0$, the sector $\beta + \pi < \arg z < \alpha + 2\pi$ contains no critical points of $\Psi_1(z)$.*

If we have $\lambda_2 \leqq 0$, *no boundary point* P *of the sector* $\beta < \arg z < \alpha + \pi$ *other than a multiple zero of* $\Psi_1(z)$ *can be a critical point unless we have* $\lambda_2 = 0$ *and all zeros and poles of* $\Psi_1(z)$ *lie on the line* OP; *if we have* $\lambda_2 \geqq 0$, *no boundary point* P *of the sector* $\beta + \pi < \arg z < \alpha + 2\pi$ *other than a multiple zero of* $\Psi_1(z)$ *can be a critical point unless we have* $\lambda_2 = 0$ *and all zeros and poles of* $\Psi_1(z)$ *lie on the line* OP.

The force at a point $z \neq 0$ due to a positive unit particle at the non-real point α_1 and to a negative unit particle at the point $\beta_1 = \bar{\alpha}_1$ has a non-zero component orthogonal to the line Oz in the sense from z toward β_1 in the direction of the arc $\alpha_1 z \beta_1$, provided α_1 or β_1 lies in a sector not containing z, the sector being NE convex with respect to the upper or lower half-plane containing z. A corresponding improvement can be made in the Lemma in this case. As an application we have

THEOREM 7. *Under the conditions of Theorem* 5, *with* $\lambda = \lambda_1 + i\lambda_2$ *and* $\beta_j = \bar{\alpha}_j$ *for every* j, *let the* α_j *lie in the sector* $\alpha \leqq \arg z \leqq \beta$, $0 < \alpha \leqq \pi/2 \leqq \beta < \pi$. *If we have* $\lambda_2 \leqq 0$, *the sector* $\beta < \arg z < 2\pi - \beta$ *contains no critical points of* $\Psi_1(z)$. *If we have* $\lambda_2 \geqq 0$, *the sector* $-\alpha < \arg z < \alpha$ *contains no critical points of* $\Psi_1(z)$. *No critical points other than multiple zeros of* $\Psi_1(z)$ *lie on the boundaries of these latter two sectors except in the case* $\lambda_2 = 0$, $\alpha = \beta = \pi/2$.

§6.4. Simply periodic functions. Let the single-valued analytic function $f(w)$ have the period $2\pi i$:

$$f(w + 2\pi i) \equiv f(w), \qquad w = u + iv. \tag{1}$$

We study [Walsh, 1947b] this function by use of the classical transformation

$$w = \log z, \tag{2}$$

under which the z-plane corresponds to each period strip $v_0 \leqq v < v_0 + 2\pi i$. The function

$$F(z) \equiv f(\log z) \tag{3}$$

is uniform, and is analytic in points z corresponding to points w in which $f(w)$ is analytic. We assume $F(z)$ to be meromorphic at every point of the extended z-plane, hence a rational function of z. When u becomes negatively infinite, the point z approaches zero, so in the w-plane the function $f(w)$ behaves in many ways as if $u \to -\infty$ corresponded to the approach of w to a single point. Likewise when $u \to +\infty$, the point z becomes infinite, and again w behaves as if approaching a single point. We introduce the convention of artificial "end-points" of a period strip. The left-hand end-point is a zero, a non-zero point of analyticity, or a pole according as we have

$$\lim_{u \to -\infty} \frac{|f(u + iv)|}{e^{mu}} = A \neq 0,$$

where the integer m is positive, zero, or negative. The right-hand end-point is a zero, a non-zero point of analyticity, or a pole according as we have

$$\lim_{u \to +\infty} \frac{|f(u + iv)|}{e^{mu}} = A \neq 0,$$

where the integer m is negative, zero, or positive. In each case $|m|$ is necessarily integral and if not zero is defined as the *order* of the zero or pole of $f(w)$; this is also the order of the corresponding zero or pole of $F(z)$ at zero or at infinity. The *order of the function* $f(w)$ is the sum of the orders of its poles in a period strip, end-points included, which is equal to the sum of the orders of its zeros in a period strip, end-points included.

§6.4.1. General theorems. Results in the w-plane may now be found by the use of the transformation (2). We shall prove

THEOREM 1. *Let the analytic function $f(w)$ have the period $2\pi i$, and have the right-hand end-point of a period strip as a pole, the only singularity in the period strip, end-points included.*

1). *If all the zeros of $f(w)$ satisfy the inequality $u \leqq u_0$ (where $u = -\infty$ is not excluded), so do all the zeros of $f'(w)$.*

2). *If all the finite zeros of $f(w)$ in a period strip satisfy the inequality $\alpha \leqq v \leqq \beta$ with $\beta - \alpha \leqq \pi$, so do all the finite zeros of $f'(w)$.*

3). *If $f(w)$ is of order n, if the left-hand end-point of a period strip is a zero of order k, and if the region $u \leqq u_0$ contains no finite zeros of $f(w)$, then the region $u \leqq u_0 + \log (k/n)$ contains no finite zeros of $f'(w)$.*

The function $F(z)$ defined by (3) is here single-valued and analytic, with a pole of order n at infinity, hence is a polynomial in z of degree n. The image of the region $u \leqq u_0$ is the region $|z| \leqq e^{u_0}$, which in 1) contains all zeros of $F(z)$ and hence of $F'(z)$ by Lucas's Theorem. It follows that all zeros of $f'(w)$ lie in the region $u \leqq u_0$; it is to be noted that we have

$$F'(z) = f'(\log z)/z, \tag{4}$$

so the zeros of $F'(z)$ and $f'(w)$ are in precise correspondence, except that $z = 0$ is a zero of $F'(z)$ of order less by unity than the left-hand end-point of a period strip of $f'(w)$. Even if $z = 0$ is not a zero of $F'(z)$, the left-hand end-point of a period strip is a zero of $f'(w)$, and satisfies the conclusion of 1).

It is of interest to remark that 1) for the case $u_0 = -\infty$ is precisely the fact that the derivative of e^{nw} has no finite zeros.

Under the transformation (2) we have $u + iv = \log |z| + i \arg z$, and in 2) the given closed strip corresponds to the closed sector $\alpha \leqq \arg z \leqq \beta$, here convex because of the relation $\beta - \alpha \leqq \pi$. The conclusion of 2) follows from Lucas's Theorem.

Under the transformation (2) we carry over known results from the z-plane to the w-plane; the new results may be expressed in terms of the geometry in the w-plane, as in parts 1) and 2), or may be expressed alternately by carrying over the geometry of the z-plane onto the w-plane; compare §§3.6.1 and 6.2.1. Thus let $\mathfrak{L}$ be the set of images in a period strip of the straight lines of the z-plane; we identify the boundary points $u + iv_0$ and $u + i(v_0 + 2\pi)$ of the strip $v_0 \leqq v < v_0 + 2\pi$. It follows from Lucas's Theorem that in the strip *any point set convex with respect to the curves* $\mathfrak{L}$ *which contains all zeros of* $f(w)$ *also contains all zeros of* $f'(w)$, *except perhaps the left-hand end-point.* This statement includes both 1) and 2).

Part 3) follows from §3.1.1 Corollary to Theorem 1.

Other results can be proved if $f(w)$ is permitted to have finite poles:

THEOREM 2. *Let the analytic function* $f(w)$ *have the period* $2\pi i$ *and be meromorphic in each period strip, end-points included.*

1). *If a line* $L{:}u =$ const *does not pass through all the zeros and poles of* $f(w)$ *but separates the zeros of* $f(w)$ *not on* L *from the poles of* $f(w)$ *not on* L, *then* L *cannot pass through a critical point of* $f(w)$ *other than a multiple zero. Consequently if each of the lines* $u = u_0$, $a < u_0 < b$, *separates all the zeros of* $f(w)$ *from all the poles of* $f(w)$, *then the region* $a < u < b$ *contains no zeros of* $f'(w)$. *If here all the zeros of* $f(w)$, k *in number in each period strip, lie in the closed region* $u \leqq a$, *then that closed region contains precisely* k *zeros of* $f(w)$ *in each period strip.*

2). *In a period strip* $S{:}v_0 \leqq v < v_0 + 2\pi$ *we identify the two boundary points having the same abscissa. Let* $L{:}v = v_1$ *and* $L'{:}v = v_1 + \pi$ *be a pair of lines in* S *on which do not lie all the finite zeros and poles of* $f(w)$ *and which separate in* S *the finite zeros of* $f(w)$ *not on* $L + L'$ *from the finite poles of* $f(w)$ *not on* $L + L'$. *Then no finite zero of* $f'(w)$ *lies on* $L + L'$ *other than a multiple zero of* $f(w)$. *Consequently if each pair of such lines respectively in the two strips* $(v_0 \leqq)v_2 < v < v_3$, $v_2 + \pi < v < v_3 + \pi(\leqq v_0 + 2\pi)$ *separates all the finite zeros of* $f(w)$ *in* S *from all the finite poles of* $f(w)$ *in* S, *then neither of these strips contains any finite zeros of* $f'(w)$.

3). *In each period strip let the region* $u \leqq u_1$ *contain* k *zeros of* $f(w)$, *the region* $(u_1<)u_2 \leqq u$ *contain all the remaining* $n - k$ *zeros of* $f(w)$, *and the region* $(u_1<)$ $u_3 \leqq u$ *contain all the* n *poles of* $f(w)$. *We set*

$$e^{u_0} = \frac{ke^{u_2+u_3} - ne^{u_1+u_2} - (n-k)e^{u_1+u_3}}{(n-k)e^{u_2} + ne^{u_3} - ke^{u_1}}.$$

If we have $u_1 < u_0 \leqq u_3$, *then no zeros of* $f'(w)$ *lie in the region* $u_1 < u < u_0$; *if we have* $u_3 < u_0 < u_2$, *then no zeros of* $f'(w)$ *lie in the region* $u_1 < u < u_3$, *in either of these cases the region* $u \leqq u_1$ *contains in each period strip precisely* k *zeros of* $f'(w)$.

Parts 1) and 2) follow from §4.2.1 Corollaries 1 and 2 to Bôcher's Theorem. Part 3) follows from §4.3 Theorem 1. As a limiting case of part 2), a consequence of §4.2.3 Theorem 2, we have the

COROLLARY. *If all finite zeros [or poles] of $f(w)$ in the period strip $S: v_0 \leqq v < v_0 + 2\pi$ lie on the line $v = v_1$ in S, and all finite poles [or zeros] of $f(w)$ in S lie on the line $v = v_1 + \pi$ in S, then all finite zeros of $f'(w)$ in S lie on those two lines. On any open segment of either line bounded by two zeros or two poles (not necessarily finite) and containing no zero or pole of $f(w)$ lies a unique zero of $f'(w)$.*

§6.4.2. Functions with symmetry. The results just established can be somewhat improved if $f(w)$ is required to possess symmetry; we formulate merely a few examples.

THEOREM 3. *Let the function $f(w)$ have the period $2\pi i$ and be meromorphic in each period strip, end-points included. Let the point $w = w_0 + \pi i$ be a zero or pole of $f(w)$ respectively whenever $w = w_0$ is a zero or pole. Let the finite zeros of $f(w)$ in the period strip $S: v_0 < v \leqq v_0 + 2\pi$ lie in the strips $S_1: (v_0 <) v_1 \leqq v \leqq v_2$, $S_1': v_1 + \pi \leqq v \leqq v_2 + \pi (\leqq v_0 + 2\pi)$, with $v_2 - v_1 < \pi/2$, and let the finite poles of $f(w)$ in S lie in the strips $S_2: (v_0 <) v_3 \leqq v \leqq v_4$, $S_2': v_3 + \pi \leqq v \leqq v_4 + \pi (\leqq v_0 + 2\pi)$, with $v_4 - v_3 < \pi/2$, where $S_1 + S_1'$ has no finite point in common with $S_2 + S_2'$. Then the four maximal strips in S which can be constructed from all quadruples of lines $v = a + k\pi/2$, $k = 0, 1, 2, 3$, which separate $S_1 + S_1'$ from $S_2 + S_2'$ in S, where we identify the two boundary points of S having the same abscissa, contain no finite zeros of $f'(w)$ in S.*

Theorem 3 is a consequence of §5.5.2 Theorem 2. Of course study of the transformed functions in the z-plane shows that $[f(w)]^2$ is periodic with period πi, so Theorem 3 can also be proved from part 2) of Theorem 2.

THEOREM 4. *Let the function $f(w)$ have the period $2\pi i$ and be meromorphic in each period strip, end-points included. Let the point $w = w_0 + \pi i$ be a zero or pole respectively of $f(w)$ whenever the point $w = w_0$ is a pole or zero of $f(w)$. If the zeros of $f(w)$ in the period strip $S: v_0 < v \leqq v_0 + 2\pi$ lie in the strip $S_1: (v_0 <) v_1 \leqq v \leqq v_2 (\leqq v_1 + \pi/2 < v_0 + \pi)$, and in the band $u_0 \leqq u \leqq u_1$, then all finite critical points of $f(w)$ lie in the regions common to that band and the two strips $v_1 \leqq v \leqq v_2$, $v_1 + \pi \leqq v \leqq v_2 + \pi$.*

Theorem 4 follows from §5.6.1 Theorem 1.

THEOREM 5. *Let the function $f(w)$ have the period $2\pi i$, be meromorphic in each period strip, end-points included, and have the modulus unity on the two lines $v = 0$ and $v = \pi$.*

1). *Let no poles of $f(w)$ lie in the strip $0 \leqq v \leqq \pi$. If no zeros or poles of $f(w)$ lie in the region $u < u_0$ [or $u > u_0$], then no finite zeros of $f'(w)$ lie in that region.*

2). *If a line $u = u_0$ separates the zeros of $f(w)$ in the strip $0 < v < \pi$ from the poles in that strip, then no zeros of $f'(w)$ lie on that line.*

3). *If all zeros of $f(w)$ in the strip $S: -\pi \leqq v \leqq \pi$ lie in the strip $S_1: (0 <) v_0 \leqq$*

$v \leqq v_1(<\pi)$, with $v_0 \leqq \pi/2 \leqq v_1$, then all finite zeros of $f'(w)$ in S lie in S_1 and in the strip $-v_1 \leqq v \leqq -v_0$.

4). *If the line $v = \pi/2$ separates the zeros in the strip $0 < v < \pi$ from the poles in that strip, then no finite zero of $f'(w)$ lies on that line.*

The logarithm of the modulus of the function $F(z)$ defined by (3) is zero when z is real, so by the Schwarz principle of reflection the zeros and poles of $F(z)$ are mutually symmetric in the axis of reals. Parts 1) and 3) follow from §5.3.1 Theorem 1, and parts 2) and 4) from §5.3.2 Theorem 2.

Functions $f(w)$ which satisfy the functional equation

$$f(w + 2\pi i) = e^{2\lambda\pi i}f(w), \qquad \lambda = \lambda_1 + i\lambda_2, \tag{5}$$

are readily treated by the present methods, for the function $f_1(w) \equiv e^{\lambda w}$ satisfies equation (5), and $f(w)/f_1(w)$ has the period $2\pi i$. In particular, the Lemma of §6.3.2 is applicable, as the reader may observe.

§6.5. Doubly periodic functions. Methods already developed enable us to study doubly periodic functions with both real and pure imaginary periods; we do not assume that these periods form a primitive period pair. We shall prove

THEOREM 1. *Let the function $f(w)$ be meromorphic at every finite point of the plane of $w = u + iv$, and possess the periods $2\pi i$ and τ, where τ is real and positive. In the strip $S: v_0 < v \leqq v_0 + 2\pi$ identify the two boundary points having the same abscissa.*

1). *Let not all the zeros and poles of $f(w)$ in S lie on the lines $L: v = v_1$ and $L': v = v_1 + \pi$ in S, and let L and L' separate in S the zeros not on $L + L'$ from the poles not on $L + L'$; then no critical point of $f(w)$ other than a multiple zero of $f(w)$ lies on L or L'. Consequently if each pair of lines L and L' respectively in the two strips $S_1: (v_0 \leqq) v_2 < v < v_3$, $S_2: v_2 + \pi < v < v_3 + \pi (\leqq v_0 + 2\pi)$, separates all the zeros of $f(w)$ in S from all the poles of $f(w)$ in S, then no critical point of $f(w)$ lies in S_1 or S_2.*

2). *If all the zeros [or poles] of $f(w)$ in S lie on the line $v = v_1$ in S and all the poles [or zeros] lie on the line $v = v_1 + \pi$ in S, then all critical points of $f(w)$ in S lie on these lines. Any open segment of either line bounded by two zeros or by two poles of $f(w)$ and containing no zero or pole of $f(w)$ contains precisely one critical point.*

3). *Let the point $w_0 + \pi i$ be a zero or pole of $f(w)$ respectively whenever $w = w_0$ is a zero or pole. Let the zeros of $f(w)$ in S lie in the strips $S_1: (v_0 \leqq)\ v_2 \leqq v \leqq v_3$, $S_1': v_2 + \pi \leqq v \leqq v_3 + \pi (\leqq v_0 + 2\pi)$, with $v_3 - v_2 < \pi/2$, and let the poles of $f(w)$ in S lie in the strips $S_2: (v_0 \leqq) v_4 \leqq v \leqq v_5$, $S_2': v_4 + \pi \leqq v \leqq v_5 + \pi (\leqq v_0 + 2\pi)$, with $v_4 - v_3 < \pi/2$, where $S_1 + S_1'$ has no finite point in common with $S_2 + S_2'$. Then the four maximal strips in S which can be constructed from all quadruples of lines $v = a + k\pi/2$, $k = 0, 1, 2, 3$, which separate $S_1 + S_1'$ from $S_2 + S_2'$ in S contain no critical points of $f(w)$.*

4). *Let the function* $f(w)$ *have the modulus unity on the two lines* $v = 0$ *and* $v = \pi$. *If all zeros of* $f(w)$ *in the strip* $S: -\pi < v \leqq \pi$ *lie in the strip* $S_1: (0<)\, v_2 \leqq v \leqq v_3(<\pi)$, *with* $v_2 \leqq \pi/2 \leqq v_3$, *then all critical points of* $f(w)$ *in* S *lie in* S_1 *and in the strip* $-v_3 \leqq v \leqq -v_2$.

We state for reference the classical results proved in §6.3.2 that if $a_1, a_2, \cdots, a_n$ are the zeros and $b_1, b_2, \cdots, b_n$ the poles of $f(w)$ in a period parallelogram (whose sides are respectively $2\pi i$ and τ), then we have

$$(1) \qquad \sum_{k=1}^{n}(a_k - b_k) = 2\pi p i + q\tau,$$

where p and q are integers.

Again we use the transformation $w = \log z$; the function $F(z) \equiv f(\log z)$ is single-valued for all values of z and meromorphic at every point of the extended z-plane other than $z = 0$ and $z = \infty$. We set $\rho = e^{\tau}(>1)$, and write the equation $f(\log z + \log \rho) \equiv f(\log z)$ as $F(\rho z) \equiv F(z)$, so $F(z)$ possesses the multiplicative period ρ. If we denote by $z = \alpha_1, \alpha_2, \cdots, \alpha_n$ the zeros and by $\beta_1, \beta_2, \cdots, \beta_n$ the poles of $F(z)$ in the ring $1 \leqq |z| < \rho$, then by (1) we may write

$$(2) \qquad \alpha_1 \cdot \alpha_2 \cdot \cdots \cdot \alpha_n/\beta_1 \cdot \beta_2 \cdot \cdots \cdot \beta_n = \rho^q,$$

where q is a suitably chosen integer. Here we can apply §6.3.1 Corollary to Theorem 1, and by a suitable change of notation if necessary, where the α_j and β_j need no longer lie in the ring $1 \leqq |z| < \rho$, replace the second member of (2) by unity.

Parts 1) and 2) now follow from §6.3.1 Theorem 2, part 3) follows from §5.5.2 Theorem 2, and part 4) follows from §6.3.1 Theorem 3.

Under the conditions of Theorem 1, the function $f(\tau w/2\pi i)$ also has the period $2\pi i$, and the theorem may also be applicable to it.

Theorem 1 extends to the case of a function $f(w)$ meromorphic at every finite point of the w-plane and satisfying the two functional equations $f(w + 2\pi i) \equiv \sigma_1 f(w)$, $f(w + \tau) \equiv \sigma_2 f(w)$, where τ is real and positive, $\rho = e^{\tau}$. We set $F(z) \equiv f(\log z)$ as before, and $F(z)$ is not necessarily uniform, having possible branch points at zero and infinity. When z traces in the positive sense a circle whose center is the origin, $F(z)$ goes into $\sigma_1 F(z)$. For the function $F_1(z) \equiv z^{\lambda} \equiv e^{\lambda \log z}$, when z traces a circle whose center is the origin, $F_1(z)$ goes into the branch $e^{2\pi\lambda i} F_1(z)$, and we may choose $\lambda = (\log \sigma_1)/2\pi i$. Then the function $F_2(z) \equiv F(z)/z^{\lambda}$ is uniform, and satisfies the equation $F_2(\rho z) \equiv (\sigma_2/\rho^{\lambda})F_2(z)$, so §6.3.2 Theorem 4 applies.

The analog of Theorem 1 is now an immediate consequence of §6.3.2 Theorems 6 and 7; its formulation is left to the reader. In proving the analog of part 3) of Theorem 1 we use the auxiliary transformation $z' = z^2$.

§6.6. General analytic functions. Cauchy's integral formula for an analytic function $F(z)$:

$$F(z) = \frac{1}{2\pi i} \int_C \frac{F(t)\, dt}{t - z} \tag{1}$$

suggests of itself the expression of the second member as the limit of a sum $\Sigma(z)$ of terms each of the form $A_k/(t_k - z)$, which is precisely the kind of sum suggesting a field of force whose positions of equilibrium we have been investigating. Thus it might be expected that the critical points of $F(z)$ can be studied by means of the critical points of $\Sigma(z)$, for if $F(z)$ is the uniform limit of $\Sigma(z)$, the critical points of $F(z)$ are the limits of those of $\Sigma(z)$. The difficulty with this method is that of characterizing A_k as to the signs of its real and pure imaginary part, for in the field of force the direction of the force depends essentially on these algebraic signs; general results can hardly be anticipated without heavy restrictions on $F(z)$.* In one general case, however, progress can be made.

Let R be a finite region bounded by two disjoint Jordan configurations (§1.1.1) C_1 and C_2, let the function $f(z)$ be analytic and different from zero in R, continuous in $R + C_1 + C_2$, and have the respective constant moduli M_1 and M_2 on C_1 and C_2, $M_1 < M_2$. We write Cauchy's formula (1) for the function $F(z) = f'(z)/f(z)$:

$$\frac{f'(z)}{f(z)} = \frac{1}{2\pi i} \int_{C_1+C_2} \frac{d[\log f(t)]}{t - z}, \tag{2}$$

where z lies in R. For the present we assume C_1 and C_2 each to consist of a finite number of disjoint analytic Jordan curves; this restriction will be removed later. The function $|f(z)|$ takes its minimum and maximum values on C_1 and C_2 respectively, hence has values in R between M_1 and M_2. The function $\log |f(z)|$ is harmonic in R and can be harmonically extended across C_1 and C_2, hence $f(z)$ is analytic on C_1 and C_2; and no critical point of $f(z)$ lies on C_1 or C_2, since neither locus $|f(z)| = M_k$ has a point in R. Under the conformal map $w = f(z)$, the region R in the neighborhood of C_1 is transformed into the neighborhood of γ_1: $|w| = M_1$ exterior to that circle in the w-plane, and the region R in the neighborhood of C_2 is transformed into the neighborhood of γ_2: $|w| = M_2$ interior to that circle. When z traces C_2 in the positive sense with respect to R, w traces γ_2 in the counterclockwise sense: $w = M_2 e^{i\theta}$ and we have $d[\log f(z)] = d[\log w] = i\,d\theta$, and $d\theta$ is positive; when z traces C_1 in the positive sense with

---

* As an illustration here, suppose $f(z)$ analytic in the closed interior of C: $|z| = 1$, a k-fold zero at 0, no other zeros interior to C, and with $\int_C |f'(z)\, dz/f(z)| \leqq 2\pi M$. We have for $z \neq 0$ interior to C

$$\frac{f'(z)}{f(z)} = \frac{k}{z} + \frac{1}{2\pi i} \int_C \frac{f'(t)\, dt}{f(t)(t - z)},$$

from which it follows that $f'(z)$ is different from zero in the annulus $0 < |z| < k/(k + M)$.

respect to R, w traces γ_1 in the clockwise sense: $w = M_1 e^{i\theta}$, and we have $d[\log f(z)] = d[\log w] = i d\theta$, and $d\theta$ is negative. Thus equation (2) can be written

$$\frac{f'(z)}{f(z)} = \frac{1}{2\pi} \int_{C_1} \frac{d\mu_1}{z - t} + \frac{1}{2\pi} \int_{C_2} \frac{d\mu_2}{z - t}, \qquad z \text{ in } R, \tag{3}$$

where $d\mu_k(t) = -d[\arg f(t)]$, which is positive on C_1 and negative on C_2. We note too the equation

$$\int_{C_1} d\mu_1 + \int_{C_2} d\mu_2 = 0,$$

for we have

$$\int_{C_1 + C_2} \frac{f'(z)\, dz}{f(z)} = 0,$$

since $f(z)$ has no zeros or poles in R.

We can now return to our original assumption on C_1 and C_2, first establishing (3) where the integrals are taken over level loci $|f(z)| = M_1'$ and $|f(z)| = M_2'$, $M_1 < M_1' < M_2' < M_2$, where each locus consists of a finite number of disjoint analytic Jordan curves in R, and we allow M_1' to approach M_1 and M_2' to approach M_2, keeping z fixed in R. These level loci and this method of proof are described in more detail in §8.1; compare also §7.1.3. During this limiting process we can allow t to approach C_1 or C_2, remaining on a curve $\arg f(z) = \text{const}$; thus the parameter t varies uniformly continuously with M_1' and M_2', and the integrals taken over the auxiliary level loci considered as Stieltjes integrals approach the integrals in (3); the first member of (3) is unchanged during this process, so (3) in its present form is established.

Equation (3) is valid whether R is finite or infinite, provided C_1 and C_2 are finite.

If $f(z)$ is not single-valued but $|f(z)|$ is single-valued, any two determinations of $\log [f(z)] = \log |f(z)| + i \arg [f(z)]$ differ merely by an additive constant, so the derivative $f'(z)/f(z)$ of $\log [f(z)]$ is single-valued. The differential $d[\log f(z)]$ is independent of the branch of the function chosen, so equations (2) and (3) remain valid.

We proceed to prove

THEOREM 1. *Let R be a region whose boundary consists of disjoint Jordan configurations C_1 and C_2. Let the function $f(z)$ be analytic and different from zero in R, continuous in $R + C_1 + C_2$, and have the constant moduli M_1 and M_2 on C_1 and C_2, $M_1 < M_2$. The function $f(z)$ may be multiple-valued provided $|f(z)|$ is single-valued in R. Let C_1 and C_2 lie in disjoint circular regions Γ_1 and Γ_2 respectively. Then Γ_1 and Γ_2 contain all critical points of $f(z)$ in R. If C_1 and C_2 consist of m_1 and m_2 components respectively, Γ_1 and Γ_2 contain $m_1 - 1$ and $m_2 - 1$ critical points respectively of $f(z)$ in R.*

Transform if necessary so that the boundary of R is finite. In equation (3) we take conjugates and replace each integral by a Riemann sum of which it is the limit. For the Riemann sums, the field of force is precisely that considered in the proof of Bôcher's Theorem (§4.2.1), except that in the present case the particles do not necessarily have integral masses. Nevertheless §1.5.1 Lemma 1 readily extends to the present case, and shows that if positive particles of total mass μ lie in a circular region Γ, then the corresponding force at a point z exterior to Γ is equal to the force exerted at z by a suitably chosen particle of mass μ in Γ. Thus (§4.1.2 Corollary to Theorem 3) no point of R exterior to Γ_1 and Γ_2 can be a position of equilibrium in the field of force for the Riemann sums, so by Hurwitz's Theorem no point of R exterior to Γ_1 and Γ_2 can be a critical point of $f(z)$.

In this proof of the first part of Theorem 1, it is perhaps more familiar to use a field of force due to particles, but it is possible to consider the conjugate of (3) to define a field of force directly, a field due to a spread of positive matter on C_1 and to an equal total mass of negative matter on C_2. For the latter field, we may study the variation in the argument of the conjugate of $f'(z)/f(z)$ as z traces a circle in R exterior to but near Γ_1, and also traces a level locus near C_1; the remainder of Theorem 1 follows by this method.

If a circle Γ separates all the points of C_1 not on Γ from all the points of C_2 not on Γ, and if not all points of C_1 and C_2 lie on Γ, then no critical point of $f(z)$ lies in R on Γ.

We shall prove also

THEOREM 2. *Let R be a region whose boundary C is a Jordan configuration, let the function $f(z)$ be analytic in R and continuous in $R + C$, and have the constant modulus M on C. The function $f(z)$ need not be single-valued in R if the modulus $|f(z)|$ is single-valued there. Let all the zeros $\alpha_1, \alpha_2, \cdots, \alpha_m$ of $f(z)$ in R, and the point set C lie in disjoint circular regions Γ_1 and Γ_2 respectively. Then Γ_1 and Γ_2 contain all critical points of $f(z)$ in R. If C consists of n components, Γ_1 and Γ_2 contain respectively $m - 1$ and $n - 1$ critical points of $f(z)$ in R.*

Theorem 2 follows from Theorem 1 merely by choosing an auxiliary variable circular region Γ_1' near Γ_1 but containing Γ_1 in its interior, and noting that a level locus $C_1 : |f(z)| = M_1$, where M_1 is positive but small, can be chosen in Γ_1'. If Γ_1' and Γ_2 are disjoint, they contain all critical points of $f(z)$ in R, and this is true for every Γ_1'. In totalling the number of critical points in Γ_1, we consider the number between C_1 and Γ_1', and also the number in points α_k.

Theorem 2 is readily proved also by modifying the field of force used in Theorem 1. If γ denotes a Jordan curve in R (assumed finite) belonging to the locus C_1, containing in its interior the zero α, of $f(z)$ of multiplicity k and containing no other zero of $f(z)$, we have for the clockwise sense on γ

$$\log f(t) \,|_\gamma = -2\pi k i;$$

as M_1 approaches zero, so also does the greatest distance from γ to α_j ; for fixed z in R the constant (nominally a function of γ)

$$\frac{1}{2\pi i}\int_\gamma \frac{d[\log f(t)]}{t - z}$$

approaches the function $k/(z - \alpha_j)$. Thus for z in R we may write

$$\frac{f'(z)}{f(z)} = \sum_{j=1}^{n} \frac{1}{z - \alpha_j} - \frac{1}{2\pi i}\int_C \frac{d[\log f(t)]}{z - t}.$$

The field of force found here by taking conjugates is due to a spread of negative matter on $C: d[\log f(t)] > 0$, and to unit positive particles at each of the points α_j.

§6.7. Analytic functions and harmonic functions. It is clear from the examples given in §6.6 that the entire theory developed in Chapters I–V can be paralleled so as to apply to suitably chosen arbitrary analytic functions instead of merely polynomials and rational functions. Moreover, any result on the critical points of an analytic function $f(z) = u(x, y) + iv(x, y)$ is also a result on the critical points of various harmonic functions, namely $u(x, y)$, $v(x, y)$, $\log | f(z) |$, etc., so the entire theory already developed can be paralleled so as to apply to suitably chosen arbitrary harmonic functions, not merely those harmonic functions trivially obtained from polynomials and other rational functions. Reciprocally, results on critical points for a harmonic function $u(x, y)$ include essentially results for its conjugate $v(x, y)$, for the analytic function $f(z) = u(x, y) + iv(x, y)$, and for the functions $\log f(z)$, $e^{f(z)}$, etc.

It is a matter of taste as to which is preferable, developing an extensive theory for analytic functions or for the related harmonic functions. Results of either kind can be obtained at once from results of the other kind. It can hardly be argued that the one category of functions is intrinsically more important than the other, although in certain cases such an argument can be made: the Blaschke product seems more important than the harmonic functions immediately related to it; harmonic measure is a harmonic function which seems more important than the analytic functions immediately related to it. Harmonic functions (such as Green's function for a multiply-connected region) have, however, the great advantage of often being single-valued even when the related analytic functions are multiple-valued, and this may mean a gain in elegance.

Henceforth in the present work we shall emphasize results on harmonic functions, without explicitly mentioning the applications to analytic functions, but it is to be understood that those applications are readily made, and are to be considered as part of the broad background.

We proceed to develop for harmonic functions a theory parallel to that of Chapters I–V.

CHAPTER VII

GREEN'S FUNCTIONS

Applications of the foregoing methods and results can be made to the study of the critical points of harmonic functions. Broadly speaking, Green's functions are analogous to polynomials, and more general harmonic functions are analogous to more general rational functions, so we commence with the study of Green's functions. Green's functions are of great importance in the study of electricity, gravitation, and hydrodynamics, as well as in potential theory, conformal mapping, and polynomial expansion problems.

§7.1. Topology. Let R be an infinite region whose boundary B is finite; by *Green's function* for R with pole at infinity we mean the function $G(x, y)$ which is harmonic at every finite point of R, continuous and equal to zero at every point of B, and such that $G(x, y) - \frac{1}{2} \log (x^2 + y^2)$ approaches a finite limit when $z = x + iy$ becomes infinite. If R is a finite or infinite region with boundary B and if O: (x_0 , y_0) is a finite point of R, Green's function for R with pole in O is the function $G(x, y)$ harmonic at every point of R other than O, continuous and equal to zero at every point of B, and such that $G(x, y) + \frac{1}{2} \log [(x, x_0)^2 + (y - y_0)^2]$ approaches a finite limit when (x, y) approaches O. Green's function for a region R is not defined at the pole, but is positive at every other point of R.

Green's function need not exist for a given region, but if existent is unique. Green's function exists for a region R and for an arbitrary choice of pole in R if the boundary of R is a Jordan configuration (§1.1.1). Green's function is invariant under inversion in a circle and under one-to-one conformal transformation of the plane, in the sense that if a region R and interior point O are transformed into a region R' and interior point O', then Green's function for R with pole in O is transformed into Green's function for R' with pole in O'.

§7.1.1. Level curves. For convenience we choose R as an infinite region whose boundary B is finite, and study Green's function $G(x, y)$ for R with pole at infinity. For the present we take B as composed of a finite number of disjoint Jordan curves. The level (equipotential) loci

$$G(x, y) = \mu > 0, \ (x, y) \text{ in } R, \tag{1}$$

have properties analogous to those of lemniscates, to be established by the same methods (§1.6).

Through a given point (x_1 , y_1) of R passes one and only one of the loci (1), namely the locus $G(x, y) = G(x_1 , y_1)$. The locus (1) consists in the neighborhood of one of its points (x_1 , y_1) of an analytic Jordan arc if (x_1 , y_1) is not a critical point of $G(x, y)$, but if (x_1 , y_1) is a critical point of order m, the locus consists of

$m + 1$ analytic Jordan arcs through (x_1, y_1) making successive angles $\pi/(m + 1)$ at (x_1, y_1). The function $G(x, y)$ vanishes on B and becomes infinite as (x, y) becomes infinite, so the locus consists of a finite number of Jordan curves interior to R separating B from the point at infinity; at most a finite number of points can be points of intersection of two or more of these curves. No point of (1) can lie in a region bounded wholly by points of (1) and points of B; no point of (1) can lie in an infinite subregion of R whose finite boundary points all belong to (1); any finite Jordan region bounded by points of (1) must contain points of B. Thus the locus (1) consists of a finite number of Jordan curves in R, mutually exterior except perhaps for a finite number of points each common to several curves; each Jordan curve belonging to (1) contains in its interior at least one of the Jordan curves of which B is composed.

The two loci $G(x, y) = \mu_1(>0)$ and $G(x, y) = \mu_2(>\mu_1)$ cannot intersect in R, and the latter separates the former from the point at infinity. When μ is small and positive, the locus (1) consists of a Jordan curve near each of the Jordan curves belonging to B. As μ increases, these Jordan curves of (1) increase monotonically in size, but the number of Jordan curves does not change unless as μ varies a locus (1) passes through one or more critical points of $G(x, y)$. If critical points of total multiplicity m lie on a locus $G(x, y) = \mu_1$ consisting of k Jordan curves, then for μ smaller than μ_1 but near μ_1 the locus (1) also consists of k Jordan curves, and these are mutually disjoint; for μ greater than μ_1 but near μ_1, the locus (1) separates the plane into m fewer regions than does $G(x, y) = \mu_1$, and hence consists of $k - m$ mutually exterior Jordan curves. If q critical points of $G(x, y)$ lie in the region $G(x, y) > \mu$ and none on (1), then (1) consists precisely of $q + 1$ mutually exterior Jordan curves. For sufficiently large μ, the locus (1) consists of a single analytic Jordan curve, which is approximately circular in shape. If the set B consists of r Jordan curves, then $G(x, y)$ has precisely $r - 1$ critical points in R, and at most $r - 1$ distinct loci (1) possess multiple points.

The loci

$$H(x, y) = \text{const}, \tag{2}$$

where $H(x, y)$ is conjugate to $G(x, y)$ in R, likewise form a family covering R; in the neighborhood of a finite point (x_1, y_1) of R not a critical point of $H(x, y)$, the locus (2) consists of a single analytic arc; in the neighborhood of an m-fold critical point (x_1, y_1) of $H(x, y)$, the locus (2) consists of $m + 1$ analytic arcs through (x_1, y_1) making successive angles $\pi/(m + 1)$ at (x_1, y_1). Through each point of R passes a locus (2) which is unique geometrically, but $H(x, y)$ is not single-valued in R, and a single geometric locus (2) corresponds to an infinity of values of the constant. Only a finite number of distinct geometric loci (2) have multiple points, namely those loci which pass through critical points of $H(x, y)$; the two functions $G(x, y)$ and $H(x, y)$ have in R identical critical points, and of the same orders for the one function as for the other. Each locus (2) consists of an arc from the point at infinity to a point of B, except that the locus (2) forks at a critical point. When z traces monotonically a locus (2), the function $G(x, y)$ varies monotonically, except that as z passes through a critical point a suitable new branch of the locus must be chosen; under those conditions z moves from the point at infinity to a point of B (or the reverse); if this procedure is not followed, and if z traces a Jordan curve from the point at infinity to a critical point and back to the point at infinity, the Jordan curve traced must separate B. At a finite point (x_1, y_1) of R not a critical point of $G(x, y)$ and $H(x, y)$, the two loci (1) and (2) are mutually orthogonal; at a finite point (x_1, y_1) of R which is a critical point of order m, the $m + 1$ branches of each locus bisect the successive angles between the branches of the other locus.

The region R, whose boundary B consists of p disjoint Jordan curves C_1, $C_2, \cdots, C_p$, can be mapped in a one-to-one manner on a region R_0 whose boundary B_0 consists of p disjoint *analytic* Jordan curves; we map R onto a region R_1 bounded by the Jordan curves C_1', $C_2', \cdots, C_p'$ by mapping the exterior of C_1 onto the exterior of a circle C_1' so that the two points at infinity correspond to each other; we map R_1 onto a region R_2 bounded by the Jordan curves C_1'', C_2'', $\cdots$, C_p'' by mapping the exterior of C_2' onto the exterior of a circle C_2'' so that the two points at infinity correspond to each other; we continue this process through p steps. After the process is completed, the final region being R_0 with boundary B_0, the invariant functions $G(x, y)$ and $H(x, y)$ are harmonic (locally) not merely at the finite interior points of R_0, but also at every point of B_0; here no critical point of $G(x, y)$ lies on B_0, for $G(x, y)$ is greater than zero in R_0 and no branch of the locus $G(x, y) = 0$ lies in R_0; thus no critical point of $H(x, y)$ lies on B_0; each point of B_0 is met by a single arc of a single geometric locus (2), and the arc cuts B_0 orthogonally. Thus in the original configuration each point of B is met by a single arc of a single geometric locus; each branch of each geometric locus (2) meets B in a single point.

If B no longer consists of a finite number of Jordan curves, the qualitative nature of each locus (1) is unchanged, but if B contains an infinity of mutually disjoint components, an infinite number of critical points of $G(x, y)$ lie in R, and the number of loci (1) with multiple points is not finite. A locus (2) which commences at infinity need not end in a single point of B, but may oscillate. Otherwise the qualitative nature of the individual loci (1) and (2) and of the totality of loci is unchanged.

Green's function with pole at infinity for the infinite region bounded by the locus (1) is obviously $G(x, y) - \mu$; the functions $H(x, y)$ and the loci (2) are the same (with properly restricted domains of definition) for the two regions.

§7.1.2. Variable regions. If R is an infinite region with finite boundary B possessing a Green's function $G(x, y)$ with pole at infinity, we denote generically by B_μ the locus (1) in R, and denote generically by R_μ the subregion $G(x, y) > \mu$ of R including the point at infinity. In view of the fact that $G(x, y) - \mu$ is Green's function for R_μ with pole at infinity, we prove

LEMMA 1. *Let R be an infinite region with finite boundary B possessing a Green's function $G(x, y)$ with pole at infinity. Let a proper subregion R' of R be an infinite* *region with finite boundary B' possessing a Green's function $G'(x, y)$ with pole at infinity. Then at every point of R' we have*

$$G(x, y) - G'(x, y) > 0.$$

The function $G(x, y) - G'(x, y)$ is harmonic at every point of R', including the point at infinity when the function is suitably defined there; the function is continuous and non-negative at every point of B', and at the points of B' interior to R (which necessarily exist) is positive, hence this function is positive throughout R'.

A related result is

LEMMA 2. *Under the conditions of Lemma 1, suppose that B' lies in the set $R + B$ exterior to R_δ. Then in R_δ we have*

$$G(x, y) - \delta < G'(x, y) < G(x, y). \tag{3}$$

Both parts of (3) follow from Lemma 1, for R' is a proper subregion of R, and R_δ is a proper subregion of R'.

Under the hypothesis of Lemma 2, for $\mu \geqq \delta$ we have in R_δ

$$G(x, y) - (\delta + \mu) < G'(x, y) - \mu < G(x, y) - \mu; \tag{4}$$

the locus B'_μ lies in R', so at any point of B'_μ we have by Lemma 1 the inequality $G(x, y) > \mu$, whence B'_μ lies in R_μ and in R_δ; at any point of B'_μ we have by (4) the inequality $G(x, y) < \delta + \mu$, so B'_μ lies exterior to $R_{\delta+\mu}$; thus *B'_μ lies between B_μ and $B_{\delta+\mu}$*; the locus B_μ lies in R', so by Lemma 1 the locus B_μ lies exterior to R'_μ, and by (4) the locus $B_{\delta+\mu}$ lies in R'_μ; thus *B'_μ separates B_μ and $B_{\delta+\mu}$*.

It is clear from Lemma 2 that if R is given, and if a sequence of infinite subregions $R^{(n)}$ with finite boundaries $B^{(n)}$ approaches R in the sense that corresponding to an arbitrary $\delta(>0)$ there exists N_δ such that $n > N_\delta$ implies that $B^{(n)}$ lies in R but not in R_δ, then Green's function $G^{(n)}(x, y)$ for $R^{(n)}$ with pole at infinity approaches $G(x, y)$ uniformly in every R_μ. It is then a consequence of (4) that for every μ the locus $B_\mu^{(n)}$ approaches B_μ uniformly.

If R is now an *arbitrary* infinite region with finite boundary B, there exists a sequence of infinite subregions $R^{(n)}$ with finite boundaries $B^{(n)}$ which approach R monotonically in the sense that $R^{(n+1)}$ contains $B^{(n)}$, and each point of R lies in some $R^{(n)}$. The regions $R^{(n)}$ may be chosen so that Green's function $G^{(n)}(x, y)$ for $R^{(n)}$ with pole at infinity exists, and then by Lemma 1 the sequence $G^{(n)}(x, y)$ is a monotonically increasing sequence at every point (x, y) of R. If Green's function $G(x, y)$ for R with pole at infinity exists, it follows from Lemma 2 that $G^{(n)}(x, y)$ approaches $G(x, y)$ uniformly in every R_μ; if Green's function for R with pole at infinity does not exist, it is still true, and follows from Harnack's Theorem (§1.1.2), that the sequence $G^{(n)}(x, y)$ converges in R to a function called the *generalized* Green's function $G(x, y)$ for R with pole at infinity, which indeed may be the infinite constant,* but otherwise is harmonic in R.

In the sequel we shall study the critical points of both Green's functions and generalized Green's functions.

§7.1.3. A formula for Green's function. In order to study in more detail the nature of Green's function and its level loci, we derive a formula [Hilbert 1897 for a simply-connected region; Walsh and Russell 1934 for a multiply-connected region] for the representation of Green's function $G(x, y)$, assumed to exist, with pole at infinity for an infinite region R whose boundary B is finite. Let $P: (x_0, y_0)$ be an arbitrary point of R, let r denote distance measured from P, let Γ be a locus B_μ passing through no critical point of $G(x, y)$ and such that P lies in R_μ, and let Γ_1 denote Γ plus a circle Γ_2 whose center is P and which contains Γ in its interior. Then we have

$$G(x_0, y_0) = \frac{1}{2\pi}\int_{\Gamma_1}\left(\log r\,\frac{\partial G}{\partial \nu} - G\,\frac{\partial \log r}{\partial \nu}\right) ds, \tag{5}$$

where ν denotes interior normal for the region bounded by Γ_1; indeed this formula is valid if $G(x, y)$ is replaced by an arbitrary function harmonic in the closed region bounded by Γ_1.

The function $G(x, y)$ is the constant μ on Γ, and P lies in the infinite region bounded by Γ, so we have

$$\int_\Gamma G\,\frac{\partial \log r}{\partial \nu}\,ds = \mu\int_\Gamma \frac{\partial \log r}{\partial \nu}\,ds = 0.$$

On Γ_2 we set $G \equiv G_1 + \log r$; the function $\log r$ is constant on Γ_2, and the function G_1 is harmonic on and exterior to Γ_2 even at infinity, so we have

$$\int_{\Gamma_2} \log r\,\frac{\partial G_1}{\partial \nu}\,ds = \log r\int_{\Gamma_2}\frac{\partial G_1}{\partial \nu}\,ds = 0;$$

the latter equation is familiar for a function harmonic in a finite region, and follows in the present situation by an inversion with P as center. By this same inversion and Gauss's mean value theorem we write

$$-\frac{1}{2\pi}\int_{\Gamma_2} G_1\frac{\partial \log r}{\partial \nu}\,ds = \frac{1}{2\pi r}\int_{\Gamma_2} G_1\,ds = G_1(\infty).$$

The value $G_1(\infty) = g$ is independent of the point P, so (5) can now be written in the form

$$G(x_0, y_0) = \frac{1}{2\pi}\int_\Gamma \log r\frac{\partial G}{\partial \nu} ds + g, \tag{6}$$

* As an example here, one may take the function (11) below and allow μ_1 to approach zero.

where g is a constant, independent of (x_0, y_0).

On $\Gamma : B_\mu$ we have $\partial G/\partial s = 0$; but Γ passes through no critical point of $G(x, y)$, so we must have $\partial G/\partial \nu \neq 0$ on Γ; the vanishing of both $\partial G/\partial s$ and $\partial G/\partial \nu$ at a point implies the vanishing of both $\partial G/\partial x$ and $\partial G/\partial y$ at that point. In R_μ we have $G(x, y) > \mu$, whence $\partial G/\partial \nu > 0$ on Γ. By way of abbreviation we write on Γ

$$(7) \qquad \frac{1}{2\pi}\frac{\partial G}{\partial \nu}\, ds = \frac{-1}{2\pi}\frac{\partial H}{\partial s}\, ds = d\sigma, \qquad 0 \leqq \sigma \leqq \sigma_1 = \int_\Gamma d\sigma;$$

here the function $H(x, y)$ is conjugate to $G(x, y)$ in R, the values of σ can be distributed over Γ continuously except for one finite jump on each of the Jordan curves composing Γ, and we have

$$2\pi\sigma_1 = \int_\Gamma \frac{\partial G}{\partial \nu}\, ds = -\int_{\Gamma_2} \frac{\partial G}{\partial \nu}\, ds = -\int_{\Gamma_2} \frac{\partial \log r}{\partial \nu}\, ds = 2\pi.$$

Thus we write (6) in the form

$$(8) \qquad G(x_0, y_0) = \int_\Gamma \log r\, d\sigma + g, \qquad 0 \leqq \sigma \leqq 1,$$

for (x_0, y_0) in R_μ, with $d\sigma > 0$.

In the particular case that R *is bounded by a Jordan configuration* (§1.1.1), the integral in (8), which is essentially a Stieltjes integral over $\Gamma : B_\mu$ of the function $\log r$ continuous on B_μ except for the values of σ each the common end-point of two σ-integrals corresponding to two different components of B_μ, can be taken over B. Indeed for sufficiently small positive μ, the locus B_μ consists of disjoint Jordan curves, in number equal to the number of components of B; the loci $H(x, y) =$ const are disjoint Jordan arcs in the closed region or regions between R and B_μ, and are terminated by B and B_μ; for (x_0, y_0), fixed in R_μ, the function $\log r$ is continuous uniformly with respect to σ as a function of μ on the totality of these Jordan arcs; the second member of (8) is a function of μ which does not change with μ, so we can allow μ to approach zero:

$$(9) \qquad G(x_0, y_0) = \int_B \log r\, d\sigma + g, \qquad 0 \leqq \sigma \leqq 1,$$

for (x_0, y_0) in R, with $d\sigma > 0$. Of course this integral is to be taken twice over a Jordan arc if R abuts on both sides.

If t denotes the running variable in (9), and we set $z = x + iy$, we see by inspection that differentiation under the integral sign is valid and that a locally single-valued function $H(x, y)$ conjugate to $G(x, y)$ in R is given by

$$H(x, y) = \int_B \arg (z - t)\, d\sigma, \qquad 0 \leqq \sigma \leqq 1,$$

where $\sigma = \sigma(t)$ and t lies on B; we have [Walsh 1948c] by differentiation of the corresponding analytic function $F(z) = G(x, y) + iH(x, y) = \int_B \log (z - t)\, d\sigma + g$, whose derivative is single-valued, the first part of

THEOREM 1. *Let R be an infinite region whose boundary is a finite Jordan configuration B, and let $G(x, y)$ be Green's function for R with pole at infinity. Then the critical points of $G(x, y)$ interior to R are the zeros interior to R of the function*

$$F'(z) = \int_B \frac{d\sigma(t)}{z - t}, \qquad 0 \leqq \sigma \leqq 1. \tag{10}$$

Consequently the critical points of $G(x, y)$ interior to R are precisely the positions of equilibrium in the field of force due to a spread σ of positive matter on B, where the matter repels with a force proportional to the mass and inversely proportional to the distance.

The second part of Theorem 1 follows merely by taking the conjugate in (10). Theorem 1 is thus the analog of Gauss's Theorem (§1.2).

We shall have occasion to use the analog of §1.5.1 Lemma 1:

LEMMA. *Let σ be a continuous spread of positive matter over a finite number of Jordan arcs J in a circular region C, and let P be a point exterior to C. Then the corresponding force at P is equal to the force at P due to the same mass concentrated at a suitable interior point of C.*

After inversion in P, this is essentially the proposition that if a distribution of mass lies in the closed interior of a circle C, so does its center of gravity; thus the conclusion is true even if the given matter consists in part of a continuous distribution and in part of discrete particles. We note that under the conditions of the Lemma the equivalent particle is *interior* to C. Moreover we can allow P to lie on the boundary of C, not on J; it is still true that the equivalent particle lies in the region C, and actually *interior* to C unless J lies on the boundary of the region C.

Theorem 1 is in accord with our general theory (§1.6.2) of fields of force, for $F'(z)$ in (10) is the logarithmic derivative of the function $e^{F(z)}$; at any point in the field the force has the direction normal to $| e^{F(z)} | =$ const, or $G(x, y) =$ const, acting in the sense of increasing $G(x, y)$, and the lines of force are the loci $\arg e^{F(z)} =$ const, or $H(x, y) =$ const. Indeed it is readily shown directly that the conjugate of $F'(z)$ is the gradient of $G(x, y)$; at an arbitrary point $z_0 = x_0 + iy_0$ of R not a critical point of $G(x, y)$, we compute dF/dz, where $dz = \epsilon\, d\nu$ ($|\epsilon| = 1$) is taken normal to the locus $G(x, y) = G(x_0, y_0)$ in the sense of increasing $G(x, y)$; we have $dF/dz = (\partial G/\partial \nu)/\epsilon$, $\partial G/\partial \nu > 0$, $\partial H/\partial \nu = 0$, whence $\arg [F'(z)] = -\arg \epsilon$, $| F'(z) | = \partial G/\partial \nu$. This proof of the conclusion that the conjugate of $F'(z)$ is the gradient of $G(x, y)$ obviously applies generally to the field defined by the conjugate of an arbitrary analytic function $F'(z)$, where $F = G + iH$.

§7.1.4. Level curves and lemniscates. The relation between level curves and lemniscates is far deeper than the analogy indicated in §7.1.1. Indeed, if we set

$$p(z) = (z - \alpha_1)(z - \alpha_2) \cdots (z - \alpha_m),$$

it may be verified at once that Green's function with pole at infinity for the region $R: | p(z) | > \mu_1$ (>0) is

$$(1/m) \log [\, | p(z) | / \mu_1]. \tag{11}$$

The region R is itself bounded by a lemniscate, and the level loci (1) are likewise lemniscates $| p(z) | =$ const.

An arbitrary level locus of Green's function can be approximated by a lemniscate [Hilbert 1897, Fekete 1933 Walsh and Russell 1934]:

THEOREM 2. *Let R be an infinite region with finite boundary B possessing a Green's function $G(x, y)$ with pole at infinity. An arbitrary level locus B_μ can be approximated by a lemniscate, in the sense that if δ is arbitrary, $0 < \delta < \mu$, there exists a lemniscate λ whose poles are exterior to $R_{\mu-\delta}$, such that the infinite region bounded by λ contains $R_{\mu+\delta}$ and is contained in $R_{\mu-\delta}$. If B_μ consists of m disjoint Jordan curves, so also does λ.*

We assume, as we may do with no loss of generality, that B consists of a finite number of disjoint Jordan curves. In equation (9) we then express the integral as the limit of a Riemann sum formed for equidistant values of σ:

$$\begin{aligned} G^{(n)}(x, y) &= g + (\log r_{1n} + \log r_{2n} + \cdots + \log r_{nn})/n, \\ G(x, y) &= \lim_{n \to \infty} G^{(n)}(x, y), \end{aligned} \tag{12}$$

where r_{kn} indicates distance to (x, y) in R measured from the point of B at which $\sigma = k/n$; the limit is uniform for (x, y) in any closed finite or infinite region in R. For sufficiently large n we have from (12)

$$G(x, y) - \delta < G^{(n)}(x, y) < G(x, y) + \delta, \tag{13}$$

for (x, y) in the closed region $R_{\mu-\delta} + B_{\mu-\delta}$. Thus we have in that closed region

$$G(x, y) - (\mu + \delta) < G^{(n)}(x, y) - \mu < G(x, y) - (\mu - \delta), \tag{14}$$

and the locus

$$G^{(n)}(x, y) = \mu, \qquad r_{1n} r_{2n} \cdots r_{nn} = e^{n(\mu - g)}, \tag{15}$$

for a suitable value of n is the lemniscate λ desired. It follows from (14) that the locus $B_{\mu-\delta}$ lies interior to the lemniscate, and that the locus $B_{\mu+\delta}$ lies exterior to the lemniscate. Thus the lemniscate λ consists at least in part of precisely one Jordan curve in each of the m annular regions composing $R_{\mu-\delta} - R_{\mu+\delta} - B_{\mu+\delta}$, provided (as we now assume) B_μ consists of m disjoint Jordan curves and δ is chosen so small that the open set $R_{\mu-\delta} - R_{\mu+\delta} - B_{\mu+\delta}$ separates into precisely m disjoint annular regions. By the general properties of lemniscates, no Jordan curve interior to the m Jordan curves of λ mentioned can belong to λ. The function $G^{(n)}(x, y)$ has no finite singularity in $R_{\mu-\delta}$, so no other Jordan curve belongs to λ. The proof is complete.

§7.2. Geometry of level loci. The fundamental result in the study of the geometric shape of level loci is the analog of Lucas's Theorem, to which we now turn.

§7.2.1. Analog of Lucas's Theorem. This analog may be given several forms, of which we choose [Walsh, 1933a]

THEOREM 1. *Let R be an infinite region whose boundary B is a finite Jordan configuration. Then all the critical points in R of Green's function $G(x, y)$ for R with pole at infinity lie in the smallest closed convex set Π containing B; no critical points lie in R on the boundary of Π unless all points of B lie on a line.*

No point interior to R but exterior to Π can be a position of equilibrium in the field of force of §7.1.3 Theorem 1, so no such point can be a critical point. No boundary point of Π in R can be a position of equilibrium unless Π degenerates to a line segment.

The simplicity of Theorem 1 and its proof is due to the choice of the finite boundary B as a Jordan configuration; if a more general boundary is admitted, a less general result appears:

COROLLARY 1. *Let R be an infinite region with finite boundary B which possesses a Green's function $G(x, y)$ for R with pole at infinity. Then all critical points of $G(x, y)$ in R lie in the smallest closed convex point set Π containing B.*

We choose an arbitrary B_μ, and denote by Π_μ the smallest convex region containing B_μ. It follows from Theorem 1 (or from §7.1.4 equation (12) and Lucas's Theorem) that the critical points of $G(x, y)$ in R_μ lie in Π_μ. The regions Π_μ decrease monotonically with μ, and by §7.1.1 approach Π as μ approaches zero, so Corollary 1 follows.

A similar limiting process yields

COROLLARY 2. *Let R be an infinite region with finite boundary B, and let the generalized Green's function $G(x, y)$ for R with pole at infinity not be the infinite constant. Then all critical points of $G(x, y)$ in R lie in the smallest closed convex point set Π containing B.*

In the limiting case of Theorem 1 critical points may lie on the boundary of Π:

COROLLARY 3. *Let R be an infinite region whose boundary B is finite and lies on the axis of reals, and let Green's function $G(x, y)$ exist for R with pole at infinity. Then all critical points of $G(x, y)$ in R lie on the smallest segment Π of the axis of reals which contains B. On any open segment of Π bounded by points of B and containing no points of B lies precisely one (a simple) critical point of $G(x, y)$.*

The first part of Corollary 3 follows from Corollary 1. Any segment of Π in R bounded by points of B contains a relative maximum of $G(x, 0)$, for $G(x, 0)$

vanishes at the end-points and is positive in the interior points; this relative maximum is a critical point of $G(x, y)$, for at this maximum of $G(x, 0)$ we clearly have $\partial G/\partial x = 0$, and symmetry shows that $\partial G/\partial y = 0$. To prove the remaining part of Corollary 3, assume a segment $\alpha < x < \beta$ to be bounded by points of B, to contain no points of B, and to contain more than one critical point of $G(x, y)$; we shall reach a contradiction. A segment S': $(\alpha <)\ \alpha' < x < \beta'\ (<\beta)$ also contains more than one critical point of $G(x, y)$. Choose μ, $0 < \mu < \min [G(\alpha', 0), G(\beta', 0)]$, so that B_μ has no multiple points; then B_μ consists of a finite number of disjoint analytic Jordan curves J. The segment S' lies in R_μ. Of course B_μ is symmetric in the axis of reals, and each component curve J must cut that axis in precisely two points and contain in its closed interior the segment joining them. On any segment S of the axis of reals in R_μ between, and bounded by points of, successive curves J, the function $G(x, y) - \mu$ is positive and continuous and vanishes at both ends of the segment, hence has on S a maximum which is a critical point of $G(x, y)$. In R_μ lie a total number of critical points equal to the number of curves J diminished by unity, so no segment S contains more than one critical point; this contradiction completes the proof of Corollary 3.

Indeed, the reasoning just given needs little modification to prove

COROLLARY 4. *Let R be an infinite region whose boundary B is finite, and consists of mutually disjoint continua each of which is symmetric in the axis of reals. Let Green's function $G(x, y)$ exist for R with pole at infinity. Then all critical points of $G(x, y)$ in R lie on the axis of reals; each finite open segment of that axis in R bounded by points of B contains precisely one (a simple) critical point, and each critical point lies on such a segment.*

The latter part of Corollary 4 follows by the method of proof of Corollary 3. If a critical point (x_1, y_1) of $G(x, y)$ were to lie in R not on the axis of reals, or not on an open finite segment of that axis in R bounded by points of B, that critical point would lie exterior to some locus $G(x, y) = \mu > 0$ consisting of a finite number q of Jordan curves each symmetric in the axis of reals; thus Green's function $G(x, y) - \mu$ for the region exterior to all those curves would have $q - 1$ real critical points and at least one other, namely at (x_1, y_1), which is impossible.

Throughout our present discussion three distinct methods suggest themselves: (i) direct use of the field of force of §7.1.3 Theorem 1, where the integral is taken over the boundary B of the given region; this is the method used in proving Theorem 1; (ii) use of the field of force of §7.1.3 Theorem 1, where the integral is now taken over an auxiliary boundary B_μ near B, and then B_μ is allowed to approach B; it is to be noted that the function $F'(z)$ whose conjugate represents the field of force is independent of μ; this method is used in the proof of Corollary 1; (iii) approximation of a level locus B_μ by a lemniscate as in §7.1.4 Theorem 2, and use of results on the critical points of a polynomial; this method can also be used in the proof of Corollary 1. We shall not discard any of these methods, for each has its own advantages.

A "method of continuity" may be useful in connection with Theorem 1 if the Jordan arcs composing B are not themselves convex:

COROLLARY 5. *Under the conditions of Theorem* 1, *let a point P of R lie in a Jordan region belonging to R bounded by a line segment belonging to the boundary of Π plus a Jordan arc belonging to B. Then P is not a critical point of $G(x, y)$.*

We can assume here that B is composed of a finite number of mutually disjoint analytic Jordan curves, for P lies on a line segment in R (say parallel to the given line segment) bounded by points of a Jordan arc of B, hence lies on a line segment in R_μ bounded by points of one of the Jordan curves composing B_μ, where μ is suitably chosen; thus P satisfies the hypothesis of Corollary 5 with R and B replaced by R_μ and B_μ.

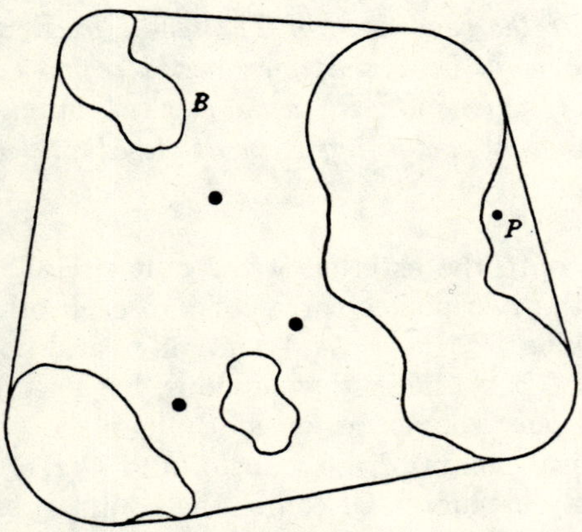

Fig. 18 illustrates §7.2.1 Corollary 5 to Theorem 1

We vary B and thus R monotonically, by replacing the given Jordan arc A_0 by a variable Jordan arc A which commences in the position of A_0 and varies continuously so as to take the position of the given line segment S_0 belonging to the boundary of Π;* the arc A is to remain in the finite Jordan region bounded by A_0 and S_0. Denote by R' the final region R. Then (§7.1.2 Lemma 2) Green's function $G(x, y)$ for the variable region R varies continuously with A. During this variation the closed convex region Π depending on R does not change with R. The critical points of $G(x, y)$ in R vary continuously with A by Hurwitz's Theorem, and by Theorem 1 none ever lies on S_0 in R. The number of components of B does not change during the process, so no critical point enters or leaves R; by the continuity of the critical points it is of course not possible for one critical point to enter R and another to leave R simultaneously. Consequently no critical point enters or leaves R', so originally P is not a critical point.

* For instance A may be chosen in the finite Jordan region bounded by A_0 and S_0 as a hypercycle joining the end-points of A_0.

The reasoning used in the proof of Corollary 5 applies also in numerous situations in the sequel, but the corresponding conclusions are mostly left to the reader.

A second proof of Corollary 5 can be given as follows. At points of B_μ chosen as before the force in the field of §7.1.3 Theorem 1 is normal to B_μ, directed into R_μ; at points of S_0 the force has a non-zero normal component directed away from the side of S_0 on which P lies. As z traces the boundary of the subregion of R_μ bounded by part of S_0 and an arc of B_μ, there is no net increase in the direction angle of the force, nor in arg $[F'(z)]$, notation of §7.1.3. Thus the subregion contains no zero of $F'(z)$ nor critical point of $G(x, y)$.

Theorem 1 is not invariant under one-to-one conformal transformation of R with the point at infinity fixed, and a new result can be obtained by any such transformation; in particular we prove

COROLLARY 6. *Under the conditions of Theorem 1 let R_1 whose boundary consists of more than one point be a simply-connected region containing R, and let $G_1(x, y)$ be Green's function for R_1 with pole at infinity. Then any locus $\Pi_1: G_1(x, y) = \mu(>0)$ in R_1 containing B in its closed interior contains all critical points of $G(x, y)$ in R.*

Map the region R_1 onto the exterior of the unit circle $|w| = 1$ so that the points at infinity in the two planes correspond to each other. The image of B lies in the closed annulus $1 \leqq |w| \leqq e^\mu$, and all critical points in the image of R of the transform of $G(x, y)$ lie in that annulus, by Theorem 1, which implies the Corollary. Indeed, any convex region in the w-plane containing the image of B contains all critical points of the transform of $G(x, y)$, and the Corollary may be correspondingly extended. Of course the boundary of R_1 may be chosen a component of B, or any continuum (not a single point) in the complement of R.

§7.2.2. Center of curvature. The property of Π expressed in Theorem 1 with reference to the location of critical points represents but a part of the importance of Π; we prove [Walsh, 1935]:

THEOREM 2. *Let R be an infinite region whose boundary B is finite and which possesses a Green's function $G(x, y)$ for R with pole at infinity, and let Π denote the smallest convex point set which contains B. If $P: (x_0, y_0)$ is an arbitrary point of R, then the normal at P to the curve $G(x, y) = G(x_0, y_0)$ in the sense of decreasing $G(x, y)$ must intersect Π. If A is the point nearest to P of intersection of this normal with Π, the radius of curvature at P of the curve $G(x, y) = G(x_0, y_0)$ is not less than PA.*

Corollary 1 to Theorem 1 is essentially included in Theorem 2, for at a critical point P of $G(x, y)$ of order m the level locus has $m + 1$ branches with tangents equally spaced. If ϵ is either positive or negative but numerically small, the locus $G(x, y) = G(x_0, y_0) + \epsilon$ near P has $m + 1$ branches, in the set of regions

$G(x, y) > G(x_0, y_0)$ or $G(x, y) < G(x_0, y_0)$ according as ϵ is positive or negative, and the normals mentioned in Theorem 2 at suitably chosen points near P are $m + 1$ half-lines through P approximately equally spaced. All of these half-lines intersect Π, for positive and negative ϵ, so P must lie in Π.

It is a consequence of Theorem 2 that for μ sufficiently large the locus $G(x, y) = \mu$ in R consists of a single Jordan curve; any such locus wholly exterior to Π has this property, and is defined by an equation $G(x, y) = \mu > \max\,[G(x, y) \text{ on } \Pi]$.

It is sufficient to prove Theorem 2 for the case that B consists of a finite number of disjoint analytic Jordan curves. Indeed, if the theorem is proved in that restricted case, to treat the more general case we need merely notice that if $P\colon(x_0, y_0)$ lies in R exterior to Π, then P lies exterior to a suitably chosen smallest convex set Π_μ containing the locus B_μ, where B_μ has no multiple points. The normal at P to the curve $G(x, y) = G(x_0, y_0)$ cuts every $\Pi_{\mu'}$, where we suppose $0 < \mu' < \mu$ and suppose $B_{\mu'}$ to have no multiple point, so this normal cuts Π; the radius of curvature at P is not less than $PA_{\mu'}$, where $A_{\mu'}$ is the point nearest P of intersection of the normal with $\Pi_{\mu'}$, so the radius of curvature is not less than PA.

The first part of Theorem 2 follows from §7.1.3 Theorem 1, assuming B a finite number of disjoint analytic Jordan curves, for if P is exterior to Π, the set Π subtends at P a certain closed sector S of angle less than π. The total force at P is not zero and is the limit of a sum of vectors terminating in P whose initial points lie in S, so the total force is represented by a vector terminating in P with its initial point in S. The line of action of this force intersects Π, and the direction of this force is known to be that of the line of force at P, orthogonal to the level locus of $G(x, y)$ through P.

To prove the second part of Theorem 2, we continue to assume B to consist of a finite number of disjoint analytic Jordan curves. In §7.1.3 equation (9) we denote by (α, β) the variable point on B, and suppose B lies in the half-plane $\beta \geqq 0$ with A real; we choose $P\colon(x_0, y_0)$ as the point $(0, b)$, $b < 0$. We have

$$r^2 = (\alpha - x)^2 + (\beta - y)^2, \qquad r\frac{\partial r}{\partial x} = x - \alpha, \qquad r\frac{\partial r}{\partial y} = y - \beta. \tag{1}$$

The usual formula for the ordinate of the center of curvature of the curve $G(x, y) = G(x_0, y_0)$ at (x_0, y_0) is

$$Y = y_0 + \frac{1 + y_0'^2}{y_0''} = y_0 - \frac{G_y(G_x^2 + G_y^2)}{G_y^2 G_{xx} - 2G_xG_yG_{xy} + G_x^2 G_{yy}}. \tag{2}$$

By means of (1) and §7.1.3 equation (9) we write the formulas for the derivatives at P; all integrals are to be taken over B:

$$\begin{aligned} G_x &= \int \frac{-\alpha}{r^2}\,d\sigma, \qquad G_y = \int \frac{b - \beta}{r^2}\,d\sigma, \qquad G_{xx} = \int \left[\frac{1}{r^2} - \frac{2\alpha^2}{r^4}\right] d\sigma, \\ G_{xy} &= \int \frac{2\alpha(b - \beta)}{r^4}\,d\sigma, \qquad G_{yy} = \int \left[\frac{1}{r^2} - \frac{2(b - \beta)^2}{r^4}\right] d\sigma. \end{aligned} \tag{3}$$

We reduce the last member of (2) to a common denominator. The combined numerator can be written

$$\begin{aligned}(4)\qquad &\left(\int \frac{b-\beta}{r^2}\,d\sigma\right)^2 \int \left[\frac{\beta}{r^2} - \frac{2b\alpha^2}{r^4}\right] d\sigma \\ &+2b\left(\int \frac{\alpha}{r^2}\,d\sigma\right)\left(\int \frac{b-\beta}{r^2}\,d\sigma\right)\int \frac{2\alpha(b-\beta)}{r^4}\,d\sigma \\ &+\left(\int \frac{\alpha}{r^2}\,d\sigma\right)^2 \int \left[\frac{\beta}{r^2} - \frac{2b(b-\beta)^2}{r^4}\right] d\sigma,\end{aligned}$$

which is the sum of two terms involving the integral of β/r^2 (and which are clearly non-negative) plus the product of $-2b$ and

$$\int \left[M\frac{\alpha}{r^2} - N\frac{b-\beta}{r^2}\right]^2 d\sigma, \qquad M = \int \frac{b-\beta}{r^2}\,d\sigma', \qquad N = \int \frac{\alpha}{r^2}\,d\sigma',$$

which is also non-negative. Hence (4) is non-negative. If the denominator in (2) is positive, we now have $Y \geqq 0$, so the radius of curvature is not less than PA.

If the denominator in (2) is negative, we compute $2b - Y$. We use the same common denominator as before; the numerator for $2b - Y$ reduces, by means of the equation $r^2 = \alpha^2 + (b-\beta)^2$, to

$$\begin{aligned}&\left(\int \frac{b-\beta}{r^2}\,d\sigma\right)^2 \int \frac{2b(b-\beta)^2 - \beta r^2}{r^4}\,d\sigma \\ &+2b\left(\int \frac{\alpha}{r^2}\,d\sigma\right)\left(\int \frac{b-\beta}{r^2}\,d\sigma\right)\int \frac{2\alpha(b-\beta)}{r^4}\,d\sigma \\ &+\left(\int \frac{\alpha}{r^2}\,d\sigma\right)^2 \int \frac{2b\alpha^2 - \beta r^2}{r^4}\,d\sigma,\end{aligned}$$

which is the sum of two terms involving the integral of $\beta r^2/r^4$ (and which are clearly non-positive) plus the product of $2b$ and

$$\int \left[N\frac{\alpha}{r^2} + M\frac{b-\beta}{r^2}\right]^2 d\sigma;$$

this product is also non-positive, so we have $2b - Y \geqq 0$, $Y \leqq 2b$, and the radius of curvature is again numerically not less than PA.

We note that the numerator of the fraction in the last member of (2) cannot vanish, for by (3) we have $G_y < 0$. Thus the center of curvature is to be considered at infinity when the denominator in (2) vanishes, so in every case the radius of curvature is not less than PA, and Theorem 2 is established.

Another form for the last part of Theorem 2 is that if B lies in a half-plane, then the center of curvature at $P\colon (x_0, y_0)$ of the curve $G(x, y) = G(x_0, y_0)$ lies either in that half-plane or in the reflection in P of that half-plane.

We add the remark, which is easily proved from the formulas already given, that as P becomes infinite the center of curvature approaches the point whose coordinates are

$$\int \alpha \, d\sigma \Big/ \int d\sigma, \qquad \int \beta \, d\sigma \Big/ \int d\sigma,$$

which is the center of gravity of the distribution σ on B and is independent of the manner in which P becomes infinite.

§7.3. Symmetry in axis of reals. Comparison of §7.2.1 Theorem 1 with Corollary 1 to that theorem shows that the former is more precise than the latter concerning critical points on the boundary of Π, but that the latter applies to less restrictive regions R than the former; however, the theorem can be readily used to prove the corollary and in that sense can be considered more general. This difference is typical of many other situations. Henceforth for simplicity we consistently state explicit results which literally are restricted, but which in reality can be applied in various more general situations, including those of the generalized Green's function.

§7.3.1. Analog of Jensen's Theorem. We establish [Walsh, 1933a]

THEOREM 1. *Let R be an infinite region whose boundary B is a finite Jordan configuration, and let $G(x, y)$ be Green's function for R with pole at infinity. Let B be symmetric in the axis of reals and let Jensen circles be constructed with diameters the segments joining all possible pairs of points of B symmetric in the axis of reals. Then every non-real critical point of $G(x, y)$ in R lies on or interior to at least one of these Jensen circles.*

We mention explicitly that these Jensen circles are not merely circles with centers on the axis of reals and which contain B in their closed interiors.

Again we use the field of force of §7.1.3 Theorem 1, defined by the integral

$$\int_B \frac{d\sigma(t)}{\bar{z} - \bar{t}}, \tag{1}$$

where $d\sigma$ is positive on B, and P: z lies in R. We write the integral (1) as an integral extended over the half B_1 of B which lies above the axis of reals, with the obvious convention for parts of B on that axis:

$$\int_{B_1} \left[\frac{1}{\bar{z} - \bar{t}} + \frac{1}{\bar{z} - t} \right] d\sigma(t); \tag{2}$$

it follows from the definition of σ in terms of the conjugate of $G(x, y)$ that $d\sigma$ taken in the positive sense on B is symmetric in the axis of reals. The integrand in (2) is precisely the force at z due to unit positive particles at t and $\bar{t}$ and is known (§1.4.1 Lemma) to be a vector which is horizontal if z lies on the Jensen circle for t and $\bar{t}$, and which has a non-zero vertical component directed away from the axis of reals if z lies exterior to that Jensen circle. Thus if P is non-real and lies exterior to all Jensen circles for B, the integrand in (2) is a vector which always has a component not zero away from the axis of reals, even if t is real, so the integral cannot vanish, P cannot be a position of equilibrium, and Theorem 1 is established.

We proved in §2.2 that the locus of points t such that a given non-real point P lies on the Jensen circle for t and $\bar{t}$ is the equilateral hyperbola H with P as vertex and the axis of reals as conjugate axis. It follows that for a non-real point P in R interior to no Jensen circle the integral (2) cannot vanish unless all Jensen circles pass through P, that is, unless every point of B lies on H.

COROLLARY. *In Theorem 1 a non-real point P in R interior to no Jensen circle can be a critical point only if all points of B lie on the equilateral hyperbola H with P as vertex and the axis of reals as conjugate axis.*

On the other hand, it is of course true that if all points of B lie on H, if for instance B is symmetric in the transverse axis of H, and if P is not a point of B, then P is a critical point of $G(x, y)$ and lies on all Jensen circles for pairs of points of B.

In connection with Jensen circles from a strictly geometric point of view, the Lemma of §3.8 yields a useful result which may be applicable to B or various subsets of B in Theorem 1:

If a point set B_1 symmetric in the axis of reals lies in the closed interior of the ellipse

$$x^2 + my^2 = m, \qquad m > 0, \tag{3}$$

then the closed interiors of all Jensen circles for pairs of points of B_1 lie in the closed interior of the ellipse

$$x^2 + (m + 1)y^2 = m + 1. \tag{4}$$

The following independent proposition is immediate:

If a point set B_1 symmetric in the axis of reals lies in the closed interior of a square S one of whose diagonals lies along the axis of reals, then the closed interiors of all Jensen circles for pairs of points of B_1 lie in the circle circumscribing S.

Let us denote generically by $J(m, \lambda)$ the convex Jordan curve formed by two arcs of the ellipse (3) and by a segment of each of the four lines with slopes $\pm\lambda(0 < \lambda)$ tangent to the ellipse, where each segment is intercepted between the point of tangency and the axis of reals. Combination of the two preceding results will yield:

If a point set B_1 symmetric in the axis of reals lies in the closed interior of $J(m, 1)$, then the closed interiors of all Jensen circles for pairs of points of B_1 lie in the closed interior of the ellipse (4).

A Jensen circle C_1 whose diameter is a vertical chord of (3) joining the points of contact of two of the tangents lies in the closed interior of (4), so any Jensen circle whose diameter is a vertical segment joining points of $J(m, 1)$ on those tangents lies in the closed interior of (4). Reciprocally, it is readily seen that the Jensen circle C_1, the curve $J(m, 1)$, and the ellipse (4) all meet on the axis of reals, from which it follows (it is convenient to study all circles with a given center) that all circles which lie in the closed interior of (4) and whose centers lie on the axis of reals are Jensen circles for pairs of points lying in the closed interior of $J(m, 1)$.

We omit the proof of the following proposition, of which the foregoing is the limiting case $\lambda = 1$; the further limiting case $m = 0$ is not excluded, and the reciprocal, in the sense just considered, is also true:

If a point set B_1 symmetric in the axis of reals lies in the closed interior of $J(m, \lambda)$, $0 < \lambda < 1$, then the closed interiors of all Jensen circles for pairs of points of B_1 lie in the closed interior of $J(m + 1, \lambda(1 - \lambda^2)^{-1/2})$.

The latter Jordan curve meets $J(m, \lambda)$ on both coordinate axes. If C_1 is a Jensen circle whose diameter is a vertical chord of $J(m, \lambda)$ joining the points of contact with the ellipse (3) of two of the tangents whose segments belong to $J(m, \lambda)$, then C_1 is tangent to the ellipse (4) at points of tangency of the ellipse and line segments belonging to $J(m + 1, \lambda(1 - \lambda^2)^{-1/2})$.

These results have obvious applications to the situation of §3.8.

§7.3.2. Number of critical points. We proceed to study the number of critical points, under the conditions of Theorem 1:

THEOREM 2. *Let R be an infinite region whose boundary B is a finite Jordan configuration symmetric in the axis of reals, let Jensen circles be constructed for all pairs of non-real symmetric points of B, and let $G(x, y)$ be Green's function for R with pole at infinity. Let α and β be points of the axis of reals in R which are not critical points of $G(x, y)$ and are not on or within any Jensen circle, and let K denote the configuration consisting of the segment $\alpha \leqq z \leqq \beta$ plus the closed interiors of all Jensen circles intersecting that segment. Let K contain precisely k components of B. Then K contains precisely $k - 1$, k, or $k + 1$ critical points of $G(x, y)$. More explicitly, this number of critical points is*

$$k - 1 \quad \textit{if} \quad \partial G(\alpha, 0)/\partial x < 0,\ \partial G(\beta, 0)/\partial x > 0;$$

$$k \quad \textit{if} \quad [\partial G(\alpha, 0)/\partial x]\cdot[\partial G(\beta, 0)/\partial x] > 0;$$

$$k + 1 \quad \textit{if} \quad \partial G(\alpha, 0)/\partial x > 0,\ \partial G(\beta, 0)/\partial x < 0.$$

We continue to use the field of force of §7.1.3 Theorem 1, and in the proof use the natural extension of the method of §2.3 Theorem 1, of which Theorem 2 is the analog. In the field represented by the conjugate of $F'(z)$, where $F(z) \equiv G + iH$, the force at any point z is (§7.1.3) orthogonal to the locus $G(x, y) = \text{const}$ through that point, in the direction ν of increasing $G(x, y)$.

At the real point $z = \alpha$, not a critical point, the locus $G(x, y) = G(\alpha, 0)$ has a vertical tangent, and the direction ν of increasing $G(x, y)$ (that is to say, the direction of the force at $z = \alpha$) is to the right if we have $\partial G(\alpha, 0)/\partial x > 0$ and to the left if we have $\partial G(\alpha, 0)/\partial x < 0$. Thus the conditions in Theorem 2 expressed in terms of the forces at α and β with reference to the interval $\alpha\beta$ are that the forces should be respectively both outward, one outward and the other inward, or both inward.

We study the variation in the direction of the force at z as z traces a Jordan curve J containing K in its interior, the curve chosen in R so close to K that no critical point or point of B lies on J or between J and K; moreover J shall start at a real point $\beta'(>\beta)$, shall remain in the upper half-plane near the boundary of K until it cuts the axis of reals in a point $\alpha'(<\alpha)$, then remain in the lower half-plane until it reaches β'. When z traces J in the positive (counterclockwise) sense, the direction angle of the force increases by 2π, 0, or -2π respectively in the various cases.

We also study the variation in the direction of the force as z traces each of k mutually exterior Jordan curves interior to J, so chosen that each curve contains precisely one component of B in its interior but contains in its closed interior no other point of B and no critical point of $G(x, y)$ in R; moreover we choose each of these k curves as part of a locus $G(x, y) = \text{const.}$ As z traces such a curve in the positive sense with respect to the region R' exterior to these curves but interior to J, that is, in the clockwise sense, the force increases its direction angle by -2π. Thus as z traces the entire boundary of R' in the positive sense, the net increase in the direction angle of the force is in the respective cases $-2(k - 1)\pi$, $-2k\pi$, or $-2(k + 1)\pi$; the net increase in the direction angle of the conjugate $F'(z)$ of the force is respectively $2(k - 1)\pi$, $2k\pi$, or $2(k + 1)\pi$; the number of critical points of $G(x, y)$ is precisely the number of zeros of $F'(z)$, and Theorem 2 follows from the Principle of Argument.

If each component of B is a segment of the axis of reals, Theorem 2 is contained in §7.2.1 Corollary 3 to Theorem 1, but the latter result includes the case of an infinity of components of B. Indeed, the method just used in the proof of Theorem 2 shows that if B in Theorem 2 is allowed to contain an infinite number of components then i) if K contains precisely k components of B, it contains precisely $k - 1$, k, or $k + 1$ critical points, and ii) if K contains an infinite number of components of B, it contains an infinite number of critical points of $G(x, y)$.

We have not emphasized the use of W-curves in the situation of Theorem 1, but the entire discussion of §2.6 is of significance here, where we consider the W-curves for all possible pairs of groups of points of B, and that discussion gives further results. We state the analogs and consequences of §2.6.3 Theorems 5 and 6:

THEOREM 3. *In the notation of* §2.6.3 *Theorem* 5, *let R be an infinite region whose boundary B is finite and symmetric in the axis of reals, and lies on S_0, S_1, and S_2. Let Green's function $G(x, y)$ for R with pole at infinity exist. Then all non-real critical points of $G(x, y)$ in R lie in $\Pi_1 + \Pi_2$.*

THEOREM 4. *In the notation of* §2.6.3 *Theorem* 6, *let R be an infinite region whose boundary B is finite and symmetric in the axis of reals, and lies on S_0, S_1,* *and S_2. Let Green's function $G(x, y)$ for R with pole at infinity exist. Then all non-real critical points of $G(x, y)$ in R lie in Π.*

Under the hypothesis of symmetry in O rather than in the axis of reals, W-curves are also important; the entire treatment of §3.6 has its counterpart here.

§7.4. Analog of Walsh's Theorem. We proceed to establish (notation of §7.1.3 Theorem 1)

THEOREM 1. *Let R be an infinite region whose boundary B is a finite Jordan configuration, the sum of two disjoint Jordan configurations B_1 and B_2. We set $m_1 = \int_{B_1} d\sigma$, $m_2 = \int_{B_2} d\sigma$, $m_1 + m_2 = 1$. If B_1 and B_2 lie respectively in the closed interiors of the circles $C_1: |z - \alpha_1| = r_1$ and $C_2: |z - \alpha_2| = r_2$, then all critical points in R of Green's function $G(x, y)$ for R with pole at infinity lie in the closed regions C_1, C_2, and*

$$C_3: |z - (m_2\alpha_1 + m_1\alpha_2)| \leqq m_2 r_1 + m_1 r_2 .$$

Let C_1 and C_2 contain respectively k_1 and k_2 components of B; if the region C_j is exterior to the other regions C_k, it contains in its closed interior precisely $k_1 - 1$, $k_2 - 1$, or 1 critical points according as $j = 1, 2$, or 3.

We use the field of force of §7.1.3 Theorem 1, and precisely the method of §1.5. A position of equilibrium z in R not in the closed interior of C_1 or C_2 is by §7.1.3 Lemma also a position of equilibrium in the field due to two particles respectively of masses m_1 and m_2, situated in the closed interiors of C_1 and C_2, exerting forces at z equal to the forces at z due to the masses on B_1 and B_2; consequently z lies in the closed interior of C_3.

§7.4.1. Numbers of critical points. The "method of continuity" in its simplest form is not adequate here to determine the numbers of critical points in the various circular regions. To be sure, we can vary B, decreasing monotonically the number of components of B, and the critical points in R vary continuously with B except as components of B vanish or meet in such a way as to reduce the total number of components of B. But as B varies, so also do m_1, m_2, and C_3; if C_3 is exterior to C_1 and C_2 for one set of values of m_1 and m_2, and if m_1 and m_2 vary, a new C_3 need not be exterior to C_1 and C_2 unless special precautions are taken. Here we shall need several lemmas.

LEMMA 1. *Let A be an annular region bounded by a fixed circle $\Gamma: r = a$ with center O and by a variable boundary Γ' interior to Γ which approaches O. Then the function $\omega(x, y)$ harmonic in A, continuous in the corresponding closed region, equal to zero on Γ and to unity on Γ' approaches zero uniformly on any closed set in the closed interior of Γ but not containing O.*

Case 1). Let Γ' be the circle $r = \alpha$. Here we write

$$\omega(x, y) = \frac{\log r - \log a}{\log \alpha - \log a}, \tag{1}$$

as may be verified at once. When α approaches zero, $\omega(x, y)$ approaches zero as described.

Case 2). Let Γ' be arbitrary, but contained in the closed region $r \leqq \alpha$; denote the second member of (1) by $\omega_1(x, y)$. On the circle $r = a$ we have $\omega(x, y) = \omega_1(x, y)$, and on the circle $r = \alpha$ we have $\omega_1(x, y) = 1 \geqq \omega(x, y)$, whence $\omega_1(x, y) \geqq \omega(x, y) \geqq 0$ in the closed region $a \geqq r \geqq \alpha$. The conclusion follows from the conclusion in Case 1).

LEMMA 2. *Let R be an infinite region whose boundary B is a variable finite Jordan configuration. Let one component Γ_1 of B vary monotonically (i.e. so that R increases monotonically) and approach a point, while the other components (assumed to exist) remain fixed. Then Green's function $G(x, y)$ for R with pole at infinity approaches uniformly on any closed fixed set S in R' Green's function $G'(x, y)$ for the region R' consisting of R plus Γ_1 plus all points separated by Γ_1 from the point at infinity.*

From §7.1.2 Lemma 1 it follows that the variable $G(x, y)$ increases monotonically on S, and that we have $G'(x, y) > G(x, y)$ in R. Let Γ_2 denote a component of B other than Γ_1, and let $\omega(x, y)$ denote the function harmonic in the annular region bounded by Γ_1 and Γ_2, continuous in the corresponding closed region, equal to unity on Γ_1 and to zero on Γ_2. We form the function

$$G(x, y) - G'(x, y) + [\max G'(x, y) \text{ on } \Gamma_1]\cdot\omega(x, y),$$

which is harmonic in R, continuous on $R + B$; on Γ_1 we have $G(x, y) = 0$ and $\omega(x, y) = 1$, so this function is non-negative there; on the remaining part of B we have $G(x, y) = G'(x, y) = 0$; this function is bounded in the neighborhood of the point at infinity, hence when properly defined is harmonic there, and is non-negative in R. Consequently on S we have

$$G'(x, y) > G(x, y) \geqq G'(x, y) - [\max G'(x, y) \text{ on } \Gamma_1]\cdot\omega(x, y);$$

the last term approaches zero uniformly on S, by Lemma 1, when the region containing the point at infinity and bounded by Γ_2 is suitably mapped onto the interior of a circle; the conclusion follows.

When the boundary components of a region vary monotonically, so do the corresponding masses of the distribution σ:

LEMMA 3. *Let R be an infinite region whose boundary B is a finite Jordan configuration B, the sum of two disjoint Jordan configurations B_1 and B_2. Let $G(x, y)$ be Green's function for R with pole at infinity; we set $m_1 = \int_{B_1} d\sigma$, $m_2 = \int_{B_2} d\sigma$, in the notation of §7.1.3. Let R be a proper subregion of R', the latter with boundary B' the sum of B_1 and a Jordan configuration B_2'. Let $G'(x, y)$ be Green's function* *for R' with pole at infinity, and set correspondingly* $m_1' = \int_{B_1} d\sigma'$, $m_2' = \int_{B_2'} d\sigma'$. *Then we have* $m_1/m_2 < m_1'/m_2'$.

The function $u(x, y) = G'(x, y) - G(x, y)$ is harmonic in R even at infinity when suitably defined there, is continuous and zero on B_1, continuous and non-negative on B_2 and positive at some points of B_2, hence is positive in R. Thus on B_1 we have $d(\sigma' - \sigma) > 0$, by the reasoning used in §7.1.3, whence $m_1' > m_1$. Of course we have $m_1 + m_2 = 1$, $m_1' + m_2' = 1$, so the lemma follows.

An integral $\int_{B_1} d\sigma_1$ simply represents the total mass on B_1 in the distribution σ, which consequently varies monotonically with B_1.

We now show that *under the conditions of Theorem* 1, *the point sets* B_1 *and* B_2 *can be varied monotonically and continuously remaining in their prescribed regions so that the ratio* $m_1{:}m_2$ *remains fixed, until both* B_1 *and* B_2 *consist of one component each.* Let a component of B_2 be varied monotonically remaining a Jordan configuration, so as to increase R monotonically; denote original and new values of $\int_{B_1} d\sigma$ and $\int_{B_2} d\sigma$ by m_1 and m_2, and m_1' and m_2' respectively. By Lemma 3 we have $m_1/m_2 < m_1'/m_2'$. Fix now B_2 at any stage and vary B_1 monotonically so as to increase R monotonically; the new variable masses m_1'' and m_2'' change in such a way that m_1''/m_2'' decreases from the value m_1'/m_2'. In this variation of B_1 (and of B_2) we vary one component at a time, reducing it to a point and then omitting it. From Lemma 2 it follows that m_1 and m_2 nevertheless vary continuously with B_1 and B_2, for such an integral as

$$m_1 = \int_{B_1} d\sigma = \frac{1}{2\pi} \int \frac{\partial G}{\partial \nu} \, ds$$

can be taken over an analytic Jordan curve or set of mutually exterior analytic Jordan curves in R *near* B_1, hence in a region S in which $G(x, y)$ and $\partial G/\partial \nu$ vary continuously even if a component of B is reduced to a point and subsequently ignored. Thus as B_1 varies, with B_2 temporarily fixed, m_1''/m_2'' decreases monotonically from the value m_1'/m_2', and when B_1 is caused to vanish completely this ratio has the value zero, hence at some intermediate stage takes the value m_1/m_2. It is therefore possible to vary B_1 and B_2 monotonically, continuously, and simultaneously so that m_1/m_2 remains fixed, until B_1 and B_2 consist each of one component, which may be chosen as small as desired.

If in Theorem 1 the numbers m_1 and m_2 are given, we choose B as the lemniscate* $|(z - \alpha_1')^{m_1}(z - \alpha_2')^{m_2}| = \mu$, where α_1' and α_2' lie interior to C_1 and C_2, and where μ is chosen so that the lemniscate consists of two Jordan curves B_1 and B_2 in C_1 and C_2 respectively. By the relation $m_1 + m_2 = 1$, Green's function for the exterior R of this lemniscate with pole at infinity is

$$\log |(z - \alpha_1)^{m_1}(z - \alpha_2)^{m_2}| - \log \mu,$$

and the hypothesis of Theorem 1 is satisfied. The conclusion of Theorem 1 is also satisfied, including the enumeration of critical points in the regions C_j, namely

---

* We use here a slight modification in the term *lemniscate*, for m_1 and m_2 are not necessarily integral.

a single critical point and that in C_3. The present Jordan curves B_1 and B_2 composing the lemniscate can be varied continuously and monotonically into any given sufficiently small Jordan configurations B_1 and B_2 interior to them and for which the ratio m_1/m_2 is the same. During this variation we keep the ratio of the masses σ on B_1 and B_2 constant, the number of critical points of Green's function remains unchanged in each of the fixed regions C_k exterior to the other regions C_j, so the original number for the small Jordan configurations B_1 and B_2 was zero, zero, or one critical point, according as $k = 1, 2,$ or 3.

During the entire original variation of the given sets B_1 and B_2 of Theorem 1, with constant m_1/m_2, the critical points of $G(x, y)$ vary continuously with B_1 and B_2, except that when B_1 or B_2 loses a component by shrinking to a point P, the function $G(x, y)$ loses a critical point; this loss must occur at P, for in the neighborhood of any other point of R the function $G(x, y)$ varies uniformly and its critical points vary continuously. The critical points remain in the closed regions C_j; during the variation none can enter or leave any region C_j disjoint from the other regions C_i except for the loss of a critical point with the loss of a component of the boundary, so the original number is that stated in the theorem. Theorem 1 is now established. If C_1, C_2, and C_3 are mutually exterior, no critical point lies on any of those circles.

§7.4.2. Numbers of critical points, alternate treatment. Still another method can be used to establish the last part of Theorem 1. In the field of force of §7.1.3 Theorem 1 we consider the force at a point z exterior to C_1 and C_2; for convenience in exposition we suppose α_1 and α_2 real, with $r_1 = r_2$; only obvious changes in phraseology are necessary to include the more general case. Suppose C_1 exterior to C_2 and C_3, and let z lie on a horizontal line L between the intersections of that line with C_1 and the intersections of that line with C_3. The force at z due to a particle of mass m_1 in C_1 and a particle of mass m_2 in C_2 is equivalent to the force at z due to a particle of mass $m_1 + m_2 = 1$ in a suitably chosen circular region $\Gamma(z)$, by §1.5.1 Lemma 2. The point z is a possible position of equilibrium for suitable positions of the given particles when and only when the point at infinity lies in $\Gamma(z)$, namely when z lies in the closed region C_3; and since $\Gamma(z)$ varies continuously with z, the circle $\Gamma(z)$ becomes a straight line when and only when z lies on the circle C_3. When z is on L near C_1, the region $\Gamma(z)$ is a finite region not containing z which cuts L on the same side (to the right or left) of z as C_1; and L cuts the four circles C_1, C_2, C_3, $\Gamma(z)$ all at the same angle. The force at z due to the two particles in C_1 and C_2, or to the distribution σ, is represented by a vector whose terminal point lies in z and whose initial point lies in the sector subtended at z by $\Gamma(z)$, a sector whose vertex is z and whose angular opening is less than π. When z lies on a common tangent to C_1 and C_2, one side of the sector lies along that tangent. Of course if z lies outside of the smallest convex region containing C_1 and C_2, the total force at z is represented by a vector with terminal point in z and initial point in the sector subtended at z by that convex region.

It is now possible to study the variation in the direction of the force at z due to the distribution σ as z traces a circle C concentric with C_1 in whose exterior C_2 and C_3 lie. It is convenient to commence with z on C on a common tangent to C_1 and C_2; when z traces C in the positive (counterclockwise) sense, the direction angle of the force at z increases by 2π. When z traces mutually exterior Jordan curves in C_1 belonging to a level locus of $G(x,.y)$ near the respective components of B_1 but enclosing those components and enclosing no critical points in R, in the positive (clockwise) sense, the direction angle of the force decreases by 2π for each of the k_1 components; compare §7.3.2. Thus for the region bounded by C and these k_1 Jordan curves, the direction angle of the force increases by $-2(k_1 - 1)\pi$, and the direction angle of $F'(z)$ increases by $2(k_1 - 1)\pi$, so $F'(z)$ has $k_1 - 1$ critical points in the region, and $G(x, y)$ has precisely $k_1 - 1$ critical points in the closed region C_1.

If the region C_2 is exterior to C_1 and C_3, the reasoning already given shows that C_2 contains precisely $k_2 - 1$ critical points of $G(x, y)$. If C_3 is exterior to C_1 and C_2, all three circles are mutually exterior; since C_1 and C_2 contain precisely $k_1 - 1$ and $k_2 - 1$ critical points, C_3 cont ains precisely one critical point.

§7.4.3. Symmetry. A great disadvantage of Theorem 1 is that m_1 and m_2 are involved, and may be difficult to determine. But if R possesses certain symmetries, such determination may be trivial:

THEOREM 2. *Let R be an infinite region whose boundary B is a finite Jordan configuration symmetric in the axis of reals. Let B lie in two circular regions C_1: $|z - \alpha| \leqq r$ and C_2: $|z - \bar{\alpha}| \leqq r$, with $|\alpha - \bar{\alpha}| \geqq 4r$. Then all critical points of Green's functions $G(x, y)$ for R with pole at infinity lie interior to C_1 and C_2 except for a single critical point on the interval $-r < z - (\alpha + \bar{\alpha})/2 < r$. If B consists of $2k$ components, each of the regions C_1 and C_2 contains $k - 1$ critical points.*

No point z of the circumference C_3: $|z - (\alpha + \bar{\alpha})/2| = r$ can be a position of equilibrium, for (§7.1.3, Lemma) at a point z exterior to C_1 and C_2 the forces due to the portions of the distribution σ on B in C_1 and C_2 are equivalent to the forces due to the same masses concentrated at points *interior* to C_1 and C_2, so (§1.5.1) z cannot lie on C_3 if we have $|\alpha - \bar{\alpha}| > 4r$. At the point $z = \frac{1}{2}(\alpha + \bar{\alpha}) + ir$ (say on C_1, if we have $|\alpha - \bar{\alpha}| = 4r$) the force due to the distribution in C_1 is equivalent to the force at a particle of equivalent mass on the circumference C_1 when and only when that distribution lies wholly on arcs of C_1; even if the portion of B in C_1 lies wholly on that circumference, the force at z due to the mass in the region C_2 is equivalent to the force due to a particle of the same mass *interior* to C_2, so z is not a position of equilibrium.

Of course the force at the point $z = \frac{1}{2}(\alpha + \bar{\alpha}) - r$ is directed to the left and that at the point $z = \frac{1}{2}(\alpha + \bar{\alpha}) + r$ to the right, so their interval contains at least one point of equilibrium. In the case $|\alpha - \bar{\alpha}| > 4r$ we prove Theorem 2 from Theorem 1; in the case $|\alpha - \bar{\alpha}| = 4r$ we use the method of proof of Theorem

1, with either method of determining the number of critical points; and in the monotonic variation of the components of B as in §7.4.1 we preserve the symmetry. The details are left to the reader.

Theorem 2 is of course the analog of §3.7 Theorem 2; similarly §3.7 Theorem 1 possesses an analog:

THEOREM 3. *Let R be an infinite region whose boundary B is a Jordan configuration symmetric in the origin O. Let B lie in two circular regions C_1:* $|z - \alpha| \leqq r$ *and C_2:* $|z + \alpha| \leqq r$, *with* $|\alpha| \geqq 2r$. *Then O is a critical point of Green's function $G(x, y)$ for R with pole at infinity, and all other critical points lie in C_1 and C_2. If B consists of $2k$ components, each of the regions C_1 and C_2 contains $k - 1$ critical points.*

The proof of Theorem 3 is similar to that of Theorem 2 and is omitted.

If in Theorem 3 we omit the restriction $|\alpha| \geqq 2r$, the conclusion requires some modification; compare §5.5.3 Theorem 5:

THEOREM 4. *Under the hypothesis of Theorem 3 modified by requiring now* $|\alpha| < 2r$, *all critical points of $G(x, y)$ except perhaps O lie in a closed finite region or regions R_1 bounded by an arc of C_1, an arc of C_2, and two arcs of an equilateral hyperbola with center O one of whose axes passes through α and $-\alpha$, the hyperbola being tangent to each of the circles C_k in two distinct points, and the arcs being so chosen that R_1 contains the closed interiors of C_1 and C_2.*

There are similar results to Theorems 3 and 4 if R has p-fold symmetry about O; we state merely a part of the analog and consequence of §5.5.6 Theorem 9:

THEOREM 5. *Let R be an infinite region whose boundary B is a Jordan configuration p-fold symmetric in O, and let B lie in the set of closed interiors S of the mutually exterior circles $C_1, C_2, \cdots, C_p$, each of which subtends an angle not greater than* $2 \sin^{-1} (1/p)$ *at O. Let the set S possess p-fold symmetry in O, and let $G(x, y)$ be Green's function for R with pole at infinity. Then all critical points of $G(x, y)$ except O lie in S.*

§7.4.4. Inequalities on masses. We have already remarked on the difficulty of applying Theorem 1 in an arbitrary situation, if m_1 and m_2 are not known. However, if m_1 and m_2 are not known exactly but inequalities on them are known, say $m_1' \leqq m_1 \leqq m_1''$ (a corresponding inequality on m_2 follows from the relation $m_1 + m_2 = 1$), then Theorem 1 can still be applied, and determines a closed region depending on m_1' and m_1'' as a locus of z for variable m_1:

$$C: |z - (1 - m_1)\alpha_1 - m_1\alpha_2| \leqq (1 - m_1)r_1 + m_1 r_2, \quad m_1' \leqq m_1 \leqq m_1'',$$

which contains all critical points of $G(x, y)$ not in the closed region C_1 or C_2. Unless C_1 contains C_2 or C_2 contains C_1, the region C is convex and bounded by segments of the common external tangents to the circles C_1 and C_2, and by an arc of each of two other circles having those same common external tangents. The case $m_1' = 0$ is not excluded, nor is the case $m_1'' = 1$. Of course the region C contains the region C_3 of Theorem 1 for each admissible m_1; if any of the regions C_1, C_2, C is disjoint from the other two of those regions, it contains precisely $k_1 - 1$, $k_2 - 1$, or 1 critical points respectively.

Simple geometric relations yielding inequalities on m_1 and m_2 are rare; we prove merely

THEOREM 6. *Let R and R' be two distinct infinite regions whose boundaries B and B' are finite Jordan configurations, respectively the sums $B = B_1 + B_2$, $B' = B_1' + B_2'$ of disjoint Jordan configurations. Let all points of B_1 be separated by B_1' from B_2, and let all points of B_2' be separated by B_2 from B_1'. Let $G(x, y)$ and $G'(x, y)$ be Green's functions for R and R' with pole at infinity, with $m_1 = \int_{B_1} d\sigma$, $m_1' = \int_{B_1'} d\sigma'$. Then we have $m_1' > m_1$.*

Each point of B_1' lies interior to R, so we have on B_1': $G(x, y) - G'(x, y) > 0$; each point of B_2 lies interior to R', so we have on B_2: $G(x, y) - G'(x, y) < 0$. Thus a locus $G(x, y) - G'(x, y) = 0$ in R and R' separates B_1 and B_1' from B_2 and B_2'. Integration over this locus if it is finite or otherwise over a neighboring level locus, of the normal derivative of the function $G(x, y) - G'(x, y)$, yields

$$\int \frac{\partial G}{\partial \nu} ds - \int \frac{\partial G'}{\partial \nu} ds < 0,$$

where ν indicates normal directed toward B_2 and B_2', and this is the inequality $m_1' > m_1$.

We have already (§5.2.1) introduced the concept of *harmonic measure* $\omega(z, B, R)$. Thus in Theorem 1 we may write

$$m_1 = \int_{B_1} d\sigma = \frac{1}{2\pi} \int_{B_1} \frac{\partial G}{\partial \nu} ds + \frac{0}{2\pi} \int_{B_2} \frac{\partial G}{\partial \nu} ds,$$

so by the classical properties of Green's function it follows that we have $m_1 = \omega(\infty, B_1, R)$. The inequality $m_1' > m_1$ just established is in essence a proposition on harmonic measure due to Carleman.

If the region R of Theorem 6 is given, in numerous cases the region R' can be chosen bounded by two circles B_1' and B_2', and indeed in numerous cases a double inequality for m_1 can be obtained by the use of two regions each bounded by two circles. If R' is bounded by the two circles

$$B_1' : |\varphi(z)| = \mu_1, \quad B_2' : |\varphi(z)| = \mu_2, \quad \varphi(z) = (z - \alpha)/(z - \beta),$$

where α and β are the null circles of the coaxal family determined by the circles B_1' and B_2', we have

$$\omega(z, B_1', R') = \frac{\log |\varphi(z)| - \log \mu_2}{\log \mu_1 - \log \mu_2}, \qquad m_1' = \omega(\infty, B_1', R') = \frac{-\log \mu_2}{\log \mu_1 - \log \mu_2}.$$

§7.4.5. Circles with collinear centers. There exists an analog of §3.3 Theorem 1:

THEOREM 7. *Let the circles $C_1, C_2, \cdots, C_n$ have collinear centers $\alpha_1, \alpha_2, \cdots, \alpha_n$ respectively, and the common radius r. Let R be an infinite region whose boundary B is a finite Jordan configuration, the sum of mutually disjoint Jordan configurations $B_1, B_2, \cdots, B_n$ lying respectively in the closed interiors of the circles C_k. Let $G(x, y)$ be Green's function for R with pole at infinity, and set $m_j = \int_{B_j} d\sigma$. Denote by $\alpha_1', \alpha_2', \cdots, \alpha_k'$ the zeros distinct from the α_j of the derivative of*

$$(z - \alpha_1)^{m_1}(z - \alpha_2)^{m_2} \cdots (z - \alpha_n)^{m_n},$$

and by C_j' the circle whose center is α_j' and radius r. Then all critical points of $G(x, y)$ in R lie in the closed interiors of the circles C_j and C_j'.

Any subset S of closed interiors of circles of the set of closed interiors of the circles $C_1 + \cdots + C_n + C_1' + \cdots + C_k'$ disjoint from the remainder of the set contains a number of critical points of $G(x, y)$ in R equal to the number of components of B in the circles C_j belonging to S minus the number of circles C_j belonging to S plus the number of circles C_j' belonging to S.

Theorem 7 can be established by the method used for Theorem 1, including the second method (§7.4.2) for the study of the numbers of critical points. Theorem 7 extends to the case of circles $C_1, C_2, \cdots, C_n$ having a common external center of similitude, without essential change of method.

Marden's Theorem (§4.5) for the case of polynomials likewise has an analog here, but like Theorem 7 has the disadvantage that $m_j = \int_{B_j} d\sigma$ may be difficult to compute. Nevertheless inequalities on the m_j may be available, and may yield some results.

§7.5. Doubly-connected regions. Thus far we have established results on the critical points of Green's function for an infinite region with pole at infinity. A linear transformation or an inversion generalizes these results. An obvious generalization of §7.2.1. Theorem 1 is

THEOREM 1. *Let R be a region whose boundary B is a Jordan configuration, and let $G(z, z_0)$ be Green's function for R with pole in z_0. Any circular region containing B but not z_0 contains all the critical points of $G(z, z_0)$ in R.*

Theorem 1 essentially asserts that no critical point lies near z_0; further results show that no critical point lies near the boundary of R:

THEOREM 2. *Let R be a doubly-connected region bounded by the Jordan curves C_1 and C_2. If $G(z, z_0)$ is Green's function for R with pole in z_0, and if the point z_0 does not lie in a given region $\omega(z, C_1, R) > \mu_1$ $(\geqq \frac{1}{2})$ or $\omega(z, C_2, R) > \mu_2$ $(\geqq \frac{1}{2})$, that region contains no critical point of $G(z, z_0)$.*

Of course $G(z, z_0)$ has precisely one critical point in R.

We map R onto a region bounded by two concentric circles, and map the universal covering surface of this annulus by means of a logarithmic transformation onto an infinite strip bounded by two parallel lines; the strip thus represents infinitely many replicas of the annulus, and the function mapping the strip onto the annulus is periodic, as is the transform of $G(z, z_0)$ in the strip. We finally map the strip onto the upper half of the w-plane so that the two end-points of the strip correspond respectively to $w = 0$ and $w = \infty$. A translation of the strip into itself corresponds to a transformation $w' = \rho w$ of this half-plane, where ρ is real. Then $G(z, z_0)$ is transformed into a function $g(w)$ harmonic in the upper half-plane except at points $\rho^n \alpha$, $n = \cdots, -1, 0, 1, 2, \cdots$, where we have $\rho > 1$, and α lies interior to the upper half-plane. The function $g(w)$ is continuous at every finite point other than O of the axis of reals and vanishes there, and satisfies the functional equation $g(\rho w) \equiv g(w)$. At each of the points $\rho^n \alpha$, the function $g(w) + \log | w - \rho^n \alpha |$ has a removable discontinuity.

The function $g(w)$ can be extended harmonically by reflection across the axis of reals, and when so extended is single-valued and harmonic at every finite point of the w-plane except the origin and the points $\rho^n \alpha$ and $\rho^n \bar{\alpha}$. In the latter points, the function $g(w) - \log | w - \rho^n \bar{\alpha} |$ has a removable discontinuity.

The analytic function $F(w)$ defined in §6.3.1 Theorem 1 having the zeros $\rho^n \alpha$ and the poles $\rho^n \bar{\alpha}$ satisfies the functional equation

$$F(\rho w) = \bar{\alpha} F(w)/\alpha,$$

from which it follows that $\log | F(w) |$ is single-valued, harmonic at all finite points except O, $\rho^n \alpha$, and $\rho^n \bar{\alpha}$, and satisfies the functional equation

$$\log | F(\rho w) | = \log | F(w) |.$$

The sum $g(w) + \log | F(w) |$ has a removable singularity at each of the points $\rho^n \alpha$ and $\rho^n \bar{\alpha}$, is harmonic (when suitably defined) at every finite point other than O, satisfies the functional equation $g(\rho w) + \log | F(\rho w) | \equiv g(w) + \log | F(w) |$, takes on the same values in any two of the closed regions $\rho^{n-1} \leqq | w | \leqq \rho^n$, and hence is identically constant. The critical points of $g(w)$ are precisely those of $\log | F(w) |$, or those of $F(w)$, and $F(w)$ is the uniform limit of a sequence of rational functions whose zeros and poles are mutually inverse in the axis of reals.

If a sector $0 \leqq \arg w < \varphi$ $(<\pi/2)$ or $(\pi/2<)$ $\varphi < \arg w \leqq \pi$ does not contain α, that sector contains no point $\rho^n \alpha$, and its complement with respect to the upper half-plane is NE convex with reference to that half-plane; consequently such a sector contains no critical points of $g(w)$.

The harmonic measure $\omega(w, B_1)$, where B_1 denotes the half-line $\arg w = 0$, with respect to the upper half-plane is seen by inspection to be $(\pi - \arg w)/\pi$, so the line $\arg w = \varphi$ is a locus $\omega(w, B_1) = \text{const}$. Similarly the harmonic measure $\omega(w, B_2)$ of B_2, the half-line $\arg w = \pi$, with respect to the upper half-plane is $(\arg w)/\pi$, so the line $\arg w = \varphi$ is a locus $\omega(w, B_2) = \text{const}$.

The functions $\omega(w, B_1)$ and $\omega(w, B_2)$ obviously are unchanged by substituting

ρw for w, hence are the transforms of the functions $\omega(z, C_1, R)$ and $\omega(z, C_2, R)$, where we assume B_1 to be the image of C_1 repeated infinitely many times, and B_2 to be the similar image of C_2. If now a region $\omega(z, C_1, R) > \mu_1$ $(\geqq \frac{1}{2})$ does not contain z_0, its image in the w-plane is a sector $0 \leqq \arg w < \varphi$ $(<\pi/2)$ not containing α and hence containing no critical point of $g(w)$; thus the original region in the z-plane contains no critical point of $G(z, z_0)$. A similar argument concerning $\omega(z, C_2, R)$ completes the proof of Theorem 2.

Theorem 2 extends to regions of higher connectivity, but the proof is more involved and is postponed to §§8.8.2 and 8.9.3. The formal statement of Theorem 2 does not exhaust the possibilities of the method, in the w-plane *any* region NE convex with respect to the upper half-plane and containing all the points $\rho^n\alpha$ also contains all critical points of $g(w)$ in the upper half-plane; Theorem 2 may be correspondingly extended.

CHAPTER VIII

HARMONIC FUNCTIONS

Green's function is analogous to a polynomial, and the pattern of the study of the location of critical points of polynomials applies also to the location of the critical points of Green's functions. Similarly, more general harmonic functions are analogous to more general rational functions, and the pattern for rational functions applies to harmonic functions more general than Green's functions. In the present chapter we study especially (i) harmonic functions defined in a region by merely two different assigned boundary values (say zero and unity) on complementary parts of the boundary and (ii) linear combinations of Green's functions for a given region. Further harmonic functions, especially linear combinations of simpler harmonic functions, are considered in Chapter IX.

§8.1. Topology. As a typical situation, we suppose R to be a region of the extended plane bounded by mutually disjoint finite Jordan curves $C_1, C_2, \cdots, C_m, D_1, D_2, \cdots, D_n$. Let the function $u(x, y)$ be harmonic in R, continuous in the corresponding closed region, equal to zero on the C_k and to unity on the D_k; this function exists. Thus $u(x, y)$ is the harmonic measure (§5.2.1) in the point (x, y) of the set $D_1 + \cdots + D_n$ with respect to the region R. The two sets of curves C_j and D_k play similar roles, for the harmonic measure in the point (x, y) of the set $C_1 + \cdots + C_m$ with respect to the region R is $1 - u(x, y)$, which has the same critical points as $u(x, y)$. This geometric configuration and the function $u(x, y)$ are essentially invariant under a conformal map of R which is one-to-one and continuous in the closed region.

§8.1.1. Level loci. Through an arbitrary point (x_1, y_1) of R passes a unique locus

$$(1) \qquad u(x, y) = \mu, \qquad 0 < \mu < 1, \qquad (x, y) \text{ in } R,$$

namely the locus $u(x, y) = u(x_1, y_1)$. In the neighborhood of any one of its points (x_1, y_1), the locus (1) consists of an analytic Jordan arc if (x_1, y_1) is not a critical point of $u(x, y)$, and consists of $q + 1$ analytic Jordan arcs through (x_1, y_1) making successive angles of $\pi/(q + 1)$ if (x_1, y_1) is a critical point of order q. Thus the locus (1) consists of a finite number of Jordan curves interior to R separating all the C_j from all the D_k; at most a finite number of points can be points of intersection of two or more of these Jordan curves. No point of (1) can lie in a subregion of R bounded wholly by points of (1) and points of the C_j, or bounded wholly by points of (1) and points of the D_k; any region bounded wholly by points of (1) must contain in its interior points of the boundary of R. Each of the Jordan curves composing the locus (1) separates at least one C_j from at least one D_k.

The given region R can be mapped conformally, and one-to-one in the *closed* region, onto a region bounded by analytic Jordan curves, by successive mappings onto the interior of a circle of the region containing R bounded by C_1, of the new region containing the image of R bounded by the image of C_2, and so on through D_n; compare §7.1.1. Various properties of $u(x, y)$ and the loci (1) are a consequence of such a map; in particular $u(x, y)$ has at most a finite number of critical points in R.

Two distinct loci $u(x, y) = \mu_1, 0 < \mu_1 < 1$, and $u(x, y) = \mu_2, 0 < \mu_1 < \mu_2 < 1$, cannot intersect in R, and the former separates the latter from the C_k; the latter separates the former from the D_j. When μ is small and positive, the locus (1) consists of m Jordan curves, one in R near each of the Jordan curves C_k and separating C_k from the curves $C_1, \cdots, C_{k-1}, C_{k+1}, \cdots, C_m, D_1, \cdots, D_n$. As μ increases, these Jordan curves vary monotonically, moving from the C_k and toward the D_j. The number of Jordan curves composing the locus (1) does not change unless the locus passes through a critical point of $u(x, y)$, and even then the number need not change if the critical point is a multiple one.* As we indicate later (§8.1.3 Theorem 2), and as may be proved by topological considerations, there are precisely $m + n - 2$ critical points of $u(x, y)$ in R, so at most $m + n - 2$ distinct loci (1) have multiple points. When μ is near unity, the locus (1) consists of n Jordan curves, one in R near each of the curves D_k and separating D_k from the curves $C_1, \cdots, C_m, D_1, \cdots, D_{k-1}, D_{k-2}, \cdots, D_n$.

The loci

$$v(x, y) = \text{const}, \qquad (x, y) \text{ in } R, \tag{2}$$

where $v(x, y)$ is conjugate to $u(x, y)$ in R, likewise form a family covering R. In the neighborhood of a point (x_1, y_1) of R on (2), the locus (2) consists of a single analytic Jordan arc if (x_1, y_1) is not a critical point of $v(x, y)$, and consists of $q + 1$ analytic Jordan arcs intersecting at (x_1, y_1) and making successive angles of $\pi/(q + 1)$ there if (x_1, y_1) is a critical point of order q; in the latter case the arcs of the locus (1) through (x_1, y_1) bisect at (x_1, y_1) successive angles between the arcs of (2). Through each point of R passes a locus (2) which is unique geometrically, but due to the multiple-valuedness of $v(x, y)$ in R each geometric locus (2) corresponds to an infinity of different values of the constant. Only a finite number (at most $m + n - 2$) of the geometric loci (2) have multiple points, namely those passing through critical points of $v(x, y)$. Each locus (2) consists of a Jordan arc joining the set of curves C_j and the set of curves D_k, except that the arc forks at a critical point; a locus (2) not passing through a critical point abuts on the boundary of R in precisely two points. As (x, y) traces

* By way of illustration here we may choose $m = n = 2$, $C_1: |z - 2| = 1$, $C_2: |z + 2| = 1$, $D_1: |z - 2i| = 1$, $D_2: |z + 2i| = 1$; the locus $u(x, y) = \frac{1}{2}$ consists of the two lines $y = \pm x$, and has as multiple points $z = 0$ and $z = \infty$. Every level locus (1) consists of precisely two Jordan curves.

monotonically an arc of (2) containing no critical point of $u(x, y)$ and $v(x, y)$, the function $u(x, y)$ changes monotonically, for we have $du = (\partial u/\partial s)\, ds = (\partial v/\partial \nu) ds$, where ν indicates normal derivative, and $\partial v/\partial \nu$ does not change sign on the arc. As (x, y) moves outward from a critical point (x_1, y_1) of order q along one of the $2(q + 1)$ Jordan arcs of the locus $v(x, y) = v(x_1, y_1)$ terminating in (x_1, y_1), $u(x, y)$ increases along $q + 1$ of those arcs and decreases along the other $q + 1$ arcs, for the $2(q + 1)$ arcs separate a neighborhood of (x_1, y_1) into $2(q + 1)$ subregions, in which the function $v(x, y) - v(x_1, y_1)$ is alternately positive and negative.

If we require merely that the C_j and D_k be mutually disjoint components of the boundary of R, not necessarily Jordan curves, and in their totality comprise that boundary, the qualitative nature of the loci (1) and (2) is not materially changed except that now a locus (2) need not abut on the boundary of R in a single point, but may oscillate. If the boundary of R is allowed to contain an infinity of mutually disjoint components, an infinite number of critical points of $u(x, y)$ and $v(x, y)$ lie in R; an infinite number of loci (1) possess multiple points.

§8.1.2. A formula for harmonic measure. We assume R finite but otherwise retain the original (§8.1) assumptions on R, and derive a formula [Walsh, 1934, 1948c] for $u(x, y)$ analogous to §7.1.3 equation (9). Choose two loci (1) with $\mu = \epsilon(>0)$ and $\mu = 1 - \epsilon(>\epsilon)$ respectively which consist of mutually disjoint analytic Jordan curves in $R: C_1', C_2', \cdots, C_m', D_1', \cdots, D_n'$, and denote by R' the subregion of R bounded by them. Let ϵ be chosen so small that all critical points of $u(x, y)$ in R lie in R'. We suppose for definiteness R' to lie interior to C_1 and C_1', and for abbreviation set $C' = C_1' + \cdots + C_m'$, $D' = D_1' + \cdots + D_n'$. If $P: (x_0, y_0)$ denotes an arbitrary point of R', and r denotes distance measured from P, we have

$$u(x_0, y_0) = \frac{1}{2\pi} \int_{C'+D'} \left(\log r \frac{\partial u}{\partial \nu} - u \frac{\partial \log r}{\partial \nu} \right) ds, \tag{3}$$

where ν indicates interior normal for the region R'. We have also the following

$$\frac{1}{2\pi} \int_{C_1'} u \frac{\partial \log r}{\partial \nu}\, ds = \frac{\epsilon}{2\pi} \int_{C_1'} \frac{\partial \log r}{\partial \nu}\, ds = -\epsilon,$$

$$\int_{C_j'} u \frac{\partial \log r}{\partial \nu}\, ds = \epsilon \int_{C_j'} \frac{\partial \log r}{\partial \nu}\, ds = 0, \qquad j > 1,$$

$$\int_{D_k'} u \frac{\partial \log r}{\partial \nu}\, ds = (1 - \epsilon) \int_{D_k'} \frac{\partial \log r}{\partial \nu}\, ds = 0.$$

Consequently (3) can be written

$$u(x_0, y_0) = \frac{1}{2\pi} \int_{C'+D'} \log r \frac{\partial u}{\partial \nu}\, ds + \epsilon. \tag{4}$$

Since $C' + D'$ passes through no critical point of $u(x, y)$, we have $\partial u/\partial \nu \neq 0$ on $C' + D'$, and indeed $\partial u/\partial \nu > 0$ on C', $\partial u/\partial \nu < 0$ on D'. Thus we may set on C'

$$\frac{1}{2\pi}\frac{\partial u}{\partial \nu}\,ds = -\frac{1}{2\pi}\frac{\partial v}{\partial s}\,ds = d\sigma, \qquad 0 \leqq \sigma \leqq \tau = \int_{C'} d\sigma,$$

and on D'

$$\frac{-1}{2\pi}\frac{\partial u}{\partial \nu}\,ds = \frac{1}{2\pi}\frac{\partial v}{\partial s}\,ds = d\sigma, \qquad 0 \leqq \sigma \leqq \tau = \int_{D'} d\sigma,$$

with $d\sigma > 0$; in each case the direction of increasing σ corresponds to the sense of description of the boundary positive with reference to the region R'. Thus (4) can be written

$$u(x_0, y_0) = \int_{C'} \log r\, d\sigma - \int_{D'} \log r\, d\sigma + \epsilon, \qquad d\sigma > 0. \tag{5}$$

The integrals in (5) are essentially Stieltjes integrals of the continuous function $\log r$. The loci $v(x, y) = \text{const}$ contain Jordan arcs joining given points of C' and given points of D' to uniquely determined points of the C_j and D_k, and $dv = \pm 2\pi\, d\sigma$ is the same between such arcs whether measured on C' and D' or on the C_j and D_k. When we allow ϵ to approach zero in (5) the function $\log r$ is continuous uniformly for all σ, and we obtain

$$u(x_0, y_0) = \int_C \log r\, d\sigma - \int_D \log r\, d\sigma, \qquad d\sigma > 0, \tag{6}$$

where $C = \sum C_j$, $D = \sum D_k$, and (6) is valid for all (x_0, y_0) in R.

Let t denote the running variable point in (6), and set $z = x_0 + iy_0$; then a locally single-valued function conjugate to $u(x, y)$ in R is given by

$$v(x, y) = \int_C \arg(z - t)\, d\sigma - \int_D \arg(z - t)\, d\sigma, \tag{7}$$

for z in R. We now set

$$f(z) = u(x, y) + iv(x, y) = \int_C \log(z - t)\, d\sigma - \int_D \log(z - t)\, d\sigma,$$

and obtain by differentiation the first part of the following theorem, under the restrictions previously mentioned on R; the function $f'(z)$ is single-valued:

THEOREM 1. *Let R be a region of the extended plane whose boundary is a finite Jordan configuration and consists of components $C_1, C_2, \cdots, C_m, D_1, \cdots, D_n$. Let the function $u(x, y)$ be harmonic in R, continuous in the corresponding closed region, equal to zero on the C_j and to unity on the D_k. Then the critical points of $u(x, y)$ interior to R are the critical points interior to R of the function $f(z)$, and we have for z in R*

$$f'(z) = \int_C \frac{d\sigma}{z-t} - \int_D \frac{d\sigma}{z-t}, \qquad d\sigma > 0, \tag{8}$$

where $C = \sum C_j$, $D = \sum D_k$.

Consequently the critical points of $u(x, y)$ *interior to* R *are precisely the positions of equilibrium in the field of force due to a spread* σ *of positive matter on* C *and a spread* $-\sigma$ *of negative matter on* D, *where the matter repels with a force proportional to the mass and inversely proportional to the distance; the total mass is zero.*

To be sure, equation (8) has been established only for a finite region R; if R is infinite with finite boundary, we choose also R' infinite with finite boundary; equation (3) is to be modified by the addition in the second member of a term involving the integral taken over a circle Γ with center (x_0, y_0) and containing in its interior the boundary of R. We have

$$\int_\Gamma \log r \frac{\partial u}{\partial \nu} ds = \log r \int_\Gamma \frac{\partial u}{\partial \nu} ds = 0,$$

$$\frac{-1}{2\pi} \int_\Gamma u \frac{\partial \log r}{\partial \nu} ds = \frac{1}{2\pi r} \int_\Gamma u \, ds = u(\infty),$$

where $u(\infty)$ is the value of $u(x_0, y_0)$ at infinity. Thus the additional term $+u(\infty)$ is to be placed in the second members of (4), (5), and (6), but equation (8) persists.

The proof of (8) remains essentially valid if R is no longer bounded by a finite number of mutually disjoint Jordan curves, but by a Jordan configuration (§1.1.1). On a Jordan arc both of whose banks (each a one-sided neighborhood) belong to R, $d\sigma$ is doubly defined.

The second part of Theorem 1 follows merely by taking conjugates in equation (8) and is the analog of §4.1.1 Theorem 1. The conjugate of $f'(z)$ is in R precisely the gradient of $u(x, y)$; compare §7.1.3.

As a matter of simplicity, we have chosen the two boundary values zero and unity for $u(x, y)$ in Theorem 1, but all the formulas undergo only minor modifications if two arbitrary distinct constant values are used. Indeed (compare §6.6) equation (8) can be proved directly when interpreted as Cauchy's integral formula for $f'(z)$, taken in the form

$$f'(z) = \frac{1}{2\pi i} \int_{C+D} \frac{f'(t)\, dt}{t - z}, \qquad z \text{ in } R;$$

on $C + D$ we write $f'(t)\, dt = d[f(t)] = i\, dv$; the integral here is to be considered a Stieltjes integral if C and D are not smooth. If $u(x, y)$ is less on C than on D, then positive matter lies on C and negative on D; the direction of the force is the direction of increasing $u(x, y)$.

Theorem 1 contains a proof of §7.1.3 Theorem 1 concerning the critical points of Green's function $G(x, y)$ for an infinite region R whose boundary is a finite Jordan configuration B. We introduce a variable auxiliary locus $B_M: G(x, y) = M$

in R, where M is large and positive, apply Theorem 1 to the region bounded by B and B_M, and allow M to become infinite; the total mass of the matter on B_M does not change with M, and as M becomes infinite B_M recedes indefinitely, so the force at an arbitrary point z of R due to the matter on B_M approaches zero with $1/M$. In fact, the total force at z is the conjugate of $f'(z)$ and is independent of M, so the force at z due to the matter on B_M is zero for every M sufficiently large; the total force at z is the force at z due merely to the spread on B.

It will be noted that in (8) the differential $d\sigma$, which except for a numerical factor is the differential of $v(x, y)$, is numerically invariant both under inversion and under one-to-one conformal transformation of R provided only Jordan configurations are involved as boundaries.

It will also be noted from (8) that if the point at infinity lies in R, the function $f'(z)$ has at least a double zero at infinity; but the point at infinity is defined (§1.1.2) as a critical point of $u(x, y)$ when and only when it is a critical point of $f(z)$, that is, a zero of $f'(z)$ of order greater than two.

The various comments in Chapter IV concerning the special nature of the point at infinity have their analogs here. As in §4.1.3, it follows that the field of force defined in Theorem 1 can be projected stereographically, in the sense that if the spread of matter $\pm\sigma$ is projected onto the sphere, and the resulting field of force on the sphere is considered due to matter repelling according to the law of inverse distance, then the direction of the field of force on the sphere is tangent to the sphere and is the stereographic projection of that in the plane; it is essential here to note that the total positive mass is equal to the total negative mass. To be sure, no direction of force in the plane is defined at infinity, but we define that direction by means of the field of force on the sphere and the stereographic projection. We also retain the convention (§4.1.1) that the point at infinity is to be considered a position of equilibrium in the plane if and only if it is a position of equilibrium on the sphere. The point at infinity is thus a position of equilibrium in the plane when and only when it is a critical point (§1.1.2) of $u(x, y)$ and $f(z)$, not a multiple zero of $f(z)$.—These comments will not be repeated, but apply in numerous situations in the sequel.

§8.1.3. Number of critical points. The field of force in Theorem 1 may be used directly in the study of critical points, or we may replace the integrals in (8) by Riemann sums, and apply results on the critical points of rational functions. We shall use both methods, but Theorem 1 is immediately useful in rendering easy the enumeration of critical points, by the methods of §§7.3.2 and 7.4.2. We prove

THEOREM 2. *Let R be a region of the extended plane whose boundary is a Jordan configuration with mutually disjoint components $C_1, C_2, \cdots, C_m, D_1, \cdots, D_n$. Let the function $u(x, y)$ be harmonic in R, continuous in the corresponding closed region, equal to zero on the C_j and to unity on the D_k. Then $u(x, y)$ has precisely $m + n - 2$ critical points in R.*

Choose the boundary of R as finite. We consider mutually disjoint auxiliary Jordan curves C'_j and D'_k in R, near the respective boundary components C_j and D_k, where each curve separates the corresponding component from the other Jordan curves, and where no critical point of $u(x, y)$ lies on these Jordan curves or between any of them and the corresponding component C_j or D_k. Denote by R' the subregion of R bounded by these Jordan curves; if R is infinite we choose the curves so that R' is infinite. Choose $C'_1 + \cdots + C'_m$ as a locus $u(x, y) = \epsilon$ and $D'_1 + \cdots + D'_n$ as a locus $u(x, y) = 1 - \epsilon$. At each point of C'_j and D'_k the force in the field of Theorem 1 is normal to the curve, in the direction of increasing $u(x, y)$. When z traces a curve C'_j or D'_k in the clockwise sense, the direction angle of the force decreases by 2π, and arg $[f'(z)]$ increases by 2π. If R' is infinite the directions on all these Jordan curves positive with respect to R' are clockwise; if R' is finite the directions positive with respect to R' are clockwise on all but one of these curves, and counterclockwise on that one. Thus when z traces all the curves in the positive sense with reference to the region R', the function arg $[f'(z)]$ increases by 2π times $m + n$ or $m + n - 2$ according as R' is infinite or finite. Theorem 2 follows.

Theorem 2 includes Macdonald's Theorem (§6.2), for under the hypothesis of the latter the locus $\log |f(z)| = -M'$ in R for M' sufficiently large consists of precisely q analytic Jordan curves, which are mutually disjoint, and one of which separates each of the q distinct zeros of $f(z)$ in R from the other zeros of $f(z)$ in R and from B. We apply Theorem 2 to the subregion R' of R bounded by B and these Jordan curves, the latter chosen so close to the respective zeros that no critical point not a multiple zero of $f(z)$ lies in R not in R'; it follows that $f(z)$ has precisely $q - 1$ critical points in R', hence has precisely $m - 1$ critical points in R, where m is the total number of zeros of $f(z)$ in R.

Theorem 2 enables us to extend Macdonald's Theorem to the case where R has connectivity $p(> 1)$; the method just used shows that if $f(z)$ is analytic in R, $|f(z)|$ is constant not zero on the boundary of R, and $f(z)$ has precisely m zeros in R, then $f(z)$ has precisely $m + p - 2$ critical points in R.

§8.2. Analog of Bôcher's Theorem. The methods just developed make almost obvious the proof [Walsh, 1934a] of

THEOREM 1. *Let R be a region of the extended plane whose boundary is a Jordan configuration with mutually disjoint components $C_1, C_2, \cdots, C_m, D_1, \cdots, D_n$. Let the function $u(x, y)$ be harmonic in R, continuous in the corresponding closed region, equal to zero on the C_j and to unity on the D_k. If a circle Γ in the closure of R separates all the points of the C_j not on Γ from all points of the D_k not on Γ, and if at least one boundary point of R does not lie on Γ, then no critical point of $u(x, y)$ lies on Γ in R.*

If disjoint circular regions C and D contain respectively all the C_j and all the D_k, they contain all critical points of $u(x, y)$ in R, and indeed contain precisely $m - 1$ and $n - 1$ critical points respectively in R.

Choose the boundary of R as finite. If z is a finite point of a circle Γ, the force at z due to the distribution of mass on the C_j is (§7.1.3 Lemma) equivalent to the force at z due to the same mass concentrated at some point in a circular region Γ' bounded by Γ and containing the C_j; this point is *interior* to Γ' unless all the C_j lie on Γ; the force at z due to the distribution of mass on the D_k is equivalent to the force at z due to the same mass (the negative of the total mass on the C_j) concentrated at some point in the circular region Γ'' bounded by Γ and containing the D_k, which is not the circular region Γ' containing the C_j; this point is *interior* to Γ'' unless all the D_k lie on Γ. Hence (§4.1.2 Corollary to Theorem 3) the total force at z has a component normal to Γ in the sense directed from Γ' to Γ'', so z is not a position of equilibrium nor a critical point of $u(x, y)$. The point at infinity is treated by a linear transformation, or by stereographic projection (§8.1.2).

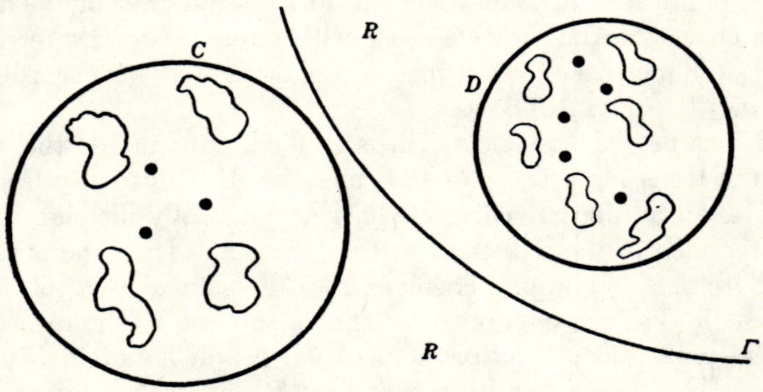

Fig. 19 illustrates §8.2 Theorem 1

That C contains precisely $m - 1$ critical points follows by the method of §8.1.3, from the study of the variation of the direction of the force as z traces Jordan curves in R near the C_j and also a finite circle Γ separating C and D. On Γ the force has a non-zero component normal to Γ from the side of Γ on which C lies toward the side of Γ on which D lies.

A second method of proof (method of continuity) of the facts concerning the numbers of critical points in C and D can be given by a simplification of the method of §7.4.1, involving monotonic variation of the C_j and D_k, and is left to the reader.

Theorem 1 is the equivalent of §6.6 Theorem 1.

We indicated in §7.2.1 that the analog of Lucas's Theorem for Green's function is not invariant under one-to-one conformal transformation and that a new result can be found by making such a transformation. The same remark applies to Theorem 1; we mention a special case. Under the conditions of Theorem 1, let C_0 and D_0 be two disjoint continua bounding a region R_1, of which R is a subregion, so that the region bounded by C_0 containing R_1 contains all the D_k, and the region bounded by D_0 containing R_1 contains all the C_k. Let $u_0(z)$ denote the function harmonic in R_1, continuous in $R_1 + C_0 + D_0$, equal to zero on

C_0 and to unity on D_0. Then any annular subregion $(0 <)\ \mu_0 < u_0(z) < \mu_1(< 1)$ of R containing no point of the C_k or D_k contains no critical point of $u(z)$. This conclusion follows merely by mapping R_1 onto an annulus.

In all of our study of fields of force, whenever (as in §§6.6, 7.3.2, 7.4.2, 8.1.3) we consider the direction of the force we are also considering the direction of the normal to the level loci. Let us prove the

Corollary. *Under the conditions of the latter part of Theorem 1, let (x_1, y_1) be a point of R not in the regions C or D. Let the circles γ_1 and γ_2 through (x_1, y_1) tangent to the circles C and D separate the interiors of the two regions C and D. Then the direction at (x_1, y_1) of the normal at (x_1, y_1) to the level locus $u(x, y) = u(x_1, y_1)$ lies interior to the angles subtended at (x_1, y_1) by the regions C and D between the circles γ_1 and γ_2. Otherwise expressed, there exists a circle γ orthogonal at (x_1, y_1) to the level locus through that point which cuts both C and D.*

The circle γ is simply the circle through (x_1, y_1) and through the two points in the regions C and D at which the masses in those regions are concentrated to yield forces at (x_1, y_1) equivalent to the forces due to the original distributions of matter.

Of course in any particular case under the latter part of Theorem 1 the circles C and D can be chosen in an infinite variety of ways; all critical points lie in two regions T_1 and T_2 whose interior points are separated by every Γ, and which are bounded by arcs of circles Γ', each circle Γ' separating all points of $\sum C_j$ not on Γ' from all points of $\sum D_k$ not on Γ'; also in the Corollary the direction at the point (x_1, y_1) (assumed not in T_1 or T_2) of the normal at (x_1, y_1) to the level locus $u(x, y) = u(x_1, y_1)$ lies in the angles subtended at (x_1, y_1) by T_1 and T_2 between the two of these circles Γ' through (x_1, y_1); compare §§4.2.1 and 4.2.5. A degenerate case is of interest here:

Theorem 2. *Let the closed arcs $C_1, C_2, \cdots, C_m, D_1, \cdots, D_n$ of a circle γ be mutually disjoint, with no two arcs C_j separating two arcs D_k on γ. Let $u(x, y)$ be the function harmonic in the extended plane except on the arcs C_j and D_k, continuous at every point of the extended plane, equal to zero on the C_j and to unity on the D_k. Then all critical points of $u(x, y)$ lie on γ; on any open arc of γ containing no point of a C_j or D_k and bounded by two points of the set $\sum C_j$ or by two points of the set $\sum D_k$ lies a unique (a simple) critical point.*

The conclusion on the location of the critical points on suitable arcs of γ is a consequence of Theorem 1, or compare §7.2.1; the conclusion on the precise number of critical points on an arc follows by the methods of §7.2.1. Indeed, those methods yield also the more general

Corollary. *Let R be a region whose boundary consists of mutually disjoint continua C_k and D_k finite or infinite in number each of which is symmetric in the* *axis of reals. Let the C_k intersect the axis in points at which z is positive, and the D_k intersect the axis in points at which z is negative. Let there exist the function $u(x, y)$ harmonic in R, continuous in the corresponding closed region, equal to zero on the C_k and to unity on the D_k. Then all critical points of $u(x, y)$ are real; each finite open segment of the axis of reals in R bounded by points of successive C_k or successive D_k contains precisely one (a simple) critical point; each finite critical point lies on such a segment.*

Here we need not require the closures of the two sets $\sum C_k$ and $\sum D_k$ to be disjoint, provided they have at most the two points $z = 0$ and $z = \infty$ in common; compare the methods of §8.6 below.

§8.3. A cross-ratio theorem. Precisely as the results of §1.5 are generalized in §7.4, the cross-ratio theorem of §4.4 can now be generalized. A difficulty in this program is that of evaluating $\int d\sigma$ over a boundary component; this integral represents the total mass spread over that boundary component. Here two methods are appropriate: i) in numerous cases majorant and minorant auxiliary harmonic functions can be found; these yield inequalities on the various masses, and (as in §7.4.4) thereby yield geometric results on critical points; ii) in cases of symmetry it may be possible to deduce equality of masses and to proceed directly with the cross-ratio theorem of §4.4.1. We do not elaborate i) further, but give an example [Walsh, 1934a] of ii):

THEOREM 1. *Let a region R of the extended plane be bounded by a finite Jordan configuration consisting of the mutually disjoint components $C_1, \cdots, C_m, D_1, \cdots, D_m, E_1, \cdots, E_n$. Let the function $u(x, y)$ be harmonic in R, continuous in the corresponding closed region, equal to zero on the C_j and the D_k and to unity on the E_k. Let the components C_j, D_k, E_k lie respectively in circular regions C', C'', C''', and let C^0 denote the locus of the point z_4 defined by the constant cross-ratio $(z_1, z_2, z_3, z_4) = 2$ when z_1, z_2, z_3 have C', C'', C''' as their respective loci.*

1). *If the entire configuration R is symmetric in some point O, and if C' is symmetric to C'' in O, and C''' symmetric in O, then all critical points of $u(x, y)$ lie in C', C'', C''', and C^0. If these four regions are mutually disjoint, then C', C'', C''' contain respectively $m - 1$, $m - 1$, $n - 1$ critical points, and the one remaining critical point lies in C^0 at O or at infinity.*

2). *If the entire configuration R is symmetric in some circle C, and C''' is symmetric in C, and if C' is symmetric to C'' in C, then all critical points of $u(x, y)$ lie in C', C'', C''', C^0. If these four regions are mutually disjoint, then C', C'', C''' contain respectively $m - 1$, $m - 1$, $n - 1$ critical points, and the one remaining critical point lies in C^0 on C.*

In both 1) and 2), if C''' is disjoint from C', C'', and C^0, then C''' contains precisely $n - 1$ critical points of $u(x, y)$.

In the actual construction of the circle C^0, §8.2 Theorem 1 is of importance; the circles (if existent) tangent to C', C'', C''', and separating the interior of C'''

from the interiors of C' and C'', are also tangent to C^0, and separate the interior of C''' from the interior of C^0; compare the situation of §4.4.4.

The fact that all critical points lie in the regions $C^{(j)}$ follows by consideration of the field of force, precisely as in §4.4.3.

The enumeration of the critical points in the various regions can be made i) by study of the change in the direction angle of the force as a point traces suitable curves, as in §7.4.2, or ii) by the method of continuity, where in varying the components of the boundaries of R we preserve the given symmetry, even if that involves varying those components in pairs instead of singly.

Despite the difficulties of determining the various masses involved, it is worthwhile to state the general cross-ratio theorem:

THEOREM 2. *Let a region R of the extended plane be bounded by a finite Jordan configuration consisting of the mutually disjoint components $C_1, \cdots, C_m$, $D_1, \cdots, D_n, E_1, \cdots, E_q$. Let the function $u(x, y)$ be harmonic in R, continuous in the corresponding closed region, equal to zero on the C_j and the D_j, and to unity on the E_k. Let us set $\mu_1 = \int_{\Sigma C_j} |d\sigma|$, $\mu_2 = \int_{\Sigma E_j} |d\sigma|$. Let the components C_j, D_k, E_k lie respectively in circular regions $\Gamma_1, \Gamma_2, \Gamma_3$, and let Γ_4 denote the locus of the point z_4 defined by the constant cross-ratio $(z_1, z_2, z_3, z_4) = \mu_2/\mu_1$ when z_1, z_2, z_3 have $\Gamma_1, \Gamma_2, \Gamma_3$ as their respective loci. Then all critical points of $u(x, y)$ in R lie in $\Gamma_1, \Gamma_2, \Gamma_3, \Gamma_4$. Any set S consisting of a number of the regions Γ_k disjoint from the remaining regions contains a number of critical points equal to the number of components of the boundary of B in S less the number of regions $\Gamma_1, \Gamma_2, \Gamma_3$ belonging to S, plus unity if Γ_4 belongs to S.*

We remark too that §4.4.4 Theorem 3 and Marden's Theorem (§4.5) also admit corresponding analogs.

§8.4. Hyperbolic plane. The results of §5.3 naturally possess an analog [Walsh, 1939, 1946] for harmonic functions:

THEOREM 1. *Let the region R be bounded by the Jordan arc or curve J_0 and a Jordan configuration B disjoint from J_0 whose components are $C_1, C_2, \cdots, C_n$. Let R be provided with a NE geometry by mapping onto the interior of a circle the region J bounded by J_0 containing R, and let Π be the smallest NE convex region in J containing B. Then Π contains all critical points in R of the function $u(x, y)$ harmonic in R, continuous in the corresponding closed region, equal to zero on J_0 and to unity on B. No critical point of $u(x, y)$ lies in R on the boundary of Π unless B lies on a NE line.*

Map temporarily onto a half-plane the region J. The function $u(x, y)$ can be extended harmonically across the boundary L of this half-plane, and when so extended is harmonic throughout the region R' bounded by the C_j and by their respective reflections C_j' in L; moreover $u(x, y)$ is continuous in the corresponding closed region, equal to minus unity on C_j'. It follows by symmetry that (notation of §8.1) for the region R', the values of $d\sigma$ are equal on corresponding arcs of C_j and C_j'.

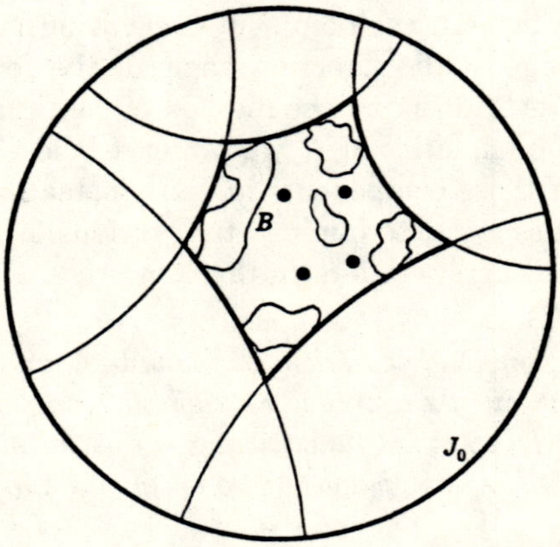

Fig. 20 illustrates §8.4 Theorem 1

In the proof of Theorem 1 we map J onto the interior of the unit circle C so that an arbitrary point of R not interior to Π is transformed into the origin O. Then B lies interior to C and its reflection B' in C lies exterior to C; both B and B' are finite. There exists a line λ through O such that all the points of both B and B' not on λ lie on the same side of λ; we assume that not all points of B and B' lie on λ. In the formula for the force at z (considered for $u(x, y)$ and the region R' bounded by B and B'):

$$\int_B \frac{d\sigma(t)}{\bar{z} - \bar{t}} - \int_{B'} \frac{d\sigma(t')}{\bar{z} - \bar{t}'}, \qquad d\sigma < 0,$$

we set $z = 0$, $t' = 1/\bar{t}$, so by the equation $d\sigma(t) = d\sigma(t')$ the force at O is

$$\int_B (t - 1/\bar{t})\, d\sigma(t),$$

which has a non-vanishing component orthogonal to λ directed from the side of λ on which B does not lie toward the opposite side. Thus O is not a position of equilibrium, and Theorem 1 is established.

It follows from §8.1.3 Theorem 2 or §8.2 Theorem 1 that $u(x, y)$ has precisely $n - 1$ critical points in R and in Π. The critical points of $u(x, y)$ in R' can of course be found by study of those in R; there are no critical points of $u(x, y)$ on C.

We shall prove also the

COROLLARY. *Under the conditions of Theorem* 1, *let* B *lie on the NE line* Γ.

Then all critical points of $u(x, y)$ in R lie on Γ; between successive components of B on Γ lies a unique (a simple) critical point.

At an arbitrary point z of Γ in R, the force is directed along Γ, as may be proved (compare §4.1.2) by an inversion with z as center. The force is orthogonal to the loci $u(x, y) = \text{const}$ in R, sensed in the direction of increasing $u(x, y)$, so at a point of Γ in R near C_j is directed toward C_j. Thus each arc of Γ between and bounded by successive components of B contains at least one critical point. Each such arc contains no more than one critical point, for there are $n - 1$ arcs and $n - 1$ critical points in R. Each critical point lies on such an arc.

THEOREM 2. *Let the region R be bounded by the Jordan arc or curve J_0 and a Jordan configuration B disjoint from J_0 whose (mutually disjoint) components are $C_1, C_2, \cdots, C_m, D_1, D_2, \cdots, D_n$. Let R be provided with a NE geometry by mapping onto the interior of a circle the region J bounded by J_0 containing R. If Γ is a NE line in R separating all the C_j from all the D_k, then Γ passes through no critical point of the function $u(x, y)$ harmonic in R, continuous in the corresponding closed region, equal to zero on J_0, to unity on the C_j, and to $-c(<0)$ on the D_k. If such a line Γ exists, all critical points of $u(x, y)$ in R lie in two NE convex regions Π_1 and Π_2 which contain respectively the C_j and D_k and which are separated by every Γ. No critical point of $u(x, y)$ lies on a NE line Γ' which separates all the points of the C_j not on Γ' from all the points of the D_k not on Γ', unless B lies on Γ'.*

The proof of Theorem 2 follows closely that of Theorem 1 and is omitted. We note that in the closure of R we have $\max u(x, y) = 1$, $\min u(x, y) = -c$; thus if C_j and D_j are analytic Jordan curves we have (notation of §8.1.2) $\partial u/\partial \nu < 0$ on C_j, $\partial u/\partial \nu > 0$ on D_j.

We have also

COROLLARY 1. *Under the conditions of Theorem 2, let a NE line Γ exist, and let B lie on a NE line Λ. Then all critical points of $u(x, y)$ in R lie on Λ on the two minimal disjoint arcs of Λ each bounded by a point of J_0 and a point of B in R, and containing respectively all the C_j and all the D_k. Between successive components C_j and between successive components D_k on Λ lies a unique critical point of $u(x, y)$.*

The fact that between successive components C_j and successive components D_k lies at least one critical point follows as in the Corollary to Theorem 1; that such a critical point is unique follows by continuous monotonic variation of those two successive components, enlarging them until they coalesce; during this variation the function $u(x, y)$ varies continuously except in the neighborhood of those components, as do the critical points of $u(x, y)$. Precisely one critical point of $u(x, y)$ disappears when the two components of B coalesce.

If J_0 in Theorem 2 is chosen as a circle whose center does not lie on B—that is to say, if J is mapped onto the interior or exterior of a circle whose center or point at infinity corresponds to an interior point of R—the function $u(x, y)$ can be extended across J_0 so as to be harmonic in the region R' bounded by the C_j, the D_k, and the reflections C_j' and D_k' of the C_j and D_k in J_0. The boundary of R' is finite. The total number of critical points of $u(x, y)$ in R' is then $2(m + n - 1)$, as follows from the method of §8.1.3 Theorem 2. Some of these critical points may lie on J_0; in particular if the C_j are symmetric to the D_j in a NE line Γ_0, with $c = 1$, the function $u(x, y)$ takes the value zero at every point of Γ_0 and of its reflection in the circle J_0; the intersections of J_0 and Γ_0 are then critical points.

The Corollary to Theorem 1 and Corollary 1 to Theorem 2 can be extended in the case of symmetry in a NE line:

Corollary 2. *Let the region R be bounded by the Jordan arc or curve J_0 and an additional boundary B disjoint from J_0 each of whose components is NE symmetric in a NE line L, NE geometry being defined for the region bounded by J_0 containing R. Let the function $u(x, y)$ be harmonic in R, continuous in the closed region, equal to zero on J_0 and to unity on B. On any segment of L in R bounded by points of B lies a unique critical point of $u(x, y)$; every critical point of $u(x, y)$ in R lies on such a segment.*

Corollary 3. *Let the region R be bounded by the Jordan arc or curve J_0 and by additional boundaries B_1 and B_2 mutually disjoint and disjoint from J_0. Let each component of B_1 and of B_2 be symmetric in a NE line L, NE geometry being defined for the region bounded by J_0 containing R. Let no segment of L bounded by points of B_1 contain a point of B_2, and no segment of L bounded by points of B_2 contain a point of B_1. Let the function $u(x, y)$ be harmonic in R, continuous in the closed region, equal to zero on J_0, to unity on B_1, and to $-c(<0)$ on B_2. Then all critical points of $u(x, y)$ in R lie on L. Any segment of L in R bounded by two points of B_1 or by two points of B_2 contains a unique critical point. Any critical point of $u(x, y)$ in R lies on such a segment except that a segment of L in R bounded by a point of J_0 and a point of B_1 or B_2 may contain a single critical point.*

Corollaries 2 and 3 follow by the methods already developed, especially those of §§7.2.1 and 5.3.2. In Corollaries 2 and 3 the components of B_1 and B_2 may be infinite in number.

Of course Theorem 1 can be expressed directly in terms of the function $u(x, y)$ harmonic in R' (notation of the proof of Theorem 1), continuous in the corresponding closed region, equal to unity on B and to minus unity on B'; this is a configuration possessing *skew-symmetry* in the circle C. A similar remark applies to Theorem 2 and the Corollaries.

If in Theorem 1 the boundary B is symmetric in a NE line but the individual components are not, we have a result intermediate between Theorem 1 and Corollary 2 to Theorem 2:

THEOREM 3. *Under the conditions of Theorem* 1 *let* B *be* (NE) *symmetric in a* NE *line* L. *Then all critical points of* $u(x, y)$ *in* R *not on* L *lie interior to the* NE *Jensen circular regions for* B *with centers on* L.

Let the arc A *of* L *be bounded by points* α *and* β *of* R, *where* α *and* β *are exterior to all* NE *Jensen circular regions for* B *and are not critical points of* $u(x, y)$, *and let the configuration* K *consist of* A *plus all closed* NE *Jensen circular regions intersecting* A. *If* K *contains* k *components of* B, *then* K *contains* $k + 1$, k, *or* $k - 1$ *critical points of* $u(x, y)$ *in* R *according as the forces at* α *and* β *are directed along* L *away from each other, both in the same sense along* L, *or both toward each other.*

Here a Jensen circular region is a subregion of J bounded by a circle whose NE center lies on L and whose NE diameter joins two points of B symmetric in L. We leave the proof of Theorem 3 to the reader; compare §§7.3 and 5.3.3.

§8.5. Other symmetries. Skew-symmetry in a circle, namely the symmetry related to the hyperbolic plane, is perhaps the most important of symmetries for a harmonic function. Nevertheless other kinds are not without interest. Indeed, the entire material of Chapter V can be applied in the study of harmonic functions.

Harmonic functions symmetric in a line or circle have been briefly considered in §8.2; we formulate only a few more of many possible results. An analog of §5.1.2 Theorem 6 does not explicitly involve symmetry:

THEOREM 1. *Let* R *be a region containing the points* $z = \pm i$, *whose boundary* B *is a Jordan configuration in an annular region* A *bounded by two circles of the coaxal family determined by the points* $z = \pm i$ *as null circles. Let* $G(z, z_0, R)$ *denote Green's function for* R *with pole in* z_0. *Then all critical points of the harmonic function* $G(z, +i, R) + G(z, -i, R)$ *lie in* A.

For the proof of Theorem 1, compare the methods of §8.9 below.

Results for harmonic functions are especially available when the corresponding results for rational functions either because of symmetry or otherwise do not explicitly involve the numbers of zeros and poles. An analog of §5.1.3 Theorem 8 is

THEOREM 2. *Let* R *be a region whose boundary* B *consists of the mutually disjoint Jordan configurations* B_1, B_2, *and* B_3, *where* B_1 *and* B_2 *are mutually symmetric in the axis of reals, and* B_3 *consists of a finite number of segments of that axis. Let* B_1 *and* B_2 *lie in the respective circular regions* C_1: $|z - bi| \leqq r$ *and* C_2: $|z + bi| \leqq r$. *If* $u(x, y)$ *is the function harmonic in* R, *continuous in* $R + B$, *equal to zero on* $B_1 + B_2$ *and to unity on* B_3, *then all critical points of* $u(x, y)$ *in* R *lie on the axis of reals plus* 1) *the closed interiors of* C_1 *and* C_2 *if we have* $b \geqq 2r$; 2) *the closed interiors of the circles of the coaxal family determined by* C_1 *and* C_2 *passing through the points* $z = \pm i[(b^2 - 2r^2)/2]^{1/2}$ *if we have* $2r > b > 2^{1/2}r$.

A further result admitting an analog here is §5.1.4 Theorem 9.

THEOREM 3. *Let R be a region whose boundary B consists of the mutually disjoint Jordan configurations B_1, B_2, B_3, where B_1 and B_2 are mutually symmetric in $C:|z| = 1$ and lie in the respective closed regions $|z| \leqq c(<1)$ and $|z| \geqq 1/c$, and B_3 is symmetric in C and lies in the closed region $b \leqq |z| \leqq 1/b$, $0 < c < b < 1$. Let $u(x, y)$ be the function harmonic in R, continuous in $R + B$, zero on $B_1 + B_2$, and unity on B_3. Then* (i) *if the inequality* (9) *of* §5.1.4 *is satisfied for $\xi = c$, no critical points of $u(x, y)$ lie in the region $c < |z| < b$;* (ii) *if inequality* (9) *of* §5.1.4 *is satisfied for some $\xi = \xi_0$, $c \leqq \xi_0 < b$, and if inequality* (10) *of* §5.1.4 *is satisfied for $\xi = \xi_0$, then no critical points lie in the region $\xi_0 < |z| < b$;* (iii) *if inequality* (10) *of* §5.1.4 *fails to be satisfied for some ξ_0, $c < \xi_0 < b$, then no critical points lie in the region $\xi_0 < |z| < b$.*

For the elliptic plane, the entire discussion of §5.4 is now of significance. We formulate but a part of the possible conclusions:

THEOREM 4. *Let R be a region of the sphere bounded by disjoint Jordan configurations B_0 and B_1, where R, B_0, and B_1 possess central symmetry. Let $u(x, y)$ be harmonic in R, continuous in $R + B_0 + B_1$, zero on B_0 and unity on B_1. Let P be an arbitrary point of the sphere, and let H denote the closed hemisphere whose pole is P.*

1). *If a great circle C through P separates all points of B_0 in H not on C from all points of B_1 in H not on C, and if at least one point of B_0 or B_1 lies in H not on C, then P is not a critical point of $u(x, y)$.*

2). *Suppose all points of B_0 in H lie exterior to the circular region C_1 containing P and bounded by a small circle of the sphere whose pole is P. Suppose all points of B_1 in H lie in the circular region C_2 interior to C_1 but not containing P. Then P is not a critical point of $u(x, y)$.*

Symmetry in the origin O, as in §5.5, also enables us to establish some special results. The analog of §5.5.2 Theorem 3 is

THEOREM 5. *Let R be a region bounded by disjoint Jordan configurations B_0 and B_1, let R, B_0, and B_1 be symmetric in O, and let the function $u(x, y)$ be harmonic in R, continuous in $R + B_0 + B_1$, zero on B_0, and unity on B_1.*

1). *If an equilateral hyperbola H whose center is O separates all points of B_0 not on H from all points of B_1 not on H, and if at least one point of B_0 or B_1 does not lie on H, then no finite critical point other than perhaps O lies on H.*

2). *If an equilateral hyperbola H_0 whose center is O contains B_0 in its closed interior and if an equilateral hyperbola H_1 whose center is O lies in the exterior of H_0, contains H_0 in its interior, and contains B_1 in its closed exterior, then between H_0 and H_1 lie no critical points of $u(x, y)$. No finite critical points other than perhaps O lie on H_0 or H_1.*

3). *If an equilateral hyperbola H_0 whose center is O contains B_0 in its closed interior, and if an equilateral hyperbola H_1 whose center is O contains in its exterior the closed interior of H_0 and contains B_1 in its closed interior, then between H_0 and H_1 lie no critical points of $u(x, y)$ other than O. No finite critical points other than perhaps O lie on H_0 or H_1.*

4). *If an equilateral hyperbola H whose center is O passes through all points of B_0 and B_1, and if on each branch of H the points of B_0 and B_1 respectively lie on two finite or infinite disjoint arcs of H, then all critical points of $u(x, y)$ other than O lie on H. On any open finite arc of H in R bounded by two points of B_0 or by two points of B_1, and not containing O, lies a unique critical point of $u(x, y)$.*

Here we expressly permit an equilateral hyperbola to degenerate into two perpendicular lines. A special case deserves mention:

5). *If B_0 lies on the axis of reals and B_1 on the axis of imaginaries, all critical points lie on those axes. Any open segment in R of either axis bounded by two points of B_0 or by two points of B_1 contains a unique critical point of $u(x, y)$.*

More generally than 5), an analog of §8.2 Corollary to Theorem 2 exists. We phrase an analog and consequence of §5.5.3 Theorem 4:

THEOREM 6. *Let (C_1, C_2) and (C_3, C_4) be two pairs of circles, each pair symmetric in O, let the two lines of centers of pairs be orthogonal, and let all the circles subtend at O the same angle, not greater than $\pi/3$. Let R be a region whose boundary consists of four mutually disjoint Jordan configurations B_1, B_2, B_3, B_4, let B_k lie in the closed interior of C_k, and let the pairs (B_1, B_2) and (B_3, B_4) be each symmetric in O. Let $u(x, y)$ be harmonic in R, continuous in the corresponding closed region, zero on B_1 and B_2, unity on B_3 and B_4. Then all finite critical points of $u(x, y)$ in R not at O lie in the regions C_k. The number of critical points in C_k is one less than the number of components of the boundary of R there.*

We omit the readily formulated theorems concerning multiple symmetry in O, and for skew symmetry in O formulate a single example from §5.6.1 Theorem 1 part 3).

THEOREM 7. *Let R be a region whose boundary consists of two disjoint Jordan configurations B_0 and B_1 mutually symmetric in O. Let B_0 and B_1 lie in the respective halves of a closed double sector S with vertex O and angular opening not greater than $\pi/2$, and in a closed annulus A bounded by circles with center O. Let $u(x, y)$ be harmonic in R, continuous in $R + B_0 + B_1$, zero on B_0 and unity on B_1. Then all critical points of $u(x, y)$ in R lie in $A \cdot S$. No critical point P lies on the boundary of $A \cdot S$ unless both B_0 and B_1 lie on a boundary circle of A passing through P or on a boundary line of S passing through P. If B_0 and B_1 consist of n components each, then each of the two closed regions common to A and S contains $n - 1$ critical points.*

The analogs of the remaining parts of Chapter V, including the interesting analog of §5.8.2 Theorem 5, are left to the reader.

§8.6. Periodic functions. Periodic harmonic functions can be studied by the methods of Chapter VI [Walsh, 1947b]:

THEOREM 1. *Let the function $U(w)$ be harmonic in an infinite region R of the w-plane, and have the period $2\pi i$ in the sense that $U(w) \equiv U(w + 2\pi i)$; whenever either of these functional values is defined, so shall the other be. Let the only boundary points of R in a specific fundamental region, a period strip with opposite boundary points identified, consist of disjoint Jordan configurations C_0 and C_1, on which $U(w)$ takes respectively the values zero and unity. Let each end-point of a period strip contain no point of C_0 or C_1, but either lie exterior to R or be a point of continuity of $U(w)$; in the latter case we assume $\lim_{u\to+\infty} U(u + iv)$ or $\lim_{u\to-\infty} U(u + iv)$ to exist in the finite sense.*

1). *Any line $u =$ const which separates C_0 from C_1 cannot pass through a critical point of $U(w)$. Consequently if each of the lines $u =$ const, $a < u < b$, separates C_0 from C_1, then the region $a < u < b$ contains no critical points of $U(w)$.*

2). *In a period strip S: $v_0 < v \leqq v_0 + 2\pi$, identify the two boundary points having the same abscissa. Then any pair of lines L: $v = v_1$ and L': $v = v_1 + \pi$ in S which separates the image of C_0 in S from the image of C_1 in S cannot pass through a finite critical point of $U(w)$. Consequently if each pair of such lines respectively in the two strips $(v_0 \leqq) v_2 < v < v_3$, $v_2 + \pi < v < v_3 + \pi (\leqq v_0 + 2\pi)$, separates the image of C_0 in S from the image of C_1 in S, then those two strips contain no finite critical points of $U(w)$.*

By means of the transformation $w = \log z$, Theorem 1 follows from §8.2 Theorem 1. Theorem 1 can be improved in various cases of symmetry; compare §§8.4 and 8.5; we leave the formulation of such results to the reader, and proceed to consider doubly periodic harmonic functions. We need a preliminary result on harmonic functions with multiplicative periods:

THEOREM 2. *Let R be a region of the z-plane which is invariant under the transformations $z = \rho^n z'$, $\rho > 1$, $n = \cdots, -1, 0, 1, 2, \cdots$. In a fundamental region R': $(0<)\ a < |z| \leqq \rho a$, let the two boundary points of R' on the same ray $\arg z =$ const be identified, and let the boundary of R in R' consist of two disjoint Jordan configurations C_0 and C_1. There exists a function $U(z)$ harmonic and bounded in R, continuous in the corresponding closed region except at $z = 0$ and $z = \infty$, and taking the value zero on C_0 and congruent Jordan configurations and the value unity on C_1 and congruent Jordan configurations. The function $U(z)$ is uniquely determined by these properties.*

Enumerate the totality of the configurations C_0, C_1, and the congruent configurations as $B_1, B_2, \cdots$, and denote by R_k the infinite region of the extended plane *bounded by $B_1, B_2, \cdots, B_k$. Denote by $U_k(z)$ the function harmonic in R_k, continuous in the corresponding closed region, and equal to $U(z)$ on the boundary of R_k. Then in any closed subregion R_0 of R not containing $z = 0$ or $z = \infty$, the sequence $U_k(z)$ approaches $U(z)$ uniformly.*

There exists a function $V_\delta(z)$ positive and harmonic in R, greater than unity in the regions $|z| < \delta$ and $|z| > 1/\delta$, and which approaches zero with δ uniformly in any closed subregion R_0 of the closure $\bar{R}$ of R not containing $z = 0$ or $z = \infty$. Transform R into a finite region S of the z'-plane, by setting $z' = 1/(z - \alpha)$ if there exists a point α exterior to R and otherwise by mapping the z-plane slit along a component of the boundary of R onto the interior of the unit circle in the z'-plane. Let $z = 0$ and $z = \infty$ correspond to $z' = z_1$ and $z' = z_2$, and let S lie interior to the circles $|z' - z_1| = d$ and $|z' - z_2| = d$. We introduce the function constructed by a well known method

$$V_\delta(z) = \frac{\log d - \log |z' - z_1|}{\log d - \log \eta} + \frac{\log d - \log |z' - z_2|}{\log d - \log \eta},$$

which for suitably chosen η depending on δ has the required properties. We return to the plane of the variable z.

Let R_0 and $\epsilon(>0)$ be given. Choose δ so that we have $V_\delta(z) < \epsilon$ in R_0, and choose N so that the configurations $B_{N+1}, B_{N+2}, \cdots$ lie exterior to R_0 and in $D: |z| < \delta$ plus $|z| > 1/\delta$. The function $U_m(z) - U_n(z)$ for $m > N, n > N$ is harmonic in R, continuous in the corresponding closed region; at any boundary point of R not in D this function vanishes, and at any boundary point of R in D has no value greater than unity or less than minus unity. Thus we have throughout R

$$-V_\delta(z) \leqq U_m(z) - U_n(z) \leqq V_\delta(z);$$

this last member is less than ϵ in R_0, so the sequence $U_n(z)$ converges uniformly in R_0. The existence of the function $U(z)$ with the requisite properties follows at once. It follows also that the critical points of $U(z)$ in R are the limits of the critical points of $U_n(z)$ in R.

Let there exist two functions $U(z)$ and $U_0(z)$ satisfying the given conditions for $U(z)$, and choose $M(\geqq 1)$ so that we have $|U(z)| \leqq M$ and $|U_0(z)| \leqq M$ in R. Let R_0 and ϵ be given as before, and choose δ so that we have $2MV_\delta(z) < \epsilon$ in R_0. The difference $U(z) - U_0(z)$ is harmonic in R, continuous in the corresponding closed region except at $z = 0$ and $z = \infty$, and zero on the boundary where defined. As z approaches $z = 0$ or $z = \infty$, no limit value of this difference is greater than $2M$ or less than $-2M$, so we have throughout R

$$-2MV_\delta(z) \leqq U(z) - U_0(z) \leqq 2MV_\delta(z);$$

this last member is less than ϵ in R_0, so $U(z)$ and $U_0(z)$ are identical in R and Theorem 2 is established.

By §8.2 Theorem 1 we now have

THEOREM 3. *Under the conditions of Theorem 2, let the two halves of a double sector with vertex O contain respectively* C_0 *and* C_1. *Then that double sector contains all critical points of* $U(z)$ *in* R.

Among the numerous cases of symmetry here, we mention the case where the boundary of R in R' consists of the Jordan configurations C_0 and C_1 each consisting of precisely one component, where both C_0 and C_1 are symmetric in the axis of imaginaries, and C_0 lies in the upper half-plane, C_1 in the lower half-plane. Precisely one critical point of $U(z)$ then lies on the axis of imaginaries between successive components congruent to C_0 and between successive components congruent to C_1; there are no other critical points of $U(z)$ in R; compare §8.2, Corollary to Theorem 2.

The exponential transformation $w = \log z$ yields from Theorem 3:

THEOREM 4. *Let the region* R *of the* w*-plane be doubly periodic with periods* $2\pi i$ *and* $\tau(>0)$, *in the sense that whenever* $w = w_0$ *lies in* R *so also does* $w = w_0 + 2\pi pi + q\tau$, *where* p *and* q *are arbitrary integers positive negative or zero. Let the boundary of* R *in any fundamental region (primitive period parallelogram with opposite boundary points identified) consist of two disjoint Jordan configurations* C_0 *and* C_1. *Then there exists a function* $u(w)$ *harmonic and doubly periodic in* R *with periods* $2\pi i$ *and* τ, *taking the values zero on* C_0 *and on congruent configurations, and unity on* C_1 *and on congruent configurations.*

In the strip $S: v_0 < v \leqq v_0 + 2\pi$ *identify the two boundary points having the same abscissa. Let the pair of lines* $L: v = v_1$ *and* $L': v = v_1 + \pi$ *in* S *separate in* S *the set* B_0 *of boundary points of* R *at which* $u(w) = 0$ *from the set* B_1 *of boundary points of* R *at which* $u(w) = 1$; *then no critical point of* $u(w)$ *in* R *lies on* L *or* L'. *Consequently if each pair of such lines in the two strips* $S_1: (v_0 \leqq)\ v_2 < v < v_3$, $S_2: v_2 + \pi < v < v_3 + \pi(\leqq v_0 + 2\pi)$, *separates* B_0 *from* B_1 *in* S, *then no critical point of* $u(w)$ *in* R *lies in* S_1 *or* S_2.

Numerous interesting special cases of Theorem 4 involving symmetry exist, and may be easily studied by the reader.

§8.7. Harmonic measure, Jordan region. The results hitherto established in the present chapter can be interpreted primarily as results on harmonic measure for a certain region, for except in §8.4 all the harmonic functions studied take only two distinct values on the boundary of a suitably chosen region. We have, however, always required the boundary values to be constant on each *component* of the boundary, a requirement which we now relinquish.

§8.7.1. Finite number of arcs. Here a fundamental theorem [Walsh, 1947] is

THEOREM 1. *Let* C *be the unit circle in the* z*-plane, and let closed arcs* $A_k: \alpha_k\beta_k$, $k = 1, 2, \cdots, n$, *of* C *be mutually disjoint; this notation is intended to imply that*

A_k is the arc of C extending in the positive (counterclockwise) sense from α_k to β_k; we assume further the points $\alpha_1, \beta_1, \alpha_2, \beta_2, \cdots, \alpha_n, \beta_n, \alpha_1 = \alpha_{n+1}, \beta_1 = \beta_{n+1}$ to lie on C in that positive order. Denote by A the set $\sum A_k$, and by $u(z)$ the harmonic measure at z with respect to the interior of C of the set A. The critical points of $u(z)$ interior to C are precisely the positions of equilibrium in the field of force due to positive unit particles situated at each point α_k and negative unit particles at each point β_k. Arcs of the NE lines $\alpha_k\alpha_{k+1}$ and $\beta_k\beta_{k+1}$ bound a closed NE convex polygon Π interior to C which contains all critical points of $u(z)$ interior to C. In the case $n = 2$, the polygon Π degenerates to a single point, which is a critical point; in every other case the critical points lie interior to Π.

Thus a point z interior to C cannot be a critical point of $u(z)$ if a NE line separates z and a point α_k or β_k from all the other points α_j and β_j.

It is to be expected as here that geometric results on critical points should be symmetric with respect to the α_k and β_k, for the function $1 - u(z)$ has the same critical points as $u(z)$, and is the harmonic measure at z of the set $C - A$ with respect to the interior of C.

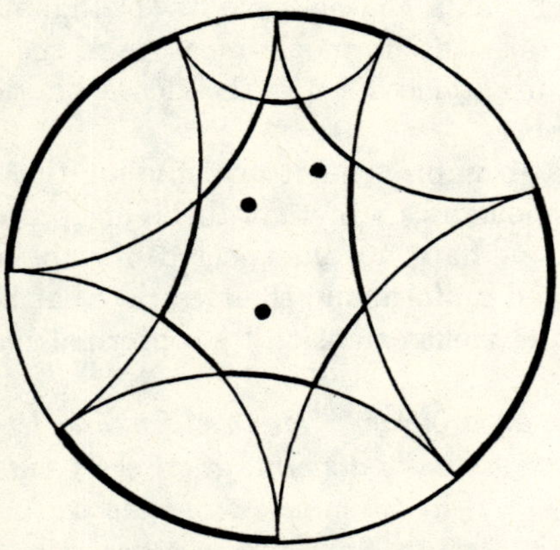

Fig. 21 illustrates §8.7.1 Theorem 1

We readily verify that the harmonic measure (§5.2.1) of A_k at z with respect to the interior of C is

$$(1) \qquad [\arg (z - \beta_k) - \arg (z - \alpha_k) - A_k/2]/\pi,$$

where A_k represents angular measure. The function (1) is the real part of the analytic function

$$(2) \qquad i[\log (z - \alpha_k) - \log (z - \beta_k) + iA_k/2]/\pi.$$

The critical points of $u(z)$ are then the critical points of the function $f(z)$, where $f(z)$ is the sum of the functions (2) for all k; we have

$$(3) \qquad f'(z) = \frac{i}{\pi} \sum_{k=1}^{n} \left[\frac{1}{z - \alpha_k} - \frac{1}{z - \beta_k} \right].$$

The first part of Theorem 1 now follows by suppressing the factor i/π and taking conjugates in (3), and the remainder of the theorem follows from §5.2.1 Theorem 1.

An alternate statement of the conclusion of Theorem 1, obvious when the point z concerned is transformed to the center of C, is that *at a critical point z of $u(z)$ interior to C, the harmonic measure of every positive arc $\alpha_k\alpha_{k+1}$ or $\beta_k\beta_{k+1}$ with respect to the interior of C is less than one-half, except in the case $n = 2$, when this harmonic measure equals one-half.*

The number of critical points of $u(z)$ interior to C is $n - 1$, as follows from §5.2.1 Theorem 1, and can also be proved by considering the variation of arg $f(z)$ as z traces a Jordan curve consisting of C modified by replacing a short arc of C in the neighborhood of each α_k and β_k by a short arc interior to C of a circle with center the α_k or β_k. This latter method, of studying the variation in the argument of an analytic function whose real or pure imaginary part is the harmonic measure, is used frequently by R. Nevanlinna in his well known treatise [1936] which profoundly investigates harmonic measure and employs it consistently as a tool in analysis.

We remark incidentally as a consequence of (1) that if the angular measure A_k approaches zero, so does the harmonic measure of the arc A_k in the point z, uniformly for z on any fixed closed set in the closed interior of C containing no point of the variable arc A_k.

Theorem 1 represents an improvement over the limiting case of §8.4 Theorem 1 as the boundary components C_k of that theorem approach fixed arcs of J_0; such an improvement is perhaps not unexpected, for the arcs A_k in the present Theorem 1 are smooth and uniform and thus restricted as boundary components.

Theorem 1 admits an obvious extension by conformal mapping:

THEOREM 2. *Let C be an arbitrary Jordan curve, and let closed arcs $A_k : \alpha_k\beta_k$, $k = 1, 2, \cdots, n$, of C be mutually disjoint, where the points $\alpha_1, \beta_1, \alpha_2, \beta_2, \cdots, \alpha_n, \beta_n, \alpha_1 = \alpha_{n+1}, \beta_1 = \beta_{n+1}$ lie in succession in the positive sense on C. Denote by A the set $\sum A_k$, and by $u(z)$ the harmonic measure at z of A with respect to the interior of C. If NE geometry is defined for the interior of C by means of a conformal map onto the interior of a circle, arcs of the NE lines $\alpha_k\alpha_{k+1}$ and $\beta_k\beta_{k+1}$ bound a closed NE convex polygon Π interior to C which contains all critical points of $u(z)$ interior to C. In the case $n = 2$, the polygon Π degenerates to a single point, which is a critical point; in every other case all critical points lie interior to Π.*

§8.7.2. Infinite number of arcs. The extension of Theorems 1 and 2 to an infinity of boundary arcs is useful and not difficult. We first establish [compare Nevanlinna, 1936] the

LEMMA. *Let $A = A_1 + A_2 + \cdots$ be the sum of an infinity of mutually disjoint open arcs A_k of the unit circle C. Then the Poisson integral*

$$(4) \qquad u(r, \varphi) = \frac{1}{2\pi} \int_A \frac{(1 - r^2)\, d\theta}{1 + r^2 - 2r \cos(\theta - \varphi)}, \qquad 0 \leqq r < 1,$$

represents a function $u(z)$ which is harmonic and bounded interior to C, and approaches the value unity when z interior to C approaches a point of A and the value zero when z approaches a point of C exterior to A.

If $u_k(z)$ is the harmonic measure of A_k with respect to the interior of C, we have

$$(5) \qquad u(z) = u_1(z) + u_2(z) + \cdots$$

interior to C, uniformly on any closed set interior to C.

Of course we have

$$(6) \qquad u_k(z) = u_k(r, \varphi) = \frac{1}{2\pi} \int_{A_k} \frac{(1 - r^2)\, d\theta}{1 + r^2 - 2r \cos(\theta - \varphi)}, \quad 0 \leqq r < 1;$$

the set A is an open set on the interval $0 \leqq \theta \leqq 2\pi$, hence measurable, so the integral in (4) exists; equation (5) follows by the complete additivity of the Lebesgue integral. The integrand in (4) and (6) is positive, so (5) is a series of positive harmonic functions in $R: 0 \leqq r < 1$; the sum of the series is not greater than unity in R:

$$1 = \int_0^{2\pi} \frac{(1 - r^2)\, d\theta}{1 + r^2 - 2r \cos(\theta - \varphi)},$$

so by Harnack's theorem (§1.1.2) convergence is uniform on any closed set in R, and $u(z)$ is harmonic in R.

Let R_n be an arbitrary closed region whose interior points are points of R and whose boundary points are interior to R except for a closed subarc A_n' of A_n. For $k \neq n$, the denominator in (6) is positive and uniformly bounded for all (r, φ) in R_n and for all θ not in A_n. Consequently the series (5) converges uniformly in R_n; each term of the series is continuous in R_n, so $u(z)$ is continuous in R_n. As z in R_n approaches a point of A_n', $u_n(z)$ approaches unity and $u_k(z)$ for $k \neq n$ approaches zero, so $u(z)$ approaches unity. Likewise if R_0 is a subregion of R whose boundary points on C consist of an arc B containing no point of the closure of A, the series (5) converges uniformly in R_0, so as z in R_0 approaches a point of B, the function $u(z)$ approaches zero. The Lemma is established.

The function $u(z)$ is defined to have the value unity at every point of A and zero at every point of $C - A$, except boundary points of A and $C - A$, and is the harmonic measure of A with respect to R.

THEOREM 3. *Let the set $A = A_1 + A_2 + \cdots$ be the sum of an infinity of disjoint open arcs A_k of the Jordan curve C, and let the complementary arcs of C be*

$B_1, B_2, \cdots$. *Let* $u(z)$ *be the harmonic measure of* A *with respect to the interior* R *of* C. *Then no critical point of* $u(z)$ *lies in the region* $\omega(z, A_j, R) > \frac{1}{2}$ *or* $\omega(z, B_k, R) > \frac{1}{2}$; *if* A_j *and* B_k *have an end-point in common, no critical point of* $u(z)$ *lies in the region* $\omega(z, A_j + B_k, R) > \frac{1}{2}$.

If no interval B_k lies adjacent to A_j, we use the former of these conclusions, which is of course weaker than the latter. The proof of Theorem 3 is immediate; it is sufficient to take C as the unit circle. If z is a point of either of the regions mentioned, so is every point of a suitable neighborhood of z. For n greater than j and k, it follows from Theorem 1 that no critical point of $\omega(z, A_1 + \cdots + A_n, R)$ lies in this neighborhood of z. Theorem 3 follows by Hurwitz's Theorem.

§8.8. Hyperbolic geometry and conformal mapping. The ideas of §8.4 together with the methods of §8.7 yield new results [Walsh 1947a] of significance.

§8.8.1. Doubly-connected region. We proceed to prove

THEOREM 1. *Let the region* R *be bounded by mutually disjoint Jordan curves* $C_1, C_2, D_1, \cdots, D_m$, *let* S *be the doubly-connected region bounded by* C_1 *and* C_2, *and let* $u(z)$ *denote the function harmonic in* R, *continuous in the corresponding closed region, equal to zero on* C_1 *and* C_2 *and to unity on the* D_k. *Then any region* $\omega(z, C_1, S) > \mu(\geqq \frac{1}{2})$ *or* $\omega(z, C_2, S) > \mu(\geqq \frac{1}{2})$ *which contains no point of* $D = \sum D_k$ *contains no critical point of* $u(z) = \omega(z, D, R)$.

As in §7.5, we map conformally the universal covering surface S^∞ of S onto a strip S_1 bounded by two parallel lines, the latter corresponding to C_1 and C_2 respectively. Each curve D_k is mapped into infinitely many Jordan curves $D_{k1}, D_{k2}, \cdots$; for fixed k, geometrically successive curves D_{kj} are congruent under a suitable translation parallel to the sides of the strip and independent of k, and the transform of $u(z)$ is correspondingly periodic. Adjacent to the sides of the strip are smaller interior strips, proper subregions of S_1, free from points of the D_{kj}.

Map now S_1 onto the interior of the unit circle C in the w-plane. The end-points of S_1 correspond to two points A and B of C; each line parallel to the sides of S_1 corresponds to a circular arc interior to C bounded by A and B. Denote by C_1' and C_2' the arcs of C terminated by A and B which are the images of C_1 and C_2 respectively, counted infinitely often. A certain closed lens-shaped region L interior to C bounded by circular arcs terminating in A and B contains all the images D_{kj}' of the curves D_{kj}; choose L as the smallest such closed region containing also the NE line AB. Then L is NE convex, and is bounded by arcs $\omega(w, C_1', |w| < 1) = \mu_1 \geqq \frac{1}{2}$, $\omega(w, C_2', |w| < 1) = \mu_2 \geqq \frac{1}{2}$. In order to apply §8.4 Theorem 1, we need merely to know that the function $U(w)$, the transform in the w-plane of $u(z)$, and which is automorphic under a group of transformations of the interior of C onto itself leaving A and B invariant, is the uniform limit as

q becomes infinite of the function $U_q(w)$ harmonic in the region R_q bounded by C and a suitably chosen *finite number* of the D'_{kj}, continuous in the corresponding closed region and taking the value zero on C and unity on the remaining boundary D'_q of R_q; of course as q increases the boundary D'_q is to increase monotonically; and each D'_{kj} shall belong to some D'_q. We write $U_q(w) = \omega(w, D'_q, R_q)$. The functions $U(w)$ and $U_q(w)$ can be reflected harmonically in C. Then §8.6 Theorem 2 applies, and this completes the proof. It will be noted that Theorem 1 determines an annular region in R adjacent to each of the boundary components C_1 and C_2, a region free from critical points of $u(z)$; compare §7.5 Theorem 2.

Of course the conclusion of Theorem 1 as stated summarizes only a part of what has been proved, for we have essentially shown that *if hyperbolic geometry is defined in S by means of a map of S^∞ onto the interior of a circle, then any NE convex region of S containing D also contains all critical points of $u(z)$.*

§8.8.2. Case $p > 2$. The method applied in Theorem 1 can also be used with a conformal map of a region of connectivity greater than two; we state at once the general result:

THEOREM 2. *Let the region R be bounded by mutually disjoint Jordan curves $C_1, C_2, \cdots, C_p, D_1, D_2, \cdots, D_m$, and let S be the region bounded by $C_1, C_2, \cdots, C_p$. Let S^∞ denote the universal covering surface of S, and define NE lines on S by a conformal map of S^∞ onto the interior of the circle $C: |w| = 1$. Then any NE convex region of S which contains all the points of $D = D_1 + D_2 + \cdots + D_m$ also contains all critical points of the function $\omega(z, D, R)$ in R. In other words if C'_k is an arc of C whose points correspond to points of C_k, then any region $\omega(w, C'_k, |w| < 1) > \mu(\geqq \frac{1}{2})$ interior to C which contains no image point of D contains no critical points of the transform of $\omega(z, D, R)$.*

We suppose $p > 2$. The NE geometry defined in S is independent of the particular choice of the conformal map of S^∞; nevertheless in the map a curve C_k corresponds not to a single arc of C but to an infinity of (open) arcs $C_{k1}, C_{k2}, \cdots$, of C, and the sequence of the arcs C_{kj} on C is a complicated one. The total angular measure of the set $C' = \sum C_{kj}$ is 2π [compare Nevanlinna 1936, p. 25]; indeed the function $\omega(z, C_1 + \cdots + C_p, S) \equiv 1$ is represented in the w-plane by the Poisson integral extended over the set C'; whenever the argument of the general function represented by the Poisson integral of unity on the set C' and zero on the complementary set approaches a point of C', the function approaches unity; since this function admits the automorphic substitutions admitted by the mapping function, it follows by considering the situation in the z-plane that the Poisson integral represents the function identically unity. On the other hand, for the value $w = 0$ the Poisson integral reduces to

$$\frac{1}{2\pi}\int_{C'} d\theta;$$

thus the angular measure of the set C' is 2π. The harmonic measure $\omega(w, C_k', |w| < 1)$, where C_k' is one of the arcs C_{kj}, is not identical with $\omega(z, C_k, S)$; the transform of the latter function onto the w-plane takes the boundary value unity not merely on C_k' but on all the arcs C_{kj}. The function $\omega(w, C_k', |w| < 1)$ is invariant under some but not all of the substitutions of the group of automorphisms of the function which maps the interior of C onto S^∞.

The set D' of images interior to C of all the curves D_k has as limit points on C an infinity of points, the complement $C - C'$ of the set C', so $C - C'$ is of measure zero. Denote by R' the image of R interior to C, and by D_{kj} the sequence of Jordan curves interior to C which are the images of the D_k. The function $\omega(z, D, R) = \omega(w, D', R')$ is the uniform limit in any closed subregion of R' of the harmonic measure $\omega(w, \Delta_q, R_q')$ in w of a finite set Δ_q of the curves D_{kj}, with respect to the subregion R_q' of the interior of C bounded by C and Δ_q, as Δ_q increases monotonically so that each D_{kj} belongs to some Δ_q; this fact is proved precisely as was the corresponding result in the proof of Theorem 1 (or of §8.6 Theorem 2), except that now we may restrict ourselves to the study of a harmonic function with a multiplicative period in a half-plane; we may employ an auxiliary function $u(w) = \omega(w, E, |w| < 1)$, where E is a set of arcs of C of total length less than a prescribed positive δ, where E covers the set $C - C'$; such a set exists because the set $C - C'$ is of measure zero; indeed the set C' is open, so the set $C - C'$ is closed, and the set E may be chosen as a finite set of arcs; when w interior to C approaches a point of E, the function $u(w)$ approaches unity; only a finite number of curves D_{kj} lie wholly or partly exterior to the point set $u(w) \geqq \frac{1}{2}$, for the closure of the set $u(w) < \frac{1}{2}$ contains no point of $C - C'$; when δ approaches zero the function $u(w)$ approaches zero uniformly on any closed set interior to C.

Since $\omega(w, D', R')$ is the uniform limit in any closed subregion of R' of the function $\omega(w, \Delta_q, R_q')$, to which we may apply §8.4 Theorem 1, it follows that any NE convex region interior to C containing all the D_{kj} contains also all critical points of $\omega(w, D', R') = \omega(z, D, R)$, so Theorem 2 follows. We remark that Theorem 2 determines an annular region in R abutting on each C_k which is free from critical points. Theorem 2 as formulated is not symmetric with respect to the C_k and D_j, but we may set $\omega(z, D, R) = 1 - \omega(z, C_1 + \cdots + C_p, R)$, and Theorem 2 applies to the latter function with the roles of the C_k and D_j interchanged.

We note that the inequality

$$\omega(z, C_k, R) = \omega(w, \textstyle\sum C_{kj}, |w| < 1) > \omega(w, C_k', |w| < 1)$$

holds interior to C, so the region $\omega(w, C_k', |w| < 1) > \frac{1}{2}$ is a subregion of the set $\omega(w, \sum C_{kj}, |w| < 1) > \frac{1}{2}$.

§8.9. Linear combinations of Green's functions. The method of conformal mapping of the universal covering surface is fruitful in numerous further cases, as we shall have occasion to indicate.

§8.9.1. Simply-connected region. Green's function for a simply-connected region R has no critical points in R, but linear combinations of such functions may have critical points:

THEOREM 1. *Let R be a Jordan region with boundary C, let $\alpha_1, \alpha_2, \cdots, \alpha_m$ be distinct points of R, and let $G(z, \alpha_k)$ be Green's function for R with pole in α_k, $k = 1, 2, \cdots, m$. The smallest NE convex region Π containing the points α_k contains all critical points in R of the function*

$$(1) \qquad u(z) = \sum_{k=1}^{m} \lambda_k G(z, \alpha_k), \qquad \lambda_k > 0.$$

In the proof we choose R bounded.

Let M be an arbitrary positive number; for M large, the locus $u(z) = M$ in R consists of m mutually exterior Jordan curves J_M, one about each of the points α_k. As M increases these curves J_M vary monotonically and shrink toward the points α_k. For given M, let Π_M denote the smallest NE convex region in R containing the curves J_M; then as M becomes infinite, the region Π_M approaches Π. If z_0 is a point of R not in Π, there exists some M for which z_0 is exterior to Π_M, so by §8.4 Theorem 1 the point z_0 is not a critical point of $u(z)$. Of course (§8.4) the function $u(z)$ has precisely $m - 1$ critical points in the region bounded by C and the J_M for M sufficiently large, and consequently has precisely $m - 1$ critical points in R.

If the points α_k lie on a NE line L for R, and if R is mapped onto the interior of a circle so that the image of L is the axis of reals, for every M sufficiently large there lies precisely one critical point of $u(z)$ on L between successive curves J_M, by §8.2 Corollary to Theorem 2. We have the

COROLLARY. *If in Theorem 1, the points α_k all lie on a NE line L, all critical points of $u(z)$ lie on the smallest segment of L containing the α_k. On each segment of L bounded by successive α_k lies a unique critical point.*

If the points α_k do not necessarily lie on a NE line L but are symmetric with respect to such a line, and if the λ_k are equal for the pairs of points which do not themselves lie on L, §8.4 Theorem 3 is applicable; we leave formulation and proof of the result to the reader.

Theorem 1 extends to the case of a linear combination of Green's functions where the coefficients are not all positive:

THEOREM 2. *Let R be a Jordan region with boundary C, let $\alpha_1, \alpha_2, \cdots, \alpha_m$, $\beta_1, \beta_2, \cdots, \beta_n$ be distinct points of R, and let generically $G(z, z_0)$ be Green's function for R with pole in z_0. If a NE line L separates all the α_j from all the β_k, no critical point of the function*

$$(2) \qquad u(z) = \sum_{k=1}^{m} \lambda_k G(z, \alpha_k) - \sum_{k=1}^{n} \mu_k G(z, \beta_k), \qquad \lambda_k > 0, \qquad \mu_k > 0,$$

lies on L. Thus if such a line L exists, all critical points of $u(z)$ in R lie in two closed NE convex regions Π_1 and Π_2 containing the α_j and β_k respectively, and every NE line separating the α_j and β_k also separates Π_1 and Π_2.

Theorem 2 follows by the method used in Theorem 1, if we consider the loci $u(z) = M$ and $u(z) = -M$, where M is large and positive, by application of §8.4 Theorem 2. A degenerate case is the

COROLLARY. *If in Theorem 2 the points α_j and β_k lie on a NE line L in such a way that the α_j and β_k lie on disjoint segments A and B of L, each segment terminating in a point of C, then all critical points of $u(z)$ in R lie on those two segments. On any open segment of L bounded by two points α_j or two points β_k and containing no α_k or β_j lies a unique critical point of $u(z)$. At most one critical point lies on an open segment of L in R terminated by C and containing no point α_k or β_k.*

Under suitable conditions, both Theorem 1 and Theorem 2 extend to an infinite number of points α_j (and β_k). Choose C as the unit circle $|z| = 1$, so we have

$$G(z, z_0) = -\log\left|\frac{z - z_0}{\bar{z}_0 z - 1}\right|. \tag{3}$$

With $\lambda_k > 0$, each term in the series

$$u(z) = \sum_{k=1}^{\infty} \lambda_k G(z, \alpha_k), \qquad \lambda_k > 0, \tag{4}$$

is positive where defined in R. We assume that every α_k is different from zero, which involves no loss of generality. The partial sums of (4) form a monotonically increasing sequence; by Harnack's Theorem a necessary and sufficient condition that (4) converge uniformly in the neighborhood of the origin, assuming such a neighborhood free from the points α_k, is the convergence of (4) for the value $z = 0$; a necessary and sufficient condition for the convergence of $\sum\lambda_k \log|\alpha_k|$ is (compare §1.6.3) the convergence of $\sum\lambda_k (1 - |\alpha_k|)$.* Thus, provided this last series converges, *Theorem 1 persists if we replace* (1) *by* (4) *and if* Π *contains all the* α_k, for if Π contains all the α_k, it contains all critical points in R of every partial sum of (4), hence by the uniformity of the convergence of $u(z)$ on every closed set in R exterior to Π, the set Π contains all critical points of $u(z)$ as defined by (4). It is similarly true that Theorem 2 persists if we replace the sums in (2) by infinite series, provided the two series $\sum\lambda_k(1 - |\alpha_k|)$ and $\sum\mu_k(1 - |\beta_k|)$ converge.

§8.9.2. Regions of arbitrary connectivity. Under the conditions of Theorem 1, the analog (§8.2) of Bôcher's Theorem is not as powerful as §8.4 Theorem 1. Nevertheless that analog has the advantages i) of not requiring conformal mapping for its application and ii) of applying to more general regions:

* If we choose any point in R other than $z = 0$ distinct from the α_k, we naturally derive an equivalent condition.

THEOREM 3. *Let R be a region bounded by a Jordan configuration B, let $\alpha_1, \alpha_2, \cdots, \alpha_m$ be distinct points of R, and let $G(z, \alpha_k)$ be Green's function for R with pole in α_k. If two disjoint circular regions C_1 and C_2 contain respectively B and all the α_k, they contain all critical points in R of $u(z)$ as defined by (1). If B contains p components, the numbers of critical points of $u(z)$ in C_1 and C_2 in R are respectively $p - 1$ and $m - 1$.*

The proof of Theorem 3, based on §8.2 Theorem 1, follows closely the proof of Theorem 1 and is omitted.

To continue a deeper study of Green's functions and their linear combinations, it is desirable to have an expression for the critical points as positions of equilibrium in a field of force.

THEOREM 4. *Let R be a region whose boundary B is a finite Jordan configuration, and let $G(z, z_0)$ be Green's function for R with pole in the finite point z_0. Then we have for z in R*

$$G(z, z_0) = \int_B \log r \, d\sigma - \log |z - z_0| \tag{5}$$

if R is finite; if R is infinite the constant $G(\infty, z_0)$ is to be added to the second member of (5); here we use the definition $d\sigma = (1/2\pi)(\partial G/\partial \nu)\, ds$ on a level locus of $G(z, z_0)$ near B, and we have $\int_B d\sigma = 1$.

We set $F(z, z_0) = G(z, z_0) + iH(z, z_0)$, where $H(z, z_0)$ is a function conjugate to $G(z, z_0)$ in R. The critical points of $G(z, z_0)$ in R are the positions of equilibrium in the field of force defined by the conjugate of

$$F'(z, z_0) = \int_B \frac{d\sigma}{z - t} - \frac{1}{z - z_0}, \tag{6}$$

namely the field of force due to the positive distribution σ on B, of total mass unity, and a unit negative particle at z_0.

By §8.1.2 equation (6) we have

$$U(z) = \int_B \log r \, d\sigma - \int_D \log r \, d\sigma, \tag{7}$$

for z in R_1, where the function $U(z)$ is harmonic in the finite region R_1 bounded by the Jordan configurations B and D, continuous in the corresponding closed region, with $U(z)$ equal to zero on B and unity on D; on level loci near B and D we have respectively

$$d\sigma = \frac{1}{2\pi} \frac{\partial U}{\partial \nu} ds, \qquad d\sigma = -\frac{1}{2\pi} \frac{\partial U}{\partial \nu} ds. \tag{8}$$

It is clear that if $U(z)$ is multiplied by any constant, the differentials $d\sigma$ defined by (8) are also multiplied by that constant, so (7) and (8) are valid for a function $U(z)$ harmonic in R_1, continuous in the corresponding closed region, zero on B and constant on D.

We now specialize $U(z)$ in (7) as the function $G(z, z_0)$ of Theorem 4, and choose D as the locus $G(z, z_0) = M$ in R, where M is large and positive. In (7) we allow M to become infinite. The function $U(z)$ and the first term in the second member are independent of M, and D approaches z_0 while $\int_D d\sigma = 1$ is constant, so the second term in the second member of (7) approaches $-\log | z - z_0 |$, and (5) is established for R finite. The proof for R infinite follows at once, by adding $U(\infty) = G(\infty, z_0)$ in the second member of (7).

The function

$$H(z, z_0) = \int_B \arg (z - t) \, d\sigma - \arg (z - z_0)$$

is not single-valued in R, but is locally single-valued and is conjugate to $G(z, z_0)$ in R. The function

$$F(z, z_0) = G(z, z_0) + iH(z, z_0) = \int_B \log (z - t) \, d\sigma - \log (z - z_0),$$

plus the term $G(\infty, z_0)$ if R is infinite, is not single-valued in R, but is locally single-valued and has the same critical points as $G(z, z_0)$ in R. Theorem 4 follows.

Theorem 3 admits a direct proof, based on the

COROLLARY. *The critical points of $u(z)$ defined by* (1) *are the positions of equilibrium in the field of force defined by the conjugate of $\sum \lambda_k F'(z, \alpha_k)$, which corresponds to a distribution of total positive mass $\sum \lambda_k$ on B, and negative particles of respective masses $-\lambda_k$ at the points α_k.*

This Corollary suggests almost of itself the application of the cross-ratio theorem (§4.4), and indeed places us in a position to develop the complete analogs of the results of Chapters IV and V. As one example we state the analog of §4.3 Corollary 3 to Theorem 1:

THEOREM 5. *Under the conditions of the Corollary to Theorem 4, let no point α_j with $j \neq k$ lie in the circle $| z - \alpha_k | = b$, and let B lie exterior to that circle. Then no critical point of $u(z)$ lies in the circle whose center is $z = \alpha_k$ and radius $\lambda_k b / (2 \sum \lambda_j - \lambda_k)$.*

We shall not formulate in detail the complete analogy of the development of Chapters IV and V, even though it now lies on the surface. A sample or two might be mentioned:

THEOREM 6. *Let R be a region whose boundary B is a finite Jordan configuration. Let the points $\alpha_1, \alpha_2, \cdots, \alpha_n$ of R lie in a circular region C_1, the points $\alpha_{n+1}, \cdots, \alpha_m$ of R lie in a circular region C_2, and let B lie in a circular region C_3. Then all critical points of $u(z)$ defined by* (1) *lie in C_1, C_2, C_3, and the circular region C_4 which is the locus of the point z_4 defined by the cross-ratio*

$$(z_1, z_2, z_3, z_4) = \sum_{k=1}^{m} \lambda_k \Big/ \sum_{k=1}^{n} \lambda_k, \tag{9}$$

when the loci of z_1, z_2, z_3 *are the regions* C_1, C_2, C_3. *If* B *consists of* p *components, then any of the regions* C_1, C_2, C_3, C_4 *disjoint from the others of those regions contains* $n - 1$, $m - n - 1$, $p - 1$, *or* 1 *critical points in* R *according as* $k = 1, 2, 3,$ *or* 4.

The first part of Theorem 6 follows at once from the Corollary to Theorem 4 by the method of §4.4.3. Here there is no difficulty in enumerating the critical points, for the various components of B can be allowed to coalesce into one component, and the points α_k in each of the regions C_1 and C_2 can be allowed to coalesce to one point, by continuous variation while remaining in their respective circular regions. During this variation we keep the λ_k fixed, so the masses do not change, and C_4 does not vary. One critical point of $u(z)$ in R is lost when two components of B coalesce, and one is lost when two distinct points α_k coalesce, so the latter part of Theorem 6 follows.

A special case of Theorem 6 is of such simplicity that it deserves explicit mention; compare §4.2.4:

COROLLARY. *Let* R *be an infinite region with finite boundary* B, *and let* B *lie in a circular region* C. *Let* z_0 *be a finite point of* R, *let* λ_1 *and* λ_2 *be positive, and let* C_0 *be the circular region found from* C *by stretching from* z_0 *in the ratio* $-\lambda_2/\lambda_1$. *Then all critical points in* R *of the function* $u(z) = \lambda_1 G(z, \infty) + \lambda_2 G(z, z_0)$ *lie in* C *and* C_0. *If those two regions are disjoint, and if* B *consists of* p *components, the circular regions contain respectively* $p - 1$ *and* 1 *critical points.*

Of course the negative sign in the ratio here means that z_0 is an internal center of similitude for C and C_0.

§8.9.3. Multiply-connected regions. The special properties of the conformal map of the universal covering surface of a doubly-connected region lead to special results:

THEOREM 7. *Let the region* R *be bounded by disjoint Jordan curves* C_1 *and* C_2, *and let* $\alpha_1, \alpha_2, \cdots, \alpha_m$ *be points of* R. *Then any region* $\omega(z, C_1, R) > \mu$ $(\geqq \frac{1}{2})$ *or* $\omega(z, C_2, R) > \mu$ $(\geqq \frac{1}{2})$ *which contains no point* α_k *contains no critical point of the function* $u(z)$ *defined by* (1).

Theorem 7 follows from §8.8.1 Theorem 1 by considering the level loci of $u(z)$.

In any given case, Theorem 7 determines an annular region adjacent to each of the curves C_1 and C_2, which is free from critical points, and Theorem 5 determines a neighborhood of each α_k which is free from critical points.

To Theorem 7 we add the

COROLLARY. *Under the conditions of Theorem 7 let all the points* α_k *lie on the locus* $C_0 : \omega(z, C_1, R) = \frac{1}{2}$. *Then all critical points of* $u(z)$ *lie on* C_0. *Any arc of* C_0

bounded by two points α_k and containing no point α_j contains precisely one critical point.

The fact that all critical points lie on C_0 is a consequence of Theorem 1 or of §8.8.1 Theorem 1. If C_0 is transformed into a straight line, the symmetry of the field of force of the Corollary to Theorem 4 shows that any arc A of C_0 bounded by two points α_k and containing no point α_j contains at least one critical point. The total number of critical points is m, as follows from §8.1.3 Theorem 2; this is also the number of arcs A, so each arc A contains precisely one critical point.

The analog of §8.8.2 Theorem 2 follows from that theorem:

THEOREM 8. *Let the region R be bounded by the mutually disjoint Jordan curves $C_1, C_2, \cdots, C_p$, let R^∞ denote the universal covering surface of R, and let NE geometry be defined in R by mapping R^∞ conformally onto the interior of the unit circle $C: |w| = 1$. Then any NE convex region containing the points α_k contains all critical points of the function $u(z)$ defined by (1). In other words if C_k' is an arc of C whose points correspond to points of C_k, then any region $\omega(w, C_k', |w| < 1) > \mu$ $(\geqq \frac{1}{2})$ interior to C which contains no image of a point α_j contains no critical points of the transform of $u(z)$.*

Theorem 8 like Theorems 1 and 7 defines a neighborhood of each curve C_k which is free from critical points, for a certain arc of C corresponds to C_k repeated infinitely often; Theorem 5 defines a neighborhood of each α_k which is free from critical points.

§8.10. Harmonic measure of arcs. In §8.7 we have studied the critical points of the harmonic measure of a finite or infinite number of boundary arcs of a Jordan region. The results and methods there developed apply by conformal mapping to the study of harmonic measure of boundary arcs of regions of higher connectivity.

§8.10.1. Doubly-connected regions. Here the simplest result is

THEOREM 1. *Let R be a region of the z-plane bounded by the disjoint Jordan curves C_1 and C_2, and let A be an arc of C_1. Let NE geometry be defined on the universal covering surface R^∞ of R by a conformal map of R^∞ onto the interior of the circle $C: |w| = 1$. Then the critical point of $\omega(z, A, R)$ does not lie in the image of the region $\omega(w, C_2', |w| < 1) > \frac{1}{2}$, where the arc C_2' is the image on C of C_2 repeated infinitely often, nor in the region $\omega(z, C_2, R) > \frac{1}{2}$. If A_k' and B_k' denote adjacent arcs of C which are respectively images of A and $C_1 - A$, the critical point of $\omega(z, A, R)$ does not lie in the image of the region $\omega(w, A_k' + B_k', |w| < 1) > \frac{1}{2}$, nor in a suitably chosen subregion $\omega(z, C_1, R) > \mu$ $(>\frac{1}{2})$.*

In the proof of Theorem 1 we use §8.7.2 Theorem 3. The image of C_1 repeated infinitely often is an arc C_1' of C, and that of C_2 repeated infinitely often is the complementary arc C_2' of C. The images of A and $C_1 - A$ are arcs A_k' and B_k' infinite in number which alternate on C_1'; when the interior of C is mapped onto the upper half of the W-plane so that C_1' and C_2' correspond to the two halves of the axis of reals, the set of arcs A_k'' (and similarly the set B_k'') which are the images of the A_k' (and the B_k'), are obtained from a single one of those arcs A_k'' (or B_k'') by a transformation $W' = \rho W$, $\rho > 1$, and its positive and negative powers. Obviously a suitable infinite sector with vertex $W = 0$ lies in the sum of all the regions $\omega[W, A_k'' + B_k'', \Im(W) > 0] > \frac{1}{2}$, where A_k'' and B_k'' are adjacent arcs; these regions are free from critical points, as is the region $\omega(w, C_2', |w| < 1) > \frac{1}{2}$, whose image in the z-plane is $\omega(z, C_2, R) > \frac{1}{2}$.

A similar proof yields the

COROLLARY. *The conclusion of Theorem* 1 *persists if A is no longer a single arc but a finite number of arcs of C_1, and A_k' and B_k' are adjacent arcs of C_1' corresponding respectively to an arc of A and an arc of $C_1 - A$; of course several critical points are now involved.*

It is not necessary to mention separately in Theorem 1 and the Corollary the function $\omega(z, A + C_2, R)$, for this function has the same critical points as $1 - \omega(z, A + C_2, R) = \omega(z, C_1 - A, R)$, and except for notation the latter function satisfies the conditions of Theorem 1 or the Corollary.

The case $p = 2$ is not excluded in §8.10.2.

§8.10.2. Arbitrary connectivity. A theorem of considerable generality [Walsh, 1947a] is

THEOREM 2. *Let R be a region of the z-plane bounded by the mutually disjoint Jordan curves $C_1, C_2, \cdots, C_p$, and let $A_1, A_2, \cdots, A_n$ be a finite number of mutually disjoint arcs of the boundary of R, with $A = \sum A_k$; in particular an arc A_k may be a curve C_j. Let the arcs A_k' be the images of the set A when the universal covering surface R^∞ of R is mapped onto the interior of $C: |w| = 1$, and let the complementary arcs of C be $B_1', B_2', \cdots$. Let L be a NE line for the interior of C which joins the end-points of an arc A_k', or the end-points of an arc B_k', or the end-points of an arc $A_j' + B_k'$, where A_j' and B_k' are adjacent arcs on C; then L separates no critical point of the function $u(z) = \omega(z, A, R) = \omega(w, \sum A_k', |w| < 1)$ from the stated arc which it spans. That is to say, in the w-plane no critical point of $u(z)$ lies in the region $\omega(w, A_k', |w| < 1) > \frac{1}{2}$, $\omega(w, B_k', |w| < 1) > \frac{1}{2}$, or $\omega(w, A_j' + B_k', |w| < 1) > \frac{1}{2}$. Thus no critical point of $u(z)$ lies in a subregion of R^∞ bounded by a NE line of R^∞ and either an arc A_k, or an arc B_k of the set of arcs complementary to A, or an arc $A_j + B_k$ consisting of an A_j plus one of the set of complementary arcs; no such critical point lies in any of the regions $\omega(z, A_k, R^\infty) > \frac{1}{2}$, $\omega(z, B_k, R^\infty) > \frac{1}{2}$, $\omega(z, A_j + B_k, R^\infty) > \frac{1}{2}$. Any NE convex region of R^∞ whose boundary contains all interior points of either A or $C - A$ contains all critical points of $u(z)$ in R.*

Theorem 2 is a consequence of §8.7.2 Theorem 3. Of course we do not assume an arc A_k' necessarily a simple image of an arc A_k; compare the proof of Theorem 1. The conclusion of Theorem 2 is symmetric with respect to the sets A and $\sum C_k - A$, and applies both to $u(z)$ and to $1 - u(z) = \omega(z, \sum C_k - A, R)$.

In Theorem 2 there is designated adjacent to every arc A_k, B_k, and $A_j + B_k$ a subregion of R which is free from critical points of $u(z)$. We can obtain adjacent to each of the Jordan curves bounding R a subregion of R free from critical points of $u(z)$. A part of the boundary of R^∞ is the curve C_1 traced continuously and infinitely often, and a corresponding arc C_1' of C consists of images of the arcs A_j on C_1 alternating with the images of the arcs B_k on C_1; the arcs A_j and B_k on C_1 are finite in number and either may be absent. A group of transformations of the interior of C onto itself leaving the mapping function of $|w| < 1$ onto R^∞ invariant transforms the arc C_1' into itself, and simply permutes the arcs A_j and B_k on C_1'. By mapping $|w| < 1$ onto a half-plane so that the image of C_1' is a half-line it is seen that a finite number of the regions $\omega(w, A_j + B_k, |w| < 1) > \frac{1}{2}$ and their images under the group mentioned wholly contain a suitably chosen region $\omega(w, C_1', |w| < 1) > \mu\ (>\frac{1}{2})$; this region corresponds to an annular subregion of R bounded in part by C_1 and free from critical points of $u(z)$.

§8.10.3. Approximating circular regions. The annular region just determined is readily approximated in a particular case:

COROLLARY 1. *Under the conditions of Theorem 2, let C_1 belong wholly to A, let the Jordan curve Γ_1 be separated by C_1 from R, and let the Jordan curve Γ_2 in R separate C_1 from $C_2 + \cdots + C_p$. Let S denote the annular region bounded by Γ_1 and Γ_2. Then no critical point of $u(z)$ lies in R in the region $\omega(z, \Gamma_1, S) > \frac{1}{2}$.*

Denote by R_1 the region bounded by $\Gamma_1, C_2, \cdots, C_p$, and denote by R_1^∞ its universal covering surface, and by S^∞ the universal covering surface of S. Then S^∞ and R^∞ are proper subregions of R_1^∞, and indeed an infinity of replicas of S^∞ lie on R_1^∞. In the inequality

$$(1) \qquad \omega(z, C_1^\infty, R^\infty) > \omega(z, \Gamma_1, S^\infty),$$

let the first member represent harmonic measure where C_1^∞ denotes the Jordan curve C_1 traced continuously and infinitely often in both senses in S but need not denote the curve C_1 in all of its occurrences as part of the boundary of R^∞. We study the functions in (1) on R_1^∞; on C_1^∞ the first member is unity and the second member is less than unity, so (1) is satisfied; on Γ_2 the first member is positive and the second member is zero, so (1) is satisfied; consequently (1) is satisfied throughout the subregion of R_1^∞ bounded by Γ_2 and C_1^∞. In the conformal map of R^∞ onto $|w| < 1$, the curve C_1^∞ corresponds to an arc of C which belongs to A, so by Theorem 2 the region $\omega(z, C_1^\infty, R^\infty) > \frac{1}{2}$ contains no critical points of $u(z)$. It follows from the proof of Theorem 1 that $\omega(z, \Gamma_1, S)$ and

$\omega(z, \Gamma_1, S^\infty)$ are identical, and follows from (1) that every point of the region $\omega(z, \Gamma_1, S) > \frac{1}{2}$ in R is a point of $\omega(z, C_1^\infty, R^\infty) > \frac{1}{2}$, so Corollary 1 is established. Of course inequality (1) is a special case of Carleman's Principle of Gebietserweiterung; compare §7.4.4.

In Corollary 1 the curve Γ_1 may also be taken identical with C_1. The corollary is especially easy to apply when Γ_1 and Γ_2 are both circles, for in that case the locus $\omega(z, \Gamma_1, S) = \frac{1}{2}$ in S is a circle of the coaxal family determined by Γ_1 and Γ_2; compare §7.4.4.

We note as in §8.8.2 an essential difference in Theorem 2 between the cases $p = 2$ and $p > 2$, for in the notation just introduced the two functions $\omega(z, C_1, R)$ and $\omega(z, C_1^\infty, R^\infty)$ are identical in the case $p = 2$, but those two functions are not identical in the case $p > 2$. Otherwise expressed, let C_1^∞ correspond to an arc A_1' of C; then in the case $p > 2$, other arcs A_k' of C also are images of C_1, but not in the case $p = 2$; in the case $p > 2$ we have $\omega(w, A_1', |w| < 1) = \omega(z, C_1^\infty, R^\infty) < \omega(z, C_1, R)$. Here $(p > 2)$ the region $\omega(z, C_1^\infty, R^\infty) > \frac{1}{2}$ is a proper subregion of the region $\omega(z, C_1, R) > \frac{1}{2}$. In both the cases $p = 2$ and $p > 2$ if an arc A_1 not an entire curve C, corresponds to arcs A_k'' of C, then we have $\omega(w, A_k'', |w| < 1) < \omega(w, \sum A_k'', |w| < 1) = \omega(z, A_1, R)$, so the image of the region $\omega(w, A_k'', |w| < 1) > \frac{1}{2}$ is a proper subregion of the region $\omega(z, A_1, R) > \frac{1}{2}$.

Corollary 1 is based on an approximation to the harmonic measure of a Jordan curve; there exists a similar result for a Jordan arc:

Corollary 2. *Under the conditions of Theorem 1, let the Jordan arc E of the boundary of R consist of an arc A_k, or an arc B_k of the set of arcs complementary to A, or an arc A, plus an adjoining arc B_k. Let a circle Ω be divided into open arcs Ω_1 and Ω_2 by the end-points of E, let Ω_1 lie in R and Ω_2 lie exterior to R. Then no critical point of $u(z)$ lies in R in the region $\omega(z, \Omega_2, \Omega) > \frac{1}{2}$, where the region Ω is the circular region containing E bounded by the circle Ω. Otherwise expressed, the NE line for the region Ω joining the end-points of E separates no critical points of $u(z)$ in Ω from Ω_1.*

We adjoin to a single sheet of R^∞ the points of the region Ω, and denote the new region thus obtained by R_1^∞. In the inequality on R_1^∞

$$\omega(z, E, R^\infty) > \omega(z, \Omega_2, \Omega), \tag{2}$$

where the first member refers to the arc E in a single sheet of R^∞, at a point of E the first member is unity and the second is less than unity, and at a point of Ω_1 in R the first member is positive and the second member is zero. Thus (2) is satisfied at every point common to the regions R and Ω. If E' denotes the image of E on C, where E' is chosen merely as a single arc, it follows from Theorem 2 that the region $\omega(w, E', |w| < 1) = \omega(z, E, R^\infty) > \frac{1}{2}$ contains no critical point of $u(z)$. It follows from (2) that the region $\omega(z, \Omega_2, \Omega) > \frac{1}{2}$ is in R_1^∞ a subregion of $\omega(z, E, R^\infty) > \frac{1}{2}$, so Corollary 2 follows.

CHAPTER IX

FURTHER HARMONIC FUNCTIONS

In the preceding chapter we studied the critical points of assigned harmonic functions of relatively simple form; in the present chapter we emphasize the range of the methods, and show how those methods extend to the study of various more complicated functions. Broadly, and with some exceptions, the functions studied in Chapter VIII are either harmonic measures of boundary components or harmonic measures of sets of boundary arcs, while the functions studied in Chapter IX are more general harmonic measures together with their linear combinations. Green's functions are constant multiples of harmonic measures in suitably chosen regions and appear in both chapters. Potentials due to assigned distributions of matter are treated in §9.9.

§9.1. A general field of force. Let R be a region whose boundary B is a finite Jordan configuration, and let the function $u(x, y)$ be harmonic in R, continuous in $B + R$. At least formally, the fields of force introduced in §§7.1.3, 8.1.2, and 8.9.2 can be broadly generalized [Walsh, 1948c]. The function $u(x, y)$ is represented by Green's formula

$$(1) \qquad u(x, y) = \frac{1}{2\pi} \int_B \left(\log r \frac{\partial u}{\partial \nu} - u \frac{\partial \log r}{\partial \nu} \right) ds + u_0, \qquad (x, y) \text{ in } R,$$

where ν indicates interior normal and where u_0 is zero or $u(\infty)$ according as R is finite or infinite. We denote by $v(x, y)$ a function conjugate to $u(x, y)$ in R, whence $(\partial u/\partial \nu)\, ds = -\, dv$, $(\partial \log r/\partial \nu)\, ds = -d[\arg (z - t)]$, and for $z = x + iy$ in R we have

$$u(x, y) = \frac{-1}{2\pi} \int_B \log r\, dv + \frac{1}{2\pi} \int_B u d[\arg (z - t)] + u_0$$

$$= \frac{-1}{2\pi} \int_B \log r\, dv + \frac{1}{2\pi} [u \arg (z - t)]_B - \frac{1}{2\pi} \int_B \arg (z - t)\, du + u_0,$$

$$v(x, y) = \frac{-1}{2\pi} \int_B \arg (z - t)\, dv - \frac{1}{2\pi} [u \log |z - t|]_B + \frac{1}{2\pi} \int_B \log |z - t|\, du,$$

$$f(z) = u + iv = \frac{-1}{2\pi} \int_B \log (z - t)\, dv - \frac{i}{2\pi} [u \log (z - t)]_B + \frac{i}{2\pi} \int_B \log (z - t)\, du + u_0,$$

$$(2) \qquad f'(z) = \frac{-1}{2\pi} \int_B \frac{dv}{z - t} - \frac{i}{2\pi} \left[\frac{u}{z - t} \right]_B + \frac{i}{2\pi} \int_B \frac{du}{z - t};$$

here $v(x, y)$ and $f(z)$ may be multiple-valued, but $f'(z)$ is single-valued.

Equation (2)* suggests the taking of conjugates and interpretation of the second member as a field of force, but the factor i requires the introduction of skew particles, as in §6.3.2, and of skew matter on B. Certain fields of force due to skew particles are important for the sequel; we prove the

LEMMA. *In the field of force due to positive and negative unit skew particles at distinct points α and β respectively, the lines of force are the circles of the coaxal family determined by α and β as null circles, and the sense of the force is counterclockwise on the circles containing α in their interiors, clockwise on those containing β in their interiors.*

This conclusion follows from the characterization (§4.1.2 Corollary to Theorem 3) of the field due to equal and opposite ordinary particles. At each point of the field of the Lemma the force is found from the force of the previous field by positive rotation through $\pi/2$.

The field of force defined as the conjugate of $f'(z)$ in (2) is thus the field due to the distribution $-dv/2\pi$ of matter on B, to skew particles at the points of discontinuity of $u(x, y)$ on B, and to the distribution $-\ du/2\pi$ of skew matter (that is, a distribution which is the limit of a suitably chosen sequence of sets of skew particles) on B. Under favorable circumstances, the signs of du and dv on various parts of B can be determined, and the distribution of skew matter can be approximated if necessary by a distribution of skew particles, so the field of force can be effectively studied with reference to positions of equilibrium, namely critical points of $u(x, y)$ and of $f(z)$.

We note that in equation (2) we have $\int_B dv = 0$, $\int_B du = [u]_B$, so the total mass (if existent) of ordinary matter is zero, as is the total mass (if existent) of skew matter.

As in the special cases previously considered, at a finite point of R *the conjugate of $f'(z)$ is the gradient of $u(x, y)$*; compare §7.1.3. Naturally this same field of force can be considered as due to a number of different distributions of matter.

Under linear transformation of z or arbitrary one-to-one conformal transformation of $R + B$, the functions u, v, $f(z)$, and the quantities dv, $[u]_B$, du are invariant, as in the direction of the force, $-\arg [f'(z)]$, namely the direction of grad $u(x, y)$. But the magnitude $|f'(z)|$ of the force is altered by a multiplicative factor, the modulus of the derivative of the mapping function.

Thus far the derivation of (1) and (2) has been formal, and requires justification, but that justification can be provided in numerous instances. It may be desirable to commence with Cauchy's integral formula for $f'(z)$, where (as in §6.6) the integral is taken over the boundary of the given region, or over a suitable modification of that boundary. Thus we consider for simplicity a positive

---

* Equation (2) is precisely the expression of $f'(z)$ by Cauchy's integral formula, including an additional term provided $u(x, y)$ is not continuous on B. For instance if $f(z)$ has a singularity of the form $i \log (z - \alpha)$ on B and if B is smooth, equation (2) may be valid when suitably interpreted, but the usual form of Cauchy's integral formula is not valid.

arc $A_0 : \alpha < \arg z < \beta$ of the unit circle $C: | z | = 1$, and study (§8.7.1) the function $u(z) = \omega(z, A_0, | z | < 1) = [\arg (z - \beta) - \arg (z - \alpha) - A_0/2]/\pi$; in the latter formula A_0 indicates arc length. If $v(z)$ is conjugate to $u(z)$ in R we set $f(z) = u(z) + iv(z)$, which is not necessarily single-valued, but $f'(z)$ is analytic and single-valued, and for z interior to C is represented by Cauchy's integral taken over a suitable Jordan curve C':

$$f'(z) = \frac{1}{2\pi i} \int_{C'} \frac{du + idv}{t - z}, \qquad | z | < 1. \tag{3}$$

The functions $u(z) + (1/\pi) \arg (z - \alpha)$ and $v(z) - (1/\pi) \log | z - \alpha |$ when suitably defined at $z = \alpha$ are harmonic at that point; we choose C' as C modified by replacing the arcs of C in the neighborhoods of α and β by small arcs interior to C of two loci $v(z) = \text{const}$. Let the former of these arcs have the end-points α' and α'' on C; let the points a_1 and a_2 be chosen in the positive order $a_1, \alpha', \alpha, \alpha'', a_2, \beta$, on C; we fasten our attention on the portion of the integral in (3) taken over the arcs $a_1\alpha'$, $\alpha'\alpha''$, $\alpha''a_2$ of C', where $\alpha'\alpha''$ is an arc of the locus $v(z) = -M$ (<0). On these three arcs we have respectively $u = \text{const}$, $v = \text{const}$, $u = \text{const}$. On the arc $\alpha'\alpha''$ the function $u(z)$ varies monotonically from $u(z) = 0$ to $u(z) = 1$, for no critical point of $u(z)$ lies on that arc. As M becomes infinite the total integral over these three arcs remains constant, by (3); for z fixed interior to C we clearly have

$$\frac{1}{2\pi i} \int_{t=\alpha'}^{t=\alpha''} \frac{du}{t - z} \to i/2\pi(z - \alpha);$$

consequently we have

$$\frac{1}{2\pi} \left(\int_{t=a_1}^{t=\alpha'} + \int_{t=\alpha''}^{t=a_2} \right) \frac{dv}{t - z} \to \frac{1}{2\pi} \int_{t=a_1}^{t=a_2} \frac{-dv}{z - t},$$

where the latter integral is improper, and is to be regarded as a kind of "principal value"; that is to say, we truncate $v(z)$ on both sides of the point $z = \alpha$ by the same value $-M$, compute the integral for the truncated $v(z)$, and define the principal value as the limit (just shown to exist) of this integral for the truncated $v(z)$. A similar discussion of the arc of C' in the neighborhood of the point $z = \beta$ now yields the formula

$$f'(z) = \frac{1}{2\pi} \int_C \frac{-dv}{z - t} + \frac{i}{2\pi(z - \alpha)} - \frac{i}{2\pi(z - \beta)}, \tag{4}$$

which is a special case of (2). Of course the particular function $u(z)$ has no critical points interior to C, but in locating critical points equation (4) is of interest in combination with similar formulas, and moreover by taking conjugates represents a field of force, the gradient of $u(z)$. This field is then due to the spread $-dv/2\pi$ on C, and to skew particles of masses $-1/2\pi$ and $1/2\pi$ at α and β. It is to be noted, however, that the total mass in this spread is not finite; care must be exercised in interpreting the integral; the force should be interpreted in terms of the definition of the improper integral; nevertheless the formula is adequate for the relatively simple applications to follow.

§9.2. Harmonic measure of arcs. It is characteristic of the general methods of §9.1 that in the final results they do not essentially involve conformal mapping.

§9.2.1. A single arc. Our first application of those methods is

THEOREM 1. *Let the region R be bounded by a finite Jordan curve J and a finite Jordan configuration B disjoint from J, and let A be an arc of J from α to β, positive with respect to R. Then the critical points of $u(z) = \omega(z, A, R)$ in R are the positions of equilibrium in the field of force due to a positive spread $-dv/2\pi$ on $B + J - A$, to a negative spread $-dv/2\pi$ on A, and to skew particles of masses $-1/2\pi$ and $1/2\pi$ at α and β respectively, where $v(z)$ is conjugate to $u(z)$ in R.*

On various occasions (§§6.6, 7.1.3, 8.1.2) we have shown how such an integral as that in §9.1 equation (3) can be taken over a Jordan arc or curve which is not necessarily smooth, by first taking the integral over an auxiliary smooth arc or curve, and then allowing this auxiliary arc or curve to approach the given one. In an integral involving $dv(t)$, we keep z fixed and allow t to vary along arcs of the loci $v(t) = \text{const}$, except in the neighborhoods of discontinuities of $v(t)$. Under the conditions of Theorem 1, this method can be used for the integral over $B + J$, where neighborhoods of α and β are cut out by loci $v(z) = \text{const}$; moreover the essential behavior of $u(z)$ and $v(z)$ in the neighborhoods of the points α and β is the same as in the case (§9.1) that B is empty and J is the unit circle; in each case $u(z)$ is unity on A and zero on $J - A$; the function $v(z)$ varies monotonically (as may be seen by studying the situation after conformal map of J onto the unit circle) on the two arcs A and $J - A$; when z in R approaches α or β, the function $v(z)$ becomes negatively or positively infinite respectively; we use auxiliary loci $v(z) = \text{const}$ near α and β passing through no critical points of $u(z)$ and separating all such critical points in R from α and β. Again we derive an improper integral over the boundary of R itself as part of the representation, whose "principal value" is to be used as before; thus for z in R we have

$$(1) \qquad f'(z) = \frac{1}{2\pi} \int_{B+J} \frac{-dv}{z - t} - \frac{i}{2\pi} \left[\frac{1}{z - \beta} - \frac{1}{z - \alpha} \right],$$

where $f(z) = u(z) + iv(z)$. At every point of A we have $-dv < 0$; at every point of $B + J - A$ we have $-dv > 0$; Theorem 1 follows.

As a consequence of Theorem 1 we prove

THEOREM 2. *Under the conditions of Theorem 1, let the circle Γ separate the interior points of A from the points of B not on Γ and from the interior points of $J - A$. Then no critical point of $u(z)$ lies in R on Γ.*

By a linear transformation transform Γ into a straight line, while J becomes a Jordan curve whose interior contains R. At an arbitrary point z of R on Γ, the force due to the distribution on A has a component orthogonal to Γ, and the force due to the distribution on $B + J - A$ has a component orthogonal to Γ in the same sense; to be sure, these distributions cannot be considered separately and in their entirety, for each distribution contains an infinite mass, and the integral in (1) is improper. Nevertheless the definition of the improper integral suggests the interpretation of the force as the finite limit of the force due to a finite spread of matter; in the limiting process defining the integral, the component orthogonal to Γ of the variable force increases monotonically. With this interpretation of the force, the senses of the two components at z orthogonal to Γ are the same. Moreover the skew particles at α and β exert forces at z which are also orthogonal to Γ and act in this same sense. Consequently z is not a position of equilibrium nor a critical point of $u(z)$.

The enumeration of the critical points $u(z)$ in R is not difficult; we study the variation in the direction angle of the force as z traces in the positive sense with respect to R level loci $u(z) = \text{const}$ in R near each component of B, and also traces a curve in R near J avoiding α and β by small circular arcs near those points; on each of the former curves, say $p - 1$ in number, the direction angle of the force decreases by 2π, and on the latter curve the direction angle of the force changes by zero. The total increase in the direction angle of the force is $-2(p - 1)\pi$, and that in the direction angle of arg $[f'(z)]$ is $2(p - 1)\pi$, so $u(z)$ has precisely $p - 1$ critical points in R. It is to be noticed that this same method applies to the enumeration of the critical points of $u(z)$ in the region R' bounded by B, $J - A$, and Γ', where Γ' is the arc $\alpha\beta$ of Γ interior to R, and the number of critical points in R' is the same as that in R:

COROLLARY. *Under the conditions of Theorem 2, no critical points of $u(z)$ lie in the Jordan region bounded by A and by the arc $\alpha\beta$ of Γ interior to R.*

§9.2.2. Several arcs. Theorems 1 and 2 can be widely extended:

THEOREM 3. *Let the boundary of a region R consist of two Jordan configurations B_1 and B_2 having no more than a finite number of common points, and let Γ be a circle. With respect to Γ let the boundaries of each of the sets $B_1 \cdot \Gamma$ and $B_2 \cdot \Gamma$ contain at most a finite number of isolated points; and let Γ separate the points of B_1 not on Γ from the points of B_2 not on Γ. Then no critical point of $u(z) = \omega(z, B_1, R)$ lies in R on Γ.*

In Theorem 3 we may allow B_1 and B_2 to have a finite number of arcs of Γ in common, provided the two banks of such arcs are distinguished and are assigned to B_1 and B_2 respectively. In the proof we take $B_1 + B_2$ as finite and Γ as a line; it is sufficient to show that no finite critical point lies on Γ.

Theorem 3 is to be proved by the use of a field of force, and two methods suggest themselves: (i) derivation of a formula analogous to (1) by the same method and (ii) addition of the formulas (1) for the functions corresponding to the individual arcs of B_1. The former method is slightly preferable, for in the latter method we need to add several different values of the differential dv, not necessarily of the same sign on each arc of $B_1 + B_2$; moreover the definition of the mproper integral in (1) is expressed in terms of $v(x, y)$, and this definition does not carry over directly to the sum of two such integrals;—but these objections are easily met, and either (i) or (ii) suffices. The new field of force corresponds to an improper integral and also to pairs of skew particles (α_k, β_k) of masses $\mp 1/2\pi$ at the end-points of arcs of B_1 which are also end-points of arcs of B_2.

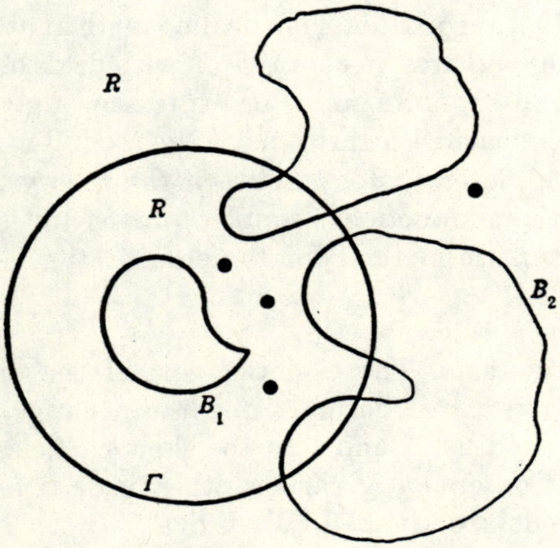

Fig. 22 illustrates §9.2.2 Theorem 3

The points α_k and β_k lie on Γ, but not necessarily in such a way that the sense $\alpha_k\beta_k$ on Γ is independent of k; for instance the order $\alpha_1, \beta_1, \alpha_2, \beta_3, \alpha_3, \beta_2$ is possible if the Jordan arcs $\alpha_1\beta_1, \alpha_2\beta_2, \alpha_3\beta_3$ of B_1 are simple arcs lying on the same side of Γ such that arcs $\alpha_2\beta_3$ and $\alpha_3\beta_2$ of B_2 lie on the opposite side of Γ. However, if A is a minimal (i.e. containing no smaller) open arc of Γ in $R + B_1 + B_2$ both of whose end-points are end-points of arcs of both B_1 and B_2, a particle α, lies at one end-point of A and a particle β_k at the other end-point; this order of particles on Γ is independent of the particular arc A considered. It follows that the points α, and the points β_k alternate on Γ when arranged in a suitable order; we consider them paired in order on Γ commencing with the particles nearest a point z in R on Γ as initial point; no two particles of a pair shall be separated by z, where z is studied as a possible position of equilibrium.

The proof of Theorem 3 now follows that of Theorem 2; details are omitted. Theorem 3 is a generalization of the analog (§8.2) of Bôcher's Theorem, and is itself a further analog of Bôcher's Theorem.

Theorems 2 and 3 are somewhat different in content if not in proof, insofar as the existence of a circle Γ of Theorem 2 ordinarily implies the existence of an infinite number of those circles, whereas in Theorem 3 it may occur that only one circle Γ can be constructed. However, under the conditions of Theorem 3 the direction of the force is approximately known near every boundary point of R, and also at every point of Γ in R. It is then quite easy to determine the number of critical points of $u(z)$ in any subregion of R bounded wholly by points of $B_1 + B_2 + \Gamma$. In particular we note the

COROLLARY. *Under the conditions of Theorem 3 a Jordan subregion of R bounded wholly by an arc of Γ and a Jordan arc of B_1 or B_2 contains no critical points of $u(z)$.*

We emphasize the advantages here of the use of the field of force represented by the conjugate of §9.2.1 equation (1), dealing with matter spread directly over given Jordan curves and arcs, in comparison with fields obtained by approximating Cauchy's integral by finite sums, or expressing a given harmonic function as the limit of a sequence of logarithms of the moduli of rational functions. Great advantages are a closer analogy between the theory for harmonic functions and that for rational functions, simpler proofs and stronger theorems, and in addition a systematic method for the enumeration of critical points of harmonic functions.

§9.2.3. Linear combinations. The essential part of the proof of Theorem 3 is that the positive matter shall lie in Γ_1 and the negative matter in Γ_2, the two closed regions bounded by Γ, and that the skew particles if any shall lie on Γ in the proper order. Consequently various other fields of force may be added, provided these conditions are satisfied. We prove

THEOREM 4. *Under the conditions of Theorem 3 let $R^{(k)}$ be a region containing R bounded by Jordan configurations $B_1^{(k)}$ and $B_2^{(k)}$ belonging to B_1 and B_2 or separated from R by B_1 and B_2 respectively, $k = 1, 2, \cdots, n$. Let $B_1^{(k)}$ and $B_2^{(k)}$ have at most a finite number of common points, and let the boundary on Γ of the intersections of Γ with $B_1^{(k)}$ and $B_2^{(k)}$ contain at most a finite number of isolated points. Then no critical point of the function*

$$\sum_{k=1}^{n} \lambda_k \, \omega(z, B_1^{(k)}, R^{(k)}), \qquad \lambda_j > 0, \tag{2}$$

lies in R on Γ.

Since R lies in $R^{(k)}$, the field of force for the function $\omega(z, B_1^{(k)}, R^{(k)})$ satisfies the essential conditions that the positive matter lies in Γ_1, the negative matter in Γ_2, and that the skew particles lie on Γ in the proper order, an order defined by choosing as initial point on Γ a point in R and therefore in every $R^{(k)}$. We add these fields of force in the proportions indicated by the λ_j in (2); the method of proof of Theorem 3 applies to the combined field of force, and yields Theorem 4. We do not exclude the possibility that $B_1^{(k)}$ and $B_2^{(k)}$ should be identical with B_1 and B_2.

In this field of force, the direction of the force is approximately known at every point z_0 of Γ in R, and also near an arbitrary point z_0 of $B_1 + B_2$ which belongs to every $B_1^{(k)} + B_2^{(k)}$. Thus the number of critical points of (2) lying in any region bounded only by such points z_0 is readily found by our present methods. As an analog of the Corollary to Theorem 3, it is true that a Jordan subregion of R bounded by an arc of Γ and a Jordan arc of B_1 or B_2 which belongs to every $B_1^{(k)}$ or $B_2^{(k)}$ contains no critical point of (2).

Under suitable conditions the total number of critical points in the given region is easily obtained. The following theorem is due to Nevanlinna [1936], who studies the function $u + iv$ and its derivative but not the present interpretation as a field of force:

THEOREM 5. *Let R be a region bounded by mutually disjoint Jordan curves $C_1, C_2, \cdots, C_p$, and let $A_1, A_2, \cdots, A_m$ be mutually disjoint closed Jordan arcs (not Jordan curves) of the curves $C_{n+1}, C_{n+2}, \cdots, C_p$. Then the total number of critical points in R of the function*

$$u(z) = \omega(z, C_1 + C_2 + \cdots + C_n + A_1 + A_2 + \cdots + A_m, R)$$

is $m + p - 2$.

Theorem 5 and its proof are valid where each C_j is a Jordan configuration with a single component, provided a suitable convention is made for the separate banks of an arc of C_j if both banks lie in R.

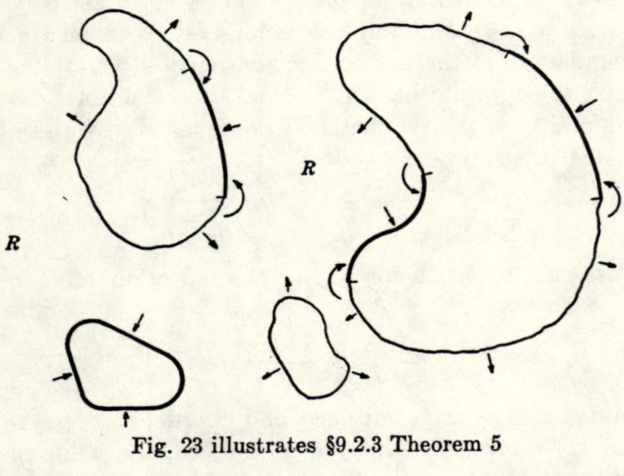

Fig. 23 illustrates §9.2.3 Theorem 5

We choose the point at infinity as an interior point of R, and denote by k_j the number of arcs A_k on C_j: $\sum k_j = m, j > n$. Let C_j' for $j \leqq n$ denote a Jordan curve in R near C_j of the locus $u(z) = 1 - \epsilon$, where ϵ is small, and where no critical point of $u(z)$ lies on C_j' or between C_j' and C_j. At a point of C_j' the force is directed toward the interior of C_j', and as z traces C_j' in the clockwise sense the direction angle of the force increases by -2π. If the curve C_j for $j > n$

has no arc A_k lying on it, we consider in R near C_j a Jordan curve C_j' of the locus $u(z) = \epsilon$, where ϵ is small, and where no critical point of $u(z)$ lies on C_j' or between C_j' and C_j. At a point of C_j' the force is directed toward the exterior of C_j', and as z traces C_j' in the clockwise sense the direction angle of the force increases by -2π. If we have $k_j > 0$, we consider a Jordan curve C_j' in R near C_j consisting of arcs $u(z) = 1 - \epsilon$ or ϵ near arcs of C_j on which $u(z) = 1$ or 0, and near end-points of arcs A_k consisting of circular arcs with those end-points as centers; no critical point of $u(z)$ shall lie on C_j' or between C_j' and C_j. On an arc of C_j' near an arc of C_j on which we have $u(z) = 1$, the force is directed toward the interior of C_j', and on an arc of C_j' near an arc of C_j on which we have $u(z) = 0$, the force is directed toward the exterior of C_j'. At the initial point of each A_k as C_j is traced in the clockwise sense is a negative skew particle, and at the terminal point of A_k is a positive skew particle. Thus when z traces C_j' in the clockwise sense, the total increase in direction angle of the force is $-2(k_j + 1)\pi$. For the entire set of curves C_j' the total change in the direction angle of the force is $-2\pi\sum_1^p(k_j + 1) = -2(m + p)\pi$, and the total change in arg $[f'(z)]$ is $2(m + p)\pi$. By the Principle of Argument the number of zeros of $f'(z)$ is $m + p$, and hence the number of critical points in R is $m + p - 2$.

§9.3. Arcs of a circle. In the use of the field of force of §9.1, the circle has an important property in addition to the inherent simplicity of the figure and its importance in conformal mapping. If a function $u(x, y)$ is harmonic in a region R bounded at least in part by an arc A of a circle C, and if $u(x, y)$ vanishes on A, then $u(x, y)$ can be extended harmonically across A so as to be harmonic at every point of A. If $u(x, y)$ takes any continuous values on an arc A_1 of C which is part of the boundary of R, the harmonic extension of $u(x, y)$ takes the negatives of those values on the opposite bank of A_1 and the values of $dv = -(\partial u/\partial \nu)\, ds$ are equal and opposite on the two banks. Consequently if we use §9.1 equation (1) for R, we need to study

$$\frac{1}{2\pi}\int_{A_1}\left(\log r\,\frac{\partial u}{\partial \nu} - u\,\frac{\partial \log r}{\partial \nu}\right) ds,$$

whereas if we use that equation for R plus the reflection of R in C, we study instead

$$-\frac{1}{\pi}\int_{A_1} u\,\frac{\partial \log r}{\partial \nu}\, ds; \tag{1}$$

here the integral is taken over A_1 but once, and in order to preserve the equality in §9.1 equation (1) the term $+u(\infty) = -u(0, 0)$, the value of $u(x, y)$ harmonically extended to infinity, may need to be added in (1). The simplification represented by (1) leads to more refined results for regions bounded at least in part by a circular arc.* If u is constant on A_1, the value of (1) is (formally)

* Naturally the method of reflection of $u(x, y)$ across an *arbitrary* analytic Jordan arc A_1 on which $u(x, y)$ is constant can be used, but if A_1 is not an arc of a circle, new singularities of $u(x, y)$ may arise not merely because of original singularities of $u(x, y)$ but also because of the failure of reflection as a one-to-one conformal transformation.

$$\frac{u}{\pi} \arg (z - t) |_{A_1}.$$

§9.3.1. Disjoint arcs. The simplest non-trivial case to which the remarks just made apply is that of the harmonic measure of the arc $A_1: \alpha_1\beta_1$ with respect to the interior of C, for in this case the methods of §9.1 reduce (1) to $[\arg (z - \beta_1) - \arg(z - \alpha_1)]/\pi$, plus a possible constant. Thus for a number of arcs A_k we have as in §9.1

THEOREM 1. *Let C be the unit circle $z = e^{i\theta}$, let A_j be the arc $\alpha_j < \theta < \beta_j$ and B_k be the arc $\gamma_k < \theta < \delta_k$, $j = 1, 2, \cdots, m$; $k = 1, 2, \cdots, n$. The critical points interior to C of the function*

$$(2) \qquad u(z) = \sum_{j=1}^{m} \lambda_j \omega(z, A_j, R) - \sum_{k=1}^{n} \mu_k \omega(z, B_k, R), \qquad \lambda_j > 0, \qquad \mu_k > 0,$$

where R is the interior of C, are the positions of equilibrium in the field of force due to skew particles of masses $+\lambda_j$ and $-\lambda_j$ in the points β_j and α_j, and of masses $+\mu_k$ and $-\mu_k$ in the points γ_k and δ_k.

Theorem 1 is in essence not greatly different from the first part of §8.7.1 Theorem 1; the present result is expressed in terms of skew particles, and the former result in terms of ordinary particles. It is clear that two fields of force due respectively to ordinary and skew particles of the same masses differ merely in that the force in the latter field is obtained from that in the former one by a rotation of $+\pi/2$. The positions of equilibrium are the same in the two fields. In general, the transition from a field due to ordinary particles to a field due to skew particles is readily made, and if the two fields are not to be combined with other fields, the field more likely to appeal to the intuition would naturally be chosen.* The W-curves for one field are identical with those for the other. In the present case we shall shortly combine the field with others, and skew particles are advantageous.

In our further study of harmonic measure we shall use the W-curves:

LEMMA. *If $\alpha_1\beta_1$ and $\alpha_2\beta_2$ are positive closed disjoint arcs of C, then the W-curves for equal and opposite skew particles at α_1 and β_1, and equal and opposite skew particles at α_2 and β_2, consist only of arcs of C if the particles at α_1 and α_2 are opposite in kind, and consist of arcs of C plus the circle K common to the two coaxal families determined by (α_1, β_1) and (α_2, β_2) as null circles if the particles at α_1 and α_2 are alike in kind.*

The Lemma is proved by inverting with α_1 as center of inversion, and of course making use of the Lemma of §9.1. The two families of lines of force for

* For instance Bôcher's Theorem seems intuitively reasonable if skew particles are used. Of course such a result as §1.5.1 Lemma 1 applies both to ordinary particles and skew particles.

pairs of particles are then two coaxal families of circles, each family determined by two null circles on the line C' which is the image of C; the centers (when existent; one circle is a straight line) of all circles of these families lie on C'; a point of tangency of two circles must lie on C', so no two lines of force have the same direction at a point not on C' unless the lines of force are identical. The two coaxal families have a unique circle in common; compare the remark at the end of §5.1.5.

THEOREM 2. *Let the closed mutually disjoint arcs $A_1, A_2, \cdots, A_m = A_0$ lie on $C: |z| = 1$ in the counterclockwise order indicated, and let Π be the closed subregion of $R: |z| < 1$ bounded by subarcs of the A_j and NE lines for R, respectively arcs in R of the circles C_j common to the two coaxal families defined by the end-points of A_j and by the end-points of A_{j+1} as null circles. Then Π contains all critical points in R of the function*

$$u(z) = \sum_{j=1}^{m} \lambda_j \omega(z, A_j, R), \qquad \lambda_j > 0. \tag{3}$$

Except in the case $m = 2$, no critical point of $u(z)$ lies on the boundary of R.

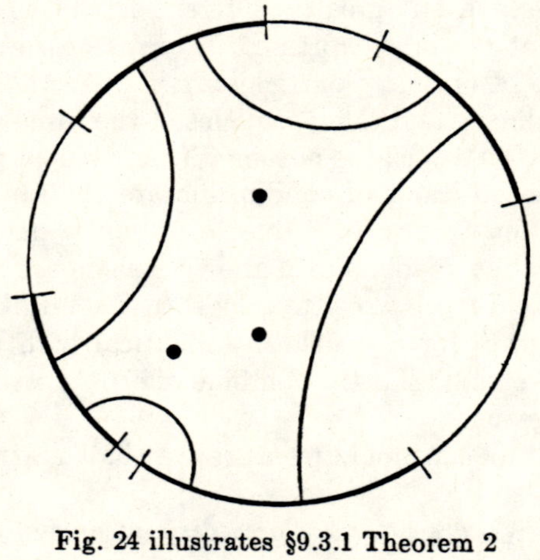

Fig. 24 illustrates §9.3.1 Theorem 2

Let z be a point of R not in Π, and suppose for definiteness no point of Π to lie between C_1 and z. Let A_k be the positive arc $\alpha_k\beta_k$, and let Γ be the circle through z of the coaxal family determined by α_1 and β_1 as null circles. In the field of force defined by Theorem 1, it follows from the Lemma that the force at z due to each pair of particles α_j, β_j $(j > 1)$ has a non-zero component orthogonal to Γ in the sense toward the side of Γ on which Π lies; the force at z due to the pair $\alpha_1\beta_1$ acts along Γ, so the total force is not zero and z is not a critical point. This proof requires only obvious modification if z lies on the boundary of Π in the case $m > 2$.

The conclusion of Theorem 2 applies also to the function

$$(4) \qquad \sum_{j=1}^{m} \lambda_j \, \omega(z, C - A_j, R) = \sum_{j=1}^{m} \lambda_j - u(z),$$

a linear combination of the harmonic measures of the arcs complementary to the A_j.

In the case $\lambda_1 = \lambda_2 = \cdots = \lambda_m = 1$, the function $u(z)$ is the harmonic measure of the set $A = A_1 + A_2 + \cdots + A_m$; Theorem 2 can also be applied to the harmonic measure of $C - A$, and yields a new closed region Π' containing the critical points of $\omega(z, C - A, R) = 1 - u(z)$, hence containing the critical points of $u(z)$ in R. The two regions Π and Π' have in common a polygon of $2m$ sides which contains the critical points of $u(z)$ in R. In the case $m = 2$ this polygon degenerates to a single point, the unique critical point of $u(z)$ in R. In the case $m > 2$, this polygon is actually larger than that of §8.7 Theorem 1, as we shall indicate. Denote by δ_1 the intersection of C_1 with A_1, and by δ_2 the intersection of C_1 with A_2; then an entire arc of the NE line $\delta_1\delta_2$ is part of the boundary of Π, and also part of the boundary of the intersection of Π and Π'. But the points α_1 and β_1 are mutually inverse in C_1, as are the points α_2 and β_2, so (as is seen by transforming C into a straight line) the NE lines $\alpha_1\alpha_2$ and $\beta_1\beta_2$ are mutually inverse in C_1, and it follows that the closed polygon of §8.7 Theorem 1 of $2m$ sides contains only a single point γ of the boundary side on $\delta_1\delta_2$ of the polygon $\Pi \cdot \Pi'$; this method applies to each side of $\Pi \cdot \Pi'$, so the former polygon is a proper subset of the latter.

Of course Theorem 2 is not contained in §8.7 Theorem 1, nor is the latter theorem contained in the former. It is remarkable however that in the case $m = 2$, $\lambda_1 = \lambda_2 = 1$ we have two sharp theorems yet formally different, one obtained by the use of skew particles and the other by the use of ordinary particles.

Theorem 2 can be generalized at once to an arbitrary Jordan region R by conformal mapping; we leave the formulation of the new result to the reader. A characterization of the circles C_j independent of conformal mapping is desirable. In the notation already introduced, the points δ_1 and δ_2 are harmonically separated by the pair (α_1, β_1), and also by the pair (α_2, β_2). Moreover, *C_1 is the NE line cutting A_1 and A_2 bisecting the angles between the NE lines $\alpha_1\alpha_2$ and $\beta_1\beta_2$*. This characterization extends to the other lines C_j.

§9.3.2. Abutting arcs. In Theorem 2 we have required the mutually disjoint arcs A_j to be closed, a requirement which we now relinquish; common endpoints are permitted:

COROLLARY. *Let the set A be the sum of open mutually disjoint arcs $A_1, A_2, \cdots, A_m$ lying on $C: |z| = 1$, in the counterclockwise order indicated and let an arc B of $C - A$ be adjacent to A_j and A_{j+1}. The NE line C_j bounds a NE half-plane whose closure contains B, where C_j is an arc of a circle belonging* *both to the coaxal family determined by the end-points of* A_j *as null circles and to the coaxal family determined by the end-points of* A_{j+1} *as null circles. This NE half-plane contains no critical point of* $u(z)$ *as defined by* (3).

The proof of the Corollary is so similar to that of Theorem 2 that it is omitted. The Corollary suggests of itself a limiting case [Walsh, 1948b]:

THEOREM 3. *Let the function* $u(z)$ *be harmonic and bounded but not identically zero interior to* $C: |z| = 1$, *and continuous also on* C *except perhaps in a finite number of points. Let* $u(z)$ *be non-negative on an arc* A *of* C *and zero on the complementary arc. Then the NE half-plane bounded by* A *and by a NE line joining the end-points of* A *contains all critical points of* $u(z)$.

On any closed set interior to C the function $u(z)$ of Theorem 3 is the uniform limit of a sequence of functions of the kind studied in the Corollary, and Theorem 3 is a consequence of the Corollary.

In Theorems 2 and 3 we have studied only non-negative functions, but more general functions can be considered:

THEOREM 4. *Let open mutually disjoint arcs* $A_1, A_2, \cdots, A_m, B_1, B_2, \cdots, B_n$ *of* $C: |z| = 1$ *follow in the counterclockwise order indicated with* $m > 1, n > 1$; *let* C_1 *be a NE line, an arc of a circle belonging both to the coaxal family determined by the end-points of* A_1 *as null-circles and the coaxal family determined by the end-points of* A_m *as null-circles, and let* C_2 *be a NE line, an arc of the corresponding circle for the end-points of* B_1 *and* B_n. *Then all critical points of* $u(z)$ *as defined by* (2) *in* $|z| < 1$ *lie interior to the two disjoint NE half-planes bounded by* C_1 *and* C_2 *respectively.*

Let z be a point not interior to either of these half-planes, let Γ be the circle through z of the coaxal family determined by C_1 and C_2, and transform z into the origin and Γ into the axis of imaginaries; we retain the original notation. For definiteness suppose the arc C_1 to lie to the right of Γ or to coincide with Γ. We use the field of force of Theorem 1. The force at z due to each pair of skew particles at the ends of an arc A_j has, by the Lemma of §9.1, a horizontal component which is either zero or acts toward the right; the same conclusion holds for the force at z due to each pair of skew particles at the end-points of an arc B_k, and the total horizontal component is not zero. Thus z is not a position of equilibrium nor a critical point of $u(z)$.

The exceptional cases under Theorem 4 deserve mention. If we have $m = n = 1$, no critical point of $u(z)$ lies in $|z| < 1$, as follows from the Lemma of §9.3.1. In the case $m > 1$, $n = 1$, the arc C_2 does not exist, but the method already used shows that all critical points of $u(z)$ lie in the NE half-plane bounded by C_1 not containing B_1; similarly in the case $m = 1, n > 1$, all critical points lie in the NE half-plane bounded by C_2 not containing A_1. The Corollary to Theorem 2 can be considered the degenerate case $n = 0$.

A limiting case of Theorem 4 is also true, proved from Theorem 4 by the method used in Theorem 3:

THEOREM 5. *Let the function $u(z)$ be harmonic and bounded but not identically zero interior to $C: |z| = 1$, and continuous on C except perhaps in a finite number of points. Let $u(z)$ be non-negative on a closed arc A of C, non-positive on a closed arc B of C disjoint from A, and zero on $C - A - B$. Let C_1 and C_2 be the NE lines joining the end-points of A and B respectively. Then all critical points of $u(z)$ in $|z| < 1$ lie in the two closed disjoint NE half-planes bounded by C_1 and C_2 respectively.*

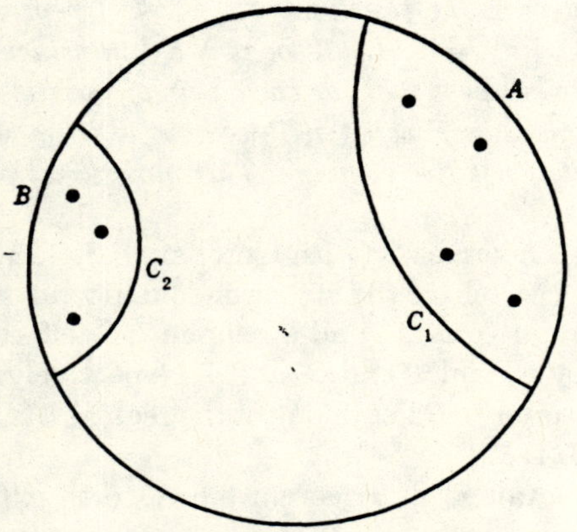

Fig. 25 illustrates §9.3.2 Theorem 5

§9.3.3. Poisson's Integral. The results of §§9.3.1 and 9.3.2 suggest the study of critical points by the use of Poisson's integral, which indeed is a special case of the formulas of §9.1. We write

$$u(r, \theta) = \frac{1}{2\pi} \int_0^{2\pi} \frac{(1 - r^2)\, U(\psi)\, d\psi}{1 + r^2 - 2r \cos (\theta - \psi)}, \tag{5}$$

where for simplicity we choose $U(\psi)$ bounded and integrable in the sense of Lebesgue on $C: |z| = 1$; the function $u(r, \theta)$ is then known to be bounded interior to C and continuous at every point of C at which $U(\theta)$ is continuous; and the Poisson integral is invariant under conformal mapping of the region $|z| < 1$ onto itself. We differentiate (5) with respect to r for the values $r = 0, \theta = 0$:

$$\frac{\partial u(0, 0)}{\partial r} = \frac{1}{\pi} \int_0^{2\pi} U(\psi) \cos \psi \, d\psi. \tag{6}$$

If we interpret the integral of $U(\psi) \cos \psi \, d\psi$ over part of C to be the *moment* at O of the corresponding values of $U(\psi)$, the vanishing of the first member of (6) expresses that the moment of the values of $U(\psi)$ on the right semicircle of C is equal and opposite to the moment of the values of $U(\psi)$ on the left semicircle. If the origin is a critical point this relation must hold for every orientation of C. Any condition sufficient to ensure the non-vanishing of this second member of (6) is sufficient to ensure that the origin is not a critical point. An obvious sufficient condition is the inequality $U(\pi/2 - \varphi) \geqq U(\pi/2 + \varphi)$, for $0 < \varphi < \pi$, with the strong inequality holding on a set of positive measure. A reformulation of this condition in terms invariant under linear transformation of $|z| < 1$ onto itself gives

THEOREM 6. *Let $u(e^{i\theta})$ be bounded and integrable in the sense of Lebesgue for $0 \leqq \theta \leqq 2\pi$, and let $u(z)$ be the function for $|z| < 1$ represented by the Poisson integral of $u(e^{i\theta})$ over $C: |z| = 1$. Let Γ be a NE line which separates C into the two arcs C_1 and C_2. We suppose that for each pair of points z_1 and z_2 of C_1 and C_2 respectively which are mutually inverse in Γ we have $u(z_1) \geqq u(z_2)$, where the strong inequality holds on a set of positive measure. Then no critical point of $u(z)$ lies on Γ.*

Theorem 6 contains Theorems 2 (with Corollary), 3, 4, 5. If in Theorem 6 we have $u(z_2) \equiv -u(z_1) \geqq 0$, where the strong inequality holds on a set of positive measure, we have $u(z) \equiv 0$ on Γ, and Theorem 6 itself can be applied to the two regions bounded by C_1 and Γ and C_2 and Γ respectively. It follows that each critical point of $u(z)$ interior to C lies in one of the regions $\omega(z, C_1, |z| < 1) > 3/4$, $\omega(z, C_2, |z| < 1) > 3/4$.

Two other results contained in Theorem 6 are worth explicit statement:

COROLLARY 1. *Let $u(e^{i\theta})$ be bounded and integrable in the sense of Lebesgue for $0 \leqq \theta \leqq 2\pi$, and let $u(z)$ be the corresponding function for $|z| < 1$ represented by Poisson's integral over $C: |z| = 1$, assumed not identically zero. Let the function $u(e^{i\theta})$ be monotonically non-decreasing on the arc $\theta_1 \leqq \theta < \theta_2$, non-negative on the arc $\theta_2 < \theta < \theta_3$, monotonically non-increasing on the arc $\theta_3 < \theta \leqq \theta_4$, and zero on the arc $\theta_4 \leqq \theta \leqq \theta_1 + 2\pi$, with $\theta_1 < \theta_2 < \theta_3 < \theta_4 < \theta_1 + 2\pi$. Let C_1 be the arc interior to C of the circle which belongs to the coaxal family determined by the points $e^{i\theta_1}$ and $e^{i\theta_2}$ as null circles, and to the family determined by the points $e^{i\theta_3}$ and $e^{i\theta_4}$ as null circles. Then all critical points of $u(z)$ interior to C lie in the open NE half-plane bounded by the NE line C_1 and bounded in part by the arc $\theta_2 < \theta < \theta_3$.*

If we relinquish the assumption $\theta_2 < \theta_3$ here, and take $\theta_2 = \theta_3$, the function $u(z)$ has no critical point interior to C; this conclusion persists even if we have in addition $\theta_4 = \theta_1 + 2\pi$; this result is the analog of, and can be proved by a limiting process from, §4.2.3 Theorem 2; compare Theorem 1.

COROLLARY 2. *Let A and B be two disjoint closed arcs of $C: |z| = 1$, let each arc A_j contain A, and each arc B_k be contained in B, where every A_j is disjoint from every B_k. Let C_1 and C_2 be the NE lines for $|z| < 1$ joining the end-points of A*

and of B respectively, and let Γ_1 be the circle in which the two circular arcs C_1 and C_2 are mutually inverse. Then if $u(z)$ is not identically constant, no critical point of the function

$$u(z) = \sum_{j=1}^{m} \lambda_j \omega(z, A_j, |z| < 1) + \sum_{k=1}^{n} \mu_k \omega(z, B_k, |z| < 1),$$

$$\lambda_j > 0, \quad \mu_k > 0, \quad \sum \lambda_j \geqq \sum \mu_k,$$

lies interior to C in the NE half-plane bounded by Γ_1 containing C_1.

If Γ is a circle of the coaxal family (C_1, Γ_1) which lies between C_1 and Γ_1, and if Γ is the axis of imaginaries, the circle Γ_1 separates C_2 and Γ; the inverse of C_1 in Γ_1 is C_2, and the inverse of C_1 in Γ is an arc which separates C_2 and Γ. The conditions of Theorem 6 are satisfied, so no critical point of $u(z)$ lies on Γ. Likewise if Γ is a circle of the coaxal family (C_1, Γ_1), and if C_1 separates Γ and Γ_1, the conditions of Theorem 6 are satisfied; Corollary 2 is established; compare §5.2.2 Corollary 3 to Theorem 2. The NE half-plane bounded by Γ_1 and containing C_1 can be defined by the inequality $\omega(z, C_1, |z| < 1) > \omega(z, C_2, |z| < 1)$.

Under the conditions of Corollary 2, there may well be other regions, say bounded by NE lines not separating A and B, which by Theorem 3 contain no critical points of $u(z)$.

Theorems 2–6 and their corollaries are of significance in the study of critical points of rational functions whose zeros and poles lie on a circle; and reciprocally, results (§5.2) on the latter topic apply to harmonic measure and other functions harmonic interior to the unit circle; we leave the elaboration of this remark to the reader. Naturally Theorems 2–6 can be generalized at once by conformal mapping so as to apply to an arbitrary Jordan region.

Even though Theorems 2–5 are contained in Theorem 6, the former methods are not to be considered obsolete, for they will be of later use.

§9.3.4. Poisson's integral as potential of a double distribution. It is well known that Poisson's integral is the expression of a potential due to a double layer distribution, so we proceed to connect Poisson's integral with our general fields of force.

If a given function $u(x, y)$ is harmonic in the interior R of the unit circle C, continuous in $R + C$, zero on C except on an arc A_1 of C, we have formally derived in (1):

$$u(x, y) = -\frac{1}{\pi} \int_{A_1} u \frac{\partial \log r}{\partial \nu} ds - u(0, 0), \tag{7}$$

which is valid for (x, y) in R. The integral in (7) may also be taken over C. If now $u(x, y)$ is harmonic in R, continuous in $R + C$, the boundary values of $u(x, y)$ may be considered as the sum of two functions $u_1(x, y)$ and $u_2(x, y)$ on C, where $u_1(x, y)$ equals $u(x, y)$ on an arc A_1 of C and is zero elsewhere on C, while $u_2(x, y)$ equals zero on A_1 and equals $u(x, y)$ elsewhere on C. By adding the representations (7) for these two functions we derive an expression for $u(x, y)$ in R:

$$u(x, y) = \frac{1}{\pi} \int_C u d\theta - u(0, 0), \tag{8}$$

where we set $(\partial \log r/\partial \nu)\, ds = -d\theta$, θ being rectilinear angle at (x, y) from the positive horizontal to a variable point of C. Equation (8) is merely an unusual form for Poisson's integral. It is readily verified, as we indicate, that (8) solves the boundary value problem and also that (8) is equivalent to the usual form of Poisson's integral.

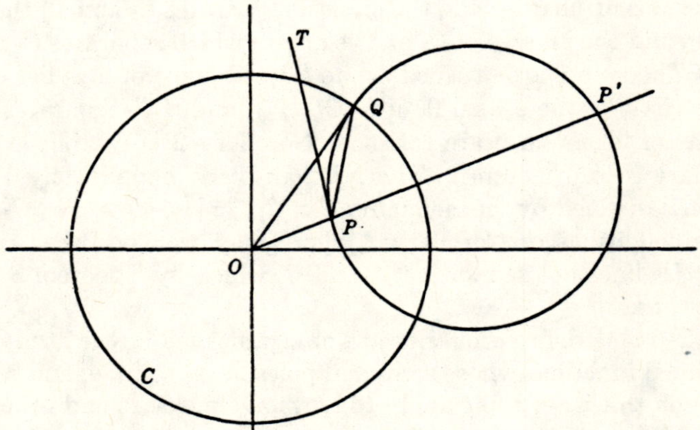

Fig. 26 illustrates §9.3.4

By Gauss's mean value theorem for harmonic functions transformed by inversion in a circle, Bôcher interprets Poisson's integral as expressing the fact that the value of $u(x, y)$ in a point P of R is the mean of the values of $u(x, y)$ on C with respect to angles measured at P, the correspondence between values of $u(x, y)$ on C and angles at P being made by circular arcs through P orthogonal to C. We prove that (8) is equivalent to Bôcher's interpretation of Poisson's integral. Let P be an arbitrary point of R, P' the inverse of P in C, Q a point of C, $P'PQ$ a circular arc necessarily orthogonal to C, and PT the tangent to this arc at P in the sense PQP'. The angles QPT and OQP are equal, being angles between the chord PQ and tangents to the arc PQ, so we have for the rectilinear angles:

$$\angle P'PQ = \angle POQ + \angle OQP = \angle POQ + \angle QPT,$$

$$2\,\angle P'PQ = \angle POQ + (\angle P'PQ + \angle QPT) = \angle POQ + \angle P'PT,$$

$$\angle P'PT = 2\,\angle P'PQ - \angle POQ. \tag{9}$$

The first member of (9) is the angle whose differential occurs in Bôcher's interpretation, and the second member is the angle whose differential occurs in (8) if we express $u(0, 0)$ by Gauss's mean value theorem, so (8) is established.

A function conjugate to $u(x, y)$ in R is

$$v(x, y) = -\frac{1}{\pi}\int_C u\, d(\log r),$$

and the corresponding function analytic in R is

$$f(z) = u + iv = -\frac{i}{\pi}\int_C u\, d[\log (z - t)] - u(0, 0)$$

$$= \frac{i}{\pi}\int_C \frac{u\, dt}{z - t} - u(0, 0).$$

The field of force is represented by the conjugate of $f'(z)$, where

$$f'(z) = \frac{1}{\pi i}\int_C \frac{udt}{(z - t)^2}, \tag{10}$$

for z in R.

If two equal and opposite ordinary particles are approximately equidistant from a fixed point, and approach that point from opposite directions while suitably increasing in mass, the field of force approaches a limit, and this limiting field of force is said to be due to a *dipole* or *doublet* (an artificial element) at the fixed point; the direction (axis) of the dipole is the direction of approach of the variable particles. Thus let α be a point of C, let points α_1 and α_2 lie on the half-line $\arg z = \arg \alpha$, $|\alpha_1| > |\alpha| > |\alpha_2|$, and let equal positive and negative particles be placed at α_1 and α_2 respectively. The corresponding field of force is the conjugate of a positive multiple of

$$\frac{1}{z - \alpha_1} - \frac{1}{z - \alpha_2} = \frac{\alpha_1 - \alpha_2}{(z - \alpha_1)(z - \alpha_2)},$$

and as α_1 and α_2 approach α a suitable positive multiple of this function approaches the function $\alpha/(z - \alpha)^2$; the conjugate of the latter function thus represents the force at z due to a dipole at α whose axis is normal to C at α. The lines of force due to a dipole are the directed circles tangent to the axis of the dipole at the dipole.

For t on C we have $\arg (dt) = \arg t + \pi/2$, so it is clear from (10) that the field of force represented by the conjugate of $f'(z)$ in (10) is the limit of the field due to a number of dipoles of suitably chosen masses on C whose axes are normal to C; that is to say, the conjugate of $f'(z)$ represents the field of force due to a double layer distribution on C. The algebraic sign of the double distribution may be considered that of $u(x, y)$, on each arc of C on which $u(x, y)$ has constant sign. We have represented in (10) the field of force, the conjugate of $f'(z)$, as due to a double distribution, so $u(x, y)$ is the potential due to the double distribution. The force is the gradient of the potential.

For the special value $z = 0$, we set $t = e^{i\theta}$, and the force as obtained from (10) reduces to

$$\frac{1}{\pi}\int u\cdot(\cos\theta + i\sin\theta)\,d\theta, \tag{11}$$

from which the horizontal and vertical components may be read off at once; compare §9.3.3 equation (6). We note too by obvious geometric properties of the field of force that such an inequality as $u(x, y) \geqq u(x, -y)$ holding on C for $y > 0$ with $u(x, y) \not\equiv u(x, -y)$ implies that at every point of the axis of reals in R the force has a non-zero vertical component. Likewise the inequality $u(x, y) \geqq u(-x, -y)$ on C for $y > 0$ with $u(x, y) \not\equiv u(-x, -y)$ implies that at 0 the force has a non-zero vertical component.

Merely for the sake of simplicity we have supposed the function $u(x, y)$ to be continuous on C, but Poisson's integral is valid, and hence (10) is valid, under much more general assumptions.

The field of a dipole can be considered not merely as the limit of the field due to equal and opposite ordinary particles, but also as the limit of the field due to equal and opposite skew particles, when the latter are approximately equidistant from a fixed point and approach that point from opposite directions while suitably increasing in mass; the axis of the dipole is then orthogonal to those directions. Thus, let α be a point of C, let α_1 and α_2 be neighboring points equidistant from α lying on C in the counterclockwise order $\alpha_2\alpha\alpha_1$, and let the positive and negative skew particles be placed at α_1 and α_2 respectively. The corresponding field of force is the conjugate of a positive multiple of

$$\frac{-i}{z-\alpha_1} - \frac{-i}{z-\alpha_2} = \frac{-i(\alpha_1-\alpha_2)}{(z-\alpha_1)(z-\alpha_2)},$$

and as α_1 and α_2 approach α a suitable positive multiple of this function approaches the function $\alpha/(z-\alpha)^2$; this is the same function as before.

Thus the field of force defined by the conjugate of (10) may be approximated by dipoles on C or by pairs of equal skew particles on C. If $u(x, y)$ is constant on an arc A of C, and if the pairs of skew particles on C are chosen all of equal mass, with internal particles on A coinciding in pairs of opposite kinds, the entire field may reduce to that due to equal and opposite skew particles at the end-points of A; compare §§8.7.1 and 9.3.1.

§9.3.5. Harmonic functions and rational functions. Still another useful formula can be derived from (7) or (8). We write formally for $z = x + iy$ interior to C

$$u(x, y) = \frac{1}{\pi}\int_C u\,d[\arg(z-t)] - u(0, 0),$$

$$u(x, y) = \frac{1}{\pi}[u\arg(z-t)]_C - \frac{1}{\pi}\int_C [\arg(z-t)]\,du - u(0, 0). \tag{12}$$

The first two terms in the second member depend essentially on the initial point of integration on C, but the first of those terms is independent of z. We have

(13) $$f(z) = u + iv = \frac{-i}{\pi}[u \log (z - t)]_C + \frac{i}{\pi}\int_C [\log (z - t)]\, du - u(0, 0),$$

where again the first two terms in the last member depend on the particular initial point of C chosen, but where the first of these terms is independent of z. The corresponding field of force is the conjugate of

(14) $$f'(z) = \frac{i}{\pi}\int_C \frac{du}{z - t}.$$

A rigorous derivation of (14) from (8) is readily given for a function $u(x, y)$ which is piecewise constant with only a finite number of discontinuities on C, hence for an arbitrary function $u(x, y)$ of bounded variation on C, for the latter can be expressed on C as the uniform limit of a sequence $u_n(x, y)$ of functions piecewise constant with only a finite number of discontinuities, such that the total variation of $u(x, y) - u_n(x, y)$ on C approaches zero.

Let us investigate in more detail the significance of (13) and (14). Denote by $U(\theta)$ the boundary values of $u(x, y)$ in (8); we assume $U(\theta)$ to be of bounded variation. The definition of the particular value $U(2\pi)$ is of no significance whatever in (8); we may set $U(2\pi) = U(0)$, or $U(2\pi) = \lim_{\epsilon\to 0} U(2\pi - \epsilon)$, or $U(2\pi) = U(0) = \frac{1}{2}\lim_{\epsilon\to 0}[U(\epsilon) + U(2\pi - \epsilon)]$. However, if we take the integral in (8) over the interval $0 \leqq \theta \leqq 2\pi$, and allow $\arg (z - t)$ to vary continuously in that interval, the particular choice of $U(2\pi)$ is of considerable significance in the individual terms of the second member of (12); to be more explicit, we write

(15) $$u \arg (z - t)\,|_C = U(2\pi) \arg (z - e^{2\pi i}) - U(0) \arg (z - e^0),$$

(16) $$\int_C [\arg (z - t)]\, du = \int_{\theta=0}^{\theta=2\pi} [\arg (z - t)]\, dU(\theta);$$

the two values $t = e^{\theta}$ for $\theta = 0$ and $\theta = 2\pi$ may properly be distinguished. If we introduce the requirement* $U(2\pi) = U(0)$, as we now do, the second member of (15) reduces to $+2\pi U(0)$, which is independent of z; moreover in the integral in (12) it is now immaterial whether for the value $t = 1$ we interpret $\arg (z - t)$ as $\arg (z - e^0)$ or $\arg (z - e^{2\pi i})$; this is the basis of our derivation of (14). However, if the requirement $U(2\pi) = U(0)$ is not fulfilled, and if the interpretation (16) is not otherwise modified, then the second member of (15) is not independent of z, and an additional term (or interpretation) is required in (14). —These comments apply in more general form in connection with §9.1 equation (2), and are the basis for the insertion of the second term in the second member of that equation.

With the conventions made, we thus have

THEOREM 7. *Let $u(x, y)$ be harmonic and bounded interior to the unit circle C, represented by the Poisson integral of boundary values u that are of bounded varia-*

* It is customary but not necessary to use the definition $U(0) = U(2\pi) = \frac{1}{2}\lim_{\epsilon\to 0}[U(\epsilon) + U(2\pi - \epsilon)]$, so that $U(\theta)$ is the boundary value of $u(x, y)$ for normal approach to C.

tion on C. Then the critical points of $u(x, y)$ interior to C are the positions of equilibrium in the field of force due to the spread of ordinary matter of mass du on C.

In Theorem 7 we suppress the factor i/π in (14), for either ordinary matter or skew matter may be used. Theorem 7 includes §8.7.1 Theorem 1, §9.3.1 Theorem 1, and various other results.

Theorem 7 makes available for use in the study of harmonic functions all the results of §5.2, and yields numerous new results for harmonic functions. Some of the results of §5.2 depend essentially on the fact that the masses involved are integral, and are therefore of limited application here, but other results are independent of the magnitude of the masses and apply at once.

We do not carry over in detail these applications of all the results of §5.2 to harmonic functions, but as a further example we state the analog of §5.2.3 Theorem 4, proved by Theorem 7:

THEOREM 8. *Let $U(z)$ be a real function defined on the axis of imaginaries, of total variation $2n$, with a finite jump of k at 0, and constant in each of the intervals $0 < |y| < b$. Let $u(z)$ be bounded and harmonic in the open right half-plane, and at every point of continuity of $U(z)$ on the axis of imaginaries take on continuously the value $U(z)$. Then $u(z)$ has no critical point in the right half-plane interior to the curve*

$$n(n-k)(x^2+y^2)^2 - kb^2[nx^2+(n-k)y^2] = 0.$$

§9.4. A circle as partial boundary. We continue our use of fields of force, where now a circle is chosen as part of the boundary of the given region.

§9.4.1. Harmonic measure. By the methods already introduced in §§9.1–9.3 we have

THEOREM 1. *Let C be the unit circle, let $A_1, A_2, \cdots, A_m$ be mutually disjoint arcs of C, and let a region R be bounded by C and by a Jordan configuration B interior to C. Let R' be the region bounded by B, by the reflection B' (assumed finite) of B in C, and by the arcs A_j, and let $u(z)$ be the harmonic measure $\omega(z, B + \sum A_j, R)$ and its harmonic extension throughout R'. Then the critical points of $u(z)$ in R' are the positions of equilibrium in the field of force due to the negative spread $-dv/2\pi$ on B, the positive spread $-dv/2\pi$ on B', where $v(x, y)$ is conjugate to $u(x, y)$ in R', and to skew particles of masses $1/\pi$ and $-1/\pi$ at the points β_j and α_j respectively, where A_j is the positive arc $\alpha_j\beta_j$.*

We introduce the notation C_j for all circles of the coaxal family determined by α_j and β_j as null circles. We shall prove

THEOREM 2. *Let C be the unit circle, let $A_1, A_2, \cdots, A_m, A_{m+1} = A_1$ be mutually disjoint closed arcs of C ordered counterclockwise, and let a region R be bounded by*

C and by a Jordan configuration B interior to C. With respect to the interior of C, let Π_B be the smallest NE convex region containing B, let Π_j be the point set consisting of all maximal arcs of the family C_j each bounded by a point of A_j and a point of Π_B, and let Π_0 be the NE convex region bounded by the m NE lines respectively common to the families C_j and C_{j+1}, $j = 1, 2, \cdots, m$. Then all critical points of $\omega(z, B + \sum A_j, R)$ in R lie in the smallest NE convex set Π containing $\sum_0^m \Pi_j$.

The set Π need not coincide with the set $\sum \Pi_j$, as may be seen for instance by choosing $m = 2$, with B small and not intersecting the NE line common to the families C_1 and C_2. Nevertheless, as we shall show, each boundary point of Π is a boundary point of at least one of the sets Π_j. Consider for instance the NE half-plane H bounded by the NE line belonging to both of the families C_1 and C_2, H bounded by no point of $A_3 + \cdots + A_m$. The boundary if any of Π interior to H consists of a maximal arc T_1 of a circle of the family C_1 which is part of the boundary of Π_1, of a maximal arc T_2 of a circle of the family C_2 which is part of the boundary of Π_2, and possible boundary points z_0 of Π_B. The set Π_1 lies in a NE half-plane bounded by the NE line of which T_1 is an arc, and also lies in a NE half-plane bounded by the NE line of which T_2 is an arc. If a boundary point z_0 of Π lies in H not on T_1 or T_2, there exists a NE half-plane containing Π_B, Π_1, and Π_2, bounded by a NE line through z_0; the point z_0 is a boundary point of Π_B, Π_1, and Π_2.

The boundary of Π_B can be considered the limiting W-curve interior to C for the groups of points of B and B'; the boundary of Π_j can be considered the limiting W-curve interior to C for the group (α_j, β_j) and groups on B and B'; the NE line belonging to the two families C_j and C_{j+1} is the W-curve for the two groups (α_j, β_j) and $(\alpha_{j+1}, \beta_{j+1})$. Thus Π is bounded wholly by W-curves and limiting W-curves, and is the smallest NE convex set containing all the W-curves.

In the proof of Theorem 2 it is sufficient to consider the origin z as a possible critical point; we assume Π to lie in the right half of the interior of C. The distributions of matter on B and B' are equal in magnitude and of opposite sign in pairs of inverse points, so the force at z due to these distributions has a non-zero component toward the right. Each NE line through z of a family C_j cuts the arc A_j in the right semicircle, so the corresponding force at z has for each j a non-zero component toward the right. Thus z cannot be a position of equilibrium. In the case $m = 2$, the set Π_0 reduces to a single NE line. In the case $m = 1$, Π_0 does not exist and Π is identical with Π_1. In the case $m = 0$, Theorem 2 reduces to §8.4 Theorem 1. No position of equilibrium lies on the boundary of Π in R unless Π reduces to a segment of a NE line; this can only occur in the case $m \leqq 2$, and B empty or a segment of a NE line; this line is necessarily Π_0 in the case $m = 2$, and contains $\Pi = \Pi_1$ in the case $m = 1$. Theorem 2 admits an obvious extension by conformal map, where C is now an arbitrary Jordan curve; the formulation is left to the reader.

It may be noted that Theorem 2 for $m = 1$ does not include §9.2.1 Theorem 2; neither result includes the other.

Under the conditions of Theorem 2, let the Jordan configuration B_k lie interior to C, and let $B_k + C$ bound a region R_k containing R in its interior; let R_k' be the region bounded by B_k and the reflection B_k' of B_k in C. We do not assume B_k to be a component or even a subset of B; if a component of B is a Jordan curve, the configuration B_k may lie interior to that curve, or B_k may consist of several components lying in or on such Jordan curves. The critical points of $\omega(z, B_k, R_k)$ in R_k are the positions of equilibrium in a field of force due to a negative spread on B_k and an equal and opposite positive spread on B_k'. The total field of force for a sum of positive multiples of such harmonic measures, plus a linear combination with positive coefficients of harmonic measures of sets B_k plus arcs A_j with respect to R_k, plus a linear combination with positive coefficients of harmonic measures of arcs A_k with respect to the interior of C, is in R qualitatively identical with the field of force considered in Theorem 2, although here the spreads of matter lie in the closed exterior of R' but need not lie on B and B'. Thus we have

THEOREM 3. *Under the conditions of Theorem 2 let the Jordan configuration B_k lie interior to C, and let $B_k + C$ bound a region R_k containing R. Then all critical points in R of the function*

$$\sum_{k=1}^{n} \lambda_k \omega(z, B_k, R_k) + \sum_{k=1}^{m} \mu_k \omega(z, B_k + \sum A_j, R_k)$$

$$+ \sum_{k=1}^{m} \nu_k \omega(z, A_k, |z| < 1), \qquad \lambda_k > 0, \mu_k > 0, \nu_k > 0,$$

lie in Π.

In $\sum A_j$ here the sum need not include all arcs A_j, and the particular arcs A_j chosen may depend on k.

There exist still further analogs of the theorems of §9.3. For instance, in Theorem 3 we may allow adjacent arcs A_k to have end-points in common, with a result analogous to §9.3.2, Corollary to Theorem 2. An analog of §9.3.2 Theorem 3 is

THEOREM 4. *Under the conditions of Theorem 3 let the function $u_1(z)$ be harmonic and bounded interior to C, continuous and non-negative on C except perhaps in a finite number of points, and zero on C except perhaps on the arcs $A_1, A_2, \cdots, A_m$. Let Π' be the smallest NE convex region whose closure contains $B + A_1 + \cdots + A_m$. Then Π' contains all critical points in R of the function*

$$u_1(z) + \sum_{k=1}^{n} \lambda_k \omega(z, B_k, R_k), \qquad \lambda_k > 0.$$

Theorem 4 may be proved from Theorem 3, for the function $u_1(z)$ can be approximated uniformly on any closed region interior to C by a linear combination with positive coefficients of harmonic measures with respect to the interior of C of arcs belonging to the set $A_1 + \cdots + A_m$.

§9.4.2. Linear combinations. In the actual application of the methods of §9.1, it is essential to know the sign of dv on various parts of the boundary of the region considered. Thus we ordinarily assume a given harmonic function $u(x, y)$ to take but two (distinct) values on the boundary; these are respectively the maximum and minimum of the function in the closed region, so the algebraic sign of $(\partial u/\partial \nu)ds$ or its substitute $\pm dv$ is known. Linear combinations of functions $u_k(x, y)$ each of which meets the conditions outlined can also be studied by these same methods, by addition of the corresponding fields of force. For various linear combinations we are able to study the force at suitable points of R and to show that some points of R are not critical points, but we are nevertheless not able to deduce the sign of $\partial u/\partial \nu$ or dv at all boundary points of R, and hence are not able to determine the total number of critical points in R; compare §9.5.1 below.

If an entire circle C is involved as part of the boundary of a given region R, it may be possible (as in §8.4 Theorem 2) to assign *three* distinct boundary values to a given function $u(x, y)$, say zero, unity, and $-c$ (<0), provided the value zero is assigned only in points of C, for in R we have $\max u = 1$, $\min u = -c$, equations which determine the algebraic sign of the matter placed on the boundary of R; reflection of R in C then yields by harmonic extension a new region R' containing R involving as new boundary values only minus unity and $+c$; on the new boundary components the algebraic sign of the spread of matter can be found at once; the circle C does not appear as part of the boundary of R'. We have by no means fully exploited these methods.

In the field of force used in the proof of Theorem 2, we have paired the particles on C in such a way that in R the force due to the pair (α_j, β_j) is directed along the circles of a coaxal family *toward* the arc (α_j, β_j). If the particles are paired as (β_j, α_{j+1}), the force in R due to the pair is directed along circles of a coaxal family *away from* the arc (β_j, α_{j+1}); we study this new interpretation of the field of force, which may alternately be obtained by reversing the usual field for the function $1 - \omega(z, B + \sum A_j, R)$. We shall prove

THEOREM 5. *Under the conditions of Theorem 2, let A_j' be the arc of C whose initial point is the terminal point of A_j and whose terminal point is the initial point of A_{j+1}. Denote by Π_0' the NE convex region analogous to Π_0 with the A_j replaced by the A_j'. If a NE line Γ separates Π_0' and B in C, then no critical point of $u(z) = \omega(z, B + \sum A_k, R)$ lies on Γ in R. Consequently all critical points of $u(z)$ in R lie in the NE convex region Π' common to all NE half-planes containing Π_0' bounded by NE lines Γ, and in a NE convex region Π'' containing B, defined as the point set common to all NE half-planes containing B bounded by NE lines Γ.*

Naturally it may occur that no line Γ exists. In the case $m = 1$ we consider

Π_0' to be empty. In the case $m = 2$ the set Π_0' is a NE line. In every case with $m > 1$, Theorem 5 contains information not contained in Theorem 2 provided Γ exists, for Γ must intersect the set Π.

The set Π'' is bounded in part by arcs of two distinct NE lines each intersecting Π_0' in a point of C, intersecting each other in some point Q. The set Π'' can be defined also as the totality of points on maximal arcs disjoint from Π_0' terminated by C and Π_B of NE lines Γ' cutting both Π_0' and Π_B. Indeed, every NE line Γ' cutting both Π_0' and Π_B is cut by every NE line Γ; the maximal arc of Γ' lies in every NE half-plane containing B bounded by Γ; every point of Π'' is joined to points of Π_0' by a NE line Γ' through Q which is cut by every NE line Γ, so every point of Π'' lies on such a maximal arc of Γ'.

To prove Theorem 5 we choose an arbitrary point z of Γ in R, and transform the region $|z| < 1$ onto itself so that z is the origin and Γ the axis of imaginaries. Let Π_0' lie in the right semicircle; then the force at z due to each pair of particles at the end-points of an interval A_j' has a non-zero component toward the left; the set B lies in the left semicircle, so the force at z due to the spread on B and on the inverse B' of B in C also has a non-zero component toward the left, so z is not a position of equilibrium nor a critical point of $u(z)$.

In Theorem 5 the set $\Pi' + \Pi''$ is bounded by limiting W-curves: arcs of NE lines through Q, which are W-curves for pairs of particles at the end-points of an arc A_j' and for pairs of points on B and B'; arcs of NE lines bounding Π_B, which are W-curves for two pairs of particles on B and B'; the boundary of Π_0', which is a W-curve for pairs of particles at the end-points of arcs A_j' and A_{j+1}'.

The proof as given establishes also the

Corollary. *Under the conditions of Theorem 5, the set $\Pi' + \Pi''$ contains all critical points in R of the function*

$$\omega(z, B, R) - \sum_{k=1}^{m} \lambda_k \omega(z, A_k', |z| < 1), \qquad \lambda_k > 0.$$

A further general result is

Theorem 6. *Let C be the unit circle, and let $A_1, A_2, \cdots, A_m, A_{m+1} = A_1, A_1', A_2', \cdots, A_{m'}', A_{m'+1}' = A_1'$ be mutually disjoint open arcs of C ordered counterclockwise. Let R be a region bounded by C and by disjoint Jordan configurations B_0 and B_0' interior to C. Let Π_0 be a NE convex region containing B_0, containing for every j the NE line belonging to both coaxal families determined by the end-points of A_j and of A_{j+1} as null-circles, and containing all arcs of circles of the family determined by the end-points of A_j joining the arc A_j with the smallest convex set containing B_0. Let Π_0' be the corresponding set for B_0' and the A_j'. Let the Jordan configurations B_k and B_k' $(k > 0)$ any of which may be empty lie interior to C, let $B_k + B_k' + C$ bound a region R_k containing R, where B_k and B_k' lie in Π_0 and Π_0' respectively. If a NE line Γ separates Π_0 and Π_0' then no critical point of*

$$u(z) = \sum_1^n \lambda_k \omega(z, B_k, R_k) + \sum_1^m \mu_k \omega(z, B_k + \sum A_j, R_k)$$

$$+ \sum_1^m \nu_k \omega(z, A_k, |z| < 1) - \sum_1^n \lambda_k' \omega(z, B_k', R_k) \tag{1}$$

$$- \sum_1^{m'} \mu_k' \omega(z, B_k' + \sum A_j', R_k) - \sum_1^{m'} \nu_k' \omega(z, A_k', |z| < 1),$$

where the coefficients $\lambda_k, \mu_k, \nu_k, \lambda_k', \mu_k', \nu_k'$ *are all positive and where* $\sum A_j$ *and* $\sum A_j'$ *need not include all* A_j *and* A_j' *and may depend on* k, *lies in* R *on* Γ; *consequently if a line* Γ *exists, all critical points of* $u(z)$ *in* R *lie in two NE convex sets* Π *and* Π' *containing* Π_0 *and* Π_0' *respectively and separated by every* Γ *which separates* Π_0 *and* Π_0'.

Even Theorem 6 may be further generalized, by permitting further arcs S_k of C on which $u_1(z)$ is non-negative and continuous except perhaps for a finite number of points, with $u_1(z)$ harmonic and bounded interior to C, and by permitting arcs S_k' of C on which $u_2(z)$ is non-negative and continuous except perhaps for a finite number of points, with $u_2(z)$ harmonic and bounded interior to C. We then require that the closure of Π_0 shall contain the S_k, and that the closure of Π_0' shall contain the S_k'; the conclusion of Theorem 6 remains valid if we add $u_1(z) - u_2(z)$ to the second member of (1).

§9.4.3. Harmonic measure, resumed. We return to the situation of Theorems 1 and 2, for although both Theorems 2 and 5 apply to that situation, and are mutually complementary, we have not yet made full use of the fact that in the field of force the magnitudes of all the particles on C are the same; Theorems 2 and 5 correspond to the methods of §9.3.1 rather than those of §8.7, and we have already pointed out (§9.3.1) that those of §8.7 and of §5.2 are in some respects more powerful.

In the field of force defined in Theorem 1 for the study of the critical points of the function

$$u(z) = \omega(z, B + \sum_1^m A_j, R), \tag{2}$$

we make the modification that all forces shall be rotated through the angle $-\pi/2$. Thus at the points α_k and β_k we have ordinary negative and positive unit particles respectively, and on B and B' we have respectively spreads of positive and negative skew matter. We prove

THEOREM 7. *Let* R *be a region bounded by the unit circle* C *and by a Jordan configuration B interior to C, and let* $A_1, A_2, \cdots, A_m, A_{m+1} = A_1$ *be mutually disjoint closed arcs* (α_j, β_j) *of C ordered counterclockwise on C. Let* Π *be the polygon of 2m sides bounded by arcs of the NE lines* $\alpha_j\alpha_{j+1}$ *and* $\beta_k\beta_{k+1}$. *A NE half-plane disjoint to* Π *bounded by a positive arc* $\beta_k\beta$ *of C and the NE line* $\beta_k\beta$ *and containing no point of B contains no critical points of* $u(z)$ *as defined by* (2); *a NE half-plane disjoint to* Π *bounded by a positive arc* $\alpha\alpha_j$ *of C and the NE line* $\alpha\alpha_j$ *and containing no point of B contains no critical point of* $u(z)$.

Let z be a point interior to C, exterior to Π, and in the first of these NE half-planes. We suppose, as can be arranged by a map of C onto itself, $z = 0$ and $\beta_j = 1$. It follows (as in §5.2.1) that the total force at z due to the particles on C has a non-zero component toward the left. The set B lies in the lower semicircle of C; if an *ordinary* positive distribution is spread on B and its negative in symmetric points of B', the force at z due to this matter has a non-zero upward component; for the distribution of *skew* matter, the force at z has a non-zero horizontal component toward the left. Thus z is not a position of equilibrium nor a critical point of $u(z)$.

The second part of Theorem 7 is similarly proved. No critical point z of $u(z)$ can lie in R on the boundary of the former half-plane considered if $m > 1$, unless $m = 2$, z lies on the boundary of II, and B lies on the NE circle $\beta_j\beta$.

In the case $m = 1$, the polygon Π does not exist and in Theorem 7 can be ignored; in the case $m = 2$, that polygon reduces to a single point.

Theorem 7 is intended to be supplementary to Theorems 2 and 5 and not to replace them. Theorem 7 is not included in those earlier theorems, as may be seen by choosing $m = 2$, $\alpha_1 = -i$, $\beta_1 = 1$, $\alpha_2 = e^{3\pi i/4}$, $\beta_2 = e^{5\pi i/4}$, with B well filling the lower semicircle of C.

We state also

COROLLARY 1. *Under the hypothesis of Theorem* 7, *the conclusion holds for the function*

$$u(z) = \omega\left(z, B + \sum_1^m A_k, R\right) + \lambda\omega\left(z, \sum_1^m A_k, |z| < 1\right), \qquad \lambda > 0.$$

It is essential here to include all the A_k in the summation.

§9.4.4. Symmetry. It is to be expected that the results established can be improved in certain cases of symmetry. A fairly obvious result is

THEOREM 8. *Let the Jordan configuration B lie interior to the unit circle C, and let* $B + C$ *bound the region R. Let every component of B be symmetric in the axis of reals, and let the disjoint arcs* A_1 *and* A_2 *of C each be symmetric in that axis. Then all critical points in R of the functions* $\omega(z, A_1 + B, R)$ *and* $\omega(z, A_1 + A_2 + B, R)$ *lie on the axis of reals. On any open segment of that axis in R bounded by points of* $A_1 + B$ *or* $A_1 + A_2 + B$ *respectively lies a unique critical point.*

On the closure of such a segment the function is continuous and takes the value unity at both ends; on the open segment the function is positive and less than unity, hence at some point has an absolute minimum for the segment. This minimum is a critical point of the given function. All critical points of the function in R are accounted for by this enumeration.

Theorem 8 admits arcs of C symmetric in the axis of reals as components of the set whose harmonic measure is considered, and is thus an extension of §8.4 Corollary 2 to Theorem 2. In proving a similar extension of §8.4 Theorem 3 we shall make use of

LEMMA 1. *Let C be the unit circle, let positive and negative unit skew particles be placed on C at α and $\bar{\alpha}$ respectively, with α in the upper half-plane, and let the non-real point z lie interior to C to the left of the NE line $\alpha\bar{\alpha}$. Then the force at z due to the particles at α and $\bar{\alpha}$ has a non-zero component toward the axis of reals in the direction of the NE vertical at z.*

Under a linear transformation of the interior of C onto itself which carries the axis of reals into itself and preserves sense on that axis, the content of this lemma is invariant; compare §4.1.2 Theorem 3, with the remark that a field due to skew particles is derived from the field for ordinary particles of the same location and masses by rotating all vectors through the angle $\pi/2$. We assume, as we may do, that z lies on the axis of imaginaries above the axis of reals. The locus of points at which the coaxal family of circles determined by α and $\bar{\alpha}$ as null circles has vertical tangents is the equilateral hyperbola whose vertices are α and $\bar{\alpha}$ (the corresponding fact holds for an arbitrary coaxal family of this kind); this hyperbola passes through the points $+i$ and $-i$ (§2.2). At the point $+i$ the force is vertically downward; the center of the circle of the family through z lies farther from the axis of reals than does $+i$ and Lemma 1 follows. Naturally if z (non-real) lies on the NE line $\alpha\bar{\alpha}$ the force at z is in the direction of the NE horizontal, and if z lies to the right of that NE line the force at z has a non-zero component directed away from the axis of reals in the direction of the NE vertical at z.

In Theorem 9 either arc A_1 or A_2 may fail to exist:

THEOREM 9. *Let R be a region bounded by the unit circle C and by a Jordan configuration B interior to C which is symmetric in the axis of reals Ox. Let closed disjoint arcs A_1 and A_2 of C each symmetric in Ox intersect Ox in the points $z = -1$ and $z = +1$ respectively. Then all non-real critical points in R of the function $u(z) = \omega(z, A_1 + A_2 + B, R)$ lie in the closed interiors of the NE Jensen circles for B and in the two closed NE half-planes bounded by A_1 and A_2 and the NE lines joining their end-points.*

Transform if necessary so that the origin is a point of R. In the usual field of force (Theorem 1), at a non-real point z of R exterior to all the Jensen circles and to both NE half-planes, the force due to the attracting matter on B and the symmetric repelling matter on B' has (§5.3.3 Lemmas 1 and 2) a non-zero component in the direction of the NE vertical at z directed toward the axis of reals; this conclusion is true for the force at z due to the particles at the end-points of A_j $(j = 1, 2)$, hence is true for the total force at z; thus z is not a position of equilibrium nor a critical point of $u(z)$.

Any segment of Ox in R bounded by points of $A_1 + A_2 + B$ contains at least one critical point of $u(z)$, for at a point of Ox near $A_1 + A_2 + B$ the force (gradient of $u(z)$) is directed toward $A_1 + A_2 + B$. This latter remark is also of significance in enumerating critical points not on Ox.

COROLLARY 1. *Let the real points α and β in R not be critical points of $u(z)$, and not lie in the NE half-planes of Theorem 9 nor in a Jensen circle. Let K be the configuration consisting of the segment $\alpha \leqq z \leqq \beta$ plus the closed interiors of all NE Jensen circles intersecting that segment, and let K contain precisely k components of B. Then K contains $k + 1$, k, or $k - 1$ critical points of $u(z)$ according as the forces at α and β are both directed outward, one directed inward and the other outward, or both directed inward.*

If the arc A_1 is not empty, let K_1 be the configuration consisting of the segment $-1 \leqq z \leqq \alpha$ plus the NE half-plane bounded by the arc A_1 and the NE line joining its end-points plus the closed interiors of all Jensen circles intersecting that segment, and let K_1 contain precisely k_1 components of B. We still assume α to satisfy the original conditions. Then K contains k_1 or $k_1 + 1$ critical points of $u(z)$ according as the force at α is directed toward the left or right.

The first part of Corollary 1 is proved from Lemma 1 by considering the variation in the direction angle of the force and hence of the function $f'(z)$ of §9.1 as z traces a Jordan curve C_1 in $R - K$ near the boundary of K, where C_1 contains K in its interior, and traces Jordan curves in R near the respective components of B in K. The second part of Corollary 1 is similarly proved by using a Jordan curve C_2 consisting of a Jordan arc in $R - K_1$ near the boundary of K_1, arcs of small circles whose centers are the end-points of A_1, and a Jordan arc near A_1 in the NE half-plane bounded by A_1 and by the NE line joining the end-points of A_1; this curve C_2 is to contain in its interior the components of B in K_1; in addition to C_2, z traces also Jordan curves in R near the components of B in K_1. It is perhaps not obvious that at a point of the arc of C_2 near A_1 the force due to the spreads on B and B' (which acts toward the interior of R) plus the force due to the skew particles at the end-points of A_1 (which acts toward the exterior of R) in sum acts toward the exterior of R, but that fact follows because $u(z)$ has the value unity on A_1 and has values less than unity in R; the total force is the gradient of $u(z)$.

It may be noted that *if C_1 and C_2 are Jordan curves interior to C and mutually symmetric in Ox, and if the point z_0 lies interior to C_1, then the closed interior of the NE Jensen circle for z_0 and $\bar{z}_0$ lies interior to the NE Jensen circle for a suitably chosen pair of points on C_1 and C_2 respectively.* By way of proof we choose z_0 on the axis of imaginaries; a point of C_1 lies on that axis farther than z_0 from the axis of reals; the Jensen circle for the latter point and its image in Ox contains in its interior the Jensen circle for z_0 and $\bar{z}_0$. This remark is of importance in the proof of

COROLLARY 2. *Under the conditions of Theorem 9 let the Jordan configuration B_k symmetric in the axis of reals lie interior to C, and let $B_k + C$ bound a region R_k containing R. Let the arc $A_1^{(k)}$ be a sub-arc of A_1, and the arc $A_2^{(k)}$ a sub-arc of A_2. Then all non-real critical points in R of the function*

$$(3)\quad \begin{aligned} u(z) &= \sum_1^n \lambda_k \omega(z, B_k, R_k) + \sum_1^n \mu_k \omega(z, A_1^{(k)} + A_2^{(k)} + B_k, R_k) \\ &\quad + \sum_1^n \nu_k \omega(z, A_1^{(k)} + A_2^{(k)}, |z| < 1), \qquad \lambda_k > 0, \mu_k > 0, \nu_k > 0, \end{aligned}$$

lie in the closed interiors of the NE Jensen circles for B and in the two closed NE half-planes bounded by A_1 and A_2 and the NE lines joining their end-points.

The first part of Corollary 1 has a precise analog here, but not the second part of Corollary 1, for here we have insufficient information on the nature of the force near A_1 and A_2 to complete the proof.

Let the function $u_1(z)$ be harmonic, non-negative, and bounded interior to C, continuous on C except perhaps for a finite number of points, zero on $C - A_1 - A_2$, monotonic non-increasing as z traces in the positive sense the part of A_1 in the lower half-plane and the part of A_2 in the upper half-plane, and symmetric in the axis of reals. Then $u_1(z)$ can be uniformly approximated on any closed set interior to C by a sum of the form

$$\sum_1^n \rho_k \omega(z, A_1^{(k)} + A_2^{(k)}, |z| < 1), \qquad \rho_k > 0,$$

so Corollary 2 remains valid if $u_1(z)$ is added to the second member of (3).

§9.4.5. Symmetry, several arcs. Lemma 1 and its application in §9.4.4 obviously hold for disjoint arcs only in the case of arcs A_1 and A_2 no more than two in number, each arc being symmetric in the axis of reals. If two disjoint arcs of C symmetric with respect to each other in that axis are objects of study, further methods must be used.

We have (§9.3.4) defined the field of a dipole as the limit of the field due to equal and opposite ordinary particles approaching a fixed point from opposite directions while suitably increasing in mass. Thus if ordinary positive particles are placed at the points $+ir$ and $-ir$, $0 < r < 1$, and ordinary negative particles at the points $+i/r$ and $-i/r$, the field is given by the conjugate of

$$\frac{1}{z - ir} + \frac{1}{z + ir} - \frac{1}{z - i/r} - \frac{1}{z + i/r} = \frac{-2z(r^2 - 1/r^2)}{(z^2 + r^2)(z^2 + 1/r^2)};$$

when r approaches unity and the masses of the particles suitably increase, this field approaches the field given by the conjugate of $z/(z^2 + 1)^2$.

The field of a dipole is also (§9.3.4) the limit of the field due to equal and opposite skew particles approximately equidistant from a fixed point and approaching that point from opposite directions while suitably increasing in mass. For instance, let ζ be a point of the first quadrant on $C: |z| = 1$, let unit positive skew particles be placed at ζ and $-\zeta$, and unit negative skew particles at $\bar{\zeta}$ and $-\bar{\zeta}$. The field of force is the product of i by the conjugate of

$$\frac{1}{z-\zeta} + \frac{1}{z+\zeta} - \frac{1}{z-\bar{\zeta}} - \frac{1}{z+\bar{\zeta}} = \frac{2z(\zeta^2 - \bar{\zeta}^2)}{(z^2 - \zeta^2)(z^2 - \bar{\zeta}^2)}.$$

The factor $\zeta^2 - \bar{\zeta}^2$ is pure imaginary, so when ζ approaches the point i and when the masses of the particles suitably increase, the field of force approaches the field given by the conjugate of $z/(z^2 + 1)^2$.

To obtain a geometric construction for this field of force, we let z_0 be an arbitrary point interior to C. The force at z_0 due to ordinary positive unit particles at the points ζ and $-\zeta$ previously considered is (§1.4.1) the force at z_0 due to a double positive particle at the point Q_1 whose distance from O is $1/|z_0|$, and such that the angles $z_0 O \zeta$ and $\zeta O Q_1$ are equal. Likewise the force at z_0 due to ordinary negative unit particles at the points $\bar{\zeta}$ and $-\bar{\zeta}$ is the force at z_0 due to a double negative particle at the point Q_2, where $OQ_2 = 1/|z_0| = OQ_1$, and the angles $z_0 O(-\bar{\zeta})$ and $(-\bar{\zeta})OQ_2$ are equal. The force at z_0 due to these four particles acts along the circle $Q_1 z_0 Q_2$, whose center lies on the half-line $\arg z = \pi - \arg z_0$. The force at z_0 due to corresponding skew particles at the given points acts along the circle through z_0 of the coaxal family of which Q_1 and Q_2 are null circles. Let ζ approach i; it follows that *the direction of the force at z_0 due to equal oppositely oriented dipoles at the points $+i$ and $-i$ is that of the circle through z_0 tangent to the half-line* $\arg z = \pi - \arg z_0$ *at the point z for which $|z| = 1/|z_0|$*.

In order to apply Theorem 1, for a positive arc $(\zeta, -\bar{\zeta})$ of C near i we need to place a positive unit skew particle at $-\bar{\zeta}$ and a negative unit skew particle at ζ; we proceed similarly for the reflection $(-\zeta, \bar{\zeta})$ of the arc $(\zeta, -\bar{\zeta})$ in the axis of reals. We now reverse the previously considered orientation of the dipoles at $+i$ and $-i$, to agree with the present genesis of the field, and proceed to derive some properties of the new field. At any point of C not $\pm i$ the force is directed toward O, and is also directed toward O at any point interior to C on the axis of reals; at any point on the axis of imaginaries interior to C the force is directed away from the axis of reals; at any point z near the origin the direction of the force is approximately that of $-\bar{z}$. A further property is established in

LEMMA 2. *In the field due to equal oppositely oriented dipoles at the points $+i$ and $-i$, the limit of the field due to skew particles oriented as for the harmonic measures of positive arcs near $+i$ and $-i$, interior to $C: |z| = 1$ the force has the direction of the NE horizontal on the axis of reals and on the circles whose arcs are the loci $\omega(z, (-i, +i), |z| < 1) = 1/4$ and $3/4$, where the arc $(-i, +i)$ indicates the positive arc on C; at a non-real point interior to both these circles the force has a non-zero component in the sense of the NE vertical directed away from the axis of reals; at a non-real point interior to C but exterior to one of these circles the force has a non-zero component in the sense of the NE vertical directed toward the axis of reals.*

At an arbitrary point z interior to C, the NE horizontal is the direction of the force due to a positive ordinary particle at $+1$, and an equal negative ordinary particle at -1, namely the direction of the conjugate of

$$\frac{1}{z-1} - \frac{1}{z+1} = \frac{2}{z^2-1}.$$

The condition that the field of the dipoles have the direction of the NE horizontal is then

$$\arg \frac{z}{(z^2+1)^2} = \arg \frac{1}{z^2-1} \qquad (\text{mod } \pi),$$

$$(4) \qquad \arg \frac{z(z^2-1)}{(z^2+1)^2} = 0 \qquad (\text{mod } \pi).$$

Equation (4) can be expressed in simpler algebraic form by substituting $z = i(1+w)/(1-w)$, $w = (z-i)/(z+i)$; equation (4) becomes

$$(5) \qquad \Re\left[\frac{w^4-1}{w^2}\right] = 0.$$

Here we set $w = \xi + i\eta$, whence

$$\frac{w^4-1}{w^2} = \frac{w^2(w^2\bar{w}^2) - \bar{w}^2}{w^2\bar{w}^2},$$

and condition (5) can be written $(\xi^2 - \eta^2)[(\xi^2+\eta^2)^2 - 1] = 0$. In the w-plane this condition is satisfied precisely on the unit circle and on the lines $\eta = \pm\xi$; in the z-plane this condition is satisfied precisely on the axis of reals and on the circles mentioned in Lemma 2. The conclusion of Lemma 2 follows from the fact that on the axis of imaginaries interior to C the force is directed away from the axis of reals; the first member of (5) reverses sign interior to C on each of the lines $\eta = \pm\xi$.

If we transform the situation of Lemma 2 by a linear transformation which leaves both the axis of reals and the interior of C invariant, the points $+i$ and $-i$ may be carried into arbitrary preassigned distinct conjugate points on C; the NE horizontal direction and content of the lemma are invariant. We shall prove

LEMMA 3. *Let the points α, β, $\bar{\beta}$, $\bar{\alpha}$ lie on C in the counterclockwise order indicated, let unit positive skew particles be placed at the points β and $\bar{\alpha}$ and unit negative skew particles at the points α and $\bar{\beta}$. Then the force at a non-real point z interior to C in either of the regions $\omega(z, \beta\bar{\beta}, |z| < 1) > 3/4$, $\omega(z, \bar{\alpha}\alpha, |z| < 1) > 3/4$ has a non-zero component in the sense of the NE vertical directed toward the axis of reals.*

We subdivide the arcs $\alpha\beta$ and $\bar{\beta}\bar{\alpha}$ indefinitely by points of subdivision which are symmetric in the axis of reals. At each interior point of subdivision we place both a unit positive and a unit negative skew particle; this procedure does not alter the original field of force. If z satisfies the conditions of Lemma 3, and if ζ is a point of the arc $\alpha\beta$, the force at z due to a pair of positive and negative skew particles on C sufficiently near to ζ and to the corresponding pair of particles at the points symmetric in the axis of reals has a non-zero component in the sense of the NE vertical directed toward the axis of reals, by Lemma 2 and by the definition of the field of a dipole; this conclusion is true for fixed z uniformly for every choice of ζ on the arc $\alpha\beta$, so Lemma 3 follows.*

We are now in a position to generalize Theorem 9:

THEOREM 10. *Let R be a region bounded by the unit circle C and by a Jordan configuration B interior to C which is symmetric in the axis of reals. Let closed disjoint arcs A_1 and A_2 each symmetric in that axis intersect the axis of reals in the points $z = -1$ and $z = +1$ respectively. Let arcs $D_j: (\alpha_j\beta_j)$, $j = 1, 2, \cdots, m$, lie on C in the positive order between A_2 and A_1, and let $\bar{D}_j$ be their reflections in the axis of reals. Then all non-real critical points in R of the function $u(z) = \omega(z, A_1 + A_2 + \sum(D_j + \bar{D}_j) + B, R)$ lie in the set consisting of the closed interiors of the NE Jensen circles for B, the NE half-planes bounded by the arcs A_j and the NE lines joining their end-points, and the m regions which are respectively the complements of the sums of the two sets $\omega(z, \beta_j\bar{\beta}_j, |z| < 1) > 3/4$, $\omega(z, \bar{\alpha}_j\alpha_j, |z| < 1) > 3/4$.*

Choose R as containing the origin. In the field of force for the function $u(z)$, the force at a non-real point z exterior to the set mentioned has a non-zero component in the sense of the NE vertical directed toward the axis of reals,—this is true for the force at z due to matter on pairs of arcs of B symmetric in the axis of reals and on their inverses, for the force due to matter on arcs of B on that axis and on their inverses, for the force due to pairs of particles at the end-points of A_1 and A_2, for the force due to quadruples of particles at the end-points of each pair D_j and $\bar{D}_j$, and for the total force at z.

Corollary 1 to Theorem 9 has an analog here, where we may suppose a configuration K either to contain or not to contain specific Jensen circles, or a NE half-plane corresponding to an arc A_j, or a region corresponding to a pair of arcs D_j and $\bar{D}_j$; we leave the formulation to the reader. Corollary 2 to Theorem 9 also has an analog:

COROLLARY. *Under the conditions of Theorem 10, let the Jordan configuration B_k lie interior to C, and let $B_k + C$ bound a region R_k symmetric in O_x containing R. Let the arc $A_j^{(k)}$ symmetric in O_x be a subarc of A_j, and let the arcs $D_j^{(k)}$ and $\bar{D}_j^{(k)}$ mutually symmetric in the axis of reals be subarcs of D_j and $\bar{D}_j$. Then the conclu-*

* Lemma 3 is not incapable of improvement; it is of course possible to be more explicit and to study directly the field of force of Lemma 3 without the use of Lemma 2; the locus of points at which the total force has a non-zero component in the direction of the NE vertical toward the axis of reals is bounded (in addition to the axis of reals) by a certain bicircular quartic, which need not degenerate.

sion of Theorem 10 *is valid for the function*

$$(6) \qquad u(z) = \sum_{k=1}^{n} \lambda_k \omega(z, A_1^{(k)} + A_2^{(k)} + \sum_j (D_j^{(k)} + \bar{D}_j^{(k)}) + B_k, R_k), \lambda_k > 0.$$

Here we expressly admit the possibility that any of the sets $A_j^{(k)}, B_k, D_j^{(k)} + \bar{D}_j^{(k)}$ may be empty. To the second member of (6) may be added any function which can be uniformly approximated on any set interior to C by a function of the type of that second member. In particular we may add a function $u_0(z)$ harmonic, non-negative, and bounded interior to C, continuous on C except perhaps for a finite number of points, zero on $C - A_1 - A_2$, monotonic non-increasing as z traces in the positive sense the part of A_1 in the lower half-plane and the part of A_2 in the upper half-plane, and symmetric in the axis of reals. We may also add a function $u_j(z)$ harmonic, non-negative, and bounded interior to C, continuous on C except perhaps for a finite number of points, zero on $C - D_j - \bar{D}_j$, and symmetric in the axis of reals.

§9.4.6. Case $p > 1$. In §§9.4.2–9.4.5 we have studied in effect the critical points, in a region R whose boundary consists of q components, of the harmonic measure of a set including proper sub-arcs of only one component; we proceed now to the harmonic measure of a more general set consisting of arcs of $p(>1)$ components, by a conformal map of the appropriate covering surface. Here the case $p = 2$ is far simpler than the case $p > 2$:

THEOREM 11. *Let R be a region of the z-plane bounded by disjoint Jordan curves C_1 and C_2 and by a Jordan configuration B in the annular region S bounded by C_1 and C_2, and let the mutually disjoint closed arcs A_{1k} and A_{2k} lie on C_1 and C_2 respectively. We set $u(z) = \omega(z, B + \sum A_{jk}, R)$.*

1). *If the set A_{1k} is empty, and if the region $\omega(z, C_1, S) > \mu$ $(\geqq \frac{1}{2})$ contains no point of B, then that region contains no critical point of $u(z)$.*

2). *Let the universal covering surface S^∞ of S be mapped onto the interior of $C: |w| = 1$. If A' and A'' are successive arcs of C which are images of arcs of the set A_{jk}, if H is a NE half-plane bounded by a NE line of the two coaxal families determined by the end-points of A' and those of A'' as null circles and bounded by an arc of C containing no complete image of an arc A_{jk}, and if H contains no point of B, then H contains no critical point of the transform of $u(z)$. If H contains points of B but if a NE half-plane H' bounded by a circle of one of those coaxal families is a subregion of H containing no point of B, then H' contains no critical point of the transform of $u(z)$.*

3). *Let R be conformally symmetric in the curve $T: \omega(z, C_1, S) = \frac{1}{2}$, and let the two arcs A_{1k} and A_{2k} be symmetric in T for every k. Any arc of T in R bounded by points of B contains at least one critical point of $u(z)$. Any critical point w of the transform of $u(z)$ not on the image of T either lies in the closed interior of a NE Jensen circle for the image of B or satisfies both inequalities $\omega(w, C_1', |w| < 1) <$ $3/4$, $\omega(w, C_2', |w| < 1) < 3/4$, where C_1' and C_2' are the arcs of C complementary to two arcs of C which are corresponding images of arcs A_{1k} and A_{2k} respectively.*

In the proof of Theorem 11, the transform of $u(z)$ in the w-plane is approximated by a harmonic function involving the harmonic measure of only a finite number of boundary components and of boundary arcs; compare §§8.7.2 and 8.8.1; we omit the details. Obviously Theorem 11 does not exhaust the immediate application of the previous methods; we add a few other remarks, likewise capable of extension.

COROLLARY 1. *If R satisfies the conditions of Theorem* 11, *if the function $u_1(z)$ is harmonic, bounded, and non-negative in R, continuous on $C_1 + C_2$ except perhaps for a finite number of points, and zero on C_1, then* 1) *is valid for the function* $u(z) = u_1(z) + \lambda\omega(z, B, R), \lambda > 0$.

COROLLARY 2. *Under the conditions of Theorem* 11 *on the region R, let the Jordan configuration B_k lie in S, and let $B_k + C_1 + C_2$ bound a region R_k containing R. Then* 1) *and* 2) *are valid for the function*

$$u(z) = \sum_1^n \lambda_k \omega(z, B_k, R_k) + \sum_1^n \mu_k \omega(z, B_k + \sum A_{jk}, R_k)$$

$$+ \sum_{j,k} \nu_k \omega(z, A_{jk}, S), \qquad \lambda_k > 0, \quad \mu_k > 0, \quad \nu_k > 0.$$

Part 3) also remains valid if R_k and $u(z)$ are conformally symmetric in T.

In numerous situations there is considerable latitude of choice of methods, for various covering surfaces can be chosen to be mapped into the interior of the unit circle C. As an illustration we formulate

COROLLARY 3. *Let the region R be bounded by a Jordan configuration whose components are $C_1, D_1, D_2, \cdots, D_n$; we set $u(z) = \omega(z, D_1 + D_2 + \cdots + D_n, R)$. Let S denote the annular region bounded by C_1 and D_1. Then any region $\omega(z, C_1, S) > \mu(\geqq \frac{1}{2})$ which contains no point of $D_2 + \cdots + D_n$ contains no critical point of $u(z)$ in R.*

Corollary 3 is to be compared especially with §8.4 Theorem 1 and §8.10.3 Corollary 1 to Theorem 2, but is presented merely as an illustration and not as a result of great generality. Corollary 3 is contained in part 1) of Theorem 11.

THEOREM 12. *Let R be a region of the z-plane bounded by mutually disjoint Jordan curves $C_1, C_2, \cdots, C_p$ and by a Jordan configuration B in the region S bounded by the C_j, and let the mutually disjoint closed arcs A_{jk} lie on C_j. Then* 2) *is valid.*

We have studied in §9.4 the critical points of the harmonic measure with respect to a region R of a set $A + B$, the sum of a set B of entire components of the boundary of R and of a set A of arcs of those components. This entire subject can be treated by approximating those arcs A by auxiliary variable boundary components interior to R and then allowing those components to approach the given arcs A. This method shows that a NE convex region containing $A + B$ contains no critical point of $\omega(z, A + B, R)$, a result analogous to those of §8.4, but this suggested method does not yield the present sharper results which are consequences of the special properties of arcs of circles with reference to the field of force. Compare §9.6.

§9.5. Green's functions and harmonic measures. Before undertaking the general study of linear combinations of Green's functions and harmonic measures, which are important in the study of various extremal problems, we consider in some detail a specific but typical illustration.

§9.5.1. A numerical example. Let R be the region $| z | < 1$, A the positive arc $(-i, i)$ of $C : | z | = 1$, and $G(z)$ Green's function for R with pole in the origin. We wish to study the critical points of $u(z) = \omega(z, A, R) + \lambda G(z)$ where λ is real and different from zero. The field of force results from unit positive and negative skew particles at $+i$ and $-i$ respectively and a particle of mass $-\lambda$ at the origin; the force is the conjugate of

$$\frac{-i}{z - i} + \frac{i}{z + i} - \frac{\lambda}{z}$$

and the critical points are the positions of equilibrium $z^2 - 2z/\lambda + 1 = 0$,

$$z = \frac{1}{\lambda} \pm \frac{(1 - \lambda^2)^{1/2}}{\lambda}. \tag{1}$$

The product of these two values of z is unity, so at most one of them lies in R. For values $0 < \lambda < 1$, precisely one critical point lies in R, and lies on the interval $0 < z < 1$; for $\lambda = 1$, the two points given by (1) coincide at $z = 1$; for $\lambda > 1$, the two values of z are conjugate imaginary with product unity, hence lie on C; the real part of z is $1/\lambda$, which is positive. For negative λ the values of z are the negatives of the values for $-\lambda$.

It is instructive to note the variation of the force on C, which of course by the Principle of Argument enables us to determine the number of critical points interior to C. We choose λ positive. On a small circle whose center is the origin the force is directed inward; near $+i$ the force is directed roughly counterclockwise about that point, and near $-i$ is directed roughly clockwise about that point. At every point of C the force due to the particle at the origin is of magnitude λ and directed toward the origin. At every point z the force due to the two skew particles is orthogonal to the circle through z, i, and $-i$; on C this force is orthogonal to C, directed outward on the right half of C and inward on the left half. On the left half of C the total force is orthogonal to C and acts inward at every point. The magnitude of the force at z due to the skew particles is $2/| z^2 + 1 |$; the denominator can be interpreted as the distance from z^2 to the point -1, which on C varies monotonically in each quadrant; the force has the value unity for $z = 1$ and increases monotonically as z moves from $z = 1$ toward the points $z = \pm i$. The total force at $z = 1$ is $1 - \lambda$, thus for values $0 < \lambda < 1$ acts toward the right, and acts outward along the normal at all points of the right half of C. For values $\lambda > 1$, the total force at $z = 1$ acts toward the left, and as z moves on C from $z = 1$ acts along the inner normal until z reaches one of the points (1); as z continues in the same sense on C the direction of the force acts along the outer normal until z reaches $\pm i$. At a point z_1 in the first quadrant and interior to C, near a point z_0 given by (1), the force due to the skew particles is directed outward normal to the circle Γ through $+i$, $-i$ and z_1, and the force due to the particle at O is directed along z_1O, so the total force has a component along Γ toward the axis of reals. When z traces clockwise a small approximate semicircle with center z_0 and interior to C, the direction angle of this force *increases* from $\arg(-z_0)$ to $\arg z_0$ approximately. It is now possible to determine also by the Principle of Argument the number of critical points of $u(z)$ interior to C in the two cases $0 < \lambda < 1$ and $\lambda > 1$. In the former case the normal derivative $\partial u/\partial \nu$ is everywhere negative on A, and in the latter case is positive on A between the points (1) and elsewhere negative on A. The additional case $\lambda < 0$ needs no separate treatment, for the force at the point z' with $\lambda = \lambda'$ is equal to the force at $-z'$ with $\lambda = -\lambda'$.

The critical points in a slightly more general case than that just treated are readily studied:

THEOREM 1. *Let R be the region $|z| < 1$, $A:(\alpha, \beta)$ a positive arc of $C:|z| = 1$, and $G(z, \gamma)$ Green's function for R with pole in the point γ of R. For $\lambda > 0$, the function $u(z) = \omega(z, A, R) + \lambda G(z, \gamma)$ has at most one critical point in R, namely on the circle through γ of the coaxal family determined by α and β as null circles, on the arc in R bounded by A and γ; for $\lambda < 0$ this function has at most one critical point in R, namely on the arc of this circle bounded by γ and a point of $C - A$.*

Under these conditions we have $G(z, \gamma) = -\log |(z - \gamma)/(\bar{\gamma}z - 1)|$, $F(z) = G + iH = -\log [(z - \gamma)/(\bar{\gamma}z - 1)]$, where H is conjugate to G, whence

$$F'(z) = \frac{1}{z - 1/\bar{\gamma}} - \frac{1}{z - \gamma}; \tag{2}$$

for $G(z, \gamma)$ the usual field of force is the conjugate of $F'(z)$, and is due to ordinary positive and negative unit particles at $1/\bar{\gamma}$ and γ respectively. To this field of force with factor λ is to be added the usual field for $\omega(z, A, R)$, namely positive and negative skew particles of masses $1/\pi$ and $-1/\pi$ at β and α respectively. Transform γ to the origin O; the lines of force for $G(z, \gamma)$ are the lines through O; the lines of force for the skew particles are the coaxal family of circles determined by α and β as null circles; these circles are orthogonal to C. If the directions of these two lines of force at a point z in R coincide, the two lines of force must coincide, hence be a single diameter of C. If z is a position of equilibrium the two forces must act in opposite senses, and z lies as indicated in Theorem 1.

It follows as in the study of the numerical example above that $u(z)$ has precisely two critical points, and the product of their moduli is unity; at most one critical point lies in R.

There are immediately available many quantitative results on the critical points of linear combinations of Green's functions and harmonic measures, by the use of fields of force and the methods of Chapters IV and V, especially by the use of §9.3.5 Theorem 7; compare the discussion of §8.9 for linear combinations of Green's functions. We shall not elaborate this remark, but turn to further results involving hyperbolic geometry.

We denote Green's function for the region R with pole in z_0 by $g(z, z_0, R)$, and if R is unambiguous by $G(z, z_0)$.

§9.5.2. General linear combinations. In the study of critical points of linear combinations of Green's functions and harmonic measures, two methods suggest themselves: i) direct consideration of the field of force, as in the proof of Theorem 1, and ii) use of the results of §§9.2–9.4. Method (i) is particularly simple for the interior of the unit circle, as the force for Green's function $G(z, \gamma)$ is the conjugate of $F'(z)$ in (2); but even if the region is not the unit circle, the method may be applicable; compare §8.9.2 Theorem 4. Method (ii) applied to such a functon as $\sum \lambda_k \omega(z, A_k, R) + \sum \mu_k G(z, \gamma_k)$, $\lambda_k > 0$, $\mu_k > 0$, involves truncating this second summation $u_2(z)$, compare §8.9.1. Let the region R be given, and let J_M be the locus $u_2(z) = M$ in R, where M is large and positive, so that J_M consists of a number of curves surrounding the points γ_k, curves which approach those points as M becomes infinite. We may write $u_2(z) = M\omega(z, J_M, R_M)$, where R_M is the region $0 < u_2(z) < M$ in R, and thereby apply our previous theorems involving the linear combination of harmonic measures, and finally allow M to become infinite. Practically all the results of §9.4 are of significance here. We suppress the proofs, and formulate but a few of the possible conclusions.

The analog of §9.4.1 Theorem 2 is

THEOREM 2. *Let C be the unit circle, let $A_1, A_2, \cdots, A_m, A_{m+1} = A_1$ be mutually disjoint closed arcs of C ordered counterclockwise, and let C_j denote the coaxal family of circles determined by the end-points of A_j as null circles. Let $a_1, a_2, \cdots, a_n$ be points interior to C, and let Π_B denote the smallest NE convex region interior to C containing the a_j. Let Π_k denote the point set consisting of all maximal arcs of the family C_k each bounded by a point of A_k and a point of Π_B, and let Π_0 denote the NE convex region bounded by the m NE lines respectively common to the families C_j and C_{j+1}, $j = 1, 2, \cdots, m$. Then all critical points of the function*

$$\sum_1^m \lambda_j \omega(z, A_j, |z| < 1) + \sum_1^n \mu_j g(z, a_j, |z| < 1), \qquad \lambda_j > 0, \quad \mu_j > 0,$$

in $R: |z| < 1$ lie in the smallest NE convex region containing $\sum_0^m \Pi_j$.

Theorem 1 is the special case $m = 1$, $n = 1$; if λ in Theorem 1 is positive, we identify the arc A of Theorem 1 with the arc A_1 of Theorem 2; if λ in Theorem 1 is negative, we identify the arc $C - A$ of Theorem 1 with the arc A_1 of Theorem 2.

COROLLARY. *Under the conditions of Theorem 2, let the function $u_1(z)$ be harmonic and bounded interior to C, continuous and non-negative on C except perhaps in a finite number of points, and zero on C except perhaps on the arcs A_j. Let Π' be the smallest NE convex region whose closure contains Π_B and the A_j. Then Π' contains all critical points in R of the function*

$$u_1(z) + \sum_1^m \lambda_j g(z, a_j, |z| < 1), \qquad \lambda_j > 0.$$

An analog of §9.4.2 Theorem 6 is

THEOREM 3. *Let C be the unit circle, and let $A_1, A_2, \cdots, A_m, A_{m+1} = A_1$, $A_1', A_2', \cdots, A_{m'}', A_{m'+1}' = A_1'$ be mutually disjoint open arcs of C ordered counter-clockwise. Let R be a region bounded by C and by disjoint Jordan configurations B_0 and B_0' interior to C, let R_0 and R_0' be the regions containing R bounded respectively by $B_0 + C$ and $B_0' + C$, and let $a_1, a_2, \cdots, a_n, b_1, b_2, \cdots, b_n$ be points of R. Let the Jordan configurations B_k and B_k' $(k > 0)$ any of which may be empty lie interior to C, and let $B_k + C$ and $B_k' + C$ bound regions R_k and R_k' containing R_0 and R_0' respectively. Let Π_0 be a NE convex region containing B_0, containing for every j the NE line belonging to both coaxal families of circles determined by the end-points of A_j and A_{j+1} as null circles, and containing all arcs of circles of the family determined by the end-points of A_j joining the arc A_j with points of the smallest convex set containing B_0 and the b_k; let Π_0' be the corresponding set for B_0', the A_j', and the a_k. If a NE line Γ separates Π_0 and Π_0', then no critical point of the function*

$$\begin{aligned} u(z) = {} & \sum_0^N \lambda_k \omega(z, B_k + \textstyle\sum A_j, R_k \cdot R_k') + \displaystyle\sum_0^N \mu_k g(z, b_k, R_k') \\ & - \sum_0^N \lambda_k' \omega(z, B_k' + \textstyle\sum A_j', R_k \cdot R_k') - \displaystyle\sum_0^N \mu_k' g(z, a_k, R_k), \end{aligned} \tag{3}$$

where the coefficients $\lambda_k, \mu_k, \lambda_k', \mu_k'$ are all positive and where $\sum A_j$ and $\sum A_j'$ may depend on k, lies in R on Γ. Consequently if a line Γ exists, all critical points of $u(z)$ in R lie in two NE convex sets Π and Π' containing Π_0 and Π_0' respectively and separated by every Γ which separates Π_0 and Π_0'.

Theorem 3 has not been previously formulated (compare §8.9) even for the case $m = m' = 0$. Theorem 3 may be further generalized by permitting further arcs S_k and S_k' of C on which $u_1(z)$ and $u_2(z)$ are respectively non-negative and continuous except perhaps for a finite number of points; we assume $u_1(z)$ and $u_2(z)$ harmonic and bounded interior to C. If the closure of Π_0 contains the S_k, and the closure of Π_0' contains the S_k', the conclusion of Theorem 3 remains valid if $u_1(z) - u_2(z)$ is added to the second member of (9).

We omit the analog of §9.4.3 Theorem 7 and its Corollary, and proceed to the analog of §9.4.5 Theorem 10:

THEOREM 4. *Let R be a region bounded by the unit circle C and by a Jordan configuration B interior to C which is symmetric in the axis of reals. Let closed disjoint arcs A_1 and A_2 each symmetric in that axis intersect the axis of reals in the points $z = -1$ and $z = +1$ respectively. Let arcs $D_j: (\alpha_j \beta_j)$, $j = 1, 2, \cdots, m$, lie on C in the positive order between A_2 and A_1, and let $\bar{D}_j$ be their reflections in the axis of reals. Let the points $a_1, a_2, \cdots, a_n$ lie in R. Then all non-real critical points in R of the function*

$$u(z) = \rho\omega(z, B, R) + \sum_{k=1}^{2} \rho_k \omega(z, A_k, |z| < 1)$$
$$+ \sum \lambda_j \omega(z, D_j + \bar{D}_j, |z| < 1) + \lambda\omega(z, A_1 + A_2 + \sum(D_j + \bar{D}_j) + B, R)$$
$$+ \sum \mu_k [g(z, a_k, |z| < 1) + g(z, \bar{a}_k, |z| < 1)],$$
$$\rho > 0, \quad \rho_k > 0, \quad \lambda_k > 0, \quad \lambda > 0, \quad \mu_k > 0,$$

lie in the set consisting of the closed interiors of the NE Jensen circles for B, the a_k, and the $\bar{a}_k$, of the NE half-planes bounded by the arcs A_j and the NE lines joining their end-points, and of the m regions which are respectively the complements of the sums of the two sets $\omega(z, \bar{\alpha}_j \alpha_j, |z| < 1) > 3/4$, $\omega(z, \beta_j \bar{\beta}_j, |z| < 1) > 3/4$.

In the case $m = 0$, A_1 and A_2 null sets, with B and the points a_k on the axis of reals, all critical points lie on that axis. Theorem 4 can be further extended; compare §9.4.5 Corollary to Theorem 10.

We state but a few of the possible results concerning Green's functions for regions of higher connectivity:

THEOREM 5. *Let R be a region of the z-plane bounded by disjoint Jordan curves C_1 and C_2, and let the mutually disjoint closed arcs A_{1k} and A_{2k} lie on C_1 and C_2 respectively. We set*

$$u(z) = \sum \lambda_k \omega(z, A_{1k}, R) + \sum \rho_k \omega(z, A_{2k}, R) + \sum \mu_k g(z, a_k, R),$$
$$\lambda_k > 0, \quad \rho_k > 0, \quad \mu_k > 0,$$

where the points a_k lie in R.

1). *If the set A_{1k} is empty, and if the region $\omega(z, C_1, R) > \mu(\geqq \frac{1}{2})$ contains no point a_k, then that region contains no critical point of $u(z)$.*

2). *Let the universal covering surface R^∞ of R be mapped onto the interior of $C: |w| = 1$. If A' and A'' are successive arcs of C which are images of arcs A_{jk}, if H is a NE half-plane bounded by a NE line of the two coaxal families determined by the end-points of A' and those of A'' as null circles and bounded by an arc of C containing no complete image of an arc A_{jk}, and if H contains no point a_k, then H contains no critical point of the transform of $u(z)$.*

Part 2) extends at once to a region R of arbitrary finite connectivity.

§9.6. Limits of boundary components. In addition to the use of Green's functions, we have hitherto studied especially i) harmonic measures of entire boundary components, as in §§8.1–8.6, and ii) harmonic measures of boundary arcs, as in §§8.7, 8.10, 9.2, 9.3. Various cases exist in which the harmonic measure of a boundary arc is more naturally treated as the limit of the harmonic measure of a variable boundary component which approaches a fixed boundary component in one or two points, rather than directly. The difference is primarily one of emphasis and method, just as frequently in the past we have found a difference in the results obtained by direct study of a given configuration and by study of that configuration after a suitable conformal transformation.

As a specific illustration here, let R be the interior of the unit circle $C: |z| = 1$, and let R_1 be the region R cut along the segment $E: 0 \leqq z \leqq 1$. We may study $\omega(z, E, R_1)$ in relation to other functions 1) by mapping R_1 onto the interior of a circle or 2) by considering without conformal map a variable region R' found by cutting R along the segment $E': 0 \leqq z \leqq 1 - \epsilon\ (< 1)$, by considering the harmonic measure $\omega(z, E', R')$, and allowing ϵ to approach zero. Naturally 1) and 2) are not intended as a complete enumeration of the possible methods, but merely to emphasize the distinctive character of the method 2) now to be used.

§9.6.1. Hyperbolic geometry. The present methods apply in the study of harmonic measures of arcs and boundary components, and their limits, and linear combinations of harmonic measures and Green's functions. We choose to present merely a few illustrations [Walsh, 1948a] and leave to the reader the full implications of the method.

THEOREM 1. *Let R be the interior of the Jordan curve C, let mutually disjoint Jordan curves or arcs $J_1, J_2, \cdots, J_m$ lie in R, and let Jordan arcs $A_1, A_2, \cdots, A_n$ disjoint from each other and from the J_k lie in $R + C$, each arc A_k having one or both end-points and no other points in common with C. Let the J_k and A_k together with suitable arcs of C bound a subregion R_1 of R. The smallest NE convex region of R whose closure contains all the J_k and A_k likewise contains all critical points in R_1 of the function $u(z) = \omega(z, \sum J_k + \sum A_k, R_1)$.*

Choose C as the unit circle: $|z| = 1$. Our use of variable harmonic functions is similar to that of §8.8.1. Let the intersections of the arcs A_k with C be the points $z_1, z_2, \cdots, z_q$, and let $z_1', z_2', \cdots, z_q'$ be auxiliary variable points near the z_k on the respective arcs. Let $\delta(> 0)$ be arbitrary, and denote by $u_k(z)$ twice the harmonic measure with respect to R of the arc of C whose center is z_k and angular measure 2δ. We suppose z_k' chosen so near z_k that on the sub-arc of A_j joining those points we have $u_k(z) > 1$. A pair of points z_k' belonging to the same arc A_j are to be joined by a Jordan arc A_j' in R exterior to R_1, and we denote the Jordan curve $A_j + A_j'$ minus arcs $z_k z_k'$ by B_j; if an arc A_j meets C in but one point z_k, we denote by B_j the subarc which remains after the arc $z_k z_k'$ is deleted from A_j. Denote by R' the variable subregion of R containing R_1 bounded by all the J_k and B_k. The inequalities

$$(1) \qquad u(z) - \sum u_j(z) \leqq \omega(z, \sum J_j + \sum B_j, R') \leqq u(z)$$

at boundary points of R_1 on the J_j or B_j and at boundary points of R_1 on C not on the A_j are obvious; the inequalities also hold in the form of limits as z in R_1 approaches any boundary point of R_1 on an arc $z_k'z_k$; thus (1) is valid throughout R_1. On any closed set in R_1 the sum $\sum u_j(z)$ approaches zero uniformly as δ approaches zero, so by (1) the function $\omega(z, \sum J_j + \sum B_j, R')$ approaches $u(z)$ uniformly. Theorem 1 is now a consequence of §8.4 Theorem 1.

A slight extension of Theorem 1 is easily given to allow arcs A_k to lie on C; here even sharper results can be obtained, as in §9.4.1 Theorem 2.

As a mere change in phraseology, any A_k in Theorem 1 may be allowed to be a Jordan curve in $R + C$ having precisely one point in common with C. The method of proof of Theorem 1 applies under certain conditions when the J_k and A_k are infinite in number; a special case here yields the

COROLLARY. *Let the region S be bounded by the disjoint Jordan curves C_1 and C_2, let mutually disjoint Jordan curves $J_1, J_2, \cdots, J_m$ lie in S, and let Jordan arcs $A_1, A_2, \cdots, A_n$ disjoint from each other and from the J_k lie in $S + C$, each arc A_k having one or both end-points and no other points in common with $C_1 + C_2$. Let the J_k and A_k together with suitable arcs of $C_1 + C_2$ bound a subregion R_1 of S. Then any region $\omega(z, C_j, S) > \mu\ (\geqq \frac{1}{2})$ which contains no point of the J_k or A_k contains no critical point of the function $\omega(z, \sum J_k + \sum A_k, R_1)$.*

As in the proof of Theorem 1 we have considered each A_k the limit of a variable boundary component, so any boundary arc can be considered the limit of a boundary component, and §8.4 Theorem 1 can be applied; results obtained by this method are not so sharp, however, as §8.7.1 Theorem 1, §8.10.1 Theorem 1, §9.3.1 Theorem 2, and numerous others; indeed these latter results can be generalized by the present method. But §9.2.1 Theorem 2 and §9.2.2 Theorem 3 can be proved by this method of variable boundary components.

A direct proof of Theorem 1, without the present use of variable boundary components, can be given by setting up the field of force for the function $u(z)$. The usual spreads of matter are placed on the J_k and A_k, with opposite spreads on their reflections in $C: |z| = 1$. It turns out that skew particles are placed at the end-points of an arc A_k on C unless only one end-point lies on C. If Theorem 1 is extended so as to allow an A_k to be a Jordan curve in $R + C$ having precisely one point in common with C, no skew particle is placed at that point.

These latter questions, relating to a Jordan curve in $R + C$ or even in R, are related to a much broader topic. If a function $f_k(z) = u_k(z) + iv_k(z)$ is analytic in a region R_k with finite boundary C_k, we consider the field of force defined by the conjugate of $f_k'(z)$, and write formally

$$(2) \qquad f_k'(z) = \frac{i}{2\pi} \int_C \frac{du_k + i dv_k}{z - t}, \qquad z \text{ in } R_k,$$

which defines a spread of ordinary and skew matter on C_k, perhaps involving ordinary and skew particles at discontinuities of $v_k(z)$ and $u_k(z)$. At a finite point z exterior to R_k, the integral in (2) has the value zero, and the force at z due to the corresponding spread of matter vanishes. Thus the critical points of the function $f_1(z)$ in R_1 can be studied as the positions of equilibrium not merely in the usual field for $f_1(z)$ in R_1, but in the sum of the fields for all the $f_k(z)$ in the regions R_k, if the regions R_j $(j > 1)$ are exterior to R_1. Some of the boundaries C_j $(j > 1)$ may coincide in whole or in part with C_1, a possibility that may enable us to simplify the spread of matter on C_1 defining the field represented by the conjugate of $f_1'(z)$. For instance it may be possible to eliminate spreads of matter on various arcs of C_1, or to eliminate various skew particles on C_1. Hitherto we have used but a single illustration of this method; reflection in a circle (§9.3) can be considered as the use of two functions defined respectively in the interior and exterior of the circle. This method is also of significance in connection with Theorem 1 and its extensions, for (notation of the proof of Theorem 1) if the arc A_k intersects C in two distinct points, and if the auxiliary Jordan arc A_k' approaches A_k, then in R_k' the function $\omega(z, \sum J_j + \sum B_j, R')$ approaches $\omega(z, A_k, R_k')$, where R_k' (if existent) denotes the Jordan subregion of R exterior to R_1 bounded by A_k and by an arc of C. The usual spread of matter on the boundary of R' for the former function approaches not the usual spread for $u(z)$ on the boundary of R_1, but a spread for $u(z)$ in R_1 plus spreads for the functions $\omega(z, A_k, R_k')$ in all the regions R_k' that exist.

§9.6.2. Comparison of hyperbolic geometries. Let at least one Jordan arc A_k in Theorem 1 have both end-points on C. Then the curve C is not uniquely determined, nor is the NE geometry of Theorem 1, for C may be altered along the arc of C bounded by the end-points of A_k not part of the boundary of R_1 without altering $u(z)$ or the truth of the hypothesis. What are the relative merits of various choices of C?

Simplicity may be a useful criterion, for in a given configuration the total or significant set of NE lines may or may not be conveniently and easily determined. It is also well to choose C so that the point set $\sum J_k + \sum A_k$ is as nearly NE convex as possible; if R is the interior of the unit circle, radial slits for the A_k are convenient. As an illustration, let C be the unit circle $|z| = 1$, let m be zero, let n be two, with A_1 and A_2 the respective segments $-1 \leqq z \leqq 0$, $\frac{1}{2} \leqq z \leqq 1$; here the NE convex region of Theorem 1 degenerates to the segment $-1 \leqq z \leqq 1$, from which it follows that the unique critical point lies on the segment $0 < z < \frac{1}{2}$; this conclusion (which of course follows by symmetry) is relatively strong, but can be improved by mapping R_1 onto the interior of the unit circle, as in §8.7.1 Theorem 1. However, if we study here not the function $u(z)$ of Theorem 1 but the function $\lambda\omega(z, A_1, R_1) + \mu\omega(z, A_2, R_1)$, $\lambda > 0$, $\mu > 0$, to which the conclusion of Theorem 1 also applies, the conclusion cannot be improved without restricting λ and μ.

If we modify the example just given by choosing C as the unit circle but with $m = 0$, $n = 3$, where the A_k are radial slits, the NE convex region determined by Theorem 1 is relatively restricted, but is less favorable than that of §8.7.1 Theorem 1.

We proceed to indicate now not by formal proof but by three explicit illustrations that under Theorem 1 *the most precise results are ordinarily obtained by choosing R as large as possible*. EXAMPLE I. Let C be the unit circle, choose $m = 0$, $n > 2$, and let A_1 and A_2 be radial slits, with no A_k meeting C on the positive arc z_1z_2 bounded by the intersections of A_1 and A_2 with C. It follows from Theorem 1 that no critical point of $u(z)$ lies in the region $\omega(z, z_1z_2, R) > \frac{1}{2}$, under the additional hypothesis that that region contains no points of $A_3 + \cdots + A_n$. Map onto the interior R' of the unit circle C': $|w| = 1$ the region R slit along the arcs A_1 and A_2, but not slit along the arcs $A_3, \cdots, A_n$; thus R_1 is mapped into a subregion R_1' of R' bounded by C' including the arcs A_1' and A_2' (images of A_1 and A_2) of C' and by the images A_k' interior to C' of the arcs A_k $(k > 2)$. If we apply Theorem 1 (slightly extended) in the w-plane, it follows that no critical point of the transform of $u(z)$ lies in the region $\omega(w, w_1w_2, R') > \frac{1}{2}$, where w_1 and w_2 are the images of z_1 and z_2, so chosen that no point of A_1' or A_2' lies on the positive arc w_1w_2. The relation $\omega(w, w_1w_2, R') < \omega(z, z_1z_2, R)$, where the points w and z correspond under the conformal map, holds at every point w in R'; for at every boundary point w of R' not on A_1' or A_2' those two harmonic measures are equal, and at interior points of A_1' and A_2' the former has the value zero and the latter is positive and continuous. Consequently the inequality $\omega(w, w_1w_2, R') > \frac{1}{2}$ implies $\omega(z, z_1z_2, R) > \frac{1}{2}$, and the region defined by the former inequality is a proper subregion of the region defined by the latter inequality;* it is more favorable to apply Theorem 1 to the region R than to the region R'; the latter is equivalent to R cut along A_1 and A_2. Geometrically, it appears that the locus $\omega(w, w_1w_2, R') = \frac{1}{2}$ is orthogonal to C' at w_1 and w_2, while the locus $\omega(z, z_1z_2, R) = \frac{1}{2}$ interpreted in the w-plane makes a zero angle with the arcs A_1' and A_2' at w_1 and w_2. Our comment on the relative merits of R and R' applies, it may be noted, only to the use of Theorem 1 itself; we have established other theorems, such as §8.7.1 Theorem 1, which yield still more favorable results.

EXAMPLE II. Again let C be the unit circle, $m = 0$, $n = 3$, Jordan arcs A_1 and A_2 (to be further restricted later) meeting C in points z_1 and z_2, and A_3 a radial slit meeting C in a point of the positive arc z_2z_1 of C. We assume no point of any A_k to lie interior to the region $\omega(z, z_1z_2, R) > \frac{1}{2}$, and it follows from Theorem 1 that no critical point of $u(z)$ lies in that region. Map onto the interior R' of the unit circle C': $|w| = 1$ the region R slit along A_3 but not slit along A_1 or A_2; thus R_1 is mapped into a subregion R_1' of R' bounded by C' including the arc A_3' of C', the image of A_3, and by Jordan arcs A_1' and A_2', the images in R' of the arcs A_1 and A_2. It follows from Theorem 1 applied in the w-plane that no critical point of the transform of $u(z)$ lies in the region $\omega(w, w_1w_2, R') > \frac{1}{2}$, where w_1 and w_2 are the images of z_1 and z_2; it is to be noticed that the inequality $\omega(w, w_1w_2, R') < \omega(z, z_1z_2, R)$ holds as in Example I

* This fact is a consequence also of a general theorem due to Nevanlinna [1936].

at every point w in R', so the region $\omega(w, w_1w_2, R') > \frac{1}{2}$ is a proper subregion of the image of $\omega(z, z_1z_2, R) > \frac{1}{2}$, from which it follows in particular that the former region contains no point of A_1' or A_2'. It also follows that it is more favorable to apply Theorem 1 to the region R than to the region R'; the latter is equivalent to R cut along A_1 and A_2.

EXAMPLE III. Let C be the unit circle, choose $m = 2$ and J_1 and J_2 mutually symmetric in the axis of reals and to be made more precise later, $n = 1$ with A_1 a segment of the axis of reals meeting C in the point $z = -1$. Choose a point z_1 of C of the first quadrant so that the NE line $z_1\bar{z}_1$ does not intersect A_1. Map onto the interior R' of the unit circle $C':|w| = 1$ the region R slit along A_1, with real points w corresponding to real points z, so that A_1 is mapped onto an arc A_1' of C' whose midpoint is $w = -1$. If the image of z_1 is w_1, the NE line $L:\omega(z, \bar{z}_1z_1, R) = \frac{1}{2}$ is transformed into an arc L' joining $\bar{w}_1$ and w_1. The region bounded by L' and arc $\bar{w}_1w_1$ of C' contains in its interior the NE line λ joining $\bar{w}_1$ and w_1. If w_2 is chosen on C' near w_1 and in the positive direction from w_1, the NE line λ' joining $\bar{w}_2$ and w_2 lies partly interior and partly exterior to the region $\omega(z, \bar{z}_1z_1, R) > \frac{1}{2}$ interpreted in the w-plane; we assume the intersection Q of L' with the axis of reals to be exterior to the NE half-plane bounded by the positive arc $\bar{w}_2w_2$. Thus small curves J_1' and J_2' mutually symmetric in the axis of reals can be drawn in R' tangent to both L' and λ', exterior to the two regions $\omega(z, \bar{z}_1z_1, R) > \frac{1}{2}$ and $\omega(w, \bar{w}_2w_2, R') > \frac{1}{2}$; the images J_1 and J_2 in the z-plane of J_1' and J_2' are chosen as boundary components of R_1. It follows that at least for points in the neighborhood of Q, it is certainly more favorable to apply Theorem 1 to the region R than to the region R'.

In Theorem 1 a given region R can often be enlarged in various ways without altering R_1 or $u(z)$ by adding to R regions adjacent to the arcs of C not part of the boundary of R_1. We may thus add arbitrary plane regions, keeping R a plane simply-connected region, or to carry this method to an extreme, may adjoin an infinitely-many-sheeted logarithmic Riemann surface along each such arc of C, with branch points at the ends of the arc. This is equivalent to mapping onto the interior of the unit circle the region R_1 plus auxiliary regions, so that except for a finite number of points the circumference of the unit circle corresponds only to that part (assumed not empty) of the boundary of R_1 on which the prescribed boundary value of $u(z)$ is zero.

§9.7. Methods of symmetry. Our primary method thus far for the study of the critical points of a harmonic function has been the use of a field of force, namely a field representing either the gradient of the given function, or the gradient of an approximating function. Any representation of a harmonic function may be applicable, however, and we have also made use (§9.3.3) of Poisson's integral and its derivatives. We proceed now to make further application of that latter method, or alternately of more elementary methods, in conjunction with reflection of the plane in a line or point; we shall give new proofs [Walsh, 1948d] and also extensions of a number of previous results.

Our previous use of the field of force has involved to some extent actual geometric loci (§§8.3, 8.9.2), but more frequently the consideration of a certain direction in which the components of various forces and of the total resultant is not zero. The present method involves essentially that latter technique, now in the use of a direction in which the directional derivative (which is the component of a force) as such is different from zero for a function, and perhaps for a number of functions and for their sum.

§9.7.1. Reflection in axis. Our first general result involves the comparison of the values of a harmonic function in points symmetric in the axis of reals:

THEOREM 1. *Denote by* Π_1 *and* Π_2 *the open upper and lower half-planes respectively. Let* $u(x, y)$ *be harmonic in a region* R *cut by the axis of reals, and let the relation*

$$(1) \qquad u(x, y) > u(x, -y), \textit{ for } (x, y) \textit{ in } \Pi_1,$$

hold whenever both (x, y) *and* $(x, -y)$ *lie in* R. *Then* $u(x, y)$ *has no critical point in* R *on the axis of reals; indeed at an arbitrary point* $(x_0, 0)$ *in* R *we have*

$$(2) \qquad \partial u(x_0, 0)/\partial y > 0.$$

Alternate sufficient conditions that $u(x, y)$ *have no critical point in* R *on the axis of reals, and indeed that* (2) *hold at an arbitrary point* $(x_0, 0)$ *of* R, *are that* R *be bounded by a Jordan configuration* B, *that* $u(x, y)$ *be harmonic and bounded in* R, *continuous in* $R + B$ *except perhaps for a finite number of points,* $u(x, y)$ *not identically equal to* $u(x, -y)$ *in any subregion of* R, *and*

1). *R symmetric in the axis of reals, with* $u(x, y) \geqq 0$ *for* (x, y) *on* $B \cdot \Pi_1$ *and* $u(x, y) \leqq 0$ *for* (x, y) *on* $B \cdot \Pi_2$.

2). *R symmetric in the axis of reals, with*

$$(3) \qquad u(x, y) \geqq u(x, -y)$$

at every point (x, y) *of* $B \cdot \Pi_1$.

3). (x, y) *lies in* R *whenever* $(x, -y)$ *lies in* $R \cdot \Pi_2$; $u(x, y) \geqq 0$ *on* $B \cdot \Pi_1$ *and* $u(x, y) = 0$ *on* $B \cdot \Pi_2$.

4). (x, y) *lies in* R *whenever* $(x, -y)$ *lies in* $R \cdot \Pi_2$; *the boundary values of* $u(x, y)$ *on* $B \cdot \Pi_1$ *are not less than* glb $[u(x, 0)$ *in* $R]$; *the boundary values of* $u(x, y)$ *on* $B \cdot \Pi_2$ *are not greater than this number.*

5). R' *is a region whose boundary* B' *is symmetric in the axis of reals, and* R *is a subregion of* R' *whose boundary in* Π_k *is denoted by* $B' \cdot \Pi_k + B_k$ $(k = 1, 2)$, *where* B_k *is disjoint from* B'; *we have* (3) *at every point* (x, y) *of* $B' \cdot \Pi_1$; *we denote by* $u_k(x, y)$ *the function harmonic and bounded in* $R' \cdot \Pi_k$ *defined by the boundary values* $u(x, y)$ *on the axis of reals and on* $B' \cdot \Pi_k$; *we suppose* $u(x, y) \geqq u_1(x, y)$ *on* B_1, $u(x, y) \leqq u_2(x, y)$ *on* B_2.

6). *R is cut by the axis of reals and* $u(x, y)$ *has boundary values unity on* $B \cdot \Pi_1$, *zero on* $B \cdot \Pi_2$.

7). *R is cut by the axis of reals; the boundary values of* $u(x, y)$ *on* $B \cdot \Pi_1$ *are not less than* lub $[u(x, 0)$ *in* $R]$; *the boundary values on* $B \cdot \Pi_2$ *are not greater than* glb $[u(x, 0)$ *in* $R]$.

As a matter of convention here, segments of the axis of reals belonging to B (or B_k) are considered to belong to both $B \cdot \Pi_1$ and $B \cdot \Pi_2$ (or $B_1 \cdot \Pi_1$ and $B_2 \cdot \Pi_2$), and boundary values may of course be different on $B \cdot \Pi_1$ and $B \cdot \Pi_2$ (or $B_1 \cdot \Pi_1$ and $B_2 \cdot \Pi_2$). The region R need not be finite. In various parts of Theorem 1 we need not assume the boundary of R to be a Jordan configuration.

The main part of Theorem 1 may be proved from §9.3.3. If $(x_0, 0)$ is a point of R on the axis of reals, a suitably chosen circle C whose center is $(x_0, 0)$ lies together with its interior in R. Inequality (1) holds at every point of C in Π_1, so we have (2).

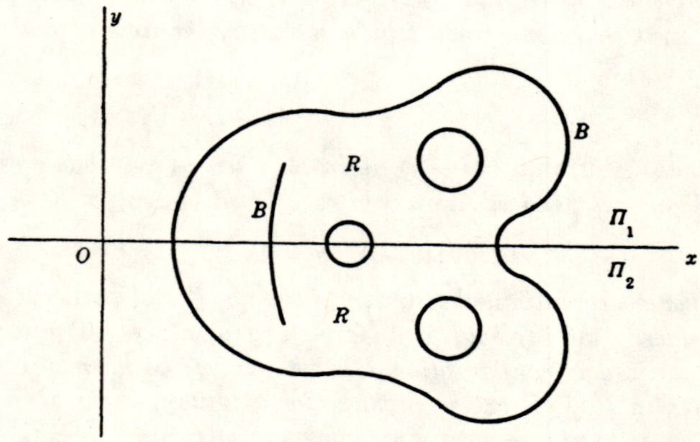

Fig. 27 illustrates §9.7.1 Theorem 1

A second proof of the main part of Theorem 1 may be given as follows. We set $U(x, y) = u(x, y) - u(x, -y)$, when both (x, y) and $(x, -y)$ lie in R, whence in Π_1 we have $U(x, y) > 0$ and in Π_2 we have $U(x, y) < 0$ if $U(x, y)$ is defined; on the axis of reals in R we have $U(x, 0) = 0$, $\partial U(x, 0)/\partial x = 0$. In the neighborhood of a critical point (x_0, y_0) of $U(x, y)$ of order k the locus $U(x, y) = U(x_0, y_0)$ consists (§1.6.1) of $k + 1$ analytic Jordan arcs through (x_0, y_0) intersecting at (x_0, y_0) at successive angles of $\pi/(k + 1)$; no arc of the locus $U(x, y) = 0$ can lie in $R \cdot \Pi_1$ or $R \cdot \Pi_2$ where $U(x, y)$ is defined, so no critical point of $U(x, y)$ lies on the axis of reals in R.* On the axis we have $\partial U/\partial x = 0$, whence $\partial U/\partial y \neq 0$; and $\partial U/\partial y > 0$ follows from the nature of $U(x, y)$ in Π_1 and Π_2. We also have $\partial U(x, 0)/\partial y = 2\partial u(x, 0)/\partial y$, so (2) follows in R.

Part 1) is contained in 2), so we proceed to the proof of 2). We set $U(x, y) = u(x, y) - u(x, -y)$, which is defined throughout R, so at every point of $B \cdot \Pi_1$ we have $U(x, y) \geqq 0$; at every point (x, y) of $B \cdot \Pi_2$ we have $U(x, y) = -U(x, -y) \leqq 0$. The function $U(x, y)$ does not vanish identically in R but

---

* Indeed, if R_1 is the interior of a circle C whose center lies on the axis of reals A and if R_1 is contained in R, it follows from §9.3.2 Theorem 3 that $U(x, y)$ has no critical point in either of the regions $\omega(z, A \cdot R_1, R_1 \cdot \Pi_1) > 1/2$ or $\omega(z, A \cdot R_1, R_1 \cdot \Pi_2) > 1/2$, hence has no critical point in the region $1/4 < \omega(z, C \cdot \Pi_1, R_1) < 3/4$.

vanishes on the axis of reals, so we have $U(x, y) > 0$ in $R \cdot \Pi_1$, $U(x, y) < 0$ in $R \cdot \Pi_2$; thus from the main part of Theorem 1 we have at any point $(x_0, 0)$ of $R: \partial U(x_0, 0)/\partial y = 2\partial u(x_0, 0)/\partial y > 0$, and (2) follows.

Part 1) contains §9.3.2 Theorem 3 and §9.3.2 Theorem 5, and part 2) contains §9.3.2 Theorem 4 and §9.3.3 Theorem 6.

To prove part 3), we denote by B_1 the reflection in the axis of reals of $B \cdot \Pi_2$; then B_1 necessarily lies in the closure of $R \cdot \Pi_1$; at every point of B_1 whether on $B \cdot \Pi_1$ or interior to R we have $u(x, y) \geqq 0$. Each point of R on the axis of reals lies in a subregion of R symmetric in the axis bounded wholly by points of B_1 and points of $B \cdot \Pi_2$, so the conclusion follows from part 2). In part 3) it is not possible to replace the requirement $u(x, y) = 0$ on $B \cdot \Pi_2$ by the weaker condition $u(x, y) \leqq 0$ on $B \cdot \Pi_2$, as we illustrate by an example. Let R be the interior of a circle C whose center lies in Π_1 and which cuts the axis of reals in two points. Let C_1 be a NE line for R cutting C in the points α_1 and α_2 in Π_2 and cutting the axis of reals in the point $(x_0, 0)$. A neighboring NE line for R through $(x_0, 0)$ also cuts the axis of reals in the point $(x_0, 0)$, and cuts C in points β_1 and β_2 in Π_2 which lie near α_1 and α_2 in the counterclockwise sense on C. The function $-\omega(z, \alpha_1\beta_1 + \alpha_2\beta_2, R)$, the negative of the sum of the harmonic measures of the positive arcs $\alpha_1\beta_1$ and $\alpha_2\beta_2$, is zero on $B \cdot \Pi_1$, non-positive on $B \cdot \Pi_2$, yet (§8.7.1 Theorem 1) has a critical point at $(x_0, 0)$ on the axis of reals. Part 3) contains a special case of part 1), and contains §9.3.2 Theorem 3.

In the proof of part 4) we set $b_2 = \text{glb}\,[u(x, 0) \text{ in } R]$. Let B_1 denote the reflection in the axis of reals of $B \cdot \Pi_2$, which necessarily lies in the closure of $R \cdot \Pi_1$. On both $B \cdot \Pi_1$ and the axis of reals in R, and hence on B_1, we have $u(x, y) \geqq b_2$, and on $B \cdot \Pi_2$ we have $u(x, y) \leqq b_2$; each point of R on the axis of reals lies in a subregion of R symmetric in the axis bounded wholly by points of B_1 and $B \cdot \Pi_2$, so the conclusion follows from part 2).

To prove part 5) we note that for (x, y) on the boundary of $R' \cdot \Pi_1$ we have $u(x, y) \geqq u(x, -y)$, both on $B' \cdot \Pi_1$ and on the axis of reals, whence $u_1(x, y) \geqq u_2(x, -y)$ for (x, y) in $R' \cdot \Pi_1$. From the relations $u(x, y) \geqq u_1(x, y)$ on B_1 and $u(x, y) \leqq u_2(x, y)$ on B_2 we deduce $u(x, y) \geqq u_1(x, y)$ in $R \cdot \Pi_1$, $u(x, y) \leqq u_2(x, y)$ in $R \cdot \Pi_2$; in sum we have for (x, y) in $R \cdot \Pi_1$ the inequalities $u(x, y) \geqq u_1(x, y) \geqq u_2(x, -y) \geqq u(x, -y)$, provided $(x, -y)$ lies in $R \cdot \Pi_2$. Throughout a suitable neighborhood of any point of R on the axis of reals we have (3) satisfied for (x, y) in Π_1, so the conclusion follows from part 2). Part 5) includes §9.4.2 Theorem 6.

Part 6) is contained in part 7), so we proceed to prove the latter. We set $b_1 = \text{lub}\,[u(x, 0) \text{ in } R]$, $b_2 = \text{glb}\,[u(x, 0) \text{ in } R]$, so we have $u(x, y) \geqq b_1 \geqq b_2$ on $B \cdot \Pi_1$, $u(x, y) \geqq b_2$ on the axis of reals in R and throughout $R \cdot \Pi_1$; we have $u(x, y) \leqq b_2$ on $B \cdot \Pi_2$, $u(x, y) \leqq b_1$ on the axis of reals in R and throughout $R \cdot \Pi_2$. We set $U(x, y) = u(x, y) - u(x, -y)$. If $U(x, y)$ is defined, at a point (x, y) of $B \cdot \Pi_1$ we have $U(x, y) \geqq 0$ and at a point of the reflection of $B \cdot \Pi_1$ we have $U(x, y) \leqq 0$; at a point (x, y) of $B \cdot \Pi_2$ we have $U(x, y) \leqq 0$ and at a point of the reflection of $B \cdot \Pi_2$ we have $U(x, y) \geqq 0$. Every point $(x, 0)$ in R lies in some subregion R' of R symmetric in the axis of reals bounded wholly by points of

$B \cdot \Pi_1$, $B \cdot \Pi_2$, and their reflections; on the boundary of R' in Π_1 we have $U(x, y) \geqq 0$ and on the boundary of R' in Π_2 we have $U(x, y) \leqq 0$. It will be noted that the relation $U(x, y) \equiv 0$ in one region R' would imply that relation in every region R', and throughout R. From part 2) we now have $\partial U(x_0, 0)/\partial y > 0$, and (2) follows. Part 6) contains §9.2.2 Theorem 3, as does part 7); thus both parts contain the analog (§8.2) of Bôcher's Theorem.

From the relation (2) in part 6) and elsewhere, if $u(x, y)$ has as boundary values only zero and unity and likewise in numerous other cases, the enumeration of the critical points in regions bounded by B and the axis of reals is immediate. We have also

COROLLARY 1. *Let* $U(x, y) = \sum_1^n \lambda_k U_k(x, y)$, $\lambda_k > 0$, *where each* $U_k(x, y)$ *is harmonic in a region* R_k *and satisfies the conditions of either the main part of Theorem* 1 *or one of the supplementary parts, except that we assume* $U(x, y)$ *not identically equal to* $U(x, -y)$ *in* $R_1 \cdot R_2 \cdot \cdots \cdot R_n$ *but do not assume every* $U_k(x, y)$ *not identically equal to* $U_k(x, -y)$ *in* R_k. *Then* $U(x, y)$ *has no critical points in* $R_1 \cdot R_2 \cdot \cdots \cdot R_n$ *on the axis of reals.*

In a Jordan region R, an arbitrary NE line behaves conformally like an axis of symmetry, for it is the image of such an axis under a suitable conformal map of R onto the interior of a circle; this fact is the foundation of some of the applications suggested. Of course Theorem 1 can be generalized by an arbitrary conformal map, where the axis of reals is replaced by an arbitrary conformal axis of symmetry, and the generalization can be expressed in invariant form. Theorem 1 and Corollary 1 are of great generality, and include various situations not previously treated, especially those in which R is multiply-connected.

It is appropriate to contrast in scope the two methods of study of critical points: (i) use of a field of force defined by a suitable spread of ordinary and skew matter on the boundary of a given region, and (ii) the present method of reflection. To be sure, the methods are not unrelated; one proof of the main part of Theorem 1 is by application of the Poisson integral and its associated field of force; both methods involve proof that the gradient of a function (which can be considered as a force) does not vanish; nevertheless there is great difference in the results obtained and in the spirit of the proofs; in (i) we use a spread of matter on the boundary of the given region defined in terms of the given function, while in (ii) the spread of matter is incidental and by no means indispensable. In delicate questions involving a symmetric (rather than skew-symmetric) configuration, such as those of §§8.3, 9.4.4, 9.4.5, method (i) seems more powerful, and there is no presumption that (ii) can be modified to yield the corresponding results. In purely quantitative questions, like those involving cross-ratios without symmetry, method (i) seems to have no competitor. In matters involving hyperbolic geometry in a simply-connected region, the two methods yield comparable results; for a Jordan region, in both (i) and (ii) we have essentially used reflection in a NE line; both methods (i) and (ii) are thus closely related to conformal axes of symmetry. For multiply-connected regions and conformal axes of symmetry, method (i) can be used (§9.2) in simple cases of harmonic measure, but method (ii) appears to be much more powerful, and seems to yield at once results beyond either the direct method (i) or that method with the help of conformal mapping of plane regions or of covering surfaces; but compare §9.9.2 below. As an illustration here we state a consequence of 2):

COROLLARY 2. *Let the region R bounded by disjoint Jordan configurations B_0, B_1, B_2, be symmetric in the axis of reals, where B_0 is symmetric in that axis and B_1 and B_2 are mutually symmetric in that axis, with B_k in Π_k. Then $\omega(z, B_1, R)$ has no critical points in R on the axis of reals.*

A second example of the advantage of method (ii) over method (i) is

COROLLARY 3. *Let the region R be cut by the axis of reals, let (x, y) lie in R whenever $(x, -y)$ lies in $R \cdot \Pi_2$, and let the points $z_1, z_2, \cdots, z_n$ lie in $R \cdot \Pi_1$. Then no critical point of $u(z) = \sum \lambda_k g(z, z_k, R)$, $\lambda_k > 0$, lies in R on the axis of reals, or (if $R \cdot \Pi_2$ is simply-connected) in $R \cdot \Pi_2$.*

That no critical points lie in R on the axis of reals is not formally included in 3), but follows by the reasoning used in proving 3); this same conclusion can be drawn even if some of the points z_i lie on the axis of reals, and also if some of the points z_k lie in Π_2, provided each such z_k is the conjugate of some z_j, and provided $u(z) \gtreqless u(\bar{z})$. The Principle of Argument applied to $R \cdot \Pi_2$ or to a suitable approximating subregion completes the proof of Corollary 3.

It is also of interest that the main part of Theorem 1 yields a simple proof of Lucas's Theorem not explicitly involving the logarithmic derivative. To establish the latter theorem it is sufficient to show that if all zeros of a polynomial $p(z)$ lie in the open upper half-plane, no critical point lies on the axis of reals; the function $u(x, y) = -\log |p(z)|$ is harmonic in the region R consisting of the extended plane except infinity and the zeros of $p(z)$; the function $u(x, y) - u(x, -y)$ is harmonic in the region $R \cdot \Pi_1$ consisting of the upper half-plane with the zeros of $p(z)$ deleted, at each of which points it becomes positively infinite, and the function approaches the boundary value zero at each finite or infinite boundary point of the upper half-plane; thus (1) is satisfied in $R \cdot \Pi_1$.

§9.7.2. Reflection in point. The primary method used in §9.7.1 is the comparison of functional values at two points mutually symmetric in the axis of reals. Similarly [Walsh, 1948d] we may compare functional values at two points mutually symmetric in the origin O:

THEOREM 2. *Denote by Π_1 and Π_2 the open upper and lower half-planes respectively. Let $u(x, y)$ be harmonic in a region R containing the origin O, and let the relation*

$$u(x, y) > u(-x, -y), \qquad (x, y) \text{ in } \Pi_1, \tag{4}$$

hold whenever both (x, y) *and* $(-x, -y)$ *lie in* R. *Then* O *is not a critical point of* $u(x, y)$.

Alternate sufficient conditions that O *not be a critical point of* $u(x, y)$ *are that* R *be bounded by a Jordan configuration* B, *that* $u(x, y)$ *be harmonic and bounded in* R, *continuous in* $R + B$ *except perhaps for a finite number of points,* $u(x, y)$ *not identically equal to* $u(-x, -y)$ *in* R, *and*

1). R *is a finite Jordan region symmetric in* O, B *is cut by the axis of reals in precisely two points, and we have*

$$u(x, y) \geqq u(-x, -y) \tag{5}$$

at every point (x, y) *of* $B \cdot \Pi_1$.

2). R *is a Jordan region containing* O, B *cuts the axis of reals in precisely two points, and* R *contains the point* $(-x, -y)$ *whenever* R *contains the point* (x, y) *with* $y < 0$; *on* $B \cdot \Pi_1$ *we have* $u(x, y) \geqq 0$ *and on* $B \cdot \Pi_2$ *we have* $u(x, y) = 0$.

3). R' *is a region whose boundary* B' *is a Jordan configuration symmetric in both coordinate axes, and* R *(containing* O*) is a subregion of* R' *whose boundary in* Π_k *is a Jordan configuration denoted by* $B' \cdot \Pi_k + B_k$, *where* B_k *is disjoint from* B'; *we denote by* $u_k(x, y)$ *the function harmonic and bounded in* $R' \cdot \Pi_k$ *defined by the boundary values* $u(x, y)$ *on the axis of reals and on* $B' \cdot \Pi_k$; *we suppose* (5) *valid on* $B' \cdot \Pi_1$, $u(x, y) \geqq u_1(x, y)$ *on* B_1, $u(x, y) \leqq u_2(x, y)$ *on* B_2.

The main part of Theorem 2 follows from §9.3.3 or §9.3.4. A suitably chosen circle C whose center is O lies together with its interior in R. Inequality (4) holds at every point of C in Π_1, so we have

$$\partial u(0, 0)/\partial y > 0. \tag{6}$$

To prove part 1) we set $U(x, y) = u(x, y) - u(-x, -y)$, so at every point of $B \cdot \Pi_1$ we have $U(x, y) \geqq 0$, and at every point of $B \cdot \Pi_2$ we have $U(x, y) \leqq 0$. The function $U(x, y)$ vanishes at O but does not vanish identically in R. If R is mapped onto the interior of the unit circle C in the w-plane so that $z = 0$ corresponds to $w = 0$, points of B symmetric in $z = 0$ correspond to points of C symmetric in $w = 0$, for rotation of R about O through the angle π is the transform of an involutory mapping of the region $|w| < 1$ onto itself with $w = 0$ fixed, so the latter mapping is a rotation about $w = 0$ through the angle π. In particular, the two points of B on the axis of reals correspond to diametrically opposite points of C. It follows (§9.3.3 or 9.3.4) that $w = 0$ is not a critical point of the transform of $U(x, y)$, and that O is not a critical point of $U(x, y)$; this conclusion can also be obtained by topological considerations, as in the second proof of the main part of Theorem 1 and as further elaborated in §9.8.1 below. Not both of the numbers $\partial U(0, 0)/\partial x$ and $\partial U(0, 0)/\partial y$ vanish. From the relations $\partial U(0, 0)/\partial x = 2\partial u(0, 0)/\partial x$, $\partial U(0, 0)/\partial y = 2\partial u(0, 0)/\partial y$ it follows that O is not a critical point of $u(x, y)$.

For simplicity we have assumed in part 1) that B cuts the axis of reals in precisely two points; it is sufficient here if the Jordan curve B is divided into two arcs, mutually symmetric with respect to O, and if for (x, y) on one of these arcs we have (5). A similar modification applies to part 2), where the end-points of the two arcs of B lie on the axis of reals, and are symmetric in O.

In the proof of part 2) we set $U(x, y) = u(x, y) - u(-x, -y)$. Each point of the reflection in O of $B \cdot \Pi_2$ is either a point of $B \cdot \Pi_1$ or lies in $R \cdot \Pi_1$; each point of $B \cdot \Pi_2$ is either a point of the reflection of $B \cdot \Pi_1$ or lies in the reflection in O of $R \cdot \Pi_1$. If $U(x, y)$ is defined, we have $U(x, y) \geqq 0$ on $B \cdot \Pi_1$ and on the reflection of $B \cdot \Pi_2$; we have $U(x, y) \leqq 0$ on $B \cdot \Pi_2$ and on the reflection of $B \cdot \Pi_1$. The point O lies in a subregion R' of R and of the reflection in O of R; R' is symmetric in O and bounded wholly by a Jordan curve consisting of points of B and of the reflection in O of B; this Jordan curve cuts the axis of reals in precisely two points; at every boundary point of R' in Π_1 we have $U(x, y) \geqq 0$ and at every boundary point of R' in Π_2 we have $U(x, y) \leqq 0$. The function $U(x, y)$ does not vanish identically in R', so by part 1) it follows that O is not a critical point of $U(x, y)$ or of $u(x, y)$. Parts 1) and 2) both contain §9.3.2 Theorem 3.

To prove part 3) we set $U_0(x, y) \equiv u(x, y) + u(-x, y)$, so $U_0(x, y)$ is defined in the region R_1 common to R and the reflection of R in the axis of imaginaries. At every point of $B' \cdot \Pi_1$ we have by (5) the inequality $U_0(x, y) \geqq U_0(x, -y)$. At every boundary point of $R \cdot \Pi_1$, whether on the axis of reals or on $B' \cdot \Pi_1$ or on B_1, and hence throughout $R \cdot \Pi_1$, we have $u(x, y) \geqq u_1(x, y)$; in particular this inequality holds on the reflection of B_1 in the axis of imaginaries. At every boundary point of $R_1 \cdot \Pi_1$ we thus have $U_0(x, y) \geqq u_1(x, y) + u_1(-x, y)$; this second member is harmonic in $R' \cdot \Pi_1$, and on the boundary of $R' \cdot \Pi_1$ takes the same values as $U_0(x, y)$. Similarly at every boundary point of $R_1 \cdot \Pi_2$ we have $U_0(x, y) \leqq u_2(x, y) + u_2(-x, y)$; this second member is harmonic in $R' \cdot \Pi_2$ and on the boundary of $R' \cdot \Pi_2$ takes the same values as $U_0(x, y)$. In order to apply part 5) of Theorem 1 and thereby to prove $0 < \partial U_0(0, 0)/\partial y = 2\partial u(0, 0)/\partial y$, which completes the proof, it remains merely to prove $U_0(x, y) \not\equiv U_0(x, -y)$ at points of R where both functions are defined. We reach a contradiction by assuming $U_0(x, y) \equiv U_0(x, -y)$; the boundary values of these functions are then equal at every point of continuity. For (x, y) on $B' \cdot \Pi_1$ from the relations $u(x, y) \geqq u(-x, -y)$, $u(-x, y) \geqq u(x, -y)$, $U_0(x, y) = u(x, y) + u(-x, y) = U_0(x, -y) = u(x, -y) + u(-x, -y)$, it follows that $u(x, y) = u(-x, -y)$, $u(x, -y) = u(-x, y)$ at every point of continuity on $B' \cdot \Pi_1$. We can now conclude: $u_1(x, y) + u_1(-x, y) \equiv u_2(-x, -y) + u_2(x, -y)$ in $R' \cdot \Pi_1$, by the definitions of $u_1(x, y)$ and $u_2(x, y)$. Unless we have $u(x, y) = u_1(x, y)$ at every point of continuity on B_1, and $u(x, y) = u_2(x, y)$ at every point of continuity on B_2, which would imply $u(x, y) \equiv u(-x, -y)$ in R contrary to hypothesis, we have at least one of the strong inequalities: $u(x, y) > u_1(x, y)$ throughout $R_1 \cdot \Pi_1$, or $u(x, y) < u_2(x, y)$ throughout $R_1 \cdot \Pi_2$, whence for (x, y) in $R_1 \cdot \Pi_1$ we have $U_0(x, y) \equiv u(x, y) + u(-x, y) >$ (or $\geqq$) $u_1(x, y) + u_1(-x, y) \equiv u_2(-x, -y) + u_2(x, -y) \geqq$ (or $>$) $u(-x, -y) + u(x, -y) \equiv U_0(x, -y)$, contrary to our assumption. This contradiction completes the proof of Theorem 2.

Part 3) includes §9.4.3 Theorem 7 and is even more general. In part 3) we have proved (6), hence if a number of functions satisfy the conditions of part 3), so also does any linear combination with positive coefficients. Theorem 2 is in some respects more powerful than Theorem 1, as is to be expected from the restricted nature of the conclusion, in applying only to a single point.

It is no accident that we have failed to establish (6) under the hypothesis of part 1):

THEOREM 3. *There exists a convex region R bounded by an analytic Jordan curve J containing O in its interior and symmetric in O, and a function* $u(x, y)$ *harmonic in R, continuous in* $R + J$, *with* $u(x, y) \equiv -u(-x, -y)$, $u(x, y) > 0$ *on* $J \cdot \Pi_1$, $u(x, y) < 0$ *on* $J \cdot \Pi_2$, *and with* $\partial u(0, 0)/\partial y < 0$.

Let E be an arbitrary ellipse (not a circle) whose center is $O: z = 0$ and whose axes are not the coordinate axes. Map the interior of E onto the interior of $C: | w | = 1$ so that the origins in the two planes correspond to each other, and so that the intersections of E with the axis of reals correspond to the intersections of C with the half-lines arg $w = 0$ and π. Let L denote the image in the z-plane of those latter half-lines, so that L cuts E orthogonally on the axis of reals, but L is not horizontal at this point of intersection. Let the direction-angle of L at O be θ, $0 \leqq \theta < \pi$. If we have $\theta \neq 0$ we choose E as J; if we have $\theta = 0$ we choose J as the image of a circle $| w | = r < 1$, where r is near unity, and where J is to be rotated slightly about O so that L as already chosen and rotating with J now cuts J on the axis of reals; the angle θ_1, $0 \leqq \theta_1 < \pi$, made with the horizontal by L at O is not zero after this rotation. It follows from a theorem due to Study that J is convex, so in the new position J is cut by the axis of reals in precisely two points.

If P_0 is a suitably chosen point on $J \cdot \Pi_1$ near the axis of reals, the image through P_0 of a half-line λ: arg w = const has a direction-angle θ_0 at O, with $+\pi < \theta_0 < 2\pi$. Let w_0 be the image in the w-plane of P_0. If α is a short arc of C whose center is w_0, and if α' is the reflection of α in $w = 0$, it follows by symmetry that the gradient at $w = 0$ of the function $\omega(w, \alpha, | w | < 1) - \omega(w, \alpha', | w | < 1) \equiv u_1(x, y)$ has the direction from $w = 0$ toward w_0. Thus the gradient of $u_1(x, y)$ at O has the direction θ_0, whence $\partial u_1(0, 0)/\partial y < 0$. If we replace $u_1(x, y)$ by a suitable approximating function $u(x, y)$ which is continuous on J and satisfies the relations $u(x, y) \equiv -u(-x, -y)$ on J and $u(x, y) > 0$ on $J \cdot \Pi_1$, we obtain a function $u(x, y)$ satisfying our requirements.

Numerous methods and results of §§9.7.1 and 9.7.2 apply without essential change to harmonic functions in n dimensions.

§9.7.3. Reflection of an annulus, Green's function. Reflection or symmetry in a line or circle, shown useful in §§9.3, 9.7.1, and 9.7.2, is also useful in other situations, as we show by two examples. Let R denote the region $(1 >) r < | z | < 1$, and let the distinct points α_1', α_2', $\cdots$, α_n' and β_1', β_2', $\cdots$, β_p' in R alternate on the axis of reals in such a way that the two sets $\alpha_1 = \alpha_1'$, $\cdots$, $\alpha_n = \alpha_n'$, $\alpha_{n+1} = 1/\beta_1'$, $\cdots$, $\alpha_m = 1/\beta_p'$ $(m = n + p)$; $\beta_1 = \beta_1'$, $\cdots$, $\beta_p = \beta_p'$, $\beta_{p+1} = 1/\alpha_1'$, $\cdots$, $\beta_m = 1/\alpha_n'$, are interlaced on the axis of reals. We study the function

$$u(z) = \sum_1^n G(z, \alpha_k', R) - \sum_1^p G(z, \beta_k', R),$$

where G represents Green's function. Since the function $u(z)$ vanishes on the boundary of R, it can be harmonically extended by reflection across both the circles $|z| = r$ and $|z| = 1$. In the annulus $r < |z| < 1/r$, the extended function $u(z)$ becomes positively infinite in the points α_k and becomes negatively infinite in the points β_k; these two sets of points are interlaced on the axis of reals. Further continued reflection of $u(z)$ in the circles $|z| = r^j$, $j = \cdots, -1, 0, 1, 2, \cdots$, defines $u(z)$ as a single-valued function harmonic in the entire plane except in the points α_k and β_k and their images, and except in the points zero and infinity, and satisfying the functional equation

$$(7) \qquad u(\rho z) \equiv u(z), \qquad \rho = 1/r^2.$$

The points where $u(z)$ becomes positively infinite and those where $u(z)$ becomes negatively infinite are interlaced on the axis of reals.

It follows from §6.3.1 Theorem 1 that there exists a function $\Psi(z)$ single-valued and meromorphic in the entire plane except in the points zero and infinity, which vanishes precisely in the points $\rho^j\alpha_k$, $j = \cdots, -1, 0, 1, 2, \cdots$, and has poles precisely in the points $\rho^j\beta_k$. Moreover we have

$$(8) \qquad \Psi(\rho z) \equiv \sigma\Psi(z), \qquad \sigma = \beta_1 \cdots \beta_m/\alpha_1 \cdots \alpha_m.$$

The function $u_1(z) \equiv u(z) + \log|\Psi(z)| - \lambda \log|z|$, $\lambda = \log\sigma/\log\rho$, is single-valued and harmonic in the entire plane except at zero and infinity and satisfies the functional equation $u_1(\rho z) \equiv u_1(z)$; thus $u_1(z)$ is identically constant, and $u(z)$ is equal to $-\log|\Psi(z)| + \lambda \log|z|$ plus a constant.

In the particular case $\sigma = 1$ we have $\lambda = 0$, and it follows from §5.2.1 Theorem 1 that *no critical point of $u(z)$ lies in a circle whose diameter is a pair of successive positive infinities or a pair of successive negative infinities of $u(z)$.*

If we no longer assume $\sigma = 1$, but make the requirement $\alpha_k > 0$, $\beta_k > 0$, it is to be noticed that λ is positive or negative according as the singularity of $u(z)$ of least modulus in the annulus $r < |z| < 1/r$ is a point α_k or a point β_k. In the field of force (§6.3.2 Lemma) corresponding to the customary finite product of which $\Psi(z)$ is the limit, plus an ordinary particle of mass $-\lambda$ at the origin O, the particle at O is then of sign opposite to that of the nearest particle at a singularity of the sequence of which $u(z)$ is the limit. For definiteness let us assume $r < \alpha_1 < \beta_1 < \alpha_2 < \beta_2 < \cdots < \beta_m < 1/r$; we then have $\log r < \log\alpha_1 < \log\beta_1 < \log\alpha_2 < \cdots < \log\beta_m < -\log r$,

$$\log\rho = -2\log r = (\log\alpha_1 - \log r) + (\log\beta_1 - \log\alpha_1) + (\log\alpha_2 - \log\beta_1) + \cdots + (\log\beta_m - \log\alpha_m) + (-\log r - \log\beta_m)$$

$$> \sum(\log\beta_k - \log\alpha_k) = \log\sigma, \qquad \log\rho > \log\sigma > 0, 1 > \lambda > 0.$$

The methods of §5.2 continue to apply. It is still true that *no critical point of* $u(z)$ *lies in a circle whose diameter is a pair of successive positive infinities or a pair of successive negative infinities of* $u(z)$. Indeed, to study the force at a non-real point P in such a circle we invert the plane in P, with O' the inverse of O and a circle C' through P the inverse of the axis of reals; the force at P due to a particle of mass $-\lambda(|\lambda| < 1)$ at O is in direction and magnitude (but not necessarily in sense) a fraction of the vector $O'P$, so this force is represented in direction and magnitude by a vector $A'B'$ which is a chord of C' parallel to $O'P$, where the points P, B', A', O' lie on C' in order in the positive sense; thus this force at P is equivalent to the force at P due to a unit particle of mass $-\lambda/|\lambda|$ at some real point A and a unit particle of mass $\lambda/|\lambda|$ at some real point B, respective inverses of A' and B'; these new unit particles together with the original ones are interlaced on the axes of reals; this conclusion is valid, although the proof requires some modification if P is real. A neighborhood of P is free from positions of equilibrium, so P is not a critical point of $u(z)$.

§9.7.4. Reflection of an annulus, harmonic measure. The method of reflection of an annulus used in §9.7.3 for the study of linear combinations of Green's functions can also be used for the study of harmonic measure. Let R be the region $(1 >)r < |z| < 1$, and let A_1 be the counterclockwise arc $(\bar{\alpha}_1, \alpha_1)$ of the circle $|z| = r$, α_1 in the upper half-plane; we study the function $u_1(z) \equiv \omega(z, A_1, R)$. Let $u_1(z)$ be extended harmonically by reflection across the circle $|z| = 1$ and across the circle $|z| = r$ on the arc complementary to A_1, and indeed by continued reflection across the circles $|z| = r^j$, $j = \cdots, -1, 0, 1, 2, \cdots$, except across the arcs denoted symbolically by $r^{2j} \cdot A_1$ that are the images of A_1. The extended function $u_1(z)$ is then harmonic and single-valued in the plane except at O and infinity, if the plane is considered to be cut along the arcs $r^{2j} \cdot A_1$. The function $u_1(z)$ takes the values -1 and $+1$ respectively on the left and right banks of each of those cuts. If we choose a particular branch of the argument, it follows that the function $u_1(z) + (1/\pi) \cdot \arg (z - \alpha_1)$ is uniform in the neighborhood of the point α_1 and if properly defined for $z = \alpha_1$ is harmonic there. Similarly the function $u_1(z) - (1/\pi) \cdot \arg (z - \bar{\alpha}_1)$ is single-valued and harmonic throughout the neighborhood of the point $\bar{\alpha}_1$. We have $u_1(\rho z) \equiv u_1(z)$, $\rho = 1/r^2$.

Let us denote by $\Psi_1(z)$ the function of §6.3.1 Theorem 1 with the zeros $\rho^j\alpha_1$ and the poles $\rho^j\bar{\alpha}_1$. We have

$$\Psi_1(\rho z) \equiv \sigma_1 \Psi_1(z), \qquad \sigma_1 = \bar{\alpha}_1/\alpha_1 .$$

The function $\Psi_1(z)$ is the uniform limit on any circle $|z| = \text{const} \neq \rho^j \cdot r$ of a sequence of rational functions each having the same number of zeros as poles interior to that circle, so $\arg \Psi_1(z)$ returns to its original value when z traces such a circle. In the plane with O and infinity deleted cut along the arcs $r^{2j} \cdot A_1$, the function $\arg \Psi_1(z)$ is single-valued, and behaves like $\arg (z - \alpha_1)$ in the neighborhood of α_1 and like $-\arg (z - \bar{\alpha}_1)$ in the neighborhood of $\bar{\alpha}_1$. The function

$u_0(z) \equiv u_1(z) + (1/\pi) \arg \Psi_1(z) + (\lambda_1/\pi) \log |z|$ is harmonic and single-valued in the cut plane, and by a suitable definition of $u_0(z)$ the cuts can be healed, leaving $u_0(z)$ harmonic in the extended plane except at O and infinity. If we now choose $\lambda_1 = -\arg \sigma_1/\log \rho$, the function $u_0(z)$ satisfies the functional equation $u_0(\rho z) \equiv u_0(z)$, hence is identically constant, so we have $u_1(z) \equiv -(1/\pi) \arg \Psi_1(z) - (\lambda_1/\pi) \log |z|$ plus a constant. The critical points of $u_1(z)$ are the critical points of the function $\arg \Psi_1(z) + \lambda_1 \log |z|$, or those of the function $z^{i\lambda_1}\Psi_1(z)$.

The algebraic sign of λ_1 is of some significance, so we investigate its value. For z in the neighborhood of the point $z = 1$, we choose $\pi < \arg (z - \alpha_1) < 2\pi$, $0 < \arg (z - \bar{\alpha}_1) < \pi$, whence $\arg [(z - \alpha_1)/(z - \bar{\alpha}_1)] = \angle \bar{\alpha}_1 z \alpha_1$, a positive angle less than 2π. With the obvious extension of this notation to the points $\rho^j\alpha_1$ and $\rho^j\bar{\alpha}_1$, we have

$$\Psi_1(z) \equiv \prod_{-\infty}^{\infty} \frac{(z - \rho^j\alpha_1)(1 - \rho^j\bar{\alpha}_1)}{(z - \rho^j\bar{\alpha}_1)(1 - \rho^j\alpha_1)}, \qquad \arg \Psi_1(1) = 0,$$

$$\arg \Psi_1(z) = \sum_{-\infty}^{\infty} [\angle(\rho^j\bar{\alpha}_1)z(\rho^j\alpha_1) - \angle(\rho^j\bar{\alpha}_1)1(\rho^j\alpha_1)],$$

$$\arg \Psi_1(\rho) = \sum_{-\infty}^{\infty} [\angle(\rho^j\bar{\alpha}_1)\rho(\rho^j\alpha_1) - \angle(\rho^j\bar{\alpha}_1)1(\rho^j\alpha_1)]$$

$$= \sum_{-\infty}^{\infty} [\angle(\rho^{j-1}\bar{\alpha}_1)1(\rho^{j-1}\alpha_1) - \angle(\rho^j\bar{\alpha}_1)1(\rho^j\alpha_1)].$$

This last expression represents precisely the angle at $z = 1$ subtended by the half-lines $\arg z = \arg \alpha_1$ and $\arg z = \arg \bar{\alpha}_1$, namely the angle $2\pi - 2 \arg \alpha_1$, $0 < \arg \alpha_1 < \pi$. However, we have here computed $\arg \Psi_1(\rho)$ for continuous motion of z from $z = 1$ to $z = \rho$ along the segment $1 \leqq z \leqq \rho$. If z moves from $z = 1$ to $z = \rho$ in the plane cut as before, we have $\arg \Psi_1(\rho) = -2 \arg \alpha_1 < 0$, $\arg \sigma_1 = \arg \Psi_1(\rho) - \arg \Psi_1(1) = -2 \arg \alpha_1$, $\lambda_1 = -\arg \sigma_1/\log \rho > 0$.

If there are counterclockwise arcs $A_k: (\bar{\alpha}_k, \alpha_k)$, $k = 1, 2, \cdots, m$, of the circle $|z| = r$, with the α_k in the upper half-plane, the critical points of the function $u(z) \equiv \sum \omega(z, A_k, R)$ are those of $z^{i\lambda}\Psi(z)$, where $\Psi(z)$ is the function of §6.3.1 Theorem 1 with the assigned zeros $r^{2j} \cdot \alpha_k$ and the assigned poles $r^{2j} \cdot \bar{\alpha}_k$, and where we have

$$\Psi(\rho z) \equiv \sigma\Psi(z), \qquad \sigma = \bar{\alpha}_1 \cdots \bar{\alpha}_m/\alpha_1 \cdots \alpha_m,$$

$$\lambda = -\arg \sigma/\log \rho, \arg \sigma = -2 \arg (\alpha_1 \cdots \alpha_m), \log \rho = -2 \log r, \lambda > 0.$$

We are now in a position to apply §6.3.2 Theorem 7. *If the* α_k *lie in the sector* $\alpha \leqq \arg z \leqq \beta$, $0 < \alpha \leqq \pi/2 \leqq \beta < \pi$, *the sector* $-\alpha < \arg z < \alpha$ *contains no critical points of* $u(z)$.*

There is no essential change here in method or result if in addition to the boundary arcs A_k of R we admit boundary arcs B_k on the circle $|z| = 1$, having

* In the case $m = 1$, §9.2.1 Theorem 2 is also of significance here.

the point $z = -1$ as midpoint. If the same sector $\alpha \leqq \arg z \leqq \beta$ contains all initial points in the upper half-plane of these new arcs, it is still true that the sector $-\alpha \leqq \arg z \leqq \alpha$ contains no critical points of the function

$$\sum \mu_k \omega(z, A_k, R) + \sum \nu_k \omega(z, B_k, R), \qquad \mu_k > 0, \qquad \nu_k > 0.$$

Results analogous to those just indicated follow under a somewhat different hypothesis from §6.3.2 Theorem 6.

The use in §§9.7.3 and 9.7.4 of reflection of an annulus as such obviously applies to much more general functions $u(z)$ than we have mentioned. Further general results here await more precise results on rational functions and on functions with a multiplicative period.

§9.8. Topological methods. In the present work our main results on the critical points of harmonic functions are geometric, in the sense that we assume properties of boundary values in various points and prove that a certain point set described geometrically in terms of the boundary contains all, or no, or a certain number of critical points. There are, however, purely topological considerations that may enable us to show that certain sets contain all or no critical points, where these sets are defined primarily in terms of the given harmonic function rather than in terms of the purely geometric configuration; compare for instance the second proof of the main part of §9.7.1 Theorem 1.

§9.8.1. Simple-connectivity. As a first further illustration here we prove

THEOREM 1. *Let the function $u(x, y)$ be harmonic but not identically zero in the region R whose boundary is the Jordan curve J, continuous on $R + J$. Let J be separated into arcs C_1 and C_2 which are disjoint except for the end-points A_1 and A_2 which they have in common, and let $u(x, y)$ be non-negative on C_1, non-positive on C_2. Then no critical point of $u(x, y)$ lies on the locus $\Gamma_0 : u(x, y) = 0$ in R.*

In the neighborhood of any point of Γ_0 not a critical point of $u(x, y)$ the locus Γ_0 consists (§1.6.1) of a single analytic Jordan arc, and in the neighborhood of a critical point (x_0, y_0) of order k, the locus consists of $k + 1$ analytic Jordan arcs intersecting at (x_0, y_0) whose tangents at (x_0, y_0) cut at successive angles of $\pi/(k + 1)$. No subregion of R can be bounded wholly by points of Γ_0, nor wholly by points of Γ_0 and arcs of J on which $u(x, y)$ vanishes identically; for $u(x, y)$ is not identically zero in R nor in any subregion of R. Thus any subregion of R bounded wholly by points of Γ_0 and of J must be bounded at least in part by points of J on which $u(x, y)$ is positive or negative or both.

An arc A of Γ_0 abutting on J must do so in a single point, assuming one end of A to lie in R. For let us map R onto the upper half-plane. The function $u(x, y)$ can be extended harmonically across the axis of reals along any interval of that axis on which $u(x, y)$ vanishes; thus at any point interior to such an interval the function $u(x, y)$ is harmonic. The arc A cannot oscillate indefinitely between the neighborhoods of two distinct points P_1 and P_2 of the axis of reals, for that would imply an infinity of mutually disjoint subarcs joining those neighborhoods, and would imply a point of R in whose neighborhood Γ_0 does not consist of a finite number of analytic Jordan arcs, or a boundary point of R interior to a segment of the axis of reals on which $u(x, y)$ vanishes identically, in whose neighborhood Γ_0 does not consist of a finite number of analytic Jordan arcs.

We assume now a point P_0 of Γ_0 (in R) to be a critical point of $u(x, y)$ of order k (> 0); we shall reach a contradiction. As we follow an arc of Γ_0 from P_0, remaining in R, the arc cannot cut itself and must terminate in some point P_1 of J; there exist $2(k + 1)$ such arcs in R commencing in P_0, no two of which can meet either in R or in a point of J. Let the arcs, extended to their first points of meeting with J, meet J in $P_1, P_2, \cdots, P_{2k+2}, P_{2k+3} = P_1$; let the notation be so chosen that those $2k + 2$ points lie on J in the counterclockwise order indicated; the function $u(x, y)$ vanishes in these points.

It is conceivable that various arcs of Γ_0 should separate P_0 from arcs of J, for instance that an arc B_0 of Γ_0 should have the same end-points as the arc B_1 of J, and that the two arcs B_0 and B_1 form a Jordan curve bounding a subregion of R whose closure does not contain P_0. Under these conditions we replace the arc B_1 of J by B_0, and we perform this operation whenever possible. We have thereby replaced arcs of J by arcs on which $u(x, y)$ takes continuously the boundary value zero. We cannot replace the whole of J by such arcs, for each arc $P_jP_0P_{j+1}$ of Γ_0 together with the arc P_jP_{j+1} of J bounds a subregion of R in which $u(x, y)$ does not vanish identically and which must be bounded in part by an arc of J on which $u(x, y)$ does not vanish identically, an arc not separated from P_0 by Γ_0. The replacement of arcs of J by arcs of Γ_0 yields a subregion R_1 of R containing P_0 bounded by a Jordan curve J_1; any arc of J on which $u(x, y)$ is positive must be separated by Γ_0 from an arc of J on which $u(x, y)$ is negative; the new boundary values on J_1 are all zero, so no pair of points of J_1 at which $u(x, y)$ is positive separates a pair of points of J_1 at which $u(x, y)$ is negative. We retain the original notation for the points P_j, chosen now to lie counterclockwise on J_1 instead of on J. The locus Γ_0 in R_1 consists precisely of the arcs P_0P_j.

Since each arc of J_1 on which $u(x, y)$ is negative is separated by Γ_0 from each arc of J_1 on which $u(x, y)$ is positive, no arc P_jP_{j+1} of J_1 can contain both points at which $u(x, y)$ is positive and points at which $u(x, y)$ is negative. At least three successive points P_{j-1}, P_j, P_{j+1} must lie on an arc $P_{j-1}P_{j+1}$ of J_1 on which $u(x, y)$ is non-negative or is non-positive. Then the function $u(x, y)$ is positive or is negative throughout the Jordan region bounded by the arc $P_{j-1}P_0P_{j+1}$ of Γ_0 and by the arc $P_{j-1}P_{j+1}$ of J_1, so the arc P_0P_j of Γ_0 cannot lie in that Jordan region, contrary to hypothesis. This contradiction completes the proof of Theorem 1.

The reasoning given also proves the

COROLLARY. *Let the function $u(x, y)$ be harmonic in the region R bounded by the Jordan curve J, continuous on $R + J$, not identically constant in R. If (x_0, y_0) is a critical point of $u(x, y)$ interior to R, then $u(x, y)$ takes the value $u(x_0, y_0)$ in at* *least four distinct points of J, and the successive alternate arcs of J bounded by these points contain each some points of J at which we have respectively $u(x, y) > u(x_0, y_0)$ and $u(x, y) < u(x_0, y_0)$.*

The requirement of continuity of $u(x, y)$ on J may be somewhat weakened in both Theorem 1 and the Corollary, for we may allow $u(x, y)$ to be bounded in R, continuous on $R + J$ except perhaps for a finite number of points. In Theorem 1 we still require $u(x, y)$ to be non-negative on C_1 and non-positive on C_2, and in the Corollary the value $u(x_0, y_0)$ is considered to be assumed by $u(x, y)$ at a point of J if $u(x, y) - u(x_0, y_0)$ changes sign there when (x, y) traces J. Compare here §9.3.3 Corollary 1 to Theorem 6.

Of course under the conditions of Theorem 1 the function $u(x, y)$ may have critical points in R. Thus let C be the unit circle, let γ_1 and γ_2 be closed disjoint arcs of C, and set $u_1(x, y) = \omega(z, \gamma_1 + \gamma_2, |z| < 1)$. The function $u_1(x, y)$ has (§8.7.1 Theorem 1) a critical point P_0 interior to C. Choose J as a Jordan curve consisting of C less small open arcs of C in the neighborhoods of the end-points of γ_1 and γ_2 plus small circular arcs interior to C. On successive arcs of J the values of $u_1(x, y)$ are assumed continuously as follows: zero, positive, zero, positive. On one of these open arcs of J on which $u_1(x, y)$ takes the value zero we assign to the function $u(x, y)$ boundary values which are numerically small but negative; elsewhere on J we assign to $u(x, y)$ the same boundary values as $u_1(x, y)$; we require that $u(x, y)$ shall be harmonic interior to J, continuous in the corresponding closed region. Then $u(x, y)$ has a critical point in R near P_0.

Theorem 1 is similar in hypothesis to §9.3.2 Theorem 5, but neither theorem is contained in the other.

§9.8.2. Higher connectivity. Analogs of Theorem 1 exist for regions of higher connectivity:

THEOREM 2. *Let the function $u(x, y)$ be harmonic but not identically zero in the annular region R bounded by the disjoint Jordan curves J^+ and J^-, continuous on $R + J^+ + J^-$. Let $u(x, y)$ be non-negative on J^+, non-positive on J^-. Then $u(x, y)$ has no critical point on the locus $\Gamma_0 : u(x, y) = 0$ in R.*

We assume that $u(x, y)$ has a critical point $P_0 : (x_0, y_0)$ on Γ_0 in R, and shall reach a contradiction.

No Jordan curve J_0 composed of arcs of Γ_0 in R can separate J^+ and J^-, or separate those curves except for a finite number of points on J_0 itself. For if such a curve J_0 exists it separates R into two or more regions each bounded by an arc of J_0 and an arc of but one of the curves J^+ and J^-; in such a region $u(x, y)$ cannot vanish identically, hence must be either positive or negative according as it is bounded in part by an arc of J^+ or J^-, and the region contains no point of Γ_0; then J_0 is the entire locus Γ_0 in R, and $u(x, y)$ has no critical point on Γ_0 in R.

No arc of Γ_0 not passing through P_0 can join J^+ and J^-; for if such an arc exists it can be considered as a cut of R; the precise method used in the proof of Theorem 1 applies and shows that no critical point can lie in R.

No more than two arcs of Γ_0 can lead either from P_0 to J^+, or from P_0 to J^-. If we assume for definiteness three such arcs to lead from P_0 to J^+, the interior points of one of these must lie in a region bounded wholly by points of Γ_0 and of J^+, a region throughout which $u(x, y)$ is necessarily positive. It follows that P_0 cannot be a critical point of order greater than unity, and that if P_0 is a critical point of order unity, one pair of arcs of Γ_0 emanating from P_0 must meet J^+ and the other pair must meet J^-. This last eventuality cannot occur, for in the neighborhood of a critical point of order one the four regions into which Γ_0 separates the plane are regions in which $u(x, y)$ is alternately positive and negative which here is impossible. This contradiction completes the proof of Theorem 2.

Under the hypothesis of Theorem 2 the function $u(x, y)$ may naturally have critical points in R not on Γ_0, as we show by an example. We set $u(x, y) = \log | z^2 - 1 | + 2$, and denote generically by Γ_μ the locus $u(x, y) = \mu$, a lemniscate. The locus Γ_2 is the Bernoullian lemniscate with a double point at the origin; the locus Γ_7 consists of a single Jordan curve containing the points $z = -1$, 0, $+1$ in its interior; the locus Γ_{-7}, like the locus Γ_0, consists of two Jordan curves interior respectively to the Jordan curves composing Γ_2. We choose J^+ as a Jordan curve composed of a segment of the line $x = q$ (> 0) to the left of the right-hand Jordan curve belonging to Γ_0, the segment bounded by points of Γ_7, plus the arc of Γ_7 which lies to the left of the line $x = q$; the function $u(x, y)$ is positive on J^+. We choose J^- as the left-hand Jordan curve belonging to Γ_{-7}, on which $u(x, y)$ is negative. The function $u(x, y)$ has a critical point in the region R bounded by J^+ and J^-, namely at the origin, but of course the critical point does not lie on Γ_0. If in Theorem 2 a critical point (x_0, y_0) lies in R, we have either on J^+ or J^- the inequality $\text{lub } u > u(x_0, y_0) > \text{glb } u$.

It is not possible to continue the direct extension of Theorems 1 and 2 to regions of higher connectivity. For instance let us consider the case of the unit circle C, with the requirement that $u(x, y)$ be non-negative on the upper half C_1 of C, non-positive on the lower half C_2 of C, harmonic in the annular region R bounded by C and by a Jordan curve B interior to C, non-negative on B. We assume $u(x, y)$ continuous in the closure of R, not identically zero. Let C_2 be divided into three successive arcs C_{21}, C_{22}, C_{23}, on which $u(x, y)$ is negative, zero, and negative respectively, and choose $u(x, y)$ as positive on C_1. For suitably chosen boundary values as indicated, the function $u_1(x, y)$ harmonic throughout the interior of C and continuous in the closure of that region vanishes interior to C at all points of a Jordan arc joining the points $z = +1$ and $z = -1$, and only there; compare Theorem 1. If now B is chosen small and near C_{23}, and if the boundary values assigned on B are positive but numerically small, the function $u(x, y)$ vanishes on a Jordan arc J_1 joining the points $z = +1$ and $z = -1$ and vanishes on a Jordan curve J_2 separating B from C_{21} and C_{23}. As B is allowed to increase in size, the values of $u(x, y)$ on B becoming larger while those on C are considered fixed, in such a way that $u(x, y)$ increases monotonically at every point of R, the arc J_1 and curve J_2 vary and meet, necessarily in a critical point of $u(x, y)$ on the locus $u(x, y) = 0$ in R. Here the conclusion of Theorem 2 is false. As B increases still further, the values of $u(x, y)$ on B continuing to increase as before, the locus $u(x, y) = 0$ in R consists of one Jordan arc separating C_1 and B from C_{21} and C_{23}.

Under restricted conditions Theorems 1 and 2 admit of extension to regions of higher connectivity. Thus let the region R be bounded by a number of mutually disjoint Jordan curves; we assume $u(x, y)$ non-negative on prescribed arcs, non-positive on the complementary arcs, harmonic and not identically zero in R, continuous in the closure of R. If a Jordan arc A of $\Gamma_0 : u(x, y) = 0$ in R separates the boundary points at which $u(x, y)$ is positive from those at which $u(x, y)$ is negative, then $u(x, y)$ has no critical point on Γ_0 in R; for in one of the subregions into which A separates R the function $u(x, y)$ is positive, and in the other subregion $u(x, y)$ is negative; neither subregion can contain a point of Γ_0.

§9.8.3. Contours with boundary values zero. Even though Theorems 1 and 2 do not extend directly to regions of higher connectivity, additional contours with assigned boundary values zero can be admitted:

THEOREM 3. *Let the function $u(x, y)$ be harmonic in the region R whose boundary consists of a Jordan curve J and a set B of Jordan curves finite in number disjoint from J and from each other, and let $u(x, y)$ be continuous and not identically zero in the closure of R. Let J be composed of the closed arcs C_1 and C_2, disjoint except for their end-points, and let $u(x, y)$ be non-negative on C_1, non-positive on C_2, and zero on B. Then no critical point of $u(x, y)$ in R lies on the locus $\Gamma_0 : u(x, y) = 0$ in R.*

If B_1 is an arbitrary one of the Jordan curves composing B, an even number of disjoint open arcs of Γ_0 have end-points on B_1, because on one side of each such arc we have $u(x, y) > 0$ and on the other side $u(x, y) < 0$. It is quite conceivable that a point of B_1 should be an end-point of several such arcs.

There exists no subregion of R bounded wholly by arcs of B and of Γ_0. If we trace monotonically arcs of Γ_0 without repetition but admitting intermediate arcs of B, we can always trace an arc terminating in J.

Assume a critical point P_0 of $u(x, y)$ of order k in R to lie on Γ_0; we shall reach a contradiction. There are $2(k + 1)$ Jordan arcs J_j of Γ_0 with P_0 as initial point which can be traced from P_0 to terminal points on J, admitting only arcs of Γ_0 plus intermediate arcs of B as necessary. No two of these arcs can intersect in the closure of R in a point other than P_0. The method of proof of Theorem 1 now applies without essential change. Jordan curves belonging to B which are not intersected by Γ_0 play no role. Arcs of Γ_0 not belonging to arcs J_j can be considered as cuts in R. As in the proof of Theorem 1 we reach a contradiction, which completes the proof of Theorem 3.

The Corollary to Theorem 1 has a corresponding analog relating to Theorem 3. We state the analog of Theorem 2:

THEOREM 4. *Let the region R be bounded by a set B consisting of a finite number of disjoint Jordan curves, and let the function $u(x, y)$ be non-negative on the Jordan curve J^+ of B, non-positive on the curve J^- of B, and zero on the remaining curves of B. Let $u(x, y)$ be harmonic but not identically zero in R and continuous in the closure of R. Then $u(x, y)$ has no critical point on the locus $\Gamma_0 : u(x, y) = 0$ in R.*

Theorem 4 may be proved by the method of proof of Theorem 2, modified as was the method of Theorem 1 to establish Theorem 3.

In Theorems 1–4 we have made the assumption $u \geqq 0$ on part of the boundary of R and $u \leqq 0$ on another part. If these inequalities are replaced by the corresponding strong inequalities $u > 0$ and $u < 0$, we obtain weaker theorems but the proofs become much simpler.

Theorems 1–4 do not extend without essential change in character, to permit both positive and negative values on two distinct boundary components of R. For instance we choose R as the region (in polar coordinates r, θ): $1/a < r < a$, with $a > 1$, and we set $u(x, y) = (r - 1/r) \sin \theta$. The locus Γ_0 in R consists of the unit circle plus the two segments $1/a < |x| < a$ of the axis of reals, and the two critical points $z = +1$ and $z = -1$ of $u(x, y)$ in R lie on Γ_0.

§9.9. Assigned distributions of matter. In the present work we have thus far discussed the critical points of harmonic functions given in terms of their boundary values. We now discuss briefly the critical points of harmonic functions defined as the potentials due to given distributions of matter.

§9.9.1. Simple layers. We have already developed detailed methods for the study of the critical points of the potential of a simple layer: for positive mass (Chapter VII), for positive and negative mass (§§8.2, 8.3), and for cases of symmetry (§§8.4, 8.5, 9.5, 9.6). Every result that we have established by use of ordinary particles or distributions can be interpreted as a result on critical points of potential due to a simple layer. We leave the details of this interpretation to the reader, and merely formulate a further analog of §4.4.3 Theorem 2 and of §8.3 Theorem 2:

THEOREM 1. *Let given circular regions C_1, C_2, C_3 contain respectively simple distributions of matter of total masses λ_1, λ_2, and $-(\lambda_1 + \lambda_2)$, where each distribution contains only positive or only negative matter. Denote by C_4 the locus of the point z_4 when C_1, C_2, C_3 are the respective loci of points z_1, z_2, z_3, with $(z_1, z_2, z_3, z_4) = (\lambda_1 + \lambda_2)/\lambda_1$. Then all critical points of the corresponding potential lie in C_1, C_2, C_3, C_4.*

Theorem 1 contains the analog of Bôcher's Theorem, obtained here by choosing C_1 and C_2 identical. A more general analog of Marden's Theorem can be formulated without difficulty.

We turn briefly to hyperbolic geometry for the interior of the unit circle C, as we have used it on various occasions. What is the most general function $u(z)$

to which applies the conclusion of §8.4 Theorem 1 that the critical points of $u(z)$ interior to C lie in the smallest NE convex region containing the boundary interior to C of a given subregion of $|z| < 1$ in which $u(z)$ is harmonic? In every case of a function $u(z)$ zero on C, continuous on C except perhaps in a finite number of points, and harmonic and positive in a subregion R of the interior of C, we have used for z in R a representation of the form

$$(1) \qquad u(z) = \int_B \log r \, d\sigma(t) - \int_{B'} \log r \, d\sigma(t) + \text{const}, \qquad r = |z - t|,$$

where B is the boundary of R interior to C and B' (assumed bounded) is the reflection of B in C; here the values of $d\sigma(t)$ are negative and are equal in corresponding points of B and B'. Equation (1) is conveniently expressed in terms of Green's function $g(z, t, |z| < 1) = -\log |z - t| + \log |z - 1/\bar{t}| + \log |t|$; again for z in R we have, by replacing σ by $-\sigma$,

$$(2) \qquad u(z) = \int_B g(z, t, |z| < 1) \, d\sigma(t) + \text{const}, \qquad d\sigma \geqq 0.$$

The function $u(z)$ is the potential due to a distribution of negative matter on B, and an equal and opposite distribution on B'; equation (2) expresses $u(z)$ in form invariant under linear transformation, for such a transformation introduces no new singularities of $u(z)$. It is clear that the conclusion of §8.4 Theorem 1 applies to any function $u(z)$ that can be expressed in form (2), with $d\sigma(t) \geqq 0$, where the integral now may be taken as a Stieltjes integral over B or even over the complement of R with respect to the interior of C; the conclusion also applies to the uniform limit in R of a sequence of such functions.

For this conclusion to apply *it is not sufficient that $u(z)$ be harmonic and positive in R, continuous in the closure of R and zero on C*, as we show by exhibiting a counterexample. Let R_1 be the annular region bounded by the circles $C: |z| = 1$ and $C_2 : |z| = r^2 < 1$, which are mutually inverse in the circle $C_1 : |z| = r$. We set $u(z) = g(z, r, R_1) + g(z, -r, R_1)$, which has precisely two critical points z_1 and z_2 interior to R_1; this set of two points z_1 and z_2 is symmetric in C_1 and in both coordinate axes, hence lies on C_1 and on the axis of imaginaries; we set $z_1 = ir$, $z_2 = -ir$. A suitably chosen circle γ orthogonal to C and exterior to C_2 contains z_1 in its interior. The locus $u(z) = \epsilon$, where ϵ is small and positive, consists of two Jordan curves in R_1 near C and C_2 respectively. We denote by Γ_1 the Jordan curve near C_2, where ϵ is chosen so small that γ lies exterior to Γ_1. We denote by Γ_2 the locus $u(z) = M$ (> 0), where M is chosen so large that the locus consists of two Jordan curves in R_1, each curve containing one of the points $z = \pm r$ in its interior, and where γ is exterior to both curves. The function $u(z)$ is harmonic and positive in the region R bounded by C, Γ_1, and Γ_2, and is zero on C, but the critical point z_1 of $u(z)$ in R does not lie in the smallest NE convex region interior to C containing the boundary of R interior to C. Of course $u(z)$ does not admit in R a representation of form (2).

§9.9.2. Superharmonic functions. In a classical paper F. Riesz [1930] relates superharmonic functions to the representation (2), so we devote some attention [Walsh, 1950] to the relation of such functions with the present development. We state for reference only a fraction of Riesz's results:

RIESZ'S THEOREM. *If the function $u(x, y)$ is superharmonic in the interior R of the unit circle, and if there exists a function harmonic in R inferior to $u(x, y)$ throughout R, then $u(x, y)$ can be represented in R in the form*

$$(3) \qquad u(x, y) = \int_R g(z, t, R)\, d\mu(e) + h(x, y), \qquad d\mu(e) \geqq 0,$$

where as t varies over R the integral is taken with respect to a suitably chosen positive additive set function $\mu(e)$ defined for t on all open sets e whose closures lie in R, and where $h(x, y)$ is harmonic in R.

As an application we prove

THEOREM 2. *Let R be the interior of a Jordan curve C, and let B be a closed set in R which together with C bounds a subregion R' of R. Let the function $u(x, y)$ be harmonic in R', superharmonic in R, continuous in the closed neighborhood of C in $R + C$, and zero on C. Then all critical points of $u(x, y)$ in R' lie in the smallest NE convex set of R containing B.*

As we may do, we choose C as the unit circle, and choose an arbitrary NE line wholly in R' as the axis of reals, with an arbitrary preassigned point of that line as the origin and with B in the upper half-plane. Equation (3) is then a consequence of Riesz's Theorem. Since both $u(x, y)$ and $h(x, y)$ are harmonic in R', it follows from (3) that the integral is harmonic in R', and follows further that we have $d\mu = 0$ in R'. Consequently the integral may be taken not over R but over an arbitrary open set in R containing $R - R'$. The function $g(z, t, R)$ has a meaning for z exterior to R; the values at points mutually inverse in C are the negatives of each other, so the integral represents a function harmonic in an annular region containing C and vanishing on C. It follows from (3) that $h(x, y)$ is continuous on $R + C$ and zero on C, hence vanishes identically in R. Thus $u(x, y)$ can be represented in a form similar to (2), and Theorem 2 follows.

If B, C, R, and R' satisfy the conditions of Theorem 2, and if a function $u(x, y)$ is harmonic in R', continuous in $R' + B + C$, zero on C and unity on $R - R'$, then $u(x, y)$ is continuous and superharmonic in R. It is sufficient to show that $u(x, y)$ is locally superharmonic in R, which is obvious for a point P of R' or a point P of $R - (R' + B)$ (the function $u(x, y)$ is itself harmonic at P) and also for a point P of B (the value of the function at P is not less than the average over any sufficiently small circumference whose center is P). Thus Theorem 2 contains §8.4 Theorem 1; moreover B in Theorem 2 need not be a Jordan configuration.

It is a further consequence of Riesz's theory that any function harmonic in a region R', continuous in the closed neighborhood of C in $R + C$, and zero on C, which can be represented by such an integral as that in (3) is superharmonic in R and thus satisfies the conditions of Theorem 2. In sum, we have here a fairly complete indication of the range of our present method, the use of a field of force defined by a potential represented by the integral in (3). In Theorem 2 we have considered only functions $u(x, y)$ vanishing on C; this particular requirement can be diminished as in §§9.4 and 9.6. However, it is to be noted that Theorem 2 is contained in part 5) of §9.7.1 Theorem 1, in which there occurs also comparison of two harmonic functions but without requiring the complete superharmonic behavior of the given function. Of course the counterexample given explicity in §9.9.1 involves a function which is not superharmonic throughout the interior of C.

The method used in Theorem 2 can be extended:

THEOREM 3. *Let Π_1 and Π_2 be the upper and lower half-planes, let R be the region $|z| < 1$, and let a subregion R' of R be bounded by the unit circle C and a Jordan configuration B not intersecting the axis of reals A. Let the function $u(x, y)$ be harmonic in R', bounded in the closed neighborhood of C in $R + C$, and continuous on C except perhaps in a finite number of points, and let $u(x, y)$ be superharmonic in $R \cdot \Pi_1$, subharmonic in $R \cdot \Pi_2$. Suppose on $C \cdot \Pi_1$ we have $u(x, y) \geqq u(x, -y)$. Then $u(x, y)$ has no critical point in R on A.*

We assume, as we may do with no loss of generality, that R' is symmetric in A and that we have $u(x, y) \equiv -u(x, -y)$ in R'; for we need merely replace a given $u(x, y)$ by the function $u_1(x, y) \equiv u(x, y) - u(x, -y)$ to obtain these conditions. Moreover we shall prove $\partial u_1(x, 0)/\partial y > 0$ on $A \cdot R$, which implies $\partial u(x, 0)/\partial y > 0$ and the theorem.

The function $U(x, y)$ which is harmonic and bounded in R and takes on the same boundary values as $u(x, y)$ on C (except perhaps in a finite number of points) vanishes on $A \cdot R$ and is inferior to $u(x, y)$ in $R \cdot \Pi_1$, so by Riesz's Theorem we have for (x, y) in $R \cdot \Pi_1$

$$u(x, y) = \int_{R \cdot \Pi_1} g(z, t, R \cdot \Pi_1)\, d\mu + h(x, y), \qquad d\mu \geqq 0, \tag{4}$$

where $h(x, y)$ is harmonic in $R \cdot \Pi_1$. The integral in (4) is taken over the open set $R \cdot \Pi_1$; throughout the region $R' \cdot \Pi_1$ the functions $u(x, y)$ and $h(x, y)$ are harmonic and $d\mu$ is zero, so the integral can be taken over any open set in $R \cdot \Pi_1$ containing the set $(R - R') \cdot \Pi_1$; the integral represents a function harmonic and continuous in the neighborhood of $C \cdot \Pi_1$ and $A \cdot R$, and vanishes on those sets, hence can be extended harmonically across them. In particular it follows that we have $h(x, y) \equiv U(x, y)$ in $R \cdot \Pi_1$, so $h(x, y)$ can be extended harmonically across A and coincides with $U(x, y)$ throughout R.

From the definitions of the functions involved we verify for t in R the identity $g(z, t, R) - g(z, \bar{t}, R) \equiv g(z, t, R \cdot \Pi_1)$, where the latter function is defined in $R \cdot \Pi_2$ by harmonic extension. Consequently equation (4) for (x, y) in $R \cdot \Pi_1$ can be extended so as to be valid throughout R:

$$u(x, y) = \int_{(R-R')\cdot\Pi_1} g(z, t, R)\, d\mu - \int_{(R-R')\cdot\Pi_2} g(z, t, R)\, d\mu + U(x, y),$$

where the integrals are taken over any disjoint open sets in R containing the respective sets $(R - R')\cdot\Pi_1$ and $(R - R')\cdot\Pi_2$. The field of force corresponding to the first two terms of the second member is the field previously considered (§8.4 Theorem 2), here due to negative matter on $(R - R')\cdot\Pi_1$ and equal positive matter on $(R - R')\cdot\Pi_2$, and (similarly to the interpretation of equation (2)) to positive matter on the inverse of $(R - R')\cdot\Pi_1$ with respect to C and to negative matter on the inverse of $(R - R')\cdot\Pi_2$ with respect to C. The function $U(x, y)$ can be treated (§9.3.4) as the potential due to a double distribution on C with $U(x, y) \equiv -U(x, -y) \geqq 0$ on $C\cdot\Pi_1$; the theorem follows.

It will be noted that Theorem 3 is related to part 5) of §9.7.1 Theorem 1, but the latter does not require the given region R to be simply-connected, nor does it require the complete superharmonic or subharmonic behavior demanded in Theorem 3.

Theorems 2 and 3 are essentially concerned with the hyperbolic (NE) plane. Similar results can be obtained, however, in the euclidean plane. We prove

Corollary 1. *Let the function $u(x, y)$ be harmonic in the closure of the annular region R bounded by disjoint circles C_1 and C_2, and let $u(x, y)$ be superharmonic in the interior of the circular region C_1 disjoint from R bounded by the circle C_1, and let $u(x, y)$ be subharmonic in the interior of the circular region C_2 disjoint from R bounded by the circle C_2. Then $u(x, y)$ has no critical points in R.*

We consider as a possible critical point an arbitrary point of R, which is chosen to be a real finite point $(x_0, 0)$, and we choose C_1 and C_2 to lie in the upper and lower half-planes respectively. By further results due to F. Riesz [1930], it follows from the superharmonic character of $u(x, y)$ interior to C_1 that we can write (either a single or a double integral may be indicated)

$$u(x, y) = \iint_{C_1} \log r\, d\mu_1(e) + u_1(x, y), \qquad d\mu_1(e) \leqq 0,$$

where the double integral is taken over the interior of the circular region C_1, and where $u_1(x, y)$ is harmonic in the closed region C_1 and at every finite point at which $u(x, y)$ is harmonic. The function $u_1(x, y)$ is subharmonic interior to C_2, so we have

$$u_1(x, y) = \iint_{C_2} \log r\, d\mu_2(e) + u_2(x, y), \qquad d\mu_2(e) \geqq 0,$$

where $u_2(x, y)$ is harmonic in the closed region C_2 and at every finite point at which $u_1(x, y)$ is harmonic; thus $u_2(x, y)$ is harmonic at every finite point. Consequently we may write

$$u(x, y) = \iint_{C_1} \log r \, d\mu_1(e) + \iint_{C_2} \log r \, d\mu_2(e) + u_2(x, y),$$

and this equation is valid at every finite point at which $u(x, y)$ is defined. In the neighborhood of the point at infinity these two integrals are equal respectively to harmonic functions plus $\frac{1}{2} \log (x^2 + y^2)$ multiplied by

$$\iint_{C_1} d\mu_1(e) \quad \text{and} \quad \iint_{C_2} d\mu_2(e);$$

these latter double integrals are numerically the quotients by 2π of the integrals of the normal derivative of $u(x, y)$ over the circles C_1 and C_2, and represent the total masses μ_1 and μ_2 in C_1 and C_2 respectively, and their sum is zero. From the fact that $u(x, y)$ is harmonic at infinity it follows that $u_2(x, y)$ is harmonic at infinity, hence identically constant. The axis of reals separates the positive distribution μ_2 from the negative distribution μ_1, so the given point $(x_0, 0)$ cannot be (§7.1.3 Lemma) a position of equilibrium in the field of force. and the Corollary is established.

The chief significance of the theorems of Riesz for the location of critical points is that they enable us to treat superharmonic functions (an additive class) by use of the field of force, and thus his theorems unify and extend our methods and results. Indeed, his representation of functions in combination with various methods developed in the present work, can be used to establish numerous general results which elude all other known methods. In a particular situation it may be difficult accurately to determine the masses involved, and such determination to be precise may require stringent hypotheses on the given function; compare for instance §9.3.5 Theorem 7. Nevertheless in particular cases the relative masses may be determinable, by symmetry or otherwise. We give a further illustration of this remark, a generalization of part 2) of §8.3 Theorem 1, and leave the proof to the reader:

COROLLARY 2. *Let C', C'', and C''' be disjoint circular regions, and let C^0 denote the locus of the point z_4 defined by the constant cross-ratio $(z_1, z_2, z_3, z_4) = 2$ when z_1, z_2, z_3 have C', C'', C''' as their respective loci. Let the function $u(z)$ be harmonic in the closed region R bounded by the circles C', C'', C''', subharmonic in C' and C'', superharmonic in C'''. Let the entire configuration be symmetric in some circle C, with C''' symmetric in C, and C' symmetric to C'' in C. Then all critical points of $u(z)$ interior to R lie in C^0.*

In connection with Corollaries 1 and 2, compare §9.9.1 Theorem 1.

§9.9.3. Double layers. Special double layer distributions have been considered hitherto on various occasions, and we shall consider now a few particular configurations.

THEOREM 4. *Let C be the unit circle, and let mutually disjoint closed arcs C_1, $C_2, \cdots, C_n$ of C contain double layer distributions all of the same orientation with* *respect to the interior of C. All critical points interior to C of the corresponding potential lie in the smallest NE convex region interior to C whose closure contains the arcs* C_k.

At any point P interior to C not interior to this convex region the total force cannot be zero, as is seen by transforming the interior of C into itself so that P is transformed into the origin; if the images of all the C_k are chosen to lie in the upper half-plane, all the forces at O due to elements of the given distribution then have non-zero vertical components in the same sense; the directions and senses of these forces are invariant under linear transformation, so Theorem 4 follows.

Theorem 4 is analogous to and indeed (§9.3.4) the equivalent of §9.3.2 Theorem 3; the corresponding analog (equivalent) of §9.3.2 Theorem 5, proved by a method similar to the one just used, is

THEOREM 5. *Let C be the unit circle, and let closed disjoint arcs* C_1 *and* C_2 *of C contain respectively double layer distributions of opposite orientations with respect to the interior of C. All critical points interior to C of the corresponding potential lie in the NE half-planes bounded respectively by* C_1 *and* C_2 *and the NE lines joining their end-points.*

Theorems 4 and 5 are invariant under arbitrary linear transformation, but not under more general conformal transformation. Rather than attempting to prove broad general theorems on the critical points of double layer distributions, we shall study certain particular cases, deriving however general methods of wide applicability. We prove

THEOREM 6. *Let* S_1 *and* S_2 *be two closed finite segments of parallel lines, and let them contain respectively double layer distributions oriented in opposite senses. Then all finite critical points of the corresponding potential lie in the set composed of the closed interiors of the circles having* S_1 *and* S_2 *as diameters, and the closed set* Π *consisting of the points of all lines cutting both* S_1 *and* S_2.

For convenience in reference we establish two lemmas.

LEMMA 1. *Let a finite circular arc S contain a double layer distribution of constant sign, and let the point P lie exterior to the circular region containing S bounded by the circle orthogonal to S at both end-points. The corresponding force at P is equal to the force at P due to a suitably chosen dipole on S with the same orientation as the original double distribution.*

The lines of force due to a dipole are (§9.3.4) the directed circles tangent to the axis of the dipole at the dipole. We omit the obvious proof of Lemma 1 when P is concyclic with S. Invert now the given configuration in the unit circle whose center is P, denote by S' the image of S and by C' (necessarily a proper circle) the image of the circle C on which S lies. Since S lies in a circular region bounded by a circle to which it is orthogonal at both end-points, and since P is exterior to that circular region, S' lies interior to a circle to which it is orthogonal at both end-points, so S' is less than a semicircle. A directed circle through P orthogonal to S at a point Q is inverted into a directed line orthogonal to S' at a point Q', and this directed line has the direction and sense (if properly chosen) of the force at P due to a dipole at Q on S. As Q varies on S and Q' varies on S', this directed line always passes through the center of C', and varies through an angle less than π. The limit of the sum of a number of vectors each representing the force at P due to a dipole on S is thus a vector through the center of C' and cutting S', which is essentially the conclusion of the lemma.

For two similarly oriented dipoles on S with axes orthogonal to S, the W-curve is the circle through them orthogonal to S. For the entire set of such dipoles on S, the W-curve of every pair lies in the closed interior of the circle orthogonal to S at both end-points; that circle may be considered as a limiting W-curve or as a W-curve for the entire set of dipoles, or for a double distribution of constant orientation on S.

LEMMA 2. *Let dipoles at distinct points Q_1 and Q_2 be parallel but oppositely oriented. The corresponding W-curve is precisely the line Q_1Q_2.*

If dipoles at distinct points Q_1 and Q_2 are parallel but similarly oriented, the corresponding W-curve is the circle whose diameter is the segment Q_1Q_2.

Let the dipoles at Q_1 and Q_2 be oppositely oriented. If P lies on the line Q_1Q_2, whether or not P lies on the finite segment Q_1Q_2, P is a center of similitude for the oriented circles through P properly tangent at Q_1 and Q_2 to the axes of the respective dipoles, and those oriented circles are improperly tangent at P. Conversely, let oriented circles properly tangent at Q_1 and Q_2 to the axes of the respective dipoles be improperly tangent at some point P. Then P is a center of similitude for those oriented circles, and points (such as Q_1 and Q_2) where the directions of those circles are oppositely parallel are collinear with P. This completes the proof of the first part of the lemma.

If the dipoles at Q_1 and Q_2 are similarly oriented, the oriented circle C whose diameter is the segment Q_1Q_2 cuts the axes of those dipoles in positive angles whose difference is π. When C is transformed by a linear transformation into a straight line C', the dipoles are transformed into parallel but oppositely oriented dipoles on C'. Since the lines of force of a dipole are invariant under linear transformation, the second part of the lemma now follows from the first part.

Lemma 2 is also readily proved by transforming to infinity one of the given dipoles; the corresponding field of force is represented by vectors all parallel to a given vector and similarly oriented.

Of course it follows from Lemma 2 that for any two dipoles (unless situated at the same point) the W-curve is the circle through them which when oriented cuts the axes of the dipoles at directed angles differing by π. Two dipoles situated at the same point have no W-curve (the force can be zero at no finite point of the plane) unless the dipoles are oppositely oriented; in the latter case any point of the plane may be a position of equilibrium, so the entire plane is to be considered the W-curve. If two dipoles lie on a circle C and their axes are tangent to C, their W-curve is either C or the circle through them orthogonal to C according as the dipoles are oppositely or similarly oriented with respect to the directed circle C.

Thanks to the lemmas, we are in a position to prove Theorem 6. Let a point P exterior to the circles having S_1 and S_2 as diameters be a position of equilibrium. By Lemma 1 the force at P due to the double distribution on S_k ($k = 1, 2$) is equal to the force at P due to a suitably chosen dipole at a point Q_k on S_k, the dipoles at Q_1 and Q_2 being oppositely oriented. It then follows by Lemma 2 that P lies on the line Q_1Q_2, hence in Π, so Theorem 6 is established. Of course Π is an infinite region bounded by two infinite polygonal lines, each consisting of three segments; thus if S_1 and S_2 are the horizontal segments A_1A_2 and B_1B_2, with A_1 and B_1 the left hand end-points, then Π is bounded by the finite segments A_1B_1 and A_2B_2, and the two lines A_1B_2 and A_2B_1 with the finite segments A_1B_2 and A_2B_1 omitted.

In Theorem 6, the original distribution can be chosen continuous or to consist of a finite number of dipoles, or a combination; in the former case no boundary point of Π exterior to the circles whose diameters are the S_k can be a critical point, nor can a point on one of those circles not interior to the other circle nor in Π be a critical point. It is to be noted that the point set mentioned in Theorem 6 is the actual *locus* of critical points where all possible double distributions, including a finite number of dipoles, are allowed.

A limiting case of Theorem 6 is that in which S_1 and S_2 are closed finite segments of the same line, and contain double layers of opposite orientation. All critical points of the corresponding potential not collinear with S_1 and S_2 lie in the closed interiors of the circles having S_1 and S_2 as diameters; this is essentially Theorem 5.

As a companion-piece to Theorem 6 we prove

THEOREM 7. *Let S_1 and S_2 be two closed finite segments of parallel lines, and let them contain respectively double layer distributions oriented in the same sense. Then all critical points of the corresponding potential lie in the set composed of the closed interiors of the six circles having as diameters the six segments joining the end-points of S_1 and S_2.*

Denote generically by (E, F) the circle whose diameter is the line segment EF. We note

LEMMA 3. *If AB is a line segment, Q a point of this segment, and P a point not on the line AB, then the circle (P, Q) lies in the two finite closed lens-shaped regions disjoint except for vertices, bounded by arcs of the two circles (P, A) and (P, B).*

All three circles pass through P and through the foot M of the perpendicular from P onto the line AB, and the conclusion follows. As Q moves on AB from A to B, the circle (P, Q) sweeps out these two lens-shaped regions. It also follows that *the closed interior of* (P, Q) *lies in the sum of the closed interiors of the circles* (P, A) *and* (P, B), for the two segments of the circle (P, Q) each bounded in part by the line segment PM lie respectively in segments of the circles (P, A) and (P, B) each bounded in part by the line segment PM. A point interior to both (P, A) and (P, B) lies interior to (P, Q).

It is a consequence of Lemma 3 that if P_1 and P_2 lie on S_1 and S_2 respectively, where S_1 and S_2 are the finite segments A_1A_2 and B_1B_2, then the closed interior of the circle (P_1, P_2) lies in the sum of the closed interiors of the four circles (A_1, B_1), (A_1, B_2), (A_2, B_1), (A_2, B_2); indeed it is a consequence of the remark just made that the closed interior of (P_1, P_2) lies in the sum of the closed interiors of the circles (P_1, B_1) and (P_1, B_2), and the latter two closed interiors lie respectively in the sum of the closed interiors of the circles (A_1, B_1) and (A_2, B_1) and in the sum of the closed interiors of the circles (A_1, B_2) and (A_2, B_2).

Theorem 7 now follows from Lemma 1 and the second part of Lemma 2, by the method of proof of Theorem 6. The point at infinity is not a critical point, as may be seen either from the field of force on the sphere or by an inversion. If the double distribution in Theorem 7 is continuous, not a finite number of dipoles, then no boundary point of the set specified is a critical point.

The set specified in Theorem 7 is not necessarily the *locus* of critical points where all possible distributions, including a finite number of dipoles, are allowed:

COROLLARY. *If all possible distributions are allowed in Theorem 7, including a finite number of dipoles, the locus consists precisely of the closed interior of* (A_1, A_2) *plus the closed interior of* (B_1, B_2); *plus the set* T_1 *consisting of the closed interiors of the four circles* (A_1, B_1), (A_1, B_2), (A_2, B_1), (A_2, B_2) *except the set* T_2 *(which may be empty) common to all four interiors.*

The locus consists of the closed interiors of the circles (A_1, A_2) and (B_1, B_2) plus the points of all circles (P_1, P_2) with P_1 on S_1 and P_2 on S_2. Each point of the closed interior of (A_1, A_2) or (B_1, B_2) belongs to the locus, for a suitably chosen W-curve passes through each such point; each point on any one of the six circles also belongs to the locus. If the point P is now interior to T_1 but exterior to T_2, it is interior to Γ_1, one of the four circles, but exterior to Γ_2, another of those four circles. However, as P_1 continuously moves along S_1 from A_1 to A_2 and P_2 moves continuously along S_2 from B_1 to B_2, the circle (P_1, P_2) moves continuously; thus the circle (P_1, P_2) can be moved continuously remaining in the locus of critical points starting in a position of coincidence with any one (say Γ_2) of the four circles and ending in a position of coincidence with any other (say Γ_1) of the four circles; consequently the variable circle (P_1, P_2) must pass through the point P interior to Γ_1 but exterior to Γ_2, so P belongs to the locus.

On the other hand, let P (not in the closed interior of (A_1, A_2) or (B_1, B_2))

now lie in T_2, assumed not empty; by the remark made in connection with Lemma 3, since P lies interior to (A_1, B_1) and to (A_1, B_2), P also lies interior to (A_1, P_2) where P_2 is an arbitrary point of S_2; similarly P lies interior to (A_2, P_2). By the same remark, since P lies interior to both (A_1, P_2) and (A_2, P_2), P lies interior to the circle (P_1, P_2), where P_1 is an arbitrary point of S_1; thus P lies on no W-curve for pairs P_1 and P_2, and is not a point of the locus; the Corollary is established.

As a limiting case of Theorem 7, the segments S_1 and S_2 may be disjoint and collinear; we have then essentially the situation of Theorem 4 with $n = 2$; the set T_2 in the Corollary is precisely the set interior to one of the four given circles.

The methods developed in the proofs of Theorems 6 and 7 apply at once in the study of arbitrary double distributions on circular arcs, as we proceed to show.

LEMMA 4. *Let P be a dipole not on a circle C, and let dipoles on C have constant orientation normal to C. Then all W-curves for P and these dipoles are circles through P and a fixed point of C.*

Lemma 4 is readily proved from Lemma 2, but we give a different proof. Transform P to infinity, so that the lines of force for the dipole P are parallel lines. One of these lines is a diameter of C, cutting C in points D and E, and is the W-curve for the dipole at P and the dipole at one (say E) of the points D and E. The force at D due to the dipole at P is opposite in sense to the force at D due to the dipole at E. Let Q be an arbitrary point of C different from D; the circle QD orthogonal to C is a line of force for the dipole at Q, and the corresponding force at D has the same direction and sense as the force at D due to the dipole at E, and has the same direction as but opposite sense to that at D due to the dipole at P. Thus D lies on the W-curve for the dipoles P and Q; it follows either from Lemma 2 or by a direct study of the fields of force of the dipoles at P and Q that the W-curve for those dipoles is a straight line through Q, so the W-curve is the line DQ. This reasoning fails if Q coincides with D, but in that case the W-curve is clearly the line tangent to C at D. Lemma 4 is established.

If we consider not dipoles to lie on C as in Lemma 4 but a double distribution of constant orientation to lie on an arc S of C, the limiting W-curves are the circles through P, D, and the end-points of S; all W-curves for the dipole P and the double distribution lie in a closed point set consisting of two lens-shaped regions bounded by these limiting W-curves.

If we modify the hypothesis of Lemma 4 by requiring P to lie on C, the W-curves are circles of a coaxal family all tangent to each other at P; if we then consider a double distribution on an arc S of C, the limiting W-curves are the circles of the coaxal family through the end-points of S.

Let there now be given two double distributions, each of constant orientation, on circular arcs $S_1 : A_1A_2$ and $S_2 : B_1B_2$ respectively. Let the circular region C_k

contain S_k and be bounded by a circle C_k orthogonal to S_k at both end-points. If a point P not in C_1 or C_2 is a position of equilibrium in the field of force, then by Lemma 1 the point P is also a position of equilibrium in the field of force due to two dipoles Q_1 and Q_2 on S_1 and S_2 respectively, with the same orientations as the given distributions. Thus P lies on the W-curve for Q_1 and Q_2. The W-curve for Q_1 and Q_2 lies in the two lens-shaped regions bounded by the W-curves for Q_1 and B_1 and for Q_1 and B_2 ; the W-curve for Q_1 and B_1 lies in the lens-shaped regions bounded by the W-curves for A_1 and B_1 and for A_2 and B_1 ; the W-curve for Q_1 and B_2 lies in the lens-shaped regions bounded by the W-curves for A_1 and B_2 and for A_2 and B_2. *Thus all positions of equilibrium lie in the regions C_1 and C_2 plus a point set Π bounded by arcs of the set of W-curves for the pairs of dipoles* (A_1, B_1), (A_1, B_2), (A_2, B_1), (A_2, B_2); *the set Π is connected and is precisely the locus of all W-curves for pairs (Q_1, Q_2), with Q_k on S_k*. This conclusion contains Theorems 6 and 7, and the Corollary to Theorem 7.

The discussion just given includes by specialization the case of a single dipole S_1 and a double distribution on a circular arc S_2. With appropriate modifications these methods apply in the study of double distributions on more general arcs. Moreover, these methods combined with those previously developed for the study of the potentials of simple layers apply to the study of potentials due to both simple and double layers.

BIBLIOGRAPHY

This bibliography is restricted to the references which have close contact with the subject-matter of the present work. A more inclusive bibliography on the zeros of polynomials, with indications of the content of the various papers, is given by Marden [1949].

ALEXANDER, J. W., II.

1915 *Functions which map the interior of the unit circle upon simple regions*. Annals of Math., vol. 17, pp. 12–22.

ANCOCHEA, G.

1945 *Sur les polynomes dont les zéros sont symmétriques par rapport à un contour circulaire*. Paris Comptes Rendus, vol. 221, pp. 13–15.

BIEBERBACH, L., AND BAUER, G.

1928 *Vorlesungen über Algebra*. Leipzig.

BIERNACKI, M.

1928 *Sur les équations algébriques contenant des paramètres arbitraires* (Thèse, Paris).

BÔCHER, M.

1904 *A problem in statics and its relation to certain algebraic invariants*. Proc. Amer. Acad. of Arts and Sciences, vol. 40, pp. 469–484.

COOLIDGE, J. L.

1916 *A treatise on the circle and the sphere*. Oxford.

CURTISS, D. R.

1919 *On the relative positions of the complex roots of an algebraic equation with real coefficients and those of its derived equation* (Abstract). Bull. Amer. Math. Soc., vol. 26, pp. 53, 61–62.

1922 *A mechanical analogy in the theory of equations*. Science, vol. 55, pp. 189–194.

DIEUDONNÉ, J.

1938 *La théorie analytique des polynomes d'une variable*. Mémorial des sciences math., fasc. 93. (Paris).

FEJÉR, L., and TOEPLITZ, O.

1925 *Problem 32 and solution*. III Abschnitt. Pólyà-Szegö, Aufgaben und Lehrsätze aus der Analysis, vol. I, (Berlin), pp. 89, 258.

FEKETE, M.

1933 *Über den transfiniten Durchmesser ebener Punktmengen* (Dritte Mitteilung). Math. Zeitschrift, vol. 37, pp. 635–646.

FORD, L. R.

1915 *On the roots of a derivative of a rational function*. Proc. Edinburgh Math. Soc., vol. 33, pp. 103–106.

GAUSS, C. F.

1816 *Lehrsatz*. Werke, vol. 3, p. 112; vol. 8, p. 32.

GONTCHAROFF, W.

1942 *Sur une extension du théorème de Gauss-Lucas*. Comptes Rendus (Doklady) de l'Académie des Sciences de l'USSR, vol. 36, pp. 39–41.

GRACE, J. H.

1902 *The zeros of a polynomial*. Proc. Cambridge Phil. Soc., vol. 11, pp. 352–357.

HEAWOOD, P. J.

1907 *Geometrical relations between the roots of* $f(x) = 0, f'(x) = 0$. Quart. Jour. Math., vol. 38, pp. 84–107.

HEINS, M. See MORSE, M.

HILBERT, D.

1897 *Über die Entwickelung einer beliebigen analytischen Funktion*. Göttinger Nachrichten, pp. 63–70.

JENSEN, J. L. W. V.

1913 *Recherches sur la théorie des équations*. Acta Math., vol. 36, pp. 181–195.

KAKEYA, S.

1917 *On zeros of a polynomial and its derivatives*. Tôhoku Mathematical Journal, vol. 11, pp. 5–16.

LAGUERRE, E.

1878 *Sur la résolution des équations numériques*. Nouvelles Annales de Math. (2), vol. 17, pp. 20–25, 97–101.

LEBESGUE, H.

1907 *Sur le problème de Dirichlet*. Palermo Rendiconti, vol. 24, pp. 371–402.

LUCAS, F.

1874 *Propriétés géométriques des fractions rationnelles*. Paris Comptes Rendus, vol. 78, pp. 271–274.

1874a *Théorèmes concernant les équations algébriques*. Ibid., vol. 78, pp. 431–433.

1879 *Géométrie des polynomes*. Jour. de l'École poly., vol. 28, cah. 46, pp. 1–33.

1888 *Statique des polynomes*. Bull. de la soc. math. de France, vol. 17, pp. 17–69.

MACDONALD, H. M.

1898 *The zeroes of the Bessel functions*. Proc. London Math. Soc., vol. 29, pp. 575–584.

MARDEN, M.

1930 *On the zeros of linear partial fractions*. Trans. Amer. Math. Soc., vol. 32, pp. 81–109.

1930a *On the zeros of certain rational functions*. Ibid., vol. 32, pp. 658–668.

1936 *On the zeros of the derivative of a rational function*. Bull. Amer. Math. Soc., vol. 42, pp. 400–405.

1949 *The geometry of the zeros of a polynomial in a complex variable*. Mathematical Surveys (Amer. Math. Soc.), vol. 3 (New York).

MONTEL, P.

1923 *Sur les modules des zéros des polynomes*. Annales de l'École normale supérieure (3), vol. 40, pp. 1–34.

1945 *Remarque sur la note précédente* [Ancochea, 1945]. Paris Comptes Rendus, vol. 221, p. 15.

MORSE, M., AND HEINS, M.

1947 *Topological methods in the theory of functions of a complex variable*. Bull. Amer. Math. Soc., vol. 53, pp. 1–14.

NAGY, J. v. S.

1918 *Über algebraische Gleichungen mit lauter reellen Wurzeln*. Jahresbericht d. deutschen Math. Verein., vol. 27, pp. 37–43.

1922 *Zur Theorie der algebraischen Gleichungen*. Ibid., vol. 31, pp. 238–251.

NEVANLINNA, R.

1936 *Eindeutige analytische Funktionen*. Berlin.

PINCHERLE, S.

1880 *Ricerche sopra una classe importante di funzioni monodrome*. Giornale di Mathematiche, vol. 18, pp. 92–136.

PORTER, M. B.

1916 *On a theorem of Lucas*. Proc. Nat. Acad. of Sciences, vol. 2, pp. 247–248.

1916a *Note on Lucas' theorem*. Ibid., vol. 2, pp. 335–336.

RIESZ, F.

1930 *Sur les fonctions subharmoniques et leur rapport à la théorie du potentiel*. Acta Math., vol. 54, pp. 321–360.

RUSSELL, H. G. See WALSH, J. L.

SIEBECK, H.

1865 *Über eine neue analytische Behandlungsweise der Brennpunkte*. Crelle's Journal, vol. 64, pp. 175–182.

Toeplitz, O. See Fejér, L.

Walsh, J. L.

1918 *On the location of the roots of the jacobian of two binary forms, and of the derivative of a rational function*. Trans. Amer. Math. Soc., vol. 19, pp. 291–298.

1920 *On the location of the roots of the derivative of a polynomial*. C. R. du congrès international des Math. Strasbourg (Toulouse).

1920a *On the location of the roots of the derivative of a polynomial*. Annals of Math., vol. 22, pp. 128–144.

1921 *On the location of the roots of the jacobian of two binary forms, and of the derivative of a rational function*. Trans. Amer. Math. Soc., vol. 22, pp. 101–116.

1921a *Sur la position des racines des dérivées d'un polynome*. Paris Comptes Rendus, vol. 172, pp. 662–664.

1922 *On the location of the roots of the jacobian of two binary forms, and of the derivative of a rational function*. Trans. Amer. Math. Soc., vol. 24, pp. 31–69.

1922a *On the location of the roots of the derivative of a polynomial*. Proc. Nat. Acad. Sciences, vol. 8, pp. 139–141.

1922b *On the location of the roots of certain types of polynomials*. Trans. Amer. Math. Soc., vol. 24, pp. 163–180.

1924 *On the location of the roots of polynomials*. Bull. Amer. Math. Soc., vol. 30, pp. 51–62.

1925 *Sur la position des racines des fonctions entières de genre zero et un*. Paris Comptes Rendus, vol. 180, pp. 2009–2011.

1933 *Note on the location of the roots of the derivative of a polynomial*. Mathematica (Cluj), vol. 8, pp. 185–190.

1933a *Note on the location of the critical points of Green's function*. Bull. Amer. Math. Soc., vol. 39, pp. 775–782.

1934 *Sur l'interpolation par fonctions rationnelles*. Paris Comptes Rendus, vol. 198, pp. 1377–1378.

1934a *Note on the location of the critical points of harmonic functions*. Proc. Nat. Acad. of Sciences, vol. 20, pp. 551–554.

1935 *Lemniscates and equipotential curves of Green's function*. Amer. Math. Monthly, vol. 42, pp. 1–17.

1936 *Note on the behavior of a polynomial at infinity*. Ibid., vol. 43, pp. 461–464.

1939 *Note on the location of zeros of the derivative of a rational function whose zeros and poles are symmetric in a circle*. Bull. Amer. Math. Soc., vol. 45, pp. 462–470.

1946 *Note on the location of the critical points of harmonic functions*. Ibid., vol. 52, pp. 346–347.

1946a *Note on the location of the zeros of the derivative of a rational function having prescribed symmetry*. Proc. Nat. Acad. of Sciences, vol. 32, pp. 235–237.

1947 *On the location of the critical points of harmonic measure*. Ibid., vol. 33, pp. 18–20.

1947a *Note on the critical points of harmonic functions*. Ibid., vol. 33, pp. 54–59.

1947b *The location of the critical points of simply and doubly periodic functions*. Duke Math. Jour., vol. 14, pp. 575–586.

1948 *On the critical points of functions possessing central symmetry on the sphere*. Amer. Jour. of Math., vol. 70, pp. 11–21.

1948a *Note on the location of the critical points of harmonic functions*. Bull. Amer. Math. Soc., vol. 54, pp. 191–195.

1948b *The critical points of linear combinations of harmonic functions*. Ibid., vol. 54, pp. 196–205.

1948c *Critical points of harmonic functions as positions of equilibrium in a field of force*. Proc. Nat. Acad. of Sciences, vol. 34, pp. 111–119.

1948d *Methods of symmetry and critical points of harmonic functions*. Ibid., vol. 34, pp. 267–271.

1948e *On the location of the zeros of the derivatives of a polynomial symmetric in the origin*. Bull. Amer. Math. Soc., vol. 54, pp. 942–945.

1950 *The location of critical points of harmonic functions*. Szeged Acta, vol. 12 (Fejér-Riesz Anniversary Volume), pp. 61–65.

WALSH, J. L. AND RUSSELL, H. G.

1934 *On the convergence and overconvergence of sequences of polynomials*. Trans. Amer. Math. Soc., vol. 36, pp. 13–28.

INDEX

Alexander, 60
Algebraic curve, iv, 19, 21, 42, 44, 79, 184
Analytic function, 2, 217, 240
Ancochea, 54
Anti-inversion, 181
Arc, Jordan, 1
Argument, Principle of, 2, 32, 54, 88, 91, 165, 180, 258, 275, 308, 312, 332, 339, 353
Automorphic function, 292

Bauer, 4
Bernoullian lemniscate, 19, 186, 363
Best possible, 8, 81, 85, 159
Bibliography, 377
Bicircular quartic, 19, 45, 68, 78, 121, 137, 147, 174, 197, 202, 324, 336
Bieberbach, 4
Biernacki, 122
Binary form, 89, 94, 97, 104
Blaschke product, 157, 223, 227
Bôcher, 13, 16, 89, 95, 97, 320
——'s Theorem, 97, 105, 116, 156, 169, 176, 184, 194, 198, 207, 211, 217, 227, 233, 239, 275, 296, 309, 313, 352, 365
Bohr, 14
Boundary point, 1

Carleman, 265, 303
Cauchy's integral formula, 236, 273, 305, 310
Chamberlin, 33
Circle, 1, 62, 98, 312
——, directed, 15, 59, 66, 111, 180, 371
Circular curve, 19, 121
—— region, 13, 58, 62, 80, 97, 103, 128, 165, 178, 186, 209, 212, 219, 238, 247, 259, 275 278, 283, 298, 302, 365, 369
Closed set, 1
Closure, 1
Coaxal family, 112
Component, 1
Configuration, Jordan, 2
Conjugate function, 4
Continuity, method of, 16, 25, 60, 98, 104, 107, 115, 124, 135, 143, 161, 168, 182, 188, 196, 221, 251, 259, 276, 279, 299
—— of zeros, 3, 42, 124
Convergence, 217, 223
Convex, 7, 14, 19, 71, 77, 80, 83, 102, 105, 138, 157, 170, 187, 193, 205, 208, 210, 222, 227, 231, 249, 256, 262, 267, 279, 289, 292 300, 325, 339, 346, 356, 366, 371
Coolidge, 112
Critical point, iii, 2, 4, 106
—— —— at infinity, 2, 4, 90, 96
—— ——, finite, 2, 4, 90
—— ——, multiplicity, 2, 4, 18
—— ——, order, 2, 4, 18
Cross-ratio, 62, 93, 109
—— Theorem, 109, 125, 278, 298, 365, 370
Cubic, 45
Curtiss, 7, 27
Curvature, 74, 82, 187, 191, 252
Curve, Jordan, 1

Degree of polynomial, 2
—— of rational function, 90
Descartes, 134, 198
Dieudonné, 122
Dipole, 321, 333, 371
Directed circle, 15, 59, 66, 111, 180, 371
Distance, NE, 168
Double distribution, 319, 321, 370
Doublet, 321, 333, 371
Doubly-connected region, 266, 292, 299, 300, 337, 343, 356, 362

Ellipse, 84, 87, 172, 203, 256, 356
Elliptic plane, 176, 284
End-point, 231
Entire function, 217
Enumeration, iii, 87, 100
Envelope, 122, 215
Equilateral hyperbola, 23, 27, 75, 82, 86, 183, 199, 209, 256, 284, 331
Equilibrium, 5, 91, 96, 274
Equivalent particle, 13, 37, 62, 169
Exponential function, 22, 220, 225, 286
Extended plane, 1
Exterior of hyperbola, 27
—— point, 1

Fejér, 8
Fekete, 248
Field of force, 5, 12, 20, 89, 93, 125, 218, 224, 237, 247, 255, 273, 280, 297, 304, 307, 313, 319, 324, 331, 345, 348, 352, 366, 368
Finite plane, 1
Ford, 99

Function, analytic, 2, 217, 240
——, automorphic, 292
——, entire, 217
——, harmonic, 4, 240, 269, 304
——, meromorphic, 2, 217, 240, 357
——, periodic, 231, 286

Gauss, iii, 5, 7
——'s mean value theorem, 245, 320
——'s Theorem, 5, 9, 15, 31, 89, 217, 247
Gebietserweiterung, 303
Gontcharoff, 222
Grace, 87
Gradient, 247, 273, 305, 321, 332, 348, 352, 356
Gravity, center of, 9, 14, 17, 21, 29, 74, 84, 127, 148, 173, 211, 217, 247, 255
Green's function, 88, 241, 276, 294, 304, 339, 344, 356, 366
—— ——, generalized, 244, 249, 255
Group, 41, 136

Harmonic function, 4, 240, 269, 304
—— conjugate, 10, 62, 125
—— measure, 140, 265, 271, 288, 294, 300, 307, 313, 324, 329, 339, 358
Harnack's Theorem, 5, 244, 291, 296
Heawood, 87
Heins, 88
Hilbert, 245, 248
Hurwitz's Theorem, 3, 5, 174, 218, 223, 239, 251, 292
Hyperbola, 203
Hyperbolic geometry, 138 157, 221, 292, 341, 346, 365
—— line, 138
—— plane, 156, 221, 279
Hypercycle, 251

Implicit function, 17, 100
Infinity, point at, 1, 14, 21, 43, 89, 92, 95, 136, 274
Interior of hyperbola, 27
—— point, 1
Interlaced points, 139, 220, 357
Inversion, 1, 12

J-circle, 127
Jensen, 9, 84
—— circle, 9, 27, 31, 45, 52, 84, 125, 136, 163, 209, 219, 222, 255, 283, 331, 336
—— ellipse, 84
——'s Theorem, 9, 24, 27, 31, 78, 84, 163, 181, 219, 255
Jordan arc, 1
—— configuration, 2, 237
—— curve, 1
—— region, 1

Kakeya, 88, 122
Klein, 176

Laguerre, 13, 25, 218, 221
Laplace's Equation, 4
Lebesgue integral, 291, 318
Lemniscate, 17, 241, 247, 261
——, Bernoullian, 19, 186, 363
——, family, 19
——, pole, 19
Level curves, 17, 91, 241, 247, 252, 269
Limit point, 1
Line of force, 20, 27, 43, 78, 91, 93, 195, 313, 371
Linear transformation, 1, 92, 126
Locus, iv, 11, 14, 24, 42, 58, 99, 103, 109, 115, 121, 128, 148, 165, 176, 178, 186, 373
——, level, 17, 91, 241, 247, 252, 269
—— of critical points, 99, 115, 121, 123
Logarithmic derivative, 5, 89
Lucas, 7, 18, 21
—— polygon, 6, 17, 20, 42, 68, 71, 159
——'s Theorem, iii, 6, 17, 40, 64, 68, 76, 81, 83, 88, 99, 157, 205, 217, 221, 232, 249, 276, 353

Macdonald, 221
——'s Theorem, 221, 275
Marden, iii, 88, 101, 121, 215, 377
——'s Theorem, 121, 266, 279, 365
Matter, 247, 273, 297, 327, 345, 352, 363
——, ordinary, 324, 329
——, skew, 305, 324, 329
Meromorphic function, 2, 217, 240, 357
Method of continuity, 16, 25, 60, 98, 104, 107, 115, 124, 135, 143, 161, 168, 182, 188, 196, 221, 251, 259, 276, 279, 299
Minkowski, 14
Moment, 317
Montel, 54, 108
Morse, 88
Multiplicity, 2, 18, 89, 147

Nagy, 25, 29, 32
n-circular (2n)-ic, 19, 121
NE, 138
Neighborhood, 1

Neumann sphere, 1, 94, 127, 176, 274, 284, 374
Nevanlinna, 290, 293, 311, 347
Non-euclidean, 138
—— distance, 168
—— geometry, 138, 279, 292, 325, 341
—— line, 138, 352
—— vertical, 163, 331, 335
Non-real point, 1

Open set, 1
Order of critical point, 2, 4, 18
—— —— pole or zero, 232
Oriented circle, 15, 59, 66, 111, 180, 371
Orthogonal trajectory, 18

Particle, 5, 89, 95
—— at infinity, 14, 21, 43, 89, 92, 95, 136, 274
——, equivalent, 13, 37, 62, 169
——, left skew, 230
——, negative, 89
——, ordinary, 230, 313, 333
——, positive, 6, 89, 334
——, right skew, 230
——, skew, 230, 305, 313, 322, 331, 333, 339, 345
——, unit, 6
Partitions, 148
Peak, 151
Period, multiplicative, 224, 236, 286, 294, 360
Periodic function, 231, 286
—— ——, doubly, 235, 286
p-hyperbola, 79, 189
Pincherle, 224
Plane, extended, 1
——, finite, 1
Point, finite, 1
—— at infinity, 1, 14, 21, 43, 89, 92, 95, 136, 274
Poisson's integral, 5, 140, 290, 293, 317, 319, 348, 352
Polynomial, 2
——, degree of, 2
Porter, 89, 101, 220
Principal value, 306
Principle of Argument, 2, 32, 54, 88, 91, 165, 180, 258, 275, 308, 312, 332, 339, 353
—— —— Maximum Modulus, 19
Product, infinite, 157, 220, 223
—— ——, convergence of, 223
Projection, stereographic, 1, 94, 127, 274

Quadratic, 27, 210
Quartic, bicircular, 19, 45, 68, 78, 121, 137, 147, 174, 197, 202, 324, 336

Radó, 156
Rational function, 89
—— ——, degree of, 90
Reciprocal equation, 167
Rectangle, 11, 52, 77, 81, 228
Reflection in line, 235, 267, 279, 346, 349, 356, 367
—— in point, 353
Region, 1
——, circular, 13, 58, 62, 80, 97, 103, 128, 165, 178, 186, 209, 212, 219, 238, 247, 259, 275, 278, 283, 298, 302, 365, 369
——, closed, 1
——, Jordan, 1
——, variable, 243, 259, 345
Riemann sum, 217, 239, 248, 274
—— surface, 18, 348
Riesz, F., 366, 369
——'s Theorem, 367
Rolle's Theorem, 5, 6, 24, 32, 73, 87, 219
Rouché's Theorem, 2, 156
Russell, 245, 248

Schwarz's principle of reflection, 235, 267, 279, 346, 349, 356, 367
Siebeck, 87
Similitude, center of, 15, 58, 67, 87, 104, 111, 120, 219, 266, 299
Simple layer, 365
Skew matter, 305, 324, 329
—— particle, 230, 305, 313, 322, 331, 333, 339 345
—— symmetry, 56, 194, 208, 282
Sphere, Neumann, 1, 94, 127, 176, 274, 284, 374
Square, 48, 72, 256
Stereographic projection, 1, 94, 127, 274
Stieltjes integral, 238, 246, 272, 366
Study, E., 356
Subharmonic function, 368
Superharmonic function, 366
Symmetry, 9, 41, 75, 80, 123, 136, 176, 181, 208, 234, 255, 263, 282, 288, 295, 330, 343
—— in origin, 75, 181, 208, 259, 284, 353, 370
—— in z and $1/z$, 202
——, method of, 348
——, p-fold, 41, 78, 189, 264
——, principle of, 235, 267, 279, 346, 349, 356, 367
——, skew, 56, 194, 208, 282

Toeplitz, 8
Topological methods, 360
Topology, 19, 241, 269, 354
Triple point, 48

Unit particle, 6, 13

Valley, 152
Variable region, 243, 259, 345
Vitali's Theorem, 224

Walsh's Theorem, 13, 26, 58, 66, 80, 103, 214, 259
W-curve, iv, 40, 135, 197, 258, 313, 325, 328, 372
Weierstrass, 225